Rethinking Agriculture

One World Archaeology Series
Sponsored by the World Archaeological Congress
Series Editors: Joan Gero, Mark Leone, and Robin Torrence

One World Archaeology volumes contain carefully edited selections of the exemplary papers presented at the World Archaeology Congress (WAC), held every four years, and intercongress meetings. WAC gives place to considerations of power and politics in framing archaeological questions and results. The organization also gives place and privilege to minorities who have often been silenced or regarded as beyond capable of making main line contributions to the field. All royalties from the series are used to help the wider work of the organization. The series is published by Left Coast Press, Inc. beginning with volume 48.

54 *Archaeology and Capitalism*, Yannis Hamilakis and Philip Duke (eds.)
53 *Living Under the Shadow*, John Grattan and Robin Torrence (eds.)
52 *Envisioning Landscapes*, Dan Hicks, Laura McAtackney, and Graham Fairclough (eds.)
51 *Rethinking Agriculture*, Tim Denham, José Iriarte, and Luc Vrydaghs (eds.)
50 *A Fearsome Heritage*, John Schofield and Wayne Cocroft (eds.)
49 *Archaeology to Delight and Instruct*, Heather Burke and Claire Smith (eds.)
48 *African Re-Genesis*, Jay B. Haviser and Kevin C. MacDonald (eds.)

Previous volumes in this series, available from Routledge:

47 Indigenous Archaeologies
46 Archaeologies of the British
45 Natural Disasters and Cultural Change
44 Material Culture
43 The Dead and Their Possessions
42 Illicit Antiquities
41 Destruction and Conservation of Cultural Property
40 Madness, Disability & Social Exclusion
39 The Archaeology of Dry Lands
38 The Archaeology of Difference
37 Time and Archaeology
36 The Constructed Past
35 Archaeology and Language IV
34 Archaeology and Language III
33 Cultural Resource Management in Contemporary Society
32 Prehistory of Food
31 Historical Archaeology
30 The Archaeology and Anthropology of Landscape
29 Archaeology and Language II
28 Early Human Behaviour in the Global Context
27 Archaeology and Language I
26 Time, Process and Structured Transformation in Archaeology
25 The Presented Past
24 Social Construction of the Past
23 Sacred Sites, Sacred Places
22 Tropical Archaeobotany
21 Archaeology and the Information Age
20 The Archaeology of Africa
19 Origins of Human Behaviour
18 From the Baltic to the Black Sea
17 The Excluded Past
16 Signifying Animals
15 Hunters of the Recent Past
14 What's New?
13 Foraging and Farming
12 The Politics of the Past
11 Centre and Periphery
10 Archaeological Approaches to Cultural Identity
9 Archaeological Heritage Management in the Modern World
8 Conflict in the Archaeology of Living Traditions
7 Animals into Art
6 The Meaning of Things
5 Who Needs the Past?
4 State and Society
3 Domination and Resistance
2 The Walking Larder
1 What Is an Animal?

Rethinking Agriculture:
Archaeological and Ethnoarchaeological Perspectives

Edited by
Tim Denham, José Iriarte and Luc Vrydaghs

Routledge
Taylor & Francis Group
LONDON AND NEW YORK

First published 2007 by Left Coast Press, Inc.
First paperback edition 2009

Published 2016 by Routledge
2 Park Square, Milton Park, Abingdon, Oxon OX14 4RN
711 Third Avenue, New York, NY 10017, USA

Routledge is an imprint of the Taylor & Francis Group, an informa business

Library of Congress Cataloging-In-Publication Data

Rethinking agriculture : archaeological and ethnoarchaeological perspective / edited by Tim Denham, José Iriarte, and Luc Vrydaghs.
p. cm.– (One world archaeology series ; 51) "Developed from a session entitled 'Inherited Models and the Denial of Prehistory: Challenging Existing Concepts of Agriculture' at the Fifth World Archaeological Congress (WAC5) in Washington DC in June 2003" – Pref. Includes bibliographical references and index.
1. Plant remains (Archaeology). 2. Ethnoarchaeology.
3. Paleoethnobotany. 4. Agriculture–Origin. 5. Plants, Cultivated–Origin. I. Denham, Tim. II. Iriarte, José, Ph. D. III. Vrydaghs, Luc.
IV. World Archaeological Congress (5th : 2003 :
Washington, DC) CC79.5.P5R48 2007 930.1–dc22
2007018694

Cover design by Joanna Ebenstein

ISBN 978-1-59874-260-2 hardcover
ISBN 978-1-59874-261-9 paperback

Contents

Preface

This book developed from a session entitled 'Inherited models and the denial of prehistory: challenging existing concepts of agriculture' at the Fifth World Archaeological Congress (WAC5) in Washington, DC, in June 2003. The purpose of the session was to bring together researchers who work on agriculture, especially early agriculture, in nonEurasian parts of the world and to assess the relevance of concepts and methods inherited from studies in Eurasia to their work. In the spirit of the session, the contributions to this book focus on three broad geographical regions: Island Southeast Asia and the Pacific, the Americas and Africa.

The editors would like to thank participants in the session who were unable, for various reasons, to contribute to this book. These include Bruce Smith, Dolores Piperno, Doug Yen, Graeme Barker, J Peter White, John Staller, Kevin Pope and Mary Pohl, Mukund Kajale, Sarah Dingalo and Alfred Tsheboeng. Several contributors to this book did not participate in the original session, but their contributions were solicited for it, including Martin Jones and Terry Brown, Elisabeth Hildebrand, Stefanie Kahlheber and Katharina Neumann, Edmond De Lange, Fiona Marshall and Roger Blench.

The session at WAC5 grew out of conversations with and support from David Harris, Dolores Piperno and Robin Torrence. The editors would like to thank them, as well as Heather Burke, Joan Gero, Mark Leone and Claire Smith, who assisted with bringing this volume to publication.

Most contributions to this volume were written in 2004. A delay in publication was beyond the control of the editors, who would like to thank Left Coast Press for agreeing to publish this volume in its original form. The editors are particularly grateful to Mitch Allen, Jennifer Collier and Erica Hill at Left Coast Press.

CHAPTER 1

Rethinking Agriculture: Introductory Thoughts

Luc Vrydaghs and Tim Denham

> ... thoughts about food in prehistory are changing and that with a proper combination of the theoretical and technical a rounded disciplinary approach is possible, which will do justice to the richness of the subject-matter (Gosden 1999: 8).

Each contributor to this book accepted an invitation to rethink agriculture, whether in terms of existing regional chronologies, in terms of techniques employed, or in terms of the concepts that frame our interpretations. In this book, new archaeological and ethnoarchaeological research on agriculture, and mostly on early agriculture, is presented for several non-Eurasian regions, including Island Southeast Asia and the Pacific, the Americas and Africa. These broad geographical regions include places often underrepresented in discussions of early agriculture, namely, New Guinea, lowland South America, and the African tropics (see Bellwood 2005 for a recent example).

Although the need to study agriculture in different parts of the world on its 'own terms' has long been recognised (Harris 1977) and reaffirmed (Piperno and Pearsall 1998), a tendency persists to evaluate agriculture across the globe using concepts, lines of evidence and methods derived from Eurasian research. However, researchers working in different regions across the globe are becoming increasingly aware of fundamental differences in the nature of, and methods employed to study, agriculture and plant exploitation in the past. Most contributions to this volume bring a conceptual and technical richness to their understanding of agriculture and plant exploitation, a richness developed through research in specific geographical, social and temporal contexts. In doing so, contributors build upon and extend the substantive, technical and theoretical deliberations of previous *One World Archaeology* books (ie Gosden and Hather 1999; Harris and Hillman 1989; Hather 1994).

In the following three sections, conceptual themes discussed by contributors to this volume, which are of general significance to the study of early agriculture, are briefly considered. These themes comprise: persistent problems with domestication-based definitions of agriculture, the diffuseness of early agricultural practices, and the importance of understanding the social contexts within which agriculture is practised. The last section of this introduction sketches how the new substantive findings reported here change our regional interpretations of early agriculture.

DEFINING AGRICULTURE: PERSISTENT PROBLEMS

Despite attempts to provide an inclusive definition, there are persistent disagreements about what constitutes agriculture (eg Harris 1989, 1996a, chapter 2). Problems of definition and corresponding archaeological visibility are most acute for differentiating early agriculture from other practices. Alternative conceptions centre on animal and plant domestication, as determined by genetic and morphological changes attributed to human agency (Harris chapter 2; cf Jones and Brown chapter 3), environmental change (Pearsall chapter 11), and social dependence (Spriggs 1996). Some have argued for the abandonment of the term altogether (Terrell et al 2003), while others have worked without an explicit definition of agriculture (Golson 1997, chapter 6). These disagreements are usually more than 'semantic' (cf Harris 1996b: 3); they are fundamental and relate to the idea of agriculture that we bring to our work and transform through our work.

None of the contributors to this volume follow the call by Terrell et al (2003) to abandon the term 'agriculture'. The word has a long tradition in several European societies from which it has colonised other areas of the world (Harris chapter 2). Although the word and its meaning are inevitably Eurocentric and not always directly transferable to other sociocultural situations (Sayre chapter 12), the term is widely used in academic and public discourse across the globe. Certainly more attention should be paid to how words, categories and concepts relevant to the study of agriculture translate between different regions and people, but this is true of many words and does not necessarily negate the value of their continued use.

If we abandon 'agriculture', surely the arena of debate will only shift onto other words and categories that replace it. Such problems are amply demonstrated by attempts to insert intermediate categories between the traditional terms of 'hunter-gatherer' and 'agriculture' – such as 'domiculture' (Hynes and Chase 1982), 'incipient agriculture' (Ford 1985), 'complex hunter-gatherers' (Zvelebil 1986), 'transitional' and 'proto-agricultural' (Yen 1989), 'wild-plant food production' (Cauvin 1997; Harris 1989, 1996b), 'hunter-horticulturalism' (Guddemi 1992), 'low-level food production' (Smith 2001) and the list goes on – as well as attempts to subsume these traditional categories using more inclusive concepts – such as '*domus*' (Hodder 1990) or 'domesticated landscapes' (Terrell et al 2003). These new terms do not fully solve the problem. The conceptual battleground shifts from 'agriculture' to alternative categorical and linguistic lines designed to demarcate or include.

Problems of definition are most acute in the demarcation of early agriculture and its differentiation from other types of practice in the past. These problems arise whether morphogenetic, environmental or social interpretations are followed. For example, many researchers, including several contributors to this volume, consider domestication, or at least a high level of dependence upon domesticates, to be a key determinant of 'agriculture'. Undoubtedly, deliberate and unintentional morphological changes in plants and animals, and more recent genetic engineering, have greatly augmented people's ability to obtain food. However, if morphogenetic changes in the past, as argued by Jones and Brown (chapter 3), have potentially been a product of genetic (often geographic) isolation of an animal or plant from its wild population (to prevent crossbreeding with wild varieties, etc), as well as a product of selective pressure, then three problems arise in trying to link the earliest agricultural practices and morphogenetic traces of initial domestication.

First, if we assume people practising a nascent form of agriculture took a species out of its original natural range, then the locus of the earliest agricultural practices may not necessarily yield the anticipated domestication signal (evident through morphogenetic change), or at least not one that is as marked as the signal in the region into which the early agriculturalists spread. In other words, the earliest practices need not yield the earliest clear evidence of morphogenetic change (consider Marshall chapter 20); only beyond the natural range can interbreeding with wild populations be prevented and reproductive isolation be assured (Jones and Brown chapter 3). Thus the clearest signals of domestication are likely to show up in the archaeobotanical or archaeozoological records at a later date and in a neighbouring region beyond the natural range of the plant or animal. According to this scenario, signals of domestication for cultivated plants and animals, whether dispersing as a result of demic expansion (eg papers in Bellwood and Renfrew 2003) or cultural interaction (eg Thomas 1996), would not be anticipated to be congruent in time or space. For example, archaeobotanical evidence of supposedly 'African crops', ie finger millet (*Eleusine coracana* L.) and sorghum millet (*Sorghum bicolor* [L.] Moench), on the Indian subcontinent are older than most reported for the African continent (Misra and Kajale 2003). Similarly, claims for the presence of *Sorghum* in Korea during the Plain Pottery Period (c 4000–2500 BP) are noteworthy (Kim et al 1978; Yim 1978). Changes through time and the sociospatial distribution of domestication signals would be animal- and plant-specific and depend upon variations in natural ranges, the abilities of species to adapt to new environmental conditions and social contexts of use (Denham 2005; Marshall chapter 20).

Secondly, and conversely, if we assume a wild plant or animal was taken out of its natural range, then signals comparable to those anticipated for domestication could be generated by what would otherwise be considered a non-agricultural practice. The claimed Pleistocene translocation of plants, most notably *Canarium indicum* (Yen 1996) and marsupials (White 2004), between New Guinea and islands of the Bismarck Archipelago (cf Spriggs 1997: 53–54) would undoubtedly result in reproductive isolation and could have, depending upon the nature of subsequent exploitation practices in different locales, yielded morphogenetic markers akin to those anticipated from agricultural dispersal (Yen 1996 on *Canarium indicum*). A similar case may pertain to the movement of starch-rich plants, most notably *Musa* bananas, taro (*Colocasia esculenta*) and yams (*Dioscorea* spp.) in the interior of New Guinea at the end of the Pleistocene and in the early Holocene (Denham and Barton 2006). Highland and lowland locales are so close together, ie within a day's walk, that people may have moved plants out of their natural ranges through adventitious exploitation, curation of reproductive portions, translocation and planting. In highland environments, the plants might have established a foothold and slowly adapted under human intervention, either sexually or somatically, to produce distinctive morphotypes and phenotypes (cf Golson chapter 6). Thus practices more generally regarded as typical of foraging behaviour, and not ordinarily acknowledged as representing agriculture, can yield distinctive macrofossil, microfossil and molecular characteristics.

Third, a similar argument pertains to the exploitation of morphogenetically altered plants and animals that are known to have radiated from a geographical region, or regions, as a result of natural dispersal or by people practicing extensive forms of plant exploitation. The presence of morphologically 'domesticated' plants and animals does not assure the presence of 'agriculture'. Such a scenario may be relevant to the dispersal of maize (*Zea mays*) after it diffused into some regions of North and South America, where it remained a minor crop and contributor to diet for extended periods. The incorporation of maize into preexisting wild plant food production strategies need not signal the advent of 'agriculture', because much depends upon the importance of maize cultivation relative to other food-yielding practices (Fritz chapter 10).

The fixation with morphogenetic changes in plants and animals as indicators of agriculture was initially derived from work in Eurasia, and particularly Southwest Asia (Harris chapter 2). However, chronological and spatial records of morphogenetic change have yet to be fully analysed for most intensively managed animals and plants in other regions, although such records are beginning to be developed (eg Harris chapter 2; Iriarte chapter 9). Of note, the reconstruction of morphogenetic transformations resulting from the human management of vegetatively propagated starch-rich plants are poorly understood. For some plants, eg the greater yam (*Dioscorea alata*), the wild ancestors of cultivated varieties are not known (Martin and Rhodes 1977), whereas for other plants, eg taro (*Colocasia esculenta*), some knowledge of wild ancestors is emerging (Matthews 1991, 1995), but there is a lack of research and knowledge tracking morphogenetic changes into the past. Additionally, there are environmental and social reasons why some plants and animals have not undergone full domestication despite millennia of cultivation and stockbreeding, eg pearl millet (*Pennisetum glaucum*; Kahlheber and Neumann chapter 17) and donkeys (*Equus* spp.) in Africa (Marshall chapter 20). The absence of morphogenetic change does not necessarily signal an absence of agriculture.

If morphogenetic signals are not to be used as definitive indicators of early agriculture, upon what types of evidence are signals of early agriculture to rest? A striking characteristic of most investigations of early and later agriculture in the regions discussed in this book is a reliance on multiple lines of evidence. These range from morphogenetic changes in plants and animals to worldviews; they do not fetishise one effect of agriculture as a diagnostic indicator of its presence, but embed archaeobotanical and archaeozoological evidence within broader environmental and social contexts.

DIFFUSENESS OF EARLY AGRICULTURE

The lack of a clear definition for agriculture, particularly early agriculture, reflects its multifaceted nature. The resultant porosity of the concept of agriculture mirrors a diffuseness in its geographical, social and temporal manifestations in the past. From such a perspective, there is less of a concern with identifying cores/peripheries and centres/non-centres; rather, we begin to see a diffuseness manifest as asynchronies and mosaics of practices related to cultivation, domestication and environmental transformation in diverse social contexts, both across space and through time (Denham 2005). Several contributors to this volume emphasise the social or environmental contexts of agriculture, as well as animal and plant exploitation practices (Denham chapter 5; Fritz chapter 10; Iriarte chapter 14; Kahlheber and Neumann chapter 17), and others do so by drawing explicitly on ethnobotanical or ethnographic information to enliven their interpretations (Bayliss-Smith chapter 7; Hildebrand chapter 15; Louwagie and Langohr chapter 8; Marshall chapter 20; Pearsall chapter 11; Sayre chapter 12).

Early agriculture in New Guinea, Africa and the Americas is characterised by a 'diffuseness' (after Bellwood's [2005] application of the term to early agriculture in the Americas). In contrast to discrete centres of early agricultural development and spread, there were mosaics of different plant and animal exploitation practices juxtaposed across space, which were variously transformed through time. Plants, technologies and other elements of material culture dispersed along local exchange pathways, as well as with movements of people. For example, mounting evidence in the Americas is showing that there was no single centre of domestication or agricultural origins (Iriarte chapter 9).

The idea of food exploitation mosaics across space, including early agriculture, based on the cultivation of varying assemblages of plants is likely to be the norm rather than the exception. Subsequent dispersal (individually or as a group) and adoption by other societies created mosaics of practices reliant on different suites of plants, many with domesticates as only a minor component. Although such scenarios are well-attested in the Americas and now in Africa, as chapters in this volume demonstrate, a similar scenario is likely to emerge with new research for New Guinea, eastern Indonesia and, potentially, the Philippines.

CONTEXTUALISING AGRICULTURE

Contextual evaluations of agriculture, often restricted to a single region, contrast markedly with panregional comparative studies, most notably the 'farming/language dispersal hypothesis' (Bellwood and Renfrew 2003;

Diamond and Bellwood 2003) and its recent incarnation, 'the early farming dispersal hypothesis' (Bellwood 2005). Contextual studies seek to understand past plant exploitation practices, including the emergence of early agriculture, as well as morphogenetic effects, within their environmental and socioeconomic contexts. Multidisciplinary information is used to understand the 'indeterminate relations' embedding local practices within macro-scale processes (Hodder 1999: 175). In contrast, comparative approaches typically present dismembered representations of the past that focus upon tracking archaeobotanical, genetic, material cultural or linguistic diagnostics through time and across space, especially those considered to be indicators of early agriculture. Macro-scale comparative hypotheses have been applied to the expansion of Indo-European language speakers (after Ammerman and Cavalli-Sforza 1984; Renfrew 1987), with subsequent applications to Austronesian and Bantu language speakers, as well as to other parts of the world (papers in Bellwood and Renfrew 2003). Several deficiencies of the comparative approach are discussed below to highlight the advantages of contextual perspectives.

First, early agriculture is not a demarcated 'all or nothing' lifestyle that can be clearly mapped across space and tracked through time (compare Bellwood 2005: 12–43 to Harris chapter 2 and Smith 2001). As indicated by the classificatory ambiguities for transitional situations between hunter-gatherer and agriculturalist in the recent and distant pasts, agriculture is a porous category and, for many, has been an optional (Kahlheber and Neumann chapter 17) or seasonal (Lévi-Strauss 1955) subsistence choice. For example, some people had access to, or incorporated domesticates as minor crops into wild plant-based cultivation strategies for centuries, and possibly millennia, without the development of agriculture (Fritz chapter 10; Kahlheber and Neumann chapter 17). Furthermore, there are now several reports of groups shifting from an agricultural economy to a hunter-gatherer economy (eg Balée 1994).

Second, the exchange of domesticated plants and animals, as well as technology, is sometimes stated to have been limited between agricultural and non-agricultural populations and, furthermore, hunter-gatherers are considered to have rarely become agriculturalists (Bellwood 2005: 12–43). From this perspective, the spread of animals, plants and technology primarily occurred due to the expansion of agricultural peoples, ie demic expansion. Several studies in this book document and hypothesise the dispersal of plants and animals without accompanying signals of population movement. Examples include the dispersals of maize to North America (Fritz chapter 10) and southern South America (Iriarte chapter 9); the plantain (*Musa* sp.) across Central Africa (De Langhe chapter 19), and the donkey across North Africa and Southwest Asia (Marshall chapter 20).

Third and following from the above, arguments hypothesising demic diffusion tend to overemphasise, despite recent provisos (eg Bellwood 2005: 2), the coherence, homogeneity and unity of 'Neolithic packages' (cf Denham 2004; Thomas 1996, 1999). As a result, cultural, genetic and linguistic markers are used as proxies for agricultural populations, ie discussions of pottery and stone artefact distributions are taken as proxies for the distribution of agricultural populations. In contrast, contextual perspectives seek to understand people and their lifestyles within their environmental and socioeconomic contexts and trace the socially mediated dispersal of individual items and practices between communities.

Fourth, from a comparative viewpoint: 'the irregularities of small-scale reality become "ironed-out"' (Bellwood 2005: 10) and are seen as isolates that do not threaten the larger model (Renfrew paraphrased in Schouse 2001: 989). However, the three points raised above would seem to suggest otherwise. Differences between contextual and comparative methods are not a function of scale; they reflect fundamental differences in how the past is represented, the role of people in the making of their own history and the way we conceive people living in the world. Comparative studies yield smoothed versions of human history largely devoid of people as agents; they also fail to account for the interrelationships between specific practices and larger-scale interpretations, which are at the core of most contributions to this book.

CHALLENGING EXISTING CONCEPTS

Most of the dominant ideas pervading the study of agriculture across the globe, particularly early agriculture, were originally developed for Eurasia. The dominance of Eurasian, often Eurocentric, perspectives reflects the long history and intensity of investigations of early agriculture in that region. Additionally, Eurasian biases to the study of agriculture cannot be divorced from broader trends in European and global history, including colonialism, nationalism, racism, neocolonialism and, more recently, globalisation (Trigger 1989). These tendencies are exemplified by the way that Eurasian-derived conceptions have become a template for understanding agriculture and, by association, non-agricultural practices across the globe.

A Eurasian template, whether culture-historic or processual, tends to prescribe components of any early agricultural practice and associated society. From Childe (1936) to Piggott (1954) to Diamond (2002), these components often include, but are not limited to: (1) clear morphogenetic changes in plants or animals, termed 'domestication'; (2) environmental transformations resulting from forest clearance for agriculture; and (3) packages of associated cultural, political and social traits. Over the last decade, this interpretative template has been questioned within the Eurasian context on conceptual, ie through postprocessual critique (Hodder 1999; Thomas 1996, 1999), and more substantive grounds (Cauvin 1997).

The interpretative framework inherited from Eurasian research has imposed conceptual and technical constraints on the study of early agriculture in other regions of the world. Conceptually, the emergence of agriculture is often marked by the appearance of animal and plant domesticates, ie animals and plants that have developed morphological and, largely by inference, genetic traits that are attributed to purposive selection by people. Explanations for the subsequent movement of agricultural peoples, plants and practices have been heavily dependant upon cultural and demic diffusionary models, which are more focussed on the 'chess game of migrations' (Renfrew 1973: 121) than on recovering the texture of social life in the past (Denham 2004; Renfrew 1973).

Technically, evidence of early agriculture in most regions of the globe has been evaluated against anticipated signals largely derived from research on cereal and animal domestications in Eurasia. Instead of the Eurasian reliance on macrofossils, advances in microfossil (ie phytoliths, pollen and starch grains) and molecular (ie DNA analysis) techniques are opening up new avenues for interpreting the human use of plants and animals in the past across the globe. For example, microfossil and molecular analyses are being used to characterise early agricultural practices and domestication in the Americas (Iriarte chapter 9). Other researchers are beginning to question pre-suppositions regarding the dependence upon and nature of domestication in relation to the study of early agriculture (eg Hather 1996; Jones et al 1996; Jones and Brown chapter 3).

In this book, the range of perspectives and evidence on agriculture cluster around three broad geographic regions: Island Southeast Asia and the Pacific, the Americas and Africa. Each broad geographical region provides evidence that challenges existing dogma and includes an area in which an editor has worked: Denham in Papua New Guinea, Iriarte in Uruguay and Vrydaghs in Cameroon. Contributions are designed to enrich our conceptual, substantive and technical understandings of agriculture in each region.

Island Southeast Asia and the Pacific

Contributions to this book on plant exploitation in New Guinea, Borneo and Easter Island have broad geographical and chronological ranges (from almost 40,000 years ago to a few hundred). Authors focus on diverse practices: non-agricultural plant exploitation that enabled permanent occupation of tropical rainforests of Borneo during the Pleistocene (Barton and Paz chapter 4); early (Denham chapter 5; Golson chapter 6) and later (Bayliss-Smith chapter 7) Holocene agriculture in highland New Guinea; and recent agriculture on Easter Island (Louwagie and Langohr chapter 8). Authors rely on diverse multidisciplinary lines of evidence to reconstruct agriculture in the past and use a diverse range of perspectives to interpret past practices.

For example, Denham's multidisciplinary research into early agriculture in New Guinea (Denham 2003, chapter 5; Denham et al 2003, 2004) built upon previous investigations by Golson and colleagues (Golson 1977, chapter 6; Golson and Hughes 1980; Hope and Golson 1985). He initially sought to address critiques of prior work that implied agriculture in New Guinea was of ultimate Southeast Asian origin (eg Spriggs 1996). Information on plant use, environmental transformations and cultivation practices addressed these critiques and showed that agriculture was practiced in the Highlands of New Guinea by at least 6950–6440 cal BP (Denham et al 2003, 2004). Despite these multidisciplinary strengths, some still qualify their discussions of early agriculture in New Guinea (Bellwood 2005: 142).

The evidence gathered during Denham's research has brought into question the dichotomous framework within which debates of early agriculture in Melanesia and elsewhere have been couched, namely 'indigenous or introduced' (Denham 2004). Instead of viewing New Guinea in isolation from areas to the west prior to Austronesian (Bellwood 1997) or Southeast Asian Neolithic (Spriggs 1989) expansion approximately 3500 cal BP years ago, the region was embedded within localised social networks (Gosden 1995) that extended from New Guinea to Indo-Malaysia and ultimately to mainland Southeast Asia. Across these regions, people were most likely engaged in various forms of plant exploitation that targeted different plants using a variety of strategies (Barton and Paz chapter 4). At present the records of these practices are sparse.

The Americas

Iriarte's research in southeastern Uruguay has shed light on early agricultural practices in another region previously considered marginal (Iriarte 2003, chapter 14; Iriarte et al 2004). His work questions the traditional view that agriculture was brought from the tropical forest around 2000 years ago by Tupi-Guarani who migrated into the diverse environments of southeastern South America inhabited by groups of foragers (eg Schmitz 1991). Iriarte (chapter 14) has shown that cultigens were introduced and integrated into preexisting local food economies earlier than expected, and that the use and manipulation of wetlands was an earlier, more important and more frequent activity than previously thought.

Recent research has revolutionised our conception of the origin and dispersal of agriculture in lowland South and Central America (see Iriarte chapter 9). Microfossil studies have revealed the presence of diverse assemblages of seed, root and tree crops, many of which are not preserved as macrofossils (Piperno and Holst 1998; Piperno and Pearsall 1998; Piperno et al 2000; Pope et al 2001; Perry 2004). The value of such approaches is clearly demonstrated by Iriarte's (chapter 14) and Perry's (chapter 13) starch analysis of soil samples and artefacts from lowland and highland locales, respectively. In addition, the reconstruction of past landscapes through the systematic collection of palaeoecological data enables archaeologists to obtain more detailed reconstructions of the natural- and human-induced changes in vegetation and climate that accompanied the development and dispersal of agriculture (Pearsall chapter 11). Such studies have the potential to detect vegetation changes associated with early swidden-type practices.

In addition to advances in microfossil, molecular and palaeoecological studies, some researchers are advancing more socially oriented interpretations (eg Hastorf 1999). Fritz (chapter 10) uses macrobotanical research to document wild plant production in the lower Mississippi Valley and considers the social contexts within which domesticates were adopted and incorporated into preexisting plant exploitation strategies. Sayre (chapter 12) uses ethnography to critique the application of the concept of 'agriculture' to traditional American societies. He considers the dichotomies underlying such attempts, such as agriculture/hunting and gathering and domesticated/wild, to be inappropriate to many social contexts in the Americas.

Africa

The identification of *Musa* phytoliths, most likely of a plantain, at Nkang, Cameroon (Mbida et al 2000, 2001; Vrydaghs 2003), has provided archaeobotanical corroboration of a general hypothesis for the dispersal of plantain across Africa. This crop was introduced to Africa from Asian and, ultimately, Melanesian sources (De Langhe chapter 19; De Langhe et al 1994–95). Various contexts with *Musa* phytoliths at Nkang also contain limited amounts of domesticated sheep and goat bones (Mbida et al 2000). Older deposits from neighbouring Nigerian sites testify to small-scale stock breeding and suggest that food production economies were well established over a broad area by 3000 years ago (Breunig 1995; Breunig et al 1996; Connah 1976, 1981). However, the interpretation of *Musa* phytoliths in Central Africa by the first millennium BC is, for some, controversial (Vansina 2004). Beside scepticism about the accuracy of phytolith analysis, these doubts are also rooted in traditional scepticism of claims for independent agricultural development in Africa (eg Bellwood 2005).

Despite recent research (see papers in van der Veen 1999; Blench and MacDonald 2000; Neumann et al 2003), theoretical discussions on early African agriculture are currently constrained by a lack of information for both forest and savanna habitats. The lack of evidence reflects research designs often notable for the lack of systematic archaeobotanical sampling.

In the case of forests, the identification of *Musa* phytoliths at Nkang (Cameroon) remains the sole evidence for prehistoric agriculture (Neumann 2005: 252; also see Van Grunderbeek and Roche chapter 16). The adoption of plantain cultivation raises questions concerning the types of practice predating the adoption of banana. Historical research indicates that people often discover, accept and transform introductions within existing traditions (Balandier 1983; Gruzinski 1999), which may account for how the banana was adopted into preexisting cultivation practices before the Bantu expansion. With the exceptions of wild resource exploitation and tentative evidence of arboriculture (Oslisly and White chapter 18; Blench chapter 21), almost nothing is known about early arboreal exploitation, especially practices involving vegetative propagation (Hildebrand chapter 15). Furthermore, for vegetative propagation and arboriculture, the transitions from gathering to transplanting to cultivation may be seamless (Denham and Barton 2006) and have limited archaeological, botanical and linguistic visibility (Blench chapter 21).

There is greater, but still limited, archaeobotanical and archaeozoological information for the savanna (Kahlheber and Neumann chapter 17; Marshall chapter 20). The most intensively investigated species are major crops today, ie *Sorghum bicolor* (sorghum), *Pennisetum glaucum* (pearl millet) and *Eleusine coracana* (finger millet). Of most significance, animal domestication appears to have systematically preceded plant domestication (Marshall chapter 20; Marshall and Hildebrand 2002). Theoretical considerations of early African agriculture should not divorce animals from plants, but consider the social contexts within which both were brought under greater human control to understand why the former preceded the latter. Of general applicability are the importance of Marshall's hypotheses (chapter 20) on the social contexts of donkey domestication to animal and plant domestication. Donkeys were not always managed in similar ways in different cultural contexts; the diversity of management practices resulted in different morphogenetic transformations developing in different places and, potentially, multiple contexts of domestication.

ACCEPTING THE CHALLENGE

The challenge set by the editors of this book was to understand past agriculture in different parts of the world on its 'own terms'. Such a challenge was intended to expose presuppositions, foster debate and reorient future research.

Each author's contribution, whether archaeological or ethnoarchaeological, sheds light on the problems of interpreting agriculture in the past. The authors also make substantive contributions to the understanding of agricultural practices in regions that have been relatively underinvestigated. Authors demonstrate a theoretical and technical openness to their subject through their willingness to challenge current orthodoxies that often constrain thinking and to adopt new techniques that illuminate investigations of earlier and later agricultural societies.

Three conceptual outcomes emerge from the contributions to this volume, and these need to be considered in terms of current debates on early agriculture. First, definitions of agriculture focussed on domestication are conceptually inadequate to characterise the emergence of early agriculture because they assume that the selective pressures exerted through deliberate management of species cause morphogenetic changes in animals and plants. Such perspectives underestimate the potential influence of genetic isolation on the emergence of distinctive genotypes and phenotypes. Genetic isolation could have occurred due to various activities, including those not ordinarily understood to be agricultural. Furthermore, such definitions are substantively insufficient because they fetishise one epiphenomenon of plant exploitation over all others, including environmental transformations and cultivation practices; thereby, they fail to represent the diversity and multifaceted nature of early agricultural practices in different parts of the world.

Second, early agricultural practices in Africa, the Americas and the New Guinea region exhibit diffuseness. In these regions, there do not seem to be clear loci of origin from which animals, plants, people and practices dispersed.

Instead, there were mosaics of different, coexisting practices across space, some of which can be characterised as 'agriculture' and some as 'non-agriculture' (see Roscoe 2002 for a recent example). Spatially variable transformations occurred through time, of which some were initiated by the differential movement of animals and plants, material culture, people and technologies among groups practicing different types of animal and plant exploitation.

Third, the development, transformation and dispersal of early agricultural practices need to be considered within broader environmental and socioeconomic contexts. Contextual approaches are preferred to an overly comparative method. Understanding the interrelationships among people and practices on the one hand, and environmental contexts and larger social, economic and political structures on the other, are fundamental to understanding any aspect of human social life, including agriculture.

ACKNOWLEDGMENTS

The authors thank José Iriarte, Robin Torrence and Peter White for constructive comments on earlier drafts of this introduction. José is also thanked for providing information on research into early agriculture in the lowland neotropics.

REFERENCES

Ammerman, A J and Cavalli-Sforza, L L (1984) *The Neolithic Transition and the Genetics of Populations in Europe*, Princeton: Princeton University Press

Balandier, G ([1957] 1983) *Afrique Ambiguë*, Paris: Presse Pocket

Balée, W (1994) *Footprints of the Forest: Ka'apor Ethnobotany—The Historical Ecology of Plant Utilization by an Amazonian People*, New York: Columbia University Press

Bellwood, P (1997) *Prehistory of the Indo-Malaysian Archipelago*, Honolulu: University of Hawaii Press

Bellwood, P (2005) *First Farmers*, Oxford: Blackwell

Bellwood, P and Renfrew, C (eds) (2003) *Examining the Farming/Language Dispersal Hypothesis*, Cambridge: McDonald Institute for Archaeological Research

Blench, R M and MacDonald, K C (eds) (2000) *The Origins and Development of African Livestock*, London: University College London Press

Breunig, P (1995) 'Gajiganna und Konduga: zur frühen besiedlung des Tschadbeckens in Nigeria', *Beiträge zur allgemeinen und vergleichenden Archäologie* 15: 3–48

Breunig, P, Neumann, K and Van Neer, W (1996) 'New research on the Holocene settlement and environment of the Chad Basin in Nigeria', *The African Archaeological Review* 13: 111–45

Cauvin, J (1997) *Naissance des divinités, naissance de l'agriculture*, Paris: Champs/Flammarion

Childe, V G (1936) *Man Makes Himself*, London: Watts

Connah, G (1976) 'The Daima sequence and the prehistoric chronology of the Lake Chad region of Nigeria', *Journal of African History* 17: 321–52

Connah, G (1981) 'Man and a lake', in *Le sol, la parole et l'écrit: 2000 ans d'histoire africaine: mélanges en hommage à Raymond Mauny*, pp 161–78, Paris: Société Française d'Histoire d'Outre-Mer

De Langhe, E, Swennen, R and Vuylsteke, D (1994–95) 'Plantain in the early Bantu world', *Azania* (Nairobi) 29–30: 147–60

Denham, T P (2003) 'The Kuk Morass: multi-disciplinary evidence of early to mid Holocene plant exploitation at Kuk Swamp, Wahgi Valley, Papua New Guinea', unpublished PhD thesis, Australian National University

Denham, T P (2004) 'The roots of agriculture and arboriculture in New Guinea: looking beyond Austronesian expansion, Neolithic packages, and indigenous origins', *World Archaeology* 36(4): 610–20

Denham, T P (2005) 'Envisaging early agriculture in the Highlands of New Guinea: landscapes, plants and practices', *World Archaeology* 37(2): 290–306

Denham, T P and Barton, H (2006) 'The emergence of agriculture in New Guinea: continuity from pre-existing foraging practices', in Kennett, D J and Winterhalder, B (eds), *Behavioral Ecology and the Transition to Agriculture*, pp 237–64, Berkeley: University of California Press

Denham, T P, Haberle, S G, Lentfer, C, Fullagar, R, Field, J, Therin, M, Porch, N and Winsborough, B (2003) 'Origins of agriculture at Kuk Swamp in the Highlands of New Guinea', *Science* 301: 189–93

Denham, T P, Haberle, S G and Lentfer, C (2004) 'New evidence and revised interpretations of early agriculture in highland New Guinea', *Antiquity* 78: 839–57

Diamond, J (2002) 'Evolution, consequences and future of animal and plant domestication', *Nature* 418: 700–07

Diamond, J and Bellwood, P (2003) 'Farmers and their languages: the first expansions', *Science* 300: 597–603

Ford, R (1985) 'The processes of plant production in prehistoric North America', in Ford, R (ed), *Prehistoric Plant Production in North America*, pp 1–18. Anthropological Paper 75, Ann Arbor: Museum of Anthropology, University of Michigan

Golson, J (1977) 'No room at the top: agricultural intensification in the New Guinea Highlands', in Allen, J, Golson, J and Jones, R (eds), *Sunda and Sahul: Prehistoric Studies in Southeast Asia, Melanesia and Australia*, pp 601–38, London: Academic Press

Golson, J (1997) 'From horticulture to agriculture in the New Guinea Highlands', in Kirch, P V and Hunt, T L (eds), *Historical Ecology in the Pacific Islands: Prehistoric Environmental and Landscape Change*, pp 39–50, New Haven and London: Yale University Press

Golson, J and Hughes, P J (1980) 'The appearance of plant and animal domestication in New Guinea', *Journal de la Société des Océanistes* 36: 294–303

Gosden, C (1995) 'Arboriculture and agriculture in coastal Papua New Guinea', *Antiquity* 69(265): 807–17

Gosden, C (1999) 'Introduction' in Gosden, C and Hather, J (eds), *The Prehistory of Food: Appetites for Change*, pp 1–10, London: Routledge

Gosden, C, and Hather, J (eds) (1999) *The Prehistory of Food: Appetites for Change*, London: Routledge

Gruzinski, S (1999) *La pensée métisse*, Paris: Fayard

Guddemi, P (1992) 'When horticulturalists are like hunter-gatherers: the Sawiyano of Papua New Guinea', *Ethnology* 31: 303–14

Harris, D R (1977) 'Alternative pathways toward agriculture', in Reed, C A (ed), *Origins of Agriculture*, pp 179–243, The Hague: Mouton

Harris, D R (1989) 'An evolutionary continuum of people-plant interaction', in Harris, D R and Hillman, G C (eds), *Foraging and Farming: The Evolution of Plant Exploitation*, pp 11–26, London: Unwin Hyman

Harris, D R (1996a) 'Domesticatory relationships of people, plants and animals', in Ellen, R and Fukui, K (eds), *Redefining Nature: Ecology, Culture and Domestication*, pp 437–63, Oxford: Berg

Harris, D R (1996b) 'Introduction: themes and concepts in the study of early agriculture', in Harris, D R (ed), *The Origins and Spread of Agriculture and Pastoralism in Eurasia*, pp 1–9, London: University College London Press

Harris, D R and Hillman, G C (eds) (1989) *Foraging and Farming: The Evolution of Plant Exploitation*, London: Unwin Hyman

Hastorf, C A (1999) 'Cultural implications of crop introductions in Andean prehistory', in Gosden, C and Hather, J G (eds), *The Prehistory of Food: Appetites for Change*, pp 35–58, New York: Routledge

Hather, J G (ed) (1994) *Tropical Archaeobotany: Applications and New Developments*, London: Routledge

Hather, J G (1996) 'The origins of tropical vegeculture: Zingiberaceae, Araceae and Dioscoreaceae in Southeast Asia', in D R Harris (ed), *The Origins and Spread of Agriculture and Pastoralism in Eurasia*, pp 538–50, London: University College London Press

Hodder, I (1990) *The Domestication of Europe: Structure and Contingency in Neolithic Societies*, Oxford: Basil Blackwell

Hodder, I (1999) *The Archaeological Process: An Introduction*, Oxford: Blackwell

Hope, G S and Golson, J (1995) 'Late Quaternary change in the mountains of New Guinea', *Antiquity* 69 (Special Number 265): 818–30

Hynes, R and Chase, A (1982) 'Plants, sites and domiculture: Aboriginal influence on plant communities', *Archaeology in Oceania* 17: 138–50

Iriarte, J (2003) 'Mid-Holocene emergent complexity and landscape transformation: the social construction of early Formative communities in Uruguay, La Plata Basin', unpublished PhD dissertation, University of Kentucky

Iriarte, J, Holst, I, Marozzi, O, Listopad, C, Alonso, E, Rinderknecht, A and Montaña, J (2004) 'Evidence for cultivar adoption and emerging complexity during the mid-Holocene in the La Plata Basin', *Nature* 432: 614–17

Jones, M K, Brown, T and Allaby, R (1996) 'Tracking early crops and early farmers: the potential of biomolecular archaeology', in D R Harris (ed), *The Origins and Spread of Agriculture and Pastoralism in Eurasia*, pp 93–100, London: University College London Press

Kim, W Y, Im, H J and Choi, M I (1978) 'The Hunamni site – a prehistoric village site on the Han river: progress reports 1976, 1977', *in Archaeological and Anthropological Papers of the Seoul National University*, volume 8, Seoul: Seoul University Museum and Department of Archaeology

Lévi-Strauss, C (1955) *Tristes tropiques*, Paris: Librairie Plon

Marshall, F and Hildebrand, E A (2002) 'Cattle before crops: the beginnings of food production in Africa', *Journal of World Prehistory* 16(2): 99–143

Martin, F W and Rhodes, A M (1977) 'Intra-specific classification of *Dioscorea alata*', *Tropical Agriculture* 54: 1–13

Matthews, P (1991) 'A possible wild type taro: *Colocasia esculenta* var. *aquatilis*', *Bulletin of the Indo-Pacific Prehistory Association* (Canberra) 11: 69–81

Matthews, P J (1995) 'Aroids and Austronesians', *Tropics* 4(2): 105–26

Mbida C, Doutrelepont, H, Vrydaghs, L, Swennen, R, Swennen, R, Beeckman, H, De Langhe, E and Maret, P de (2001) 'First archaeological evidence of banana cultivation in Central Africa during the third millennium before present', *Vegetation History and Archaeobotany* 10: 1–6

Mbida, C, Van Neer, W, Doutrelepont, H and Vrydaghs, L (2000) 'Evidence for banana cultivation and animal husbandry during the first millennium BC in the forest of southern Cameroon', *Journal of Archaeological Science* 27: 151–62

Misra, V N and Kajale, M D (eds) (2003) *Introduction of African Crops into South Asia*, Pune: Indian Society for Prehistoric and Quaternary Studies, Deccan College Post-Graduate and Research Institute

Neumann, K (2005) 'The romance of farming: plant cultivation and domestication in Africa', in Stahl, A B (ed), *African Archaeology: A Critical Introduction*, pp 249–75, Malden, MA: Blackwell Publishing

Neumann, K, Butler, E A and Kahlheber, S (eds) (2003) *Food, Fuel and Fields: Progress in African Archaeobotany*. Africa Praehistorica 15, Cologne: Heinrich-Barth-Institut

Perry, L (2004) 'Starch analyses reveal the relationship between tool type and function: an example from the Orinoco Valley of Venezuela', *Journal of Archaeological Science* 31: 1069–81

Piggott, S (1954) *The Neolithic Cultures of the British Isles*, Cambridge: Cambridge University Press

Piperno, D R and Holst, I (1998) 'The presence of starch grains on prehistoric stone tools from the humid neotropics: indications of early tuber use and agriculture in Panama', *Journal of Archaeological Science* 25: 765–76

Piperno, D R and Pearsall, D M (1998) *The Origins of Agriculture in the Lowland Neotropics*, San Diego: Academic Press

Piperno, D R, Ranere, J A, Holst, I and Hansell, P (2000) 'Starch grains reveal early root crop horticulture in the Panamanian tropical forest', *Nature* 407: 894–97

Pope, K O, Pohl, M D E, Jones, J G, Lentz, D L, Von Nagy, C, Vega, F J and Quitmyer, I R (2001) 'Origin and environmental setting of ancient agriculture in the lowlands of Mesoamerica', *Science* 292: 1370–73

Renfrew, C (1973) *Before Civilization: The Radiocarbon Revolution and Prehistoric Europe*, Harmondsworth: Penguin Books

Renfrew, C (1987) *Archaeology and Language: The Puzzle of Indo-European Origins*, London: Jonathan Cape

Roscoe, P (2002) 'The hunters and gatherers of New Guinea', *Current Anthropology* 43(1): 153–62

Schmitz, P I (1991) 'Migrantes da Amazonia: a tradiçao Tupiguarani', in Kern, A (ed), *Arqueologia Pré-histórica do Rio Grande do Sul*, pp 295–330, Porto Alegre, Brazil: Mercado Aberto

Schouse, B (2001) 'Spreading the word, scattering the seeds', *Science* 294: 988–89

Smith, B D (2001) 'Low-level food production', *Journal of Archaeological Research* 9: 1–43

Spriggs, M (1989) 'The dating of the Island Southeast Asian Neolithic: an attempt at chronometric hygiene and linguistic correlation', *Antiquity* 63: 587–613

Spriggs, M (1996) 'Early agriculture and what went before in Island Melanesia: continuity or intrusion?', in Harris, D R (ed), *The Origins and Spread of Agriculture and Pastoralism in Eurasia*, pp 524–37, London: University College London Press

Spriggs, M (1997) *The Island Melanesians*, Oxford: Blackwell

Terrell, J E, Hart, J P, Barut, S, Cellinese, N, Curet, A, Denham, T P, Haines, H, Kusimba, C M, Latinis, K, Oka, R, Palka, J, Pohl, M E D, Pope, K O, Staller, J E and Williams, P R (2003) 'Domesticated landscapes: the subsistence ecology of plant and animal domestication', *Journal of Archaeological Method and Theory* 10(4): 323–68

Thomas, J (1996) 'The cultural context of the first use of domesticates in continental Central and Northwest Europe', in Harris, D R (ed), *The Origins and Spread of Agriculture and Pastoralism in Eurasia*, pp 310–22, London: University College London Press

Thomas, J (1999) *Understanding the Neolithic*, London: Routledge

Trigger, B G (1989) *A History of Archaeological Thought*, Cambridge: Cambridge University Press

van der Veen, M (ed) (1999) *The Exploitation of Plant Resources in Ancient Africa*, New York: Kluwer Academic/Plenum Publishers

Vansina, J (2004) 'Bananas in Cameroon c 500 BCE? Not proven', *Azania* (Nairobi) 38: 174–76

Vrydaghs, L (2003) Studies in opal phytoliths: methods and identification criteria, unpublished PhD thesis, Ghent University

White, J P (2004) 'Where the wild things are: prehistoric animal translocation in the circum New Guinea Archipelago', in Fitzpatrick, S M (ed), *Voyages of Discovery: The Archaeology of Islands*, pp 147–64, Westport: Praeger

Yen, D E (1989) 'The domestication of environment', in Harris, D R and Hillman, G C (eds), *Foraging and Farming: The Evolution of Plant Exploitation*, pp 55–75, London: Unwin Hyman

Yen, D E (1996) 'Melanesian arboriculture: historical perspectives with emphasis on the genus *Canarium*', in Evans, B R, Bourke, R M and Ferrar, P (eds), *South Pacific Indigenous Nuts*, pp 36–44, Canberra: ACIAR

Yim, H (1978) 'The Hunam-ni dwelling site, report 4', in *Archaeological and Anthropological Papers of the Seoul National University*, vol 8, Seoul: Seoul University Museum and Department of Archaeology

Zvelebil, M (1986) *Hunters in Transition: Mesolithic Societies of Temperate Eurasia and the Transition to Farming*, Cambridge: Cambridge University Press

CHAPTER 2

Agriculture, Cultivation and Domestication: Exploring the Conceptual Framework of Early Food Production

David R Harris

INTRODUCTION

Progress in scientific understanding depends on the interaction of theory and data, with theory provoking the search for evidence and the evidence in turn modifying theory. This applies as much to the investigation of past as to presently observable phenomena, with the important difference that the experimental method is much less readily applied to past situations that cannot be observed directly. Notwithstanding the axiom that the present is the key to the past, direct evidence is much less accessible to the historical sciences, and dependence on surrogate data is accordingly much greater.

This problem is well-illustrated by efforts to understand how agriculture emerged and eventually became the dominant mode of human subsistence. Evidence of past agricultural production, in the form of well-preserved, accurately identified and directly dated remains of crop plants and domestic animals, is sparse. Most of it relates to those parts of the world where it has been systematically sought by archaeologists, notably in such presumed 'centres of origin' as the Southwest Asian 'Fertile Crescent', central China and Mexico. Other types of archaeologically observable evidence of early agriculture, such as ancient fields, irrigation and drainage works, crop-storage structures and agricultural tools, occur quite widely, but they generally relate to later rather than very early phases in the development of agricultural systems.

Evidence relevant to the origins and early development of agriculture can also be derived from biogeographical, genetic, palaeoenvironmental, historical, ethnographic and present-day ecological studies, but inferences drawn from these sources are often less direct and conclusive than those based on archaeological data. For example, although genetic studies have revealed the ancestry of many crops and domestic animals, assumptions about the areas of origin of their wild progenitors have often been based on the present distribution of the wild forms without regard for environmental changes during and before the Holocene and the possibility that the ranges of the progenitor species may have been different when they were domesticated. Genetic analysis is an indispensable source of evidence of the ancestry of domesticated species and, when combined with biogeographical data, can indicate their probable areas of origin, but without corroborative palaeoenvironmental evidence, inferences from genetic and biogeographical data alone are seldom conclusive. Palaeoenvironmental, principally palaeoecological, data are also a valuable source of evidence of past agricultural practices. For example, the potential of fossil pollen, phytolith and charcoal evidence to demonstrate past forest clearance for shifting (swidden) cultivation has been well demonstrated in Papua New Guinea (Haberle 1994) and Panama (Piperno 1994), but it is often difficult to determine whether changes in forest composition revealed by the analysis of pollen grains and microscopic particles of charcoal are the result of former cultivation or of natural environmental factors.

Historical records and ethnographic observations can also provide valuable data relevant to early agriculture, but they too suffer from limitations. Historical accounts are usually sketchy, seldom provide comprehensive descriptions of past agricultural practices, and mostly date to recent historical periods. The value of ethnographic evidence is likewise limited, by its recency and by the difficulties inherent in using the detailed knowledge it can provide of 'traditional', premodern agricultural systems to interpret the fragmentary direct evidence available of much earlier, prehistoric forms of agricultural production. It is seldom possible to link archaeological with ethnographic evidence directly across such large gaps in time, but it is sometimes feasible. This is well illustrated by the Kuk project in highland Papua New Guinea (Denham chapter 5, Golson chapter 6 and Bayliss-Smith chapter 7), although even there the direct connection between the ethnographic and the archaeological data relates only to the more recent phases of agricultural evolution. In general, great caution is needed when models based on ethnographic or historical data are used to interpret prehistoric agriculture.

The same caution applies to the use of inferences derived from present-day ecological observations. However, ecological data are more widely available than historical or ethnographic evidence and, especially when relevant palaeoenvironmental data also exist, they can generate well-founded interpretations of exiguous archaeological evidence. The potential value of such ecological modelling is exemplified by its application to the interpretation of the plant remains recovered from Epipalaeolithic levels at the very early agricultural site of Tell Abu Hureyra in Syria (Hillman 2000).

I have referred above, without qualification, to 'agriculture', 'agricultural production' and 'agricultural systems'. But such terms, and others used in common parlance and in academic discourse on early agriculture, such as 'cultivation', 'domestication', 'horticulture', 'husbandry' and 'pastoralism', have multiple meanings. It is not only the sparse and geographically uneven sources of available evidence that hamper investigation of the beginnings of agriculture; the subject is also held back by conceptual and terminological confusion. This is partly due to the multidisciplinary nature of the enterprise, which has introduced into the discourse numerous terms that carry prior connotations from their disciplines of origin. For many years this semantic confusion has militated against analytical precision in our thinking about how and why agriculture emerged.

Agriculture and domestication are prime examples of imprecise 'catch-all' concepts that create confusion because users of them tend to assume that others share the same, usually intuitive and seldom explicitly stated, understanding of what they mean. The published literature on the emergence and early development of agriculture is characterised by a plethora of general, often ill-defined terms. 'Agriculture' is frequently qualified with prefixes and adjectives the intended meaning of which is not always made clear, such as 'proto', 'incipient', 'developed', 'intensive' and 'extensive'. Likewise, 'domestication' is qualified with adjectives such as 'incidental', 'specialised', 'agricultural' (Rindos 1984: 153), 'behavioural', 'cultural', 'full', 'complete' (Zvelebil 1986: 174; 1995: 98), 'cultural', 'economic', 'social' and 'symbolic' (Hodder 1990: 31–39). The literature on early agriculture contains many other descriptive terms with variable connotations, for example (ordered alphabetically) agronomy (Yen 1989: 57), arboriculture (Gosden 1995), cultivation, domiculture (Chase 1989; Hynes and Chase 1982), farming, food production, gardening, herding, horticulture, husbandry, pastoralism (Clutton-Brock 1989) and transhumance (Delano Smith 1979: 239–56). Faced with this impressive yet confusing array, the need for conceptual clarification is clear enough – but exceedingly difficult to achieve. In the next section, I examine the derivation and use of three overarching concepts – domestication, agriculture and cultivation – as well as several of the more specific descriptive terms, and suggest how their meanings might most appropriately be refined. I do not cherish the illusion that this semantic exercise will lead to a consensus as to how these terms should in future be used, but I hope that it will foster greater awareness of the need for those engaged in the study of early agriculture to define their terminology more precisely.

TERMS OF DEBATE

The confusion created by the multiplicity of terms used in discussions of early agriculture is compounded by their use both as descriptors of specific agricultural activities and as more abstract labels for general categories of food production, such as horticulture and pastoralism, that are in reality more complex and variable than their labels suggest. This tendency is particularly evident in usage of the broad, overlapping terms 'domestication', 'agriculture' and 'cultivation', examined below. We start with domestication because it is central to any discussion of the conceptual dimensions of 'agriculture' and 'cultivation', and because different roles are accorded to it as a criterion for separating hunting and gathering from agriculture.

Domestication

The derivation of 'domestic' from *domus,* the Latin for house, is well known. According to the *Oxford English Dictionary* (1971) the word entered the English language via the French *domestique* and referred at first to a person's state of belonging to a home or household. By the seventeenth century in England its meaning had been extended to include cultivated plants and tame animals cared for by people and living in or near human habitations. It was not until the nineteenth century, and especially following the publication in 1868 of Charles Darwin's *The Variation of Animals and Plants under Domestication,* that morphological, behavioural and later genetic change became an integral part of what came to be regarded as the orthodox (morphogenetic) concept of plant and animal domestication. However, while biologists thought of domestication as a dynamic process, historians and archaeologists tended to regard it more as a series of past events that had brought new forms of plant and animal into existence. Frederick Zeuner was one of the earliest scientists working in an archaeological institution (the University of London's Institute of Archaeology) to treat domestication as a biological process and to relate it to archaeological and historical evidence. He distinguished five 'stages of intensity' of animal domestication, from loose human contacts with free-breeding animals to the planned development of breeds and the extermination of wild ancestors; he also proposed a chronological sequence in which animals had probably been domesticated (Zeuner 1963).

By the 1960s several archaeologists had become interested in the processes of plant and animal domestication. They began to question the previously rigid conceptual distinction between hunter-gatherers dependent on wild plants and animals and agriculturalists dependent on crops and domestic livestock. One result of this shift was the formulation of the concepts of 'pre-domestication cultivation' (Helbaek 1960; Hillman 1975) and 'pre-domestication animal husbandry' (Jarman et al 1982: 51–52).

In the 1980s more comprehensive evolutionary models of the domestication process, which sought to define its place in the long continuum of human, plant and animal relationships, were proposed by Rindos (1984: 152–66), Ford (1985) and Harris (1989). All three focused on plant exploitation and recognised the existence of forms of 'food production' involving cultivation, which occupied an intermediate position between complete dependence on wild sources of food (hunter-gatherer foraging) and predominant dependence on domesticated crops and livestock (agriculture). However, the role each author accorded to domestication varied. Both Ford and I made the presence of domesticated plants a criterion for separating 'agriculture' from early forms of cultivation, which included small-scale tending, planting and harvesting of plants that had undergone little or no morphological change from the wild state. These early forms of cultivation were termed by Ford 'incipient agriculture' and 'gardening' and by me 'wild plant-food production with minimal tillage' and 'cultivation with systematic tillage'. Rindos applied the term 'domestication' to the whole of what he regarded as a coevolutionary process based on symbiotic relationships between plants and humans that spanned all human existence. He proposed three categories of domestication: 'incidental', 'specialised' and 'agricultural', the last of which can be equated with morphogenetic domestication. The authors of all three models emphasised that they were dealing with a continuous evolutionary process and that the purpose of the categories they defined was to describe more precisely plant–people relationships, not to devise explanatory schemes. Rindos (1984: 152) did not claim that his model represented stages in the development of all agricultural systems, and I stated explicitly that mine was not unidirectional and deterministic (Harris 1989: 12, 18). The idea of a series of stages was, however, inherent in Ford's model, which envisaged incipient agriculture, gardening and field agriculture as having progressively succeeded one another (Ford 1985: 2–7).

In 1996 I published a fuller discussion of domesticatory relationships of people, plants and animals which included a revised version of my 1989 scheme for plants and also a parallel evolutionary scheme for animals (Harris 1996: figures 15.1, 15.2). The main difference between the two plant schemes is that the more recent one represents domestication as a long, drawn-out process that began well before the establishment of agricultural systems, now defined as those in which (morphogenetically) domesticated crops are the predominant source of food. Thus the model allows for the presence, but not the dominance, of some domesticated plants in pre- and non-agricultural subsistence systems, in addition to cultivated wild plants that have undergone little or no morphological (phenotypic) or genetic change as a result of human intervention in their reproduction. The boundary between 'agriculture' and 'wild plant-food production' is therefore less absolute than it was in the 1989 scheme and more recognition is given to subsistence systems that are neither strictly agricultural nor wholly based on the exploitation of wild plants. The same is true of my 1996 scheme for animal exploitation, in which I defined a series of activities by which humans intervened in the ecology and reproductive biology of animals, and grouped them into three main categories: predation, protection and domestication. Systems of protection include taming, protective herding and free-range management; the domestication threshold is crossed when groups of animals reproducing under human control become genetically isolated from their wild relatives – a process that was often facilitated in the past by fortuitous or deliberate segregation of the breeding stock, resulting in their geographical isolation from wild populations.

Ethnographic and historical evidence indicates that such 'in-between' systems of animal protection and plant management that are neither foraging nor farming have existed in many parts of the world without developing into agricultural systems, but much less attention has been paid to them by archaeologists than to the investigation of prehistoric hunter-gatherer and agricultural societies. They remain largely invisible in the archaeological record and deserve more attention from archaeobotanists and archaeozoologists than they have so far received, although it is difficult to establish criteria by which to recognise them archaeologically.

Increasing recognition of the complexity and variability of systems of premodern plant and animal exploitation has led some students of early agriculture and hunter-gatherer subsistence to question the validity of the orthodox definition of domestication as a process that necessarily involves morphogenetic change and its use as a defining criterion of agriculture. The latter issue was keenly debated at the symposium in 2003 that gave rise to this book. I believe that this use of the term domestication remains valid for two main reasons. First, inadvertent and deliberate intervention by people in the reproduction of culturally selected plants and animals, leading to their genetic isolation from wild progenitors and dependence on sustained human care for their long-term survival, represents a fundamental change in human subsistence which vastly increased the density of populations that could be supported, by means of agriculture, per unit area of food-producing land. The second, more pragmatic reason is that changes diagnostic of morphogenetic domestication, such as the large seeds and non-brittle rachises of cereal crops, are often detectable in organic remains recovered from archaeological sites and can thus provide direct evidence of the emergence of agriculture. The objection can be raised that this skews the archaeobotanical record by overemphasising the past importance in food-producing economies of cereals and other crops with hard seeds or fruits because they tend to survive better in archaeological contexts than the soft tissues of roots, tubers, leafy vegetables and many fruits. This was generally the case in past archaeological investigations of early agriculture, and it remains a potential difficulty today, but micromorphological techniques such as analyses of parenchyma, phytoliths and starch grains now offer new methods with which to identify the remains of root and tuber, as well as seed, crops (Harris 2006). As these techniques are further developed and more widely applied in varied archaeological contexts, the bias towards cereals and other crops with hard seeds that are often preserved in a charred state will diminish.

This brief discussion of the concept of domestication leads to the conclusion that, in its morphogenetic sense it remains a valuable criterion by which to define, and recognise in the archaeological record, systems of food production that have an agricultural component, ie when at least some domesticated taxa are present. However, whether a society is designated 'agricultural' as opposed to occupying a position somewhere between agriculture and hunter-gatherer foraging is a matter of the *relative* importance of domesticated and wild plants and animals in the subsistence economy. This leads us on to a consideration of the second general term of debate – agriculture and, subsumed within it, pastoralism and horticulture.

Agriculture

The English word 'agriculture' derives from two Latin roots: *ager*, a field, and *colo*, to cultivate. They were combined in the Latin *agricultura*, meaning tillage of the land. The word 'agriculture', in the sense of tillage of fields to raise crops, was in use in both French and English by the early seventeenth century. The term was also closely linked with the concept of 'husbandry', via the Latin *agricola*, meaning 'husbandman'. In modern usage 'agriculture' (unlike 'farming') is sometimes restricted to the cultivation of crops and excludes the raising of domestic animals, but more usually it implies both. Indeed, the *Oxford English Dictionary* (1971) defines agriculture as 'The science and art of cultivating the soil, including the allied pursuits of gathering in the crops and rearing live stock [*sic*]; tillage, husbandry, farming (in the widest sense)'.

Most students of agricultural history regard livestock raising as part of 'agriculture'. However, we should remember that this view stems from Eurasian traditions of 'agro-pastoralism' whereby in Southwest Asia goats, sheep and later cattle became fully integrated with the cultivation of cereal and pulse crops (Harris 1998, 2002), and in central China water buffalos became an integral part of the system of wet-rice (padi) cultivation (Glover and Higham 1996: 428; Underhill 1997: 141–44). Such integration, in which the animals provide food, fertilizer and traction, did not occur in other regions of early agriculture where domestic livestock were present but not fully incorporated into the indigenous systems of seed and root crop cultivation, for example the central Andes and northern tropical Africa. In yet other regions domestic animals were only linked to crop production as consumers, for example turkeys in Mexico and pigs in highland New Guinea. Although prehistoric and early historic agricultural systems varied widely in the extent to which domestic animals were incorporated into the production process, it is appropriate to regard livestock raising as part of 'agriculture', provided that we are careful to distinguish and define particular systems of agricultural production within the overarching concept. There are many types of such systems, for example shifting (swidden) cultivation and irrigation agriculture in its various manifestations, which need not be detailed here, but pastoral systems call for comment because the terms used to describe forms of pastoralism are often vague and confusing.

Pastoralism

Derived from the Latin *pastor*, a herdsman or shepherd, 'pastoralism' refers broadly to systems of mobile livestock pasturing in which domesticated ungulates, principally sheep, goats, cattle, horses, donkeys, camels (including the domesticated South American camelids, llama and alpaca) and reindeer, are raised to provide meat, milk (and its secondary products yoghurt, butter and cheese), hides, hair and wool, and for riding, traction and load carrying. All pastoralists depend in part on crop products for their food and usually also for supplementary fodder for their animals, but the spatial and temporal scales of their movements vary greatly. These range from short-distance daily movements of flocks and herds to and from pastures near the settlements (diurnal grazing) to long-distance seasonal movements to distant pastures (transhumance) and to the most specialised manifestation of the pastoral life (nomadic pastoralism). Nomadic pastoralists own and largely depend on their animals for food and they move from camp to camp and from year to year with their livestock, only occasionally trading with (or raiding) settled agricultural communities. Nomadic pastoralism was limited historically, and probably prehistorically, to: the arid lands of northern and eastern Africa, where the focus was on camels or cattle with sheep and goats; the deserts of mid-latitude Asia where it was on camels, horses, sheep and goats; and high-latitude Eurasia where reindeer pastoralism developed. The nomadic pastoralists of Asia and Africa routinely obtained grain and other agricultural products from farmers, but they lived apart from agricultural communities.

Transhumance differs from nomadic pastoralism in that it seldom if ever involves the movement of entire communities. It normally refers to the seasonal movements of selected members of agricultural villages with their domestic flocks and herds, most commonly to summer pastures in the mountains. It is therefore inappropriate to apply the term, as some archaeologists have, to seasonal movements of hunter-gatherers, such as the inferred altitudinal migrations of Preceramic forager bands in the northern Peruvian Andes (Lynch 1971) or the residential shifts of Desert Culture peoples of the western Great Basin in North America (Davis 1963).

Horticulture

'Horticulture' is another term that has more than one connotation and is variously, and often loosely, employed in the literature on traditional agricultural systems and the evolution of agriculture. In the same way that nomadic pastoralism can be regarded either as part of agriculture in the broad sense (as it is here) or as a separate form of food production, so too can horticulture. Horticulture, and the synonymous gardening or garden cultivation, derive from *hortus*, the Latin word for garden, in contrast to field (*ager*). The two terms imply differences both in the scale of cultivation and in the kinds of plants grown. Normally gardens are smaller than fields and are more structurally and floristically diverse. They contain a greater variety of plant forms and taxa, especially of perennial trees and shrubs and, unlike most fields, they tend to be cultivated throughout the year. The difference in scale is widely recognised in the literature on early agriculture, for example in the German distinction between *Ackerbau*, or large-scale field agriculture, and *Gartenbau*, or small-scale garden cultivation.

The terms 'horticulture' and 'gardening' have also been used to denote systems of crop cultivation in some parts of the world, principally in Melanesia and the Pacific Islands, that combine the growing of (usually annual) root and cereal crops in fields with the raising of (mainly perennial) tree, shrub and herbaceous crops in gardens. This obscures the valuable descriptive distinction between field and garden cultivation and in effect treats horticulture as a general term synonymous with agriculture. In fact most ethnographically and historically described systems of agriculture combine elements of garden and field cultivation. Although this may be particularly evident in Melanesia and the Pacific (and elsewhere in the tropics), it does not, in my view, justify replacing agriculture with horticulture as the preferred general term.

If use of the term 'horticulture' is restricted to its literal, narrower meaning of garden cultivation, it can usefully be reserved to denote a distinctive form, and subdivision, of agriculture. But there is a further contrast to consider, which stems from differences in the ecology of gardens and fields. Studies of house (or home) gardens in the tropics within village settlements and in nearby outlying areas have shown that species diversity tends to be much higher in gardens than in fields, and that many of the plants growing in the gardens are adventitious, spontaneously occurring wild or weedy species rather than deliberately planted crops. In contrast, swidden clearings and other types of field characteristically have lower ratios of wild/weed taxa to planted crops, even in within-field mixed-crop (polycultural) systems of cultivation (Coomes and Ban 2004; Kimber 1973, 1978; Padoch and de Jong 1991). The gardens are valued not only for the produce of the crops cultivated in them, but because many of the wild and weedy plants are also sources of attractive and useful products such as fruits, flowers, edible leaves, fibres, dyes, medicines, construction materials and firewood (Kimber 1978: Lamont et al 1999: 316; Landauer and Brazil 1990; Padoch and de Jong 1991: 169; Terra 1954). The semiwild ecosystems of the gardens, with their varied life forms, floristic diversity and frequently disturbed soil, favour hybridisation and the appearance of novel varieties, which are readily noticed and sometimes propagated by observant owners who visit their gardens frequently, often daily.

As sources of both domestic and useful wild plants, house gardens straddle the boundary between my definitions of 'agriculture' and 'wild plant-food production' (Harris 1996: figure 15.1), Ford's (1985: figure 1.1) division between 'cultivation' and 'domestication' and Smith's (2001a: figure 7) boundary between 'agriculture' and 'low-level food production with domesticates'. Indeed, it may well be, as I have previously argued (Harris 1973: 398–401), that house gardens were early and active locales of plant domestication. I think it highly probable that many taxa, especially of tropical trees and shrubs, were initially grown on small, frequently tended plots close to dwellings where they were reproduced vegetatively or from seed, underwent cultural selection and became domesticates. However, I do not mean to imply that horticulture, in the sense of garden cultivation, represents a general evolutionary stage that everywhere preceded field cultivation, although it may well have done so in many situations. It comprises a distinct small-scale system of agricultural production whereby food and other valued products are obtained not only from domesticated crops but also from wild and weedy plants. Although some archaeobotanists have presented putative evidence for very early garden cultivation, for example in the Early Jomon period in Japan (Crawford 1992: 19–20), it merits more archaeobotanical research than it has so far received, despite the difficulty of establishing criteria for its archaeological recognition.

Cultivation

The term 'cultivation' is even more general, and its usage is more varied, than either agriculture or domestication. Derived from the Latin verb *colo*, meaning to cultivate or till the land, and more directly from the medieval Latin *cultivare*, it is limited, in its narrowest and most literal connotation, to tillage of the soil for crop production. In common parlance, too, it usually refers to the growing of domesticated crops for food and other purposes, but its meaning has expanded from its earlier limited connotation to include, in addition to tillage, a range of other activities that promote plant growth, such as land clearance, planting, sowing, weeding, harvesting, soil drainage and irrigation. Even the deliberate burning of vegetation to aid clearance, add nutrients to the soil, and promote the growth of useful plants can be regarded as within the scope of cultivation, most obviously in the context of swidden cultivation (Harris 1972). All these activities can be, and have been, applied to wild as well as domestic plants, and although, in everyday use, the term 'cultivation' is normally restricted to agricultural contexts, it can refer with equal justification to non-agricultural contexts of wild-plant exploitation. It is a useful, if very general, term that can be applied across the continuum of plant–people interaction, all the way from situations in which people depend principally on wild-plant products but enhance productivity by tending some species, to developed systems of agriculture focused on domesticated crops.

Although the array of techniques encompassed by the term 'cultivation' can be regarded both as a prelude to, and an integral part of, agriculture, we should not interpret their practice in non-agricultural contexts as a transitional process leading (unless arrested by external factors) to the development of agriculture. The cultivation, by so-called hunter-gatherer groups without agriculture, of plants that have not been changed morphogenetically by cultural selection is well documented ethnographically and historically. Thus, many examples of cultivation practices have been reported among tribal peoples in North America, Australia and elsewhere whose societies were apparently stable, enduring and manifestly not 'on the road to agriculture' (eg Allen 1974; Bean and Lawton 1973; Ellen 1988; Harris 1984; Keeley 1995; Lawton et al 1976; Shipek 1989; Steward 1930; Yen 1989). Similarly, in the sphere of animal exploitation, such practices as taming, protective herding, and free-range management of undomesticated species – which have often been regarded as transitional states between the hunting of wild animals and the raising of domesticated ones – should be regarded as enduring practices in their own right (eg Aikio 1989; Ingold 1980; Serpell 1989; Simoons 1968).

Societies characterised by such practices engage in what Bruce Smith, in an original and timely exploration of the conceptual and actual 'middle ground' between hunting-fishing-gathering and agriculture, has labelled 'low- level food production'. Smith argues that they are 'distinctively, qualitatively different from pre-Holocene hunter-gatherers on the one hand and agriculturalists on the other' (Smith 2001a: 15, 33) – a judgment that I strongly endorse. Indeed, this middle ground, which we perceive mainly in the ethnographic and historical records, remains elusive archaeologically and deserves much more research attention from archaeologists than it has so far received.

The main purpose of the preceding sections has been to contribute to the theme of this book by exploring and refining the principal concepts that are embedded in our thinking about the origins and early development of agriculture. But this alone is not enough. We need to move beyond what some may regard merely as a semantic exercise towards a clearer conceptual framework for thinking about the diversity of past human subsistence. In the remaining sections of this chapter my aim is to contribute to this framework in two ways: by commenting on spatial scales in the study of early agriculture and by suggesting that 'biotic resource specialisation' has value as a concept for the comparative analysis of human subsistence.

SCALES OF ENQUIRY

Study of the origins and early development of agriculture is conducted at varied spatial scales that range from local, often single-site based investigations, through regional-scale analyses to worldwide comparative syntheses, but the relation between the aims and the scales of enquiry are often not clearly specified. The need for greater clarity is particularly acute when worldwide comparisons are attempted, because they inevitably depend both on what is, at the world scale, a largely fortuitous spatial distribution of areas and sites where relevant investigations have been undertaken, and on datasets of uneven quality that are seldom strictly comparable. Nevertheless, worldwide syntheses need to be attempted if we wish to understand, at the global level, how agriculture emerged. It may be objected that trajectories towards and into agriculture in different areas and ecological contexts have been so varied and distinctive that it is unrealistic to seek higher-level global, or even continental, explanations; that we should be satisfied with more local 'contingent' explanations that account for particular transitions from hunting-gathering to agriculture. But the two approaches, one emphasising the value of detailed local investigations and the other the need for worldwide comparisons, are (in my view) interdependent and equally valid. Nor does the comparative approach imply a commitment to a single 'universalist' explanation – only that continental and worldwide comparisons are needed if we are ever fully to comprehend the multiple subsistence pathways followed by human groups as they increasingly intervened in the life cycles of plants and animals through cultivation, taming and domestication.

This point can be illustrated by referring again, by way of example, to the two intensively studied early sites that I have already mentioned, which have revealed strongly contrasted pathways to agriculture: Tell Abu Hureyra in the Euphrates Valley in Syria (Moore et al 2000) and Kuk in highland Papua New Guinea (Denham et al 2003, 2004; Golson 1977, 1989, 1997). Whereas at Abu Hureyra there is evidence that cereal cultivation began under the impact of the cold, dry conditions of the early postglacial Younger Dryas climatic phase (Hillman 1996, 2000; Hillman et al 2001), at Kuk a very different trajectory was followed where the small-scale 'garden' cultivation of such root and fruit crops as taro and banana appears to have been a prelude to more extensive wetland cultivation of domesticated taro, and later sweet potato, as staple crops. This summary statement greatly simplifies two complex histories of agricultural evolution, but it serves to emphasise how important it is that local trajectories are studied in depth, not only for their own interest but also for the light they can throw on larger-scale patterns of agricultural evolution that may be observed in other environments and cultural settings. Unfortunately, the scope for such comparisons is at present restricted by the small number of sites and areas of the world where modern techniques of recovery, analysis and dating of plant and animal remains have yielded reliable data on subsistence change.

At the world scale, investigation and discussion of early agricultural and preagricultural subsistence have been hampered by the pervasive influence of the concept of centres of origin. First adumbrated by Alphonse de Candolle (1855, 1882, 1884), later formulated and elaborated by Nicolai Vavilov (1926) and modified by, among others, Jack Harlan (1971), the concept has, by focusing on the so-called primary centres or nuclear areas of agricultural origins, skewed interpretations of the evidence and discouraged new research in other regions of the world, including vast areas in Africa, South and Southeast Asia and South America that Harlan misleadingly designated 'non-centres'. As a framework for the comparative worldwide analysis of the emergence of agriculture the concept has, I believe, outlived its usefulness (Harris 1990). It is now being replaced by the more complex conception of multiple areas of local domestication that recent research has begun to reveal, for example in parts of tropical Africa (D'Andrea et al 2001; Neumann 1999), India (Fuller et al 2004; Harvey et al 2005), and the Americas (Pearsall 1992; Piperno and Pearsall 1998; Piperno et al 2000; Smith 1992, 2001b). Also, there is a growing realisation that within some of the regions conventionally regarded as primary centres, for example Southwest Asia and China, domestication and the early development of agricultural systems were much more spatially and functionally diverse than had been thought when the 'Fertile Crescent' and 'North China' were regarded simply as large undifferentiated regions in which agriculture originated (eg Bar-Yosef and Meadow 1995; Cohen 1998; Crawford and Shen 1998; Garrard 1999; Harris 2002; Lu 1999; Willcox 1998). In 1990 I suggested that it was time to decouple the concept of world centres of origin from investigations of the beginnings of agriculture and to 'focus instead on the evolutionary history of individual crops and regional crop associations' (Harris 1990: 15), a conceptual change that is now well underway.

Increasing acceptance of the concept of multiple areas of local domestication does, however, bring to the fore a related question of process and spatial scale that is of great significance in the prehistory of agriculture: how to identify, among the many local areas of domestication, those from which agricultural systems, with their crops and in some cases domestic animals, spread. The former concept of a small number of primary centres of origin led logically to the assumption that agriculture had spread to the rest of the cultivable world from them, but this view is no longer justified. It needs to be replaced by more refined hypotheses as to why some early agricultural systems spread and others did not.

Differences in the capacity of agricultural systems to provide an adequate human diet without substantial reliance on wild food resources is, I suggest, one factor that merits close attention. In a preliminary analysis, I have examined the expansion capacity of several systems by comparing broadly the nutritional status of their assemblages of domesticated plants and animals (Harris 2003). The analysis was based on the idea of 'core regions' of early agriculture, as opposed to the traditional concept of a few primary centres where agriculture was believed to have originated independently, and earlier than anywhere else, and from where it was assumed to have spread to most of the rest of the world. The concept of core regions is less deterministic and more responsive to new evidence than that of primary centres. Core regions are identified by synthesising currently available archaeological, palaeoenvironmental, biogeographical, genetic and ethnohistorical data in order to infer where distinctive suites of crops, and in some areas also domestic animals, formed early (but not necessarily the earliest) agricultural systems, some of which spread extensively while others did not. In my preliminary analysis (2003) I only presented evidence from five such regions (western Southwest Asia, central China, Mexico, the Andean highlands and northern tropical Africa), and in so doing I was keenly aware of the inadequacies of the heterogeneous sources of data on which the analysis depended. Finer-scale analysis of more adequate, especially archaeological, data from a larger number of smaller regions is needed to develop and test the hypothesis, but the preliminary exercise I undertook appears to confirm that differences in the nutritional basis of early agricultural systems do correlate with whether or not they spread. The main (here oversimplified) conclusion is that the systems which had the greatest potential to expand territorially into new environments, and to eliminate or absorb preexisting systems of wild-food procurement and production, were those that included cereals and pulses among their staple crops, especially when the crops were combined with livestock raising in systems of agro-pastoral production.

BIOTIC RESOURCE SPECIALISATION

Consideration of contrasts in the extent to which different agricultural systems spread leads to the last general concept I wish to discuss: 'biotic resource specialisation'. As the more expansive agricultural systems spread, their crops and domestic animals progressively replaced and reduced the populations of many of the wild and domesticated taxa that were previously exploited by the indigenous inhabitants of the regions into which these systems expanded. As a result of this process, biotic resource use tended to become more specialised in the regions that received the new domesticates and the overall range of resources used became narrower.

The concept of biotic resource specialisation is valuable because it encompasses the entire spectrum from wild-food procurement (hunter-gatherer foraging) through the management of selected plants and animals to enhance their productivity in systems of wild-food production, to the genetic isolation and morphogenetic domestication of particular taxa, some of which eventually became the mainstay of (agricultural) systems of crop and livestock production (Figure 2.1). It thus offers an overarching framework for the comparative analysis of long-term trends in human subsistence at local, regional, continental and world scales.

Viewed from this perspective, we can envisage the evolution of human subsistence through the Holocene as a trend towards greater specialisation in biotic resource use, however temporally punctuated and spatially irregular it

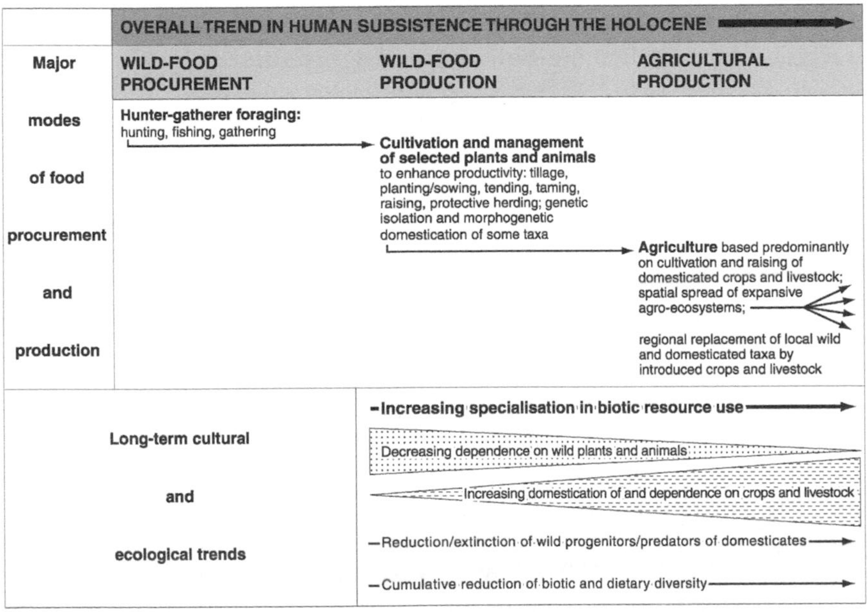

Figure 2.1 A schematic representation of major modes of food procurement and production, and long-term trends in human subsistence, through the Holocene. The three modes are shown as separate entities for analytical purposes, but the reality they represent is made up of mixed subsistence systems in which the balance between food procurement and production varied greatly, both spatially and temporally. The overall trend from hunter–gatherer foraging to agricultural production is envisaged as one of increasing specialisation in the use of biotic resources, and does *not* imply that the gradual shift from foraging to agriculture was a unidirectional process through a sequence of predetermined developmental stages.

may have been. The gradual shift from dependence on wild to dependence on domesticated plants and animals for food culminated in recent centuries with the complete reliance of almost the entire human population on the products of agriculture. Today, a mere 30 crops account for 95% of plant-derived energy in the human food supply, over half of which is provided by maize, rice and wheat. This long-term trend of increasing crop specialisation can, conversely, be characterised as a cumulative reduction of biotic and dietary diversity over time, which led, during the twentieth century, to the estimated loss (according to the United Nations Food and Agricultural Organisation) of 75% of crop-plant genetic diversity.

CONCLUSION

The either/or conceptual dichotomy of hunter-gatherers versus agriculturalists is deeply embedded in the discourse on early agriculture but it is a gross oversimplification of past human subsistence. We should no longer allow it to constrain our thinking, or go on trying to fit data to it. Instead, we should concentrate on how to investigate most effectively the resource spectra on which past human groups depended for their livelihood, including the 'richly heterogeneous' societies that occupied the 'vast and largely uncharted regions' between hunting-gathering and agriculture (Smith 2001a: 33, 24).

In the past two decades, the potential for such investigation has been greatly enhanced by an expansion in the range of techniques with which plant and animal remains can be analysed and data on past human subsistence obtained. The most novel advances have occurred since the early 1990s in molecular biology through analyses of modern and ancient biomolecules, principally DNA, proteins and lipids (Evershed et al 1999; Fankhauser 1994; Hedges and Sykes 1992; Jones and Brown 2000). For example, analysis of lipids from organic residues preserved in pottery is providing direct evidence of prehistoric use of animal milk (Copley et al 2003), and stable-isotope analysis of ancient human bone is yielding valuable new data on past subsistence and diet (Richards and Hedges 1999a, 1999b; Richards et al 2003). There have also been novel developments and applications in micro-morphological techniques, such as parenchyma, starch-grain and phytolith analysis, that make possible the identification of very small fragmentary and amorphous organic materials (Bowdery 1999; Haslam 2004; Hather 1991, 1994, 2000; Iriarte et al 2004; Loy 1994; Mbida et al 2001; Pearsall and Piperno 1993; Perry 2002, 2004; Piperno and Holst 1998; Piperno et al 2000; Piperno and Stothert 2003; Rosen 1999; Therin et al 1999). By employing all these novel techniques, as well as continuing to use more conventional archaeo-botanical and archaeozoological methods of analysis, and by thinking more clearly about the conceptual categories and terminology we employ, we can expect gradually to gain a better understanding of the roles of 'cultivation', 'domestication' and 'agriculture' in the diversity and evolution of past human subsistence.

REFERENCES

Aikio, P (1989) 'The changing role of reindeer in the life of the Sámi', in Clutton-Brock, J (ed), *The Walking Larder: Patterns of Domestication, Pastoralism, and Predation*, pp 169–84. London: Unwin Hyman

Allen, H (1974) 'The Bagundji of the Darling Basin: cereal gatherers in an uncertain environment', *World Archaeology* 5: 309–22

Bar-Yosef, O and Meadow R H (1995) 'The origins of agriculture in the Near East', in Price, T D and Gebauer, A B (eds), *Last Hunters, First Farmers: New Perspectives on the Prehistoric Transition to Agriculture*, pp 39–94, Santa Fe, NM: School of American Research Press

Bean, L J and Lawton, H W (1973) 'Some explanations for the rise of cultural complexity in native California with comments on proto-agriculture and agriculture', in Lewis, H (ed), *Patterns of Indian Burning in California: Ecology and Ethnohistory*, pp v–xlvii, Ramona, CA: Ballena Press

Bowdery, D (1999) 'Phytoliths from tropical sediments: reports from Southeast Asia and Papua New Guinea', *Bulletin of the Indo-Pacific Prehistory Association* (Canberra) 18: 159–68

Chase, A K (1989) 'Domestication and domiculture in northern Australia: a social perspective', in Harris, D R and Hillman, G C (eds), *Foraging and Farming: The Evolution of Plant Exploitation*, pp 42–54, London: Unwin Hyman

Clutton-Brock, J (1989) 'Introduction to pastoralism', in Clutton-Brock, J (ed), *The Walking Larder: Patterns of Domestication, Pastoralism, and Predation*, pp 115–18, London: Unwin Hyman

Cohen, D J (1998) 'The origins of domesticated cereals and the Pleistocene-Holocene transition in East Asia', *The Review of Archaeology* 19(2): 22–29

Coomes, O T and Ban, N (2004) 'Cultivated plant species diversity in home gardens of an Amazonian peasant village in northeastern Peru', *Economic Botany* 58: 420–34

Copley, M S, Berstan, R, Dudd, S N, Docherty, G, Mukherjee, A J, Straker, V, Payne, S and Evershed, R P (2003) 'Direct chemical evidence for widespread dairying in prehistoric Britain', *Proceedings of the National Academy of Sciences* (Washington, DC) 100: 1524–29

Crawford, G W (1992) 'Prehistoric plant domestication in East Asia', in Cowan, C W and Watson, P J (eds), *The Origins of Agriculture: An International Perspective*, pp 7–38, Washington, DC: Smithsonian Institution Press

Crawford, G W and Shen, C (1998) 'The origins of rice agriculture: recent progress in East Asia', *Antiquity* 72: 858–66

D'Andrea, A C, Klee, M and Casey, J (2001) 'Archaeobotanical evidence for pearl millet (*Pennisetum glaucum*) in sub-Saharan West Africa', *Antiquity* 75: 341–48

Davis, E L (1963) 'The Desert Culture of the western Great Basin: a lifeway of seasonal transhumance', *American Antiquity* 29: 202–12

de Candolle, A (1855) *Géographie botanique raisonnée ou, Exposition des faits principaux et des lois concernant la distribution géographique des plantes de l'epoque actuelle*, 2 vols, Paris: Victor Masson

de Candolle, A (1882) *Origine des plantes cultivées*, Paris: Germer Baillière

de Candolle, A (1884) *Origin of Cultivated Plants*, London: Kegan Paul, Trench & Co

Delano Smith, C (1979) *Western Mediterranean Europe: A Historical Geography of Italy, Spain and Southern France since the Neolithic*, London: Academic Press

Denham, T P, Haberle, S G and Lentfer, C (2004) 'New evidence and revised interpretations of early agriculture in highland New Guinea', *Antiquity* 78: 839–57

Denham, T P, Haberle, S G, Lentfer, C, Fullagar, R, Field, J, Therin, M, Porch, N and Winsborough, B (2003) 'Origins of agriculture at Kuk Swamp in the Highlands of New Guinea', *Science* 301: 189–93

Ellen, R (1988) 'Foraging, starch extraction and the sedentary lifestyle in the lowland rainforest of central Seram', in Ingold, T, Riches, D and Woodburn, J (eds), *Hunters and Gatherers*, vol 1, pp 117–34, Oxford: Berg

Evershed, R P, Dudd, S N, Charters, S, Mottram, H, Stott, A W, Raven, A, van Bergen, P F and Bland, H A (1999) 'Lipids as carriers of anthropogenic signals from prehistory', *Philosophical Transactions of the Royal Society of London*, Series B–Biological Sciences 354: 19–31

Fankhauser, B (1994) 'Protein and lipid analysis of food residues', in Hather, J G (ed), *Tropical Archaeobotany: Applications and New Developments*, pp 227–50, London: Routledge

Ford, R I (1985) 'The processes of plant production in prehistoric North America', in Ford, R I (ed), *Prehistoric Food Production in North America*, pp 1–18. Anthropological Paper no 75, Ann Arbor: Museum of Anthropology, University of Michigan

Fuller, D Q, Korisettar, R, Venkatasubbaiah, P C and Jones, M K (2004) 'Early plant domestications in southern India: some preliminary archaeobotanical results', *Vegetation History and Archaeobotany* 13: 115–29

Garrard, A (1999) 'Charting the emergence of cereal and pulse domestication in South-west Asia', *Environmental Archaeology* 4: 67–86

Glover, I C and Higham, C F W (1996) 'New evidence for early rice cultivation in South, Southeast and East Asia, in Harris, D R (ed), *The Origins and Spread of Agriculture and Pastoralism in Eurasia*, pp 413–41, London: University College London Press

Golson, J (1977) 'No room at the top: agricultural intensification in the New Guinea Highlands', in Allen, J, Golson, J and Jones, R (eds), *Sunda and Sahul: Prehistoric Studies in Southeast Asia, Melanesia and Australia*, pp 601–38, London: Academic Press

Golson, J (1989) 'The origins and development of New Guinea agriculture', in Harris, D R and Hillman, G C (eds), *Foraging and Farming: The Evolution of Plant Exploitation*, pp 678–87, London: Unwin Hyman

Golson, J (1997) 'From horticulture to agriculture in the New Guinea Highlands', in Kirch, P V and Hunt, T L (eds), *Historical Ecology in the Pacific Islands: Prehistoric Environmental and Landscape Change*, pp 39–50, New Haven: Yale University Press

Gosden, C (1995) 'Arboriculture and agriculture in coastal Papua New Guinea', *Antiquity* 69: 807–17

Haberle, S (1994) 'Anthropogenic indicators in pollen diagrams: problems and prospects for late Quaternary palynology in New Guinea', in Hather, J G (ed), *Tropical Archaeobotany: Applications and New Developments*, pp 172–201, London: Routledge

Harlan, J R (1971) 'Agricultural origins: centers and noncenters', *Science* 174: 468–74

Harris, D R (1972) 'Swidden systems and settlement', in Ucko, P J, Tringham, R and Dimbleby, G W (eds), *Man, Settlement and Urbanism*, pp 245–62, London: Duckworth

Harris, D R (1973) 'The prehistory of tropical agriculture: an ethnoecological model', in Renfrew, C (ed), *The Explanation of Culture Change: Models in Prehistory*, pp 391–417, London: Duckworth

Harris, D R (1984) 'Ethnohistorical evidence for the exploitation of wild grasses and forbs: its scope and archaeological implications', in van Zeist, W and Casparie, W A (eds), *Plants and Ancient Man: Studies in Palaeoethnobotany*, pp 63–69, Rotterdam: Balkema

Harris, D R (1989) 'An evolutionary continuum of people–plant interaction', in Harris, D R and Hillman, G C (eds), *Foraging and Farming: The Evolution of Plant Exploitation*, pp 11–26, London: Unwin Hyman

Harris, D R (1990) 'Vavilov's concept of centres of origin of cultivated plants: its genesis and its influence on the study of agricultural origins', *Biological Journal of the Linnean Society* (London) 39: 7–16

Harris, D R (1996) 'Domesticatory relationships of people, plants and animals', in Ellen, R and Fukui, K (eds), *Redefining Nature: Ecology, Culture and Domestication*, pp 437–63, Oxford: Berg

Harris, D R (1998) 'The origins of agriculture in Southwest Asia', *The Review of Archaeology* 19(2): 5–11

Harris, D R (2002) 'Development of the agro-pastoral economy in the Fertile Crescent during the Pre-Pottery Neolithic period', in Cappers, R T J and Bottema, S (eds), *The Dawn of Farming in the Near East*, pp 67–83. Studies in Early Near Eastern Production, Subsistence and Environment 6, Berlin: Ex Oriente

Harris, D R (2003) 'The expansion capacity of early agricultural systems: a comparative perspective on the spread of agriculture', in Bellwood, P and Renfrew, C (eds), *Examining the Farming/Language Dispersal Hypothesis*, pp 31–39, Cambridge: McDonald Institute for Archaeological Research

Harris, D R (2006) 'The interplay of ethnographic and archaeological knowledge in the study of past human subsistence in the tropics', *Journal of the Royal Anthropological Institute* 12 (Special Issue): 63–78

Harvey, E L, Fuller, D, Pal, J and Gupta, M C (2005) 'Early agriculture in the Neolithic Vindhyas (North-Central India)', in Franke-Vogt, U and Weisshaar, H-J (eds), *South Asian Archaeology, 2003*, pp 329–34, Aachen: Linden Soft

Haslam, M (2004) 'The decomposition of starch grains in soils: implications for archaeological residue analysis', *Journal of Archaeological Science* 31: 1715–34

Hather, J G (1991) 'The identification of charred archaeological remains of vegetative parenchymatous tissues', *Journal of Archaeological Science* 18: 661–75

Hather, J G (1994) 'The identification of charred root and tuber crops from archaeological sites in the Pacific', in Hather, J G (ed), *Tropical Archaeobotany: Applications and New Developments*, pp 51–64, London: Routledge

Hather, J G (2000) *Archaeological Parenchyma*, London: Archetype Publications

Hedges, R E M and Sykes, B C (1992) 'Biomolecular archaeology: past, present and future', in Pollard, A M (ed), *New Developments in Archaeological Science: Proceedings of the British Academy* 77: 267–83

Helbaek, H (1960) 'The palaeoethnobotany of the Near East and Europe', in Braidwood, R J and Howe, B (eds), *Prehistoric Investigations in Iraqi Kurdistan*, pp 99–118. Studies in Ancient Oriental Civilization 31, Chicago: University of Chicago Press

Hillman, G C (1975) 'The plant remains from Tell Abu Hureyra: a preliminary report', in Moore, A M T (ed), 'The excavation of Tell Abu Hureyra in Syria: a preliminary report', *Proceedings of the Prehistoric Society* 41: 50–77

Hillman, G C (1996) 'Late Pleistocene changes in wild plant-foods available to hunter-gatherers of the northern Fertile Crescent: possible preludes to cereal cultivation', in Harris, D R (ed), *The Origins and Spread of Agriculture and Pastoralism in Eurasia*, pp 159–203, London: University College London Press

Hillman, G C (2000) 'The plant food economy of Abu Hureyra 1: the Epipalaeolithic', in Moore, A M T, Hillman, G C and Legge, A J (eds), *Village on the Euphrates: From Foraging to Farming at Abu Hureyra*, pp 327–99, New York: Oxford University Press

Hillman, G, Hedges, R, Moore, A, Colledge, S and Pettitt, P (2001) 'New evidence of late glacial cereal cultivation at Abu Hureyra on the Euphrates', *The Holocene* 11: 383–93

Hodder, I (1990) *The Domestication of Europe*, Oxford: Blackwell

Hynes, R A and Chase, A K (1982) 'Plants, sites and domiculture: Aboriginal influence on plant communities in Cape York Peninsula', *Archaeology in Oceania* 17: 38–50

Ingold, T (1980) *Hunters, Pastoralists and Ranchers: Reindeer Economies and Their Transformations*, Cambridge: Cambridge University Press

Iriarte, J, Holst, I, Marozzi, O, Listopad, C, Alonso, E, Rinderknecht, A and Montana, J (2004) 'Evidence for cultivar adoption and emerging complexity during the mid-Holocene in the La Plata Basin', *Nature* 432: 614–17

Jarman, M R, Bailey, G N and Jarman, H N (eds) (1982) *Early European Agriculture: Its Foundations and Development*, Cambridge: Cambridge University Press

Jones, M and Brown, T (2000) 'Agricultural origins: the evidence of modern and ancient DNA', *The Holocene* 10: 769–76

Keeley, L H (1995) 'Protoagricultural practices among hunter-gatherers: a cross-cultural survey', in Price, T D and Gebauer, A B (eds), *Last Hunters, First Farmers: New Perspectives on the Prehistoric Transition to Agriculture*, pp 243–72, Santa Fe, NM: School of American Research Press

Kimber, C T (1973) 'Spatial patterning in the dooryard gardens of Puerto Rico', *Geographical Review* 63: 6–26

Kimber, C T (1978) 'A folk content for plant domestication: or the dooryard garden revisited', *Anthropological Journal of Canada* 16: 2–11

Lamont, S R, Eshbaugh, W H and Greenberg, A M (1999) 'Species composition, diversity, and use of home gardens among three Amazonian villages', *Economic Botany* 53: 312–26

Landauer, K and Brazil, M (eds) (1990) *Tropical Home Gardens*, Tokyo: United Nations University Press

Lawton, H W, Wilke, P J, De Decker, M and Mason, W M (1976) 'Agriculture among the Paiute of Owens Valley', *Journal of California Anthropology* 3: 13–50

Loy, T H (1994) 'Methods in the analysis of starch residues on prehistoric stone tools' in Hather, J G (ed), *Tropical Archaeobotany: Applications and New Developments*, pp 86–114, London: Routledge

Lu, T L D (1999) 'The transition from foraging to farming in China', *Bulletin of the Indo-Pacific Prehistory Association* (Canberra) 18: 77–80

Lynch, T F (1971) 'Preceramic transhumance in the Callejón de Huaylas, Peru', *American Antiquity* 36: 139–48

Mbida, C M, Doutrelepont, H, Vrydaghs, L, Swennan, R L, Swennan, R J, Beeckman, H, de Langhe, E and Maret, P de (2001) 'First archaeological evidence of banana cultivation in central Africa during the third millennium before present', *Vegetation History and Archaeobotany* 10: 1–6

Moore, A M T, Hillman, G C and Legge, A J (2000) *Village on the Euphrates: From Foraging to Farming at Abu Hureyra*, New York: Oxford University Press

Neumann, K (1999) 'Early plant food production in the West African Sahel: new evidence', in van der Veen, M (ed), *The Exploitation of Plant Resources in Ancient Africa*, pp 73–81, New York: Kluwer Academic

Padoch, C and de Jong, W (1991) 'The house gardens of Santa Rosa: diversity and variability in an Amazonian agricultural system', *Economic Botany* 45: 166–75

Pearsall, D M (1992) 'The origins of plant cultivation in South America', in Cowan, C W and Watson, P J (eds), *The Origins of Agriculture: An International Perspective*, pp 173–205, Washington, DC: Smithsonian Institution Press

Pearsall, D M and Piperno, D R (eds) (1993) *Current Research in Phytolith Analysis: Applications in Archaeology and Paleoecology*. MASCA Research Papers in Science and Archaeology 10, Philadelphia: University of Pennsylvania Museum

Perry, L (2002) 'Starch granule size and the domestication of manioc (*Manihot esculenta*) and sweet potato (*Ipomoea batatas*)', *Economic Botany* 56: 335–49

Perry, L (2004) 'Starch analyses reveal the relationship between tool type and function: an example from the Orinoco Valley of Venezuela', *Journal of Archaeological Science* 31: 1069–81

Piperno, D R (1994) 'Phytolith and charcoal evidence for prehistoric slash-and-burn agriculture in the Darien rain forest of Panama', *The Holocene* 4: 321–25

Piperno, D R and Holst, I (1998) 'The presence of starch grains on prehistoric tools from the humid neotropics: indications of early tuber use and agriculture in Panama', *Journal of Archaeological Science* 25: 765–76

Piperno, D R and Pearsall, D M (1998) *The Origins of Agriculture in the Lowland Neotropics*, San Diego: Academic Press

Piperno, D R and Stothert, K E (2003) 'Phytolith evidence for early Holocene *Cucurbita* domestication in southwest Ecuador', *Science* 299: 1054–57

Piperno, D R, Ranere, A J, Holst, I and Hansell, P (2000) 'Starch grains reveal early root crop horticulture in the Panamanian tropical forest', *Nature* 407: 894–97

Richards, M P and Hedges, R E M (1999a) 'Stable isotope evidence for similarities in the types of marine foods used by late Mesolithic humans at sites along the Atlantic coast of Europe', *Journal of Archaeological Science* 26: 717–22

Richards, M P and Hedges, R E M (1999b) 'A Neolithic revolution? New evidence of diet in the British Neolithic', *Antiquity* 73: 891–97

Richards, M P, Pearson, J A, Molleson, T I, Russell, N and Martin, L (2003) 'Stable isotope evidence of diet at Neolithic Çatalhöyük, Turkey', *Journal of Archaeological Science* 30: 67–76

Rindos, D (1984) *The Origins of Agriculture: An Evolutionary Perspective*, Orlando: Academic Press

Rosen, A (1999) 'Phytolith analysis in Near Eastern archaeology', in Pike, S and Gitin, S (eds), *The Practical Impact of Science on Near Eastern and Aegean Archaeology*, pp 86–92, London: Archetype Publications

Serpell, J (1989) 'Pet-keeping and animal domestication: a reappraisal', in Clutton-Brock, J (ed), *The Walking Larder: Patterns of Domestication, Pastoralism, and Predation*, pp 10–30, London: Unwin Hyman

Shipek, F C (1989) 'An example of intensive plant husbandry: the Kumeyaay of southern California', in Harris, D R and Hillman, G C (eds), *Foraging and Farming: The Evolution of Plant Exploitation*, pp 159–70, London: Unwin Hyman

Simoons, F J (1968) *A Ceremonial Ox of India: The Mithan in Nature, Culture and History*, Madison: University of Wisconsin Press

Smith, B D (ed) (1992) *Rivers of Change: Essays on Early Agriculture in Eastern North America*, Washington, DC: Smithsonian Institution Press

Smith, B D (2001a) 'Low-level food production', *Journal of Archaeological Research* 9: 1–43

Smith, B D (2001b) 'Documenting plant domestication: the consilience of biological and archaeological approaches', *Proceedings of the National Academy of Sciences* (Washington, DC) 98: 1324–26

Steward, J H (1930) 'Irrigation without agriculture', *Papers of the Michigan Academy of Sciences, Arts and Letters* 12: 149–56

Terra, T T A (1954) 'Mixed-garden horticulture in Java', *Malaysian Journal of Tropical Geography* 4: 33–43

Therin, M, Fullagar, R and Torrence, R (1999) 'Starch in sediments: a new approach to the study of subsistence and land use in Papua New Guinea', in Gosden, C and Hather, J (eds), *The Prehistory of Food: Appetites for Change*, pp 438–62, London: Routledge

Underhill, A P (1997) 'Current issues in Chinese Neolithic archaeology', *Journal of World Prehistory* 11: 103–60

Vavilov, N I (1926) *Studies on the Origins of Cultivated Plants*, Leningrad: Institut Botanique Appliqué et d'Amélioration des Plantes

Willcox, G (1998) 'Archaeobotanical evidence for the beginnings of agriculture in Southwest Asia', in Damania, A B, Valkoun, J, Willcox, G and Qualset, C O (eds), *The Origins of Agriculture and Crop Domestication*, pp 25–38, Aleppo: ICARDA

Yen, D E (1989) 'The domestication of environment', in Harris, D R and Hillman, G C (eds), *Foraging and Farming: The Evolution of Plant Exploitation*, pp 55–75, London: Unwin Hyman

Zeuner, F E (1963) *A History of Domesticated Animals*, London: Hutchinson

Zvelebil, M (1986) 'Mesolithic societies and the transition to farming: problems of time, scale and organisation', in Zvelebil, M (ed), *Hunters in Transition: Mesolithic Societies of Temperate Eurasia and Their Transition to Farming*, pp 167–88, Cambridge: Cambridge University Press

Zvelebil, M (1995) 'Hunting, gathering or husbandry? Management of food resources by the late Mesolithic communities of temperate Europe', in Campana, D (ed), *Before Farming: Hunter-Gatherer Society and Subsistence*, pp 79–104. MASCA Research Papers in Science and Archaeology 12 (supp), Philadelphia: University of Pennsylvania Museum

CHAPTER 3

Selection, Cultivation and Reproductive Isolation: A Reconsideration of the Morphological and Molecular Signals of Domestication

Martin Jones and Terry Brown

INTRODUCTION

While Childe's model of a Neolithic Revolution has continued to stimulate a great deal of research into the beginnings of agriculture, much of that research has gone on to unpack and separate the core components of his revolutionary transition, dispersing them to distinct episodes in different places and different millennia. One component, which has retained a central position in the debate, is the morphological change seen in plant and animal remains, commonly treated as the most direct signal of 'domestication'. Like many other concepts that are widely employed in this debate, the term 'domestication' is used in a variable manner that shifts, even within individual publications. Its meaning has ranged between a social relationship (Zeuner 1963), a form of practice (Ingold 1996), an evolutionary process (Rindos 1984) and a specific stage in that process where morphological change is visible in the plant or animal concerned (Harris and Hillman 1989). In this final usage, 'domestication' is a character associated with the visible evolutionary response of the prey in terms of heritable morphological traits, in contrast to 'agriculture', which refers to the actions and selection pressure imposed by the human predator. In this paper, we look critically at that presumed association and ask the question of what precisely is the relationship between a change in human behaviour and the morphological and molecular changes in the plants and animals they exploit, and we explore the implications for our understanding of the origins and spread of farming. The great majority of domesticated species are plant taxa, which consequently dominate the following argument. However, the central themes apply similarly to animal taxa, which are also drawn into the argument at appropriate points.

WHAT ARE THE MORPHOLOGICAL CHANGES?

The principal morphological changes associated with domestication can be divided into two groups: those related to some form of size change and those relating to a transformation in predator/prey dependency. The first group can be subdivided into shrinkage and enlargement or, to draw terms from evolutionary biology, 'dwarfism' and 'gigantism'.

Size Change

Many domesticated animals are smaller than their wild progenitors, an evolutionary trajectory referred to as 'dwarfism'. The now extinct wild cattle (aurochsen), for example, were enormous beasts, and some domesticated cattle are only a fraction of their presumed body weight (cf Clutton-Brock 1999). In a number of domesticated animals, size reduction is especially evident in the skull and its component parts; horns, teeth and mandibles are foreshortened and reduced. Such size reduction in many species serves as a primary indicator of domestication. In food plants conversely, a recurrent feature is 'gigantism', in which the seed, fruit, or caryopsis, and sometimes the whole plant, is measurably, and sometimes substantially, larger than its wild progenitor. Enlarged seeds and fruits are widespread among cereal, pulse and fruit crops, and enlarged whole plants particularly characteristic of some cereals (cf Zohary and Hopf 2000).

Predator/Prey Dependency

In the life cycle of a typical domesticated species, the human predator figures with far greater prominence than is the case in the wild progenitor's life cycle. That is at the heart of the notion of domestication as a consequence of human agency/control/domination of nature. The morphological correlates of an enhanced predator-prey, or human-species, interdependency may include a reduced ability to compete in the wild, to avoid predation in the wild, or to reproduce in the wild. Some of these correlates overlap with size changes. So, for example, smaller animals may have greater difficulty competing for patchy or mobile food resources than their larger wild counterparts, and reduced or vestigial horns and teeth may greatly diminish their defensive abilities in the absence of human protection. Such changes may be observed, for example, in domesticated goats and pigs (Clutton-Brock 1999). There is also a separate set of features among plants.

There are several ways in which wild plants avoid predation, including toxins, lignified 'armour' and spines, all of which are widespread in nature and repeatedly lost on evolutionary trajectories to domestication. Many domesticated legumes, such as many of the Old World *Vicia* spp., have reduced toxicity (cf Berger et al 2003); the domestication of yams may be accompanied by a reduction in spines, a process that may still be observed in contemporary Benin and Nigeria in the ongoing domestication of *Dioscorea* spp. (Mignouna and Dansi 2003; Vernier et al 2003); and, the lignified armour of many grain crops has evolved a softer, thinner or looser form, as observed in naked barleys and wheats (Zohary and Hopf 2000). Each of these transformations would render the domesticates more vulnerable to predation in the absence of human protection. The most striking dependency is in relation to the reproductive cycle, and this is particularly marked in some plants that have more or less lost their ability to reproduce in the absence of their human predator. The cultivated oat, *Avena sativa*, for example, has completely lost the intricate self-seeding mechanism of its wild ancestor which involved a combination of shattering rachis and a twisted, limb-like awn (Zohary and Hopf 2000).

In plants, reproduction can be either vegetative or sexual, and a steadily increasing component of the global human diet has been taken from annual plants that can only reproduce by sexual means. The sexual cycle is heavily dependant on effective pollination, seed dispersal, dormancy and germination, and in wild plants, each of these is finely triggered by maturational and environmental signals. Within domestication, such triggers have repeatedly become disabled. Seed dispersal mechanisms are disabled such that the human predator becomes the only viable dispersal route, and dormancy and germination triggers are disabled such that all planted seeds either germinate or die in the season they are sown. In archaeological terms, some forms of dormancy may be discernible, for example in seed-coat thickness, but as a general rule, the most directly visible transformations are in mechanisms of seed dispersal. In cereals, for example, the transition from a dispersing brittle rachis to a non-dispersing tough rachis may be directly visible in the surviving rachis fragments (Zohary and Hopf 2000). Similarly, a twisted fragment of naturally dehiscent legume pod may be directly distinguished from an untwisted, non-dehiscent domesticated pod (Ladizinski 1979).

In terms of what is archaeologically visible, some combination of size change, loss of defensive structure, and disablement of reproductive mechanism is grouped together under a morphological definition of domestication. In evolutionary and genetic terms, there is no intrinsic reason why these different changes need to coincide (cf Nesbitt 2002), although they are often treated as marking a common transformation. Furthermore, their relationship with the selection pressure arising from human agency need not be straightforward and unchanging.

HOW CLEAR IS THE GENETIC BASIS OF THESE CHANGES?

Morphological changes associated with domestication can be broadly subdivided into discrete polymorphisms and continuous variables. The discrete polymorphisms occur in a fixed number of states. For example, rapid pod dehiscence in peas and lentils disappears as a result of a single gene change during domestication (Zohary and Hopf 2000). A continuous variable can occur at any value within a particular range, and the size changes tend to be continuous in this way, as evident in the length and breadth measurements of the seeds and caryopses of numerous crop species (Zohary and Hopf 2000) and limb bone measurements in a number of animal species (Clutton-Brock 1999). As a general rule, if a morphological feature is directly controlled by one or a very small number of gene loci, then it will display discrete polymorphism. Continuous variables arise from one of a number of more complex scenarios, possibly involving a number of genetic loci that together control a quantitative trait, and/or combining genetic, nutritional and environmental influences in ways that are often difficult to disentangle. In the particular case of archaeological evidence, this complexity is further exacerbated by taphonomic factors, such as processing (eg seed sorting), selective culls (eg small animal capture), and preservation. While discrete polymorphisms often act as clear genetic markers, and in an increasing number of instances markers whose genomic basis is understood, the genetic contribution to continuous variables will typically remain less certain.

Unfortunately for archaeologists, among the morphological changes associated with domestication, discrete polymorphisms are in the minority. There are a few cases of size change that are discrete and may be directly related to genetic change. These include the gigantism that may arise from polyploidy in such crops as tobacco and New World cotton (cf Zohary and Hopf 2000). Another form of gigantism that is not exactly discrete, but may nevertheless result from some fairly specific genetic changes, is that of maize. The enlarged cob results from a transformation in the pattern of rachis branching (Jaenicke-Després et al 2003). However, the genetic signal behind most forms of size change connected to domestication is likely to be complex and difficult to define. Moreover, the complex interplay of additional contributors to morphological and genetic change, including physiological, developmental, environmental and taphonomic factors, is generally poorly understood. For all these reasons, alongside the archaeologically discernible instances of evolutionary coupling between humans and their food species, there may be a large number of equally significant couplings that remain archaeologically invisible.

THE GEOGRAPHY OF DOMESTICATION

At around the time of Childe's first published arguments about the origins and dispersal of agriculture from the basis of material culture, the plant geneticist Nikolai Vavilov was charting crop genetic diversity as a potential signpost of where domestication had first occurred (Childe 1928; Vavilov 1926). These two authors set out the framework within which much subsequent research into agricultural origins has been constructed. Vavilov's maps indicate target areas for exploration, and Childe's work indicates what to look for. Harris (1990) has pointed out that there is an element of circular argument in this; the idea that Vavilov's centres of diversity are indeed the points of origin is reinforced because other regions are not subjected to similar interrogation. Much subsequent work has explored trajectories of domestication beyond these centers. Research into the principal world crops today, wheat and rice, has drawn attention to the status of the periphery. In the case of wheat, George Willcox, working with Cauvin et al (1998), has drawn attention to an interesting pattern in the archaeobotanical record for emmer and einkorn wheats in Southwest Asia. In their domesticated form, in the middle Pre-Pottery Neolithic B (PPNB), these wheats each appear in regions of the Fertile Crescent from which their wild counterpart is absent, in other words, just beyond the periphery of the wild distribution. Nesbitt (2002) has argued that the most secure assemblages of morphologically transformed wheat first appear in Southeast Turkey, rather than the central areas of the Fertile Crescent, and indeed morphological domestication appears at an early date at the site of Mylouthia in Cyprus (Peltenburg et al 2000, 2001).

Rice, too, first appears in domesticated form at, or possibly beyond, the periphery of its wild ancestral range. A focus on the heart of the wild ancestral range had taken earlier researchers to Thailand and East India, where domesticated rice spreads from the 4th millennium BC onwards. However, the Middle Yangtse Valley has sites that are thousands of years older still (Sato 1997). The prominence of the periphery in these geographical patterns draws our attention to the actual mechanism of evolutionary change.

These observations resonate with the 'edge effect' discussed by Flannery (1969, 1973). Flannery was thinking in terms of adaptive human behaviour. It is the edge where he surmised that intensive care of the prey would be cost-effective in evolutionary terms. At the core of the wild prey's distribution he argued that the wild harvest is perfectly adequate without the investment of farming effort. Since his publication, other authors, such as Bender (1978), Hayden (1990) and Hodder (1990), have explored social and cognitive rather than purely economic drivers. However, another way of approaching the apparent edge effect is to focus on the actual mechanisms of evolutionary change.

MECHANISMS OF EVOLUTIONARY CHANGE

It has been commonly presumed that morphological domestication has been directly driven by sentient human action – that the physical changes observed in plants and animals have a one-to-one correspondence with human control over the breeding cycle, a transition from a passive to an active engagement with nature. This presumption has also been queried or qualified by various authors. First, the recognition that active human engagement and control may coexist without morphological change in the prey species has led to the concept of 'pre-domestication cultivation' recognised at various early sites in the Fertile Crescent (Hillman et al 2001; Willcox 1995, 2002). Second, the possibility that morphological change could result from more passive predator-prey coupling is a theme that has often been explored. For example, Zeuner (1963) and Rindos (1984) each point out that coevolution between predator and prey, in which each facilitates in some manner the reproductive cycle of the other, is widespread in nature, not just among sentient animals, but also among species in which sentient agency is a meaningless concept. Rindos is mostly concerned with changes in predator/prey dependency, but a similar argument can be developed for dwarfism and gigantism, evolutionary trends which are also widespread in nature, often in the context of island populations. Islands highlight a key element of the evolutionary equation, and that element is reproductive isolation.

The kind of wholesale shift in morphology that distinguishes domesticated from wild populations requires two things: variation in selection pressure and reproductive isolation. The core interaction that characterises 'agriculture' and 'farming' in their various forms is the care of the female of the prey, in either words, the assistance of the maternal parent during the reproductive cycle. While we are familiar with strict control of sources of male pollen/sperm in contemporary agriculture, we have no reason to assume that to have been a universal trait. If a human predator changes the selection pressure by so caring for the female of the prey, then this may enable mutational traits, such as reduced 'armour' in plants and animals and disabled dispersal mechanisms in plants, to persist. However, if the original route to reproductive success remains available to the plant or animal, then the change in selection pressure will only add diversity to the gene pool – it will not in itself lead to the eradication of the older traits. Take for example an imagined care and tending of a wild barley patch, protecting the parent plant from competition and assisting the germination and growth of the offspring. If this takes place in a landscape in which wild barley is widespread, and with which there is a certain amount of genetic interchange through pollen and grain mobility, then we might expect tough and brittle rachis forms to persist for some considerable period of time. This is not an issue of the intensity of human care for the prey, simply the degree of reproductive isolation, which in this imagined case is low. However, in transferring that human care, even in a *less labour-intensive* form to areas in which the wild barley patches are more scattered, then the greater reproductive isolation may lead to *greater* morphological change in the population.

It should be stressed that reproductive isolation can be achieved in many ways other than through spatial isolation. We probably know insufficient about past human predator/prey relationships to consider in detail other mechanisms of isolation. There is however scattered genetic and morphological evidence of varying levels of isolation, for example of introgression between wolf and dog (Uerpmann 1996). Introgression is commonly observed between cultivated and wild grain chenopods in America (Pickersgill 1989), has been used to account for the high chloroplast DNA diversity in domesticated maize (Doebley 1990), and may conceivably account for some of the brittle rachis/tough rachis mixtures in the Epipalaeolithic and Pre-Pottery Neolithic of Israel and Syria (cf Nesbitt 2002).

The central point to make is that selection pressure and reproductive isolation are both implicated in a wholesale genetic shift such as is observed in morphological domestication. While archaeological interest, stemming from a concern with human agency and control, has emphasised selection pressure, our actual observations of domestication may be more directly driven by reproductive isolation. The rather teleological concept of 'pre-domestication cultivation' implying an incipient, transitional episode, may indeed relate to something more substantial and less directional: the separate chronologies of human ecological agency and reproductive isolation. To reiterate a point, it is perfectly possible that a more intensive care and management of the prey species, at the core of that species' distribution, would result in a less prominent population shift in morphology, than less intensive care and management at the periphery of that species' distribution.

If, as we suggest, reproductive isolation played a rather more significant and complex role in domestication than previously suspected, then we might anticipate seeing a legacy of this isolation in the genetic features of domesticated plants and animals. We therefore turn our attention to the extent to which the available information on molecular markers supports our hypothesis.

MOLECULAR MARKERS

Although the use of ancient biomolecules is growing, most of the work on molecular markers relevant to domestication involves living specimens (Jones and Brown 2000). One approach is to use variations between molecular markers to construct family trees or phylogenies, displayed in the form of a bifurcating network, with each branch point in the tree indicating a specific instance of reproductive isolation in the evolutionary history of the specimens being studied. These phylogenetic techniques are robust when applied to *species*, which display complete reproductive isolation. They are less appropriate for the comparison of *populations* of a single species, as any interbreeding between the populations invalidates the assumptions that underlie the tree-building procedures, and under these circumstances spurious relationships and branch points may be inferred, leading to erroneous conclusions about evolutionary history. Instead, populations with the capacity to interbreed are studied by methods that examine population substructure to infer degrees of reproductive isolation and of mixing between groups. In more sophisticated analyses, the evolutionary changes that are inferred can be related to the current and past geographical distributions of the subpopulations.

Today, most molecular studies use DNA markers as these are highly informative and can be typed rapidly in a large number of specimens. Additionally there has been, since the 1960s, a significant field of research based on protein variations, and this research has considerably enhanced our understanding of domestication. In some ways, protein variations are equivalent to morphological traits, both being the physiological consequence of expressed genes. Proteins have an advantage over morphology in being the direct products of gene expression, and in theory might allow direct exploration of an interesting gene, such as one associated with dormancy. In practice, however, the protein phylogenies that have been used have generally had a fairly arbitrary relationship with domestication and have served primarily as lineage or population markers.

DNA has greater utility as a molecular marker because different regions (loci) within the genome are subject to different degrees of selective pressure.

Hence, one can distinguish the evolutionary histories of loci subject to selection during domestication (eg those genes directly responsible for key morphological changes) from those 'neutral' loci that are unaffected by selection and whose evolution has therefore been influenced primarily by population effects, including factors such as reproductive isolation. Our ability to fully exploit the range of variables within the genome of any domesticated species is limited by our understanding of that genome. One major block to progress is the difficulty in identifying genes involved in domestication, even when a complete DNA sequence of the genome is available. It is also necessary to analyze the variation of neutral loci in the context not only of the domesticates but also of related wild populations, which proves problematic in cases such as cattle where the wild progenitor is extinct. Nevertheless, progress is being made in the application of molecular studies to domestication, especially with the following types of marker.

Mitochondrial DNA (mtDNA)

Loci are present on the small DNA molecules located in the energy-generating organelles of animal and plant cells. In most species, the mtDNA is inherited only through the maternal line and, in mammals, is not subject to the recombination between maternal and paternal DNAs that occurs with the chromosomes present in the nucleus. These special features of mtDNA simplify the process by which evolutionary information is extracted, and the rapid rate at which mutations (the driving force behind evolutionary change) accumulate in mtDNA means that these loci display sufficient variability to be informative when closely related subpopulations are studied. Techniques for using mtDNA have been pioneered by biologists interested in recent human evolutionary history, and these techniques have been adapted for use with domesticated mammals, greatest progress having been made with cattle (Loftus et al 1999). Plants also have DNA within their mitochondria and in their second maternally inherited organelle, the chloroplast responsible for photosynthetic function. Unfortunately, plant mtDNA evolves in a more complex manner, and the chloroplast genome has generally evolved too slowly to provide information useful in the timescale of domestication. Nonetheless, some informative patterns have been derived from chloroplast DNA for rice and maize (Chen et al 1993; Matsuoka et al 2002).

Nuclear Loci Subject to Selection During Domestication

These are known in some species, and the study of these loci is starting to have a major impact on our understanding of the genetic basis to morphological and physiological changes. For example, the substantial morphological differences in maize from its wild progenitor, teosinte, has prompted extensive research of this type. This work has identified a number of genes involved in domestication of maize, and three of these – for plant architecture, storage protein synthesis and starch production – have been studied in ancient DNA from preserved maize specimens (Jaenicke-Després et al 2003). One conclusion from this work is that, in maize at least, the morphological and biochemical changes associated with domestication did not all occur at the same time; selection for the 'domesticated' variants in starch production were still not complete as recently as 2000 years ago. These results suggest that the transition from the 'wild' to the 'domesticated' phenotype was a more drawn-out process than is apparent when morphology alone is used as the indicator of domestication.

Neutral Loci

Neutral loci are not subject to selective pressure during domestication and are known in most species. Much work is being done with loci called 'microsatellites', or 'short tandem repeats', which consist of short DNA sequences that are repeated a variable number of times at a particular position in a genome. In humans, these loci form the basis to the genetic profiling tests that forensic scientists use both to assign specific fingerprints to individuals and to establish parentage or kinship in legal disputations. In population biology, microsatellites are now extensively used to reveal substructures within interbreeding populations. This approach is now being applied to domesticated species, in particular to cereal crops (for examples with wheat, see Akkaya and Buyukunal-Bal 2004; Huang et al 2002; Prasad et al 2000).

Large DNA Datasets

DNA datasets can be obtained through methods that examine many loci at once. Such methods can avoid the sampling errors that may occur when a study takes account of only a single or small group of loci. The most important of these methods is amplified fragment length polymorphism (AFLP) analysis, in which variations at several hundreds or even thousands of loci can be typed in one experiment (eg Schwarz et al 2000; Vuylsteke et al 2000). The great advantage of AFLP and related analyses is that they are applicable to little studied species in which the DNA sequence variation is only understood to a limited degree. Their weakness is that, because the loci being typed are not fully characterised, those subject to selective pressure cannot be distinguished from neutral loci. In essence, the analysis provides a general sweep of the overall variation in a genome, and hence is less applicable to sophisticated examination of domesticated populations.

MOLECULAR EVIDENCE AND DOMESTICATION

Molecular information is not sufficiently well developed for any species to provide anything more than incomplete and approximate indication of the evolutionary events underlying domestication. However, for those species and groups of species for which the molecular data are most detailed, the picture that emerges has a complexity that cannot be accounted for by conventional views of domestication. The data are most complete for Southwest Asian cereals, and it is on these that we will focus.

In the years leading up to the mid-1990s, a substantial body of molecular evidence was accumulated that suggested that wheats and barley had followed a relatively non-complex evolutionary trajectory culminating in their domestication (Zohary 1996). These data could be interpreted as indicating that each crop is monophyletic, meaning that the modern crop has arisen from a single domestication event, with all modern specimens descended from the single wild population that became domesticated. Although it is difficult or impossible to apply any rigorous dating method to the molecular data, a coincidence was naturally assumed between the domestication event inferred from the molecular data and the time of appearance in the archaeological record of the first specimens displaying the morphology associated with domestication. Hence the molecular data supported the view that the selective pressure that led to the morphological changes was the key genetic driver in domestication (Zohary 1996).

These analyses were not incorrect. However, in retrospect it is clear that their underlying basis leads to incomplete conclusions. That underlying basis is the quite reasonable assumption that to study the genetics of domestication one must study those molecular markers most closely associated with the domestication process. Hence, great weight was placed on comparisons between the seed protein contents of wild and domesticated plants and on the genetics of traits such as plant architecture, seed dispersal and seed dormancy (Zohary 1996). These are the various phenotypes that are likely to be subject to the most intensive selective pressure during domestication, and it is hence unsurprising that the story they tell fits with the morphological interpretation and places emphasis on selection.

A major step forward was provided subsequent to the mid-1990s by the application of large AFLP datasets to the domestication issue (Badr et al 2000; Heun et al 1997; Özkan et al 2002). By accessing at least some component of those parts of the genome not subject to selection, such studies should provide an independent test of the monophyletic hypothesis. The fact that these data appeared to show that einkorn, barley and the hulled and hard tetraploid wheats were each domesticated on a single occasion at discrete locations in Southwest Asia therefore attracted great interest. However, these studies are problematic, as computer modeling of artificial populations has shown that the phylogenetic methods that were used are inappropriate for AFLP data and can lead to a monophyletic origin being inferred when one does not exist (Allaby and Brown 2003). A reworking of these data using a more valid population biology approach does not obviously show that the domesticated crop has a single origin (Allaby and Brown 2004; Salamini et al 2004).

Indeed, the reworking of the AFLP datasets shows that modern domesticated populations of einkorn, barley and tetraploid wheat each have a surprisingly large amount of genetic diversity. This is not a new discovery.

Since the 1980s it has been known that the mitochondrial and chloroplast DNAs of domesticated barley have similar degrees of variability to those of wild populations (Holwerda et al 1986). Our own studies of microsatellite loci have shown that landraces of emmer wheat display a substantial proportion of the diversity present in wild emmer populations (Isaac 2003) and that landraces of einkorn not only have extensive genetic diversity but also appear to be related to wild populations from several different parts of their natural range (S. Thaw and T. Brown, unpublished results).

All of these studies suggest that the genetic uniformity inferred from examination of those loci subject to selection is only part of the picture, specifically that part relating to the morphological changes that gave rise to the domesticated forms. The other part of the picture, that painted by neutral loci, is that each of the cereal crops originating in Southwest Asia displays remarkable genetic diversity, more than one anticipates could arise during the 10,000 years that the crop has evolved postdomestication. If this view is correct, then one possibility is that the diversity began to arise *before* morphological domestication. For this to be possible, the plants that eventually became domesticated must have existed as a distinct population, with some degree of reproductive isolation, for a period prior to the onset of morphological change. This interpretation coincides precisely with the notion of 'predomestication cultivation', which is recognised at various early sites in the Fertile Crescent (cf Hillman et al 2001; Willcox 1995, 2002). Management of the crop as a distinct population, though one into which there is some gene flow from neighbouring wild plants, is entirely consistent with the molecular data. If cultivation of the predomesticated crop spread, then the crop might acquire a range of molecular markers that, under modern scrutiny, appear to ally it with many parts of the natural range, exactly as we see with microsatellite loci in einkorn. If the spread of cultivation eventually reaches the edge of the wild range then, or indeed a region in which the patchiness and fragmentation of the wild populations is enhanced, then a different dynamic comes into play. As we describe above, the higher degree of reproductive isolation would result in persistence of the domestication traits, the crop then becoming morphologically domesticated. The fully domesticated crop therefore emerges when the predomesticated population goes beyond the edge of the wild range and becomes reproductively isolated.

CONCLUSION

In this paper, we have emphasised the difference between concepts such as 'agriculture' that relate to what people do, and concepts such as 'domestication' that relate to the evolutionary response of plant and animal species to the selection pressure imposed by those human actions. Our aim has not been to further classify and contain such terms; indeed, one of the important features of the enquiry is that what people do, and how other species respond, are both extremely variable. We have instead aimed to clarify the connections between the human stimuli and the evolutionary responses in other species.

Our consideration of the range of attributes used as genetic proxies of 'domestication' leads to the conclusion that, in both morphological and molecular markers, there are a range of processes in play – processes that need not necessarily coincide. There are some that provide a broad overview of the essential distinctions between 'domesticated' and 'wild' species, and some valuable pointers as to where and when some form of transformation happened to the species in question. There are other attributes that relate more closely to those transformations, revealing evolutionary responses to a combination of varied selection pressure and reproductive isolation. It is our view that the stronger signal comes from reproductive isolation. In other words, the starting point may be an early stage of agricultural spread rather than actual agricultural origin. It may not be reasonable to anticipate that the key changes in human ecological behaviour, and the changes in selection pressure consequent upon them, might be translated immediately into the kind of wholesale changes we can find, either in the archaeological record, or in the molecular diversity of modern genomes.

The episode referred to as 'pre-domestication cultivation' may be far more than a transitional episode, and indeed reflect a highly sustainable state within an optimising/stable, rather than a maximising/expansive, human ecosystem. As other authors have made clear, we do have archaeological means of detecting direct human intervention into the life cycle of prey, for example the sheen on flint sickles indicative of grass-head harvesting (Unger-Hamilton 1989), the suppression of perennial competitors (Willcox 1995), and the birth of animals within human settlements (Uerpmann 1996). It should be no surprise that these need not coincide with either morphological or molecular variation, variations which reflect instead the connected but distinct process of reproductive isolation, a process more prominent in the context of agricultural expansion than origin.

REFERENCES

Akkaya, M S and Buyukunal-Bal, E B (2004) 'Assessment of genetic variation of bread wheat varieties using microsatellite markers', *Euphytica: International Journal of Plant Breeding* 135: 179–85

Allaby, R G and Brown, T A (2003) 'AFLP data and the origins of domesticated crops', *Genome* 46: 448–53

Allaby, R G and Brown, T A (2004) 'Reply to the comment by Salamini et al on "AFLP data and the origins of domesticated crops" ', *Genome* 47: 621–22

Badr, A, Müller, K, Schäfer-Pregl, R, El Rabey, H, Effgen, S, Ibrahim, H H, Pozzi, C, Rohde, W and Salamini, F (2000) 'On the origin and domestication history of barley (*Hordeum vulgare*)', *Molecular Biology and Evolution* 17: 499–510

Bender, B (1978) 'Gatherer-hunter to farmer: a social perspective', *World Archaeology* 10: 204–19

Berger, J D, Robertson, L D and Cocks, P C (2003) 'Agricultural potential of Mediterranean grain and forage legumes: 2) anti-nutritional factor concentrations in the genus *Vicia*', *Genetic Resources and Crop Evolution* 50(2): 201–12

Cauvin, J, Cauvin, M-C, Helmer, D and Willcox, G (1998) 'L'homme et son environnment au Levant Nord entre 30,000 et 7,500 BP', *Paléorient* 23: 51–69

Chen, W B, Nakamura, I, Sato, Y I and Nakai, H (1993) 'Distribution of deletion type in CpDNA of cultivated and wild rice', *Japanese Journal of Genetics* 68: 597–603

Childe, V G (1928) *The Most Ancient East: The Oriental Prelude to European Prehistory*, London: Kegan Paul

Clutton-Brock, J (1999) *A Natural History of Domesticated Mammals*, 2nd edition, Cambridge: Cambridge University Press

Doebley, J (1990) 'Molecular evidence and the evolution of maize', in Bretting, P K (ed), *New Perspectives on the Origin and Evolution of New World Domesticated Plants*, *Economic Botany* 44: 6–28

Flannery, K V (1969) 'Origins and ecological effects of early domestication in Iran and the Near East', in Ucko, P J and Dimbleby, G W (eds), *The Domestication and Exploitation of Plants and Animals*, pp 73–100, London: Duckworth

Flannery, K V (1973) 'The origins of agriculture', *Annual Review of Anthropology* 2: 271–310

Harris, D R (1990) 'Vavilov's concept of centres of origin of cultivated plants: its genesis and its influence on the study of agricultural origins', *Biological Journal of the Linnaean Society* (London) 39: 7–16

Harris, D R and Hillman, G C (1989) 'Introduction', in Harris, D R and Hillman, G C (eds), *Foraging and Farming: The Evolution of Plant Exploitation*, pp 1–8, London: Unwin Hyman

Hayden, B (1990) 'Nimrods, piscators, pluckers and planters: the emergence of food production', *Journal of Anthropological Archaeology* 9: 31–69

Heun, M, Schäfer-Pregl, R, Klawan, D, Castagna, R, Accerbi, M, Borghi, B and Salamini, F (1997) 'Site of einkorn wheat domestication identified by DNA fingerprinting', *Science* 278: 1312–14

Hillman, G, Hedges, R, Moore, A, Colledge, S and Pettitt, P (2001) 'New evidence of late glacial cereal cultivation at Abu Hureyra on the Euphrates', *The Holocene* 11: 383–93

Hodder, I (1990) *The Domestication of Europe*, Oxford: Blackwell

Holwerda, B C, Jana, S and Crosby, W L (1986) 'Chloroplast and mitochondrial DNA variation in *Hordeum vulgare* and *Hordeum spontaneum*', *Genetics* 114: 1271–1291

Huang, X O, Borner, A, Roder, M S and Ganal, M W (2002) 'Assessing genetic diversity of wheat (*Triticum aestivum* L.) germplasm using microsatellite markers', *Theoretical and Applied Genetics* 105: 699–702

Ingold, T (1996) 'Growing plants and raising animals: an anthropological perspective on domestication', in Harris, D R (ed), *The Origins and Spread of Agriculture and Pastoralism in Eurasia*, pp 12–24, London: University College London Press

Isaac, A (2003) 'Using microsatellites to investigate the domestication and spread of emmer wheat', unpublished PhD thesis, University of Manchester Institute of Science and Technology

Jaenicke-Després, V, Buckler, E S, Smith, B D, Gilbert, M T P, Cooper, A, Doebley, J and Pääbo, S (2003) 'Early allelic selection in maize as revealed by ancient DNA', *Science* 302: 1206–08

Jones M K and Brown, T A (2000) 'Agricultural origins: the evidence of modern and ancient DNA', *The Holocene* 10: 775–82

Ladizinski, G (1979) 'Seed dispersal in relation to the domestication of Middle East legumes', *Economic Botany* 33: 284–89

Loftus, R T, Ertugrul, O, Harba, A H, El-Barody, M A, MacHugh, D E, Park, S D and Bradley, D G (1999) 'A microsatellite survey of cattle from a centre of origin: the Near East', *Molecular Ecology* 8(12): 2015–22

Matsuoka, Y, Vigouroux, Y, Goodman, M M, Sanchez, J, Buckler, E and Doebley, J (2002) 'A single domestication for maize shown by multilocus microsatellite genotyping', *Proceedings of the National Academy of Sciences* (Washington, DC) 99: 6080–84

Mignouna, H D and Dansi, A (2003) 'Yam (*Dioscorea* ssp.) domestication by the Nago and Fon ethnic groups in Benin', *Genetic Resources and Crop Evolution* 50(5): 519–28

Nesbitt, M (2002) 'When and where did domesticated cereals first occur in Southwest Asia?', in Cappers, R T J and Bottema, S (eds), *The Dawn of Farming in the Near East*, pp 113–32, Berlin: Ex Oriente

Özkan, H, Brandolini, A, Schäfer-Pregl, R and Salamini, F (2002) 'AFLP analysis of a collection of tetraploid wheats indicates the origin of emmer and hard wheat domestication in southeast Turkey', *Molecular Biology and Evolution* 19: 1797–801

Peltenburg, E, Colledge, S, Croft, P, Jackson, A, McCartney, C and Murray, M A (2000) 'Agropastoral colonization of Cyprus in the 10th millennium BP: initial assessments', *Antiquity* 74: 844–53

Peltenburg, E, Colledge, S, Croft, P, Jackson, A, McCartney, C and Murray, M A (2001) 'Neolithic dispersals from the Levantine corridor: a Mediterranean perspective', *Levant* 33: 35–64

Pickersgill, B (1989) 'Cytological and genetical evidence on the domestication and diffusion of crops within the Americas', in Harris, D R and Hillman, G C (eds), *Foraging and Farming: The Evolution of Plant Exploitation*, pp 426–39, London: Unwin Hyman

Prasad, M, Varshney, R K, Roy, J K, Balyan, H S and Gupta, P K (2000) 'The use of microsatellites for detecting DNA polymorphism, genotype identification and genetic diversity in wheat', *Theoretical and Applied Genetics* 100: 584–92

Rindos, D (1984) *The Origins of Agriculture: An Evolutionary Perspective*, New York: Academic Press

Salamini, F, Heun, M, Brandolini, A, Özkan, H and Wunder, J (2004) 'Comment on "AFLP data and the origins of domesticated crops"', *Genome* 47: 615–20

Sato, Y I (1997) 'Cultivated rice was born in the middle and lower Yangtse River', *Nikkei Science* 1: 32–42

Schwarz, G, Herz, M, Huang, X Q, Michalek, W, Jahoor, A, Wenzel, G and Mohler, V (2000) 'Application of fluorescence-based semi-automated AFLP analysis in barley and wheat', *Theoretical and Applied Genetics* 100: 545–51

Uerpmann, H P (1996) 'Animal domestication: accident or intention?', in Harris, D R (ed), *The Origins and Spread of Agriculture and Pastoralism in Eurasia*, pp 227–37, London: University College London Press

Unger-Hamilton, R (1989) 'The Epi-Palaeolithic southern Levant and the origins of cultivation', *Current Anthropology* 30(1): 88–103

Vavilov, N I (1926) *Studies of the Origin of Cultivated Plants*, Leningrad: Institut Botanique Appliqué et d'Amélioration des Plantes

Vernier, P, Orkwor, G C and Dossou, A R (2003) 'Studies on yam domestication and farmers' practices in Benin and Nigeria', *Outlook on Agriculture* (Oxford) 32(1): 35–41

Vuylsteke, M, Mank, R, Brugmans, B, Stam, P and Kuiper, M (2000) 'Further characterization of AFLP data as a tool in genetic diversity assessments among maize (*Zea mays* L.) inbred lines', *Molecular Breeding* 6(3): 265–76

Willcox, G (1995) 'Wild and domestic cereal exploitation: new evidence from Early Neolithic sites in the northern Levant and south eastern Anatolia', *Arx: World Journal of Prehistoric and Ancient Studies* (Barcelona) 1: 9–16

Willcox, G (2002) 'Geographical variation in major cereal components and evidence for independent domestication events in the Western Asia', in Cappers, R T J and Bottema, S (eds), *The Dawn of Farming in the Near East*, pp 133–40, Berlin: Ex Oriente

Zeuner, F E (1963) *A History of Domesticated Animals*, London: Hutchinson

Zohary, D (1996) 'The mode of domestication of the founder crops of Southwest Asian agriculture', in Harris, D R (ed), *The Origins and Spread of Agriculture and Pastoralism in Eurasia*, pp 142–58, London: University College London Press

Zohary, D and Hopf, M (2000) *Domestication of Plants in the Old World*, 3rd edition, Oxford: Oxford University Press

CHAPTER 4

Subterranean Diets in the Tropical Rain Forests of Sarawak, Malaysia

Huw Barton and Victor Paz

INTRODUCTION

Recent archaeological studies at Niah Cave, Sarawak, Malaysia have demonstrated that small bands of hunter-gatherers were using the cave and the surrounding lowland tropical rain forest from at least 45,000 years ago (Barker et al 2002a, 2003). These people were using a simple stone-tool assemblage composed of unretouched flakes, primarily of quartzite, produced from small pebble cores (Barton 2005a; Zuraina 1982: 64–65). These rain forest environments have been described as relatively inaccessible to hunter-gatherers (Bailey et al 1989; Bailey and Headland 1991; Headland 1987) as most of the stored energy exists in the form of inedible woody tissue. Edible plant and animal species are widely dispersed spatially and temporally, and the foraging and processing costs of many plant foods, many of which contain toxic compounds, are argued to be too high to support foraging populations (Bailey et al 1989: 61). Specific aspects of this argument have been criticised by illustrating weaknesses in its ecological portrayal of rain forests and their potential to support nomadic groups (Colinvaux and Bush 1991; Piperno and Pearsall 1998: 77). Of most significance are the potential importance of certain plant foods, such as sago and yams, to a preagricultural diet (Brosius 1993; Ellen 1988), the abundance of sources of protein in some rain forests (Piperno and Pearsall 1998), and suggestions that early hunter-gatherers were unintentionally or intentionally manipulators of rain forest, creating favourable distributions of plant and animal resources necessary for long-term human use of this type of environment (Denham and Barton 2006; Hladik et al 1993: 128; Piperno and Pearsall 1998: 78; Puri 1997: 207).

A study of preserved macro and micro plant remains from the site of Niah Cave further challenges arguments about the inaccessibility of rain forest to human foragers. Not only have we found a wide range of carbohydrate-rich plant species at this site, including yam tubers and aroid rhizomes, but we have recovered several species known to have very high levels of natural toxins. The presence of these species indicates a broad diet breadth that included high-cost food items and suggests that people knew how to process these plants and neutralise their toxic compounds. This finding has important implications for understanding the nature of forager diet and settlement patterns within the Pleistocene tropical forests of Southeast Asia.

RAIN FORESTS OF BORNEO

It is naive to view Southeast Asian rain forests as homogeneous, with a broadly similar resource distribution throughout. For example, ethnobotanical surveys conducted during the nineteenth and twentieth centuries on the Malay Peninsula and the large island of Borneo indicate that, historically, foragers across the South China Sea may well have depended upon a different range of wild carbohydrate staples. Yams were more important on Peninsula Malaysia (Carey 1976; Dentan 1991: 438; Endicott 1984; Kuchikura 1987, 1993; Schebesta 1954, 1973) and the Philippines (Eder 1978; Paz 2001), whereas edible palm pith (sago) was more important on Borneo, Indonesia and the islands to the east (Brosius 1993; Dentan 1991: 434; Ellen 1988; Kedit 1982; Langub 1989; Spencer 1966). The history of plant use in this region is complex and defies simple assessments of resource use in times before the adoption and spread of agriculture during the mid-Holocene and introductions of cultivars from the seventeenth century onwards (Ellen 1988: 123). The unknown extent of plant movement during prehistoric times is yet another factor that adds further complexity to a historic interpretation of plant use and settlement patterns in this region (Yen 1998).

Mammal fauna on Borneo includes about 185 species and, unlike the smaller islands, Borneo is also home to several large ungulates including the Asian elephant (*Elephas maximus*), Asian two-horned rhinoceros (*Dicerorhinus sumatrensis*), tembadu (*Bos javanicus*), Sambar deer (*Cervus unicolor*) and the bearded pig (*Sus barbatus*) (Payne and Francis 1998). Elephant, two-horned rhinoceros and tembadu are locally extinct in Sarawak as a result of hunting pressure and habitat modification; these species persist as small scattered populations in Sabah and Kalimantan (Payne and Francis 1998: 293–94, 300). The Pleistocene fauna of Borneo is also known to have included several large mammals now extinct: the tiger (*Panthera tigris*), Javan rhinoceros (*Rhinoceros sondaicus*), the Malay tapir (*Tapirus indicus*) and the now extinct giant pangolin (*Manis palaeojavanica*) (Cranbrook 2000: 82). Dated remains of the Javan rhinoceros from Madai caves in Sabah indicate that it persisted until the beginning of the Holocene (Cranbrook 2000: 80). The extirpation of the tiger is less certain as the remains were recovered from mixed deposits at Madai, ranging in age from c 10,000 BP to c 3000 BP (Cranbrook 2000: 79). Several specimens of Malay tapir were recovered from stratified contexts at Niah Cave and argued to be at least 8000 years old (Medway 1960: 358). However, given the current understanding of the complexities of sediment formation at Niah, any age estimations must be treated with extreme caution; it appears the Holocene sequence may have been very shallow across much of the West Mouth, invalidating previous attempts (Harrisson 1967: 185) to formulate an age-depth curve at this site. With these uncertainties in defining the terminal occurrence of extinct species in Borneo, it seems pointless, at this stage, to speculate as to the particular agent or agents that caused their demise. It seems sufficient to indicate that hunter-gatherers in Borneo had access to a wide range of large and medium-sized game and did not encounter a landscape of small game, low in diversity, as was the case for colonists on most of the smaller islands of the archipelago (MacKinnon et al 1997).

SITE CONTEXT AND CHRONOLOGY

The island of Borneo is the third largest island in the world (c 700,000 sq km) and straddles the equator from about 4°S to 7°N. The landscape is predominantly below 1000 m with a range of mountains in the northern core of the island rising between 1000 and 2000 m. The present climate is tropical superwet (Richards 1996) with rainfall averages ranging from 2000 mm to extreme highs of 7000 mm (Hazebroek 2001). Temperatures are relatively even throughout the year (22–32°C) with increasing climatic seasonality experienced in the south and southeast of the island (Richards 1996).

Niah Cave is situated approximately 30 m above sea level on the edge of a broad coastal plain that extends seawards and now forms part of the submerged Sunda Shelf. The West Mouth of the Niah Cave complex is a massive cavernous opening that is suspended some 15 m above the valley floor near the base of a limestone formation called the Gunung Subis. The Subis towers 394 m above the lowland rain forest, its sheer craggy white walls bare of vegetation until the summit, where a rich limestone rain forest drapes precipitously overhead. The valley bottom directly in front of the cave mouth lies at a lower elevation of about 15 m above sea level and is a mixture of small flat areas of limestone bedrock with very shallow soil and deep and narrow mazelike trenches.

Today the cave is reached by a wooden boardwalk that leads from the Niah Cave National Park Museum situated on the banks of the river Niah about 3.5 km from the site. While the walk to the cave is now made with relative ease, in the past access from the river meant traversing areas of lowland swampy ground and a steep ascent over sharp limestone at the foot of the Subis which would have involved hand-over-hand climbing. Access to the cave would have presented some difficulties for humans, and the rocky terrain would have been a barrier for large terrestrial animals and possibly, medium-sized game as well. Plant material falling into the site via non-anthropogenic processes will almost solely have consisted of species that prefer limestone habitat and can persist within skeletal soils and the precipitous nature of the rock formations.

Occupation deposits within the West Mouth were once extensive and filled a large area at the left-hand side of the cave mouth (Figure 4.1). These sediments were almost entirely removed by the excavation team led by Tom and Barbara Harrisson between 1954 and 1967 (B Harrisson 1967; T Harrisson 1957, 1958, 1959, 1970) leaving several small standing sections across the site

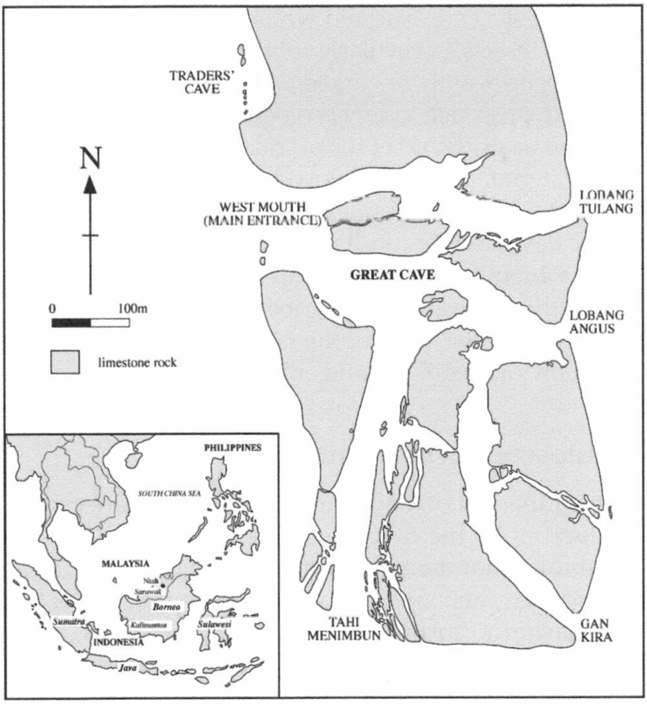

Figure 4.1 Site location (lower inset) and plan of cave.

and particularly against the rear wall of the cave. Tom Harrisson followed an aggressive excavation strategy leaving little for those to follow and nothing at all in some of the smaller cave mouths. The Niah Caves Project (NCP) under the direction of Professor Graeme Barker, University of Cambridge, was initiated as a four-year study (2000–2003) of the remaining sediment sections and trench walls to reinterpret the stratigraphy and chronology of the site, to reappraise the environmental and behavioural record within the sediments and to provide new geomorphological interpretations of deposit formation (Barker et al 2001, 2002a, 2002b, 2003). The material discussed here is derived from sampling some of these remaining blocks of sediment and from the excavation of a surviving sediment plinth in the 'Hell Trench'.

The sediments discussed cover an age range from the late Pleistocene to the early Holocene from around 45,000 BP to 8500 BP, a timespan of some 36,500 years of prehistory. While the sedimentary sequences are now well understood, detailed chronologies are not available for all deposits and thus the discussion of results is framed in general terms regarding the nature of plant utilisation at the site.

Deposit Formation

The history of deposit formation is varied across the West Mouth. The majority of sediments are either in situ or reworked bird and bat guano. This often has the appearance of a chocolate and white-coloured layer cake with vertical disturbances from insects, particularly robber wasps. In some areas of the site this material has been heavily reworked by human activity forming a uniform brown-coloured deposit rich in archaeological evidence; Areas A and B, and to some extent the upper levels of the original Hell trenches, would have consisted of this material. Within the lower levels of the Hell trench system deposit formation was more complex, involving major sediment movements from within the cave itself, deposits introduced by colluvial processes from outside the cave mouth, and a coincident phase of fluvial deposition that occurred during the earliest known period of occupation (Barker et al 2002b: 153). The NCP excavations have for the first time produced contextually controlled information for Niah Cave and enabled the reappraisal of earlier work.

The Late Pleistocene Environment

An important component of the argument about rain forest foraging at Niah concerns the location of the cave relative to the coast and the nature of the environment throughout the long history of its use. The Sunda Shelf is one of the largest tropical shelf areas linking the major islands of Sumatra, Java, Borneo and Palawan to form the landmass of the Last Glacial Maximum (LGM) defined as 'Sundaland'. To the east, narrow deepwater straits separated Sundaland from most of the Philippines and the combined landmass of Papua New Guinea and Australia known as 'Sahul'. Over the last 50,000 years there have been major changes in climate, vegetation and sea level in the tropical region of Southeast Asia that must have had direct implications for human inhabitants of the region.

The distance of Niah from the coast has varied over the last 50,000 years from a maximum of about 130 km during the LGM to possibly 0 km during the high sea stand of the mid-Holocene (Figure 4.2). As the sea level retreated following the last interglacial high stand, there were several periods of general stability. Between 110,000 and 75,000 years ago, sea level was about −50 m below modern values, dropping to about −80 m between 55,000 and 30,000 years ago (Hanebuth and Stattegger 2003a) leaving the cave between 110 km and 80 km from the coast. When sea levels were at −50 m, extensive land bridges still connected the Malay Peninsula, Sumatra, Java and Borneo and at depths below −75 m the configuration of the Sunda Shelf was very similar to that of the LGM (Voris 2000). During the LGM, sea level had reached its minimum level at about −120 m below modern values (Chappell et al 1996; Hanebuth and Stattegger 2003a). The Sunda Shelf appears to have remained tectonically stable since the mid-Tertiary (Lambeck et al 2002; Tija and Liew 1996) which suggests that in the absence of localised tectonic activity, Niah Cave may have lain at least 100 km from the coast from c 55,000 years ago until the period of rapid sea-level rise beginning around 14,600 years ago (Hanebuth et al 2000).

The climate appears to have been cooler and drier during the LGM with air temperatures perhaps as low as 6 to 7°C below current values (Flenley

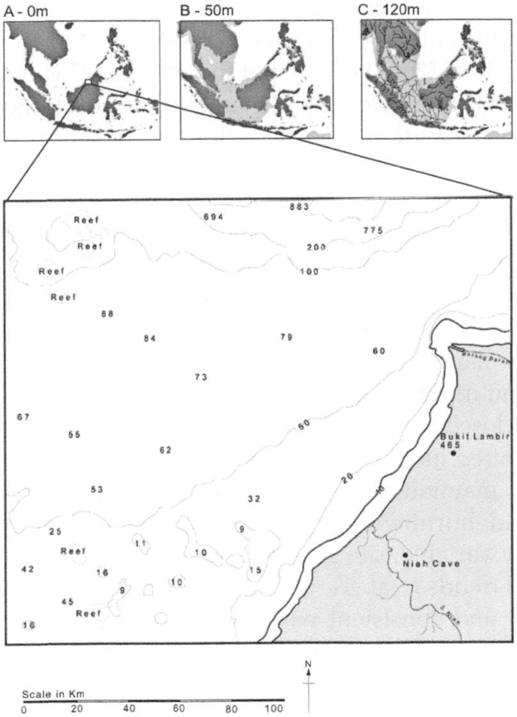

Figure 4.2 Sea level showing (a) extent of exposed Sunda Shelf at 0 m, −50 m and −120 m; (b) modern sea-floor bathymetry.

1998: 49). At this time, expanded glaciers covered the high mountains of Indonesia and New Guinea (Peterson et al 2002); and alpine and montane forest treelines were lower (Flenley 1998; Sun et al 2000; Taylor et al 2001). Flenley (1998) postulates an extreme lowering of the lower montane rain forest boundary from around 1000 m to around 50 m or 150 m above present sea level based on changed UV levels and atmospheric CO_2. Such environmental conditions may have created unique distributions of plant and animal species as montane and lowland rain forest co-occurred at low elevation, producing a forest type with no modern analogue (Anshari et al 2001: 226). Cranbrook (2000) in his review of the faunal remains from Niah postulated that humans using the cave in the Late Glacial period were foraging in 'a mosaic of closed forest alternating with scrub, bush, or lakes, or large rivers'. Rainfall may have been reduced by 30 to 50% in some parts of the Sunda Shelf (Kershaw et al 2001) but a strengthened winter monsoon is also argued to have led to increased rainfall along the exposed Sunda Shelf north of Borneo (Bird et al 2004: 150). These conditions may have created a generally drier seasonal forest or savannah grassland corridor in southern Borneo whilst maintaining moist tropical rain forest in northern Borneo and the island of Sumatra (Bird et al 2004; Heaney 1991). A pollen core from a lowland peat swamp in West Kalimantan indicates lowered temperatures at around 30 kya and a more open environment during the LGM, but that climate was sufficiently wet at times to permit peat formation (Anshari et al 2001: 225). Sun et al (2000) argue that during the LGM much of the lowland areas of the Sunda Shelf were wet enough to support large stands of lowland tropical rain forest with mangroves along river mouths and the coastline. A series of sediment cores from the mouth of the North Sunda River ('Molengraff River') dating to around 17,000 cal BP also shows the existence of mangroves, coastal swamps and interior regions supporting grassy marshlands (Hanebuth and Stattegger 2003b).

A more open, fractured type of rain forest is likely to have favoured many edible plant species, particularly those of the yam family, Dioscoreaceae, and other tuberous plants. These species may have occurred in higher densities in this period than in the mid-Holocene when the forests became more homogeneous and 'closed' as global climates warmed. A more open landscape may also have favoured indirect strategies of plant management, such as burning to increase or maintain forest gaps and edge habitats. Direct evidence of human-induced burning in the Southeast Asian and Melanesian tropics appears highly variable, site specific and difficult to tease out of larger-scale environmental trends (Haberle et al 2001; Maloney 1985). Across most of this region, a clear and consistent pattern of human-induced burning does not occur until c 4000 BP (Haberle et al 2001). Though humans are clearly implicated in some early landscape modification in locales such as the Baliem Valley, Irian Jaya, by 32,000 BP (Haberle et al 2001: 265), charcoal from a Late Glacial pollen core in West Kalimantan, Borneo only shows a clear anthropogenic signature as late as 1400 years BP (Anshari et al 2001: 226).

Due to the shallow gradient of the Sunda Shelf, the rate of landward transgression of the South China Sea during periods of sea level rise was substantial. Between 21,000 cal BP and 11,000 cal BP the average lateral rate of transgression was about 50 m yr^{-1} and as much as 450 m yr^{-1} during Meltwater Pulse 1A between 14,600 and 14,300 cal BP, when the sea rose rapidly by 16 m (Hanebuth et al 2000; Hanebuth and Stattegger 2003b), or possibly by as much as 25 m (Lambeck et al 2002). The rising seas separated Borneo from the peninsula shortly before 12,000 years ago and the large islands of Java and Sumatra by around 10,000 years ago (Bird et al 2004). Such high rates of water movement must have had major impacts on the coastal ecology of Sundaland, with many plant and animal communities adversely affected because they were unable to maintain pace with such rapidly changing environmental conditions. Depending upon the degree and nature of ecological changes resulting from sea level change, an unstable coastline with varied and unpredictable changes in resource distribution may not have represented good environments for human foragers until rates slowed after 14,000 years ago (Steinke et al 2003). The period of high sea stand around 5,000 years ago (Tija 1996) may have flooded the lowland areas surrounding the limestone subis containing Niah Cave (Barker et al 2001), with as yet unknown consequences for humans in the region. Modern marine and sedimentary conditions were not reestablished on the central part of the Sunda Shelf until 9,000 years ago and 8,000–7,000 years ago respectively (Steinke et al 2003). As Terrell (2002) has argued for the northern coasts of New Guinea, the late Pleistocene coastlines of northern Sundaland were not necessarily ideal habitats for hunter-gatherers and may not have been intensively utilised until after sea levels stabilised in the mid-Holocene.

The summer monsoon may have returned in a punctuated rather than gradual manner from around 13,000 years ago with varying effects on vegetation and climate across Southeast Asia (Hope et al 2004). Low-elevation rain forest expanded from 14,000 to 11,000 years ago in Sumatra and Java, but may have been later farther north in Thailand (Penny 2001) and seasonality less pronounced until around 7,500 years ago (Maxwell and Liu 2002). Later Holocene records indicate far less climatic instability than in the preceding glacial transition, but anthropogenic factors including burning, tree clearance and agricultural expansion complicate interpretations of climate change (Hope et al 2004).

MACRO AND MICROPLANT REMAINS

Macro and microbotanical work carried out at Niah has to date been limited, but the scope for this type of analysis seems great. Macroplant was recovered during the 1977 field excavations undertaken by Zuraina (1982) and by the Niah Caves Project (NCP) between 2000 and 2003 (Barker et al 2001, 2002a, 2003). The scope of both projects has been deliberately conservative and limited the amount of material removed during investigation. As a consequence these excavations have targeted small surviving sediment baulks and sections widely dispersed around the cave rather than concentrated efforts in one area (Figure 4.3). Zuraina (1982) only recovered botanical material from the area of Hell Trench near the face of the West Mouth from several small soundings in and around the area that the Harrisson team uncovered the human skull. The NCP botanical analysis has involved the experimental analysis of starch in sediments undertaken by Barton (2005b) and of several targeted analyses of sediment conducted by Paz (2001). The results of these studies are summarised below and indicate the potential for future work.

Cave sediments have yielded a wide range of wood and fibre materials including coffin wood, bamboo caskets, mats of pandan fibre, and woven fabric from the Neolithic cemetery (B Harrisson 1967: 189). Potential plant food remains include nut exocarp, seeds and charred parenchyma tissue from the interior tissues of tubers and fruits (Paz 2001). Starch granules from plant tubers and rhizomes have also been recovered from a wide range of contexts across the site (Barton 2005b) and new work on phytoliths is also showing promise (Kealhofer pers comm 2004). The conditions for organic preservation at Niah Cave seem good and may be in part related to the peculiar characteristics of guano chemistry at the site and the relatively dry and temperate climate of the cave mouth.

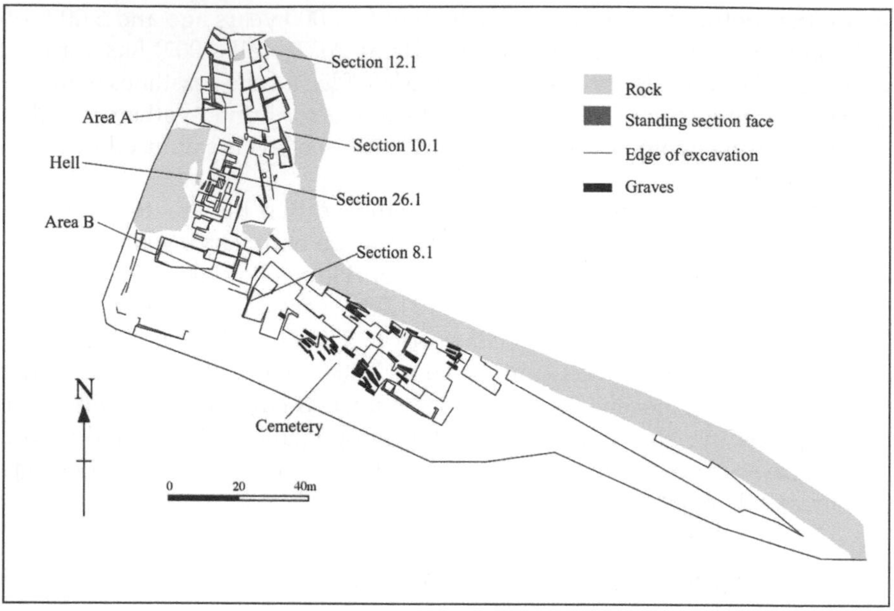

Figure 4.3 Plan of excavation area, Niah Cave West Mouth.

The macro and microplant remains have all come from the area of the West Mouth that have been defined as Areas A and B by the NCP, an area that Tom Harrisson broadly referred to as the 'occupation sector' (B Harrisson 1967: 128). Radiocarbon dates from this part of the cave have now confirmed that the remaining sedimentary sequences containing evidence of human habitation date from c 45,000 BP to the early Holocene at c 8500 BP (Barker et al 2002a). Later Holocene sequences are conspicuously absent here and appear to have been completely removed during the earlier Harrisson excavation of the 1950s or by guano collection within the cave mouth during the early part of the twentieth century. The steep rocky terrain immediately in front of the cave mouth would have made access difficult to species that were not adept climbers. It is most likely that humans are responsible for the presence of starch from tuberous species here; without doubt humans are the source of the carbonised tubers, fruit and nut fragments. Another possible vector for some starches, particularly the Araceae or aroids, could be pigs (*Sus barbatus*), as this species is known to favour shallow rooting tuberous plants, but whether they could deal with the acridity of unprocessed aroids is unknown; most yams (Dioscoreaceae) are too deeply rooted to be available as pig food. However, starch granules are routinely recovered from modern human faecal matter (Englyst et al 1992) and have recently been recorded in a good state of preservation within human and dog coprolites that are up to 500 years old (Horrocks et al 2004). In an extremely pessimistic case, some of the tuberous starches could have entered the sediments via the butchery of these animals on site, but given the quantity of additional material that is derived from human activity, such as the charred tuber remains, an anthropogenic origin seems most parsimonious for the aroids.

METHODS

Macro: Plant Recovery

In the laboratory, samples of 4.5 to 9 litres of sediment matrix per context were manually floated in plastic buckets, 5 to 7 litres at a time. The contents of the sediment matrix were collected by pouring off the floating light fraction onto a 0.3 mm laboratory test sieve. This fraction was then placed inside a drying cabinet. The heavy fraction of each sample was wet sieved to remove the fine sediments and afterwards air dried. A random quarter of each heavy fraction sample was selected using a riffle box and sorted for botanic material. Both heavy and light fractions were sorted by particle size using a nest of laboratory test sieves with apertures of four, two, and 0.5 mm. The materials were sorted using a low power stereomicroscope with magnification ranging from 10x to 50x. The identification of botanical macroremains followed procedures outlined in Paz (2001) and the parenchymatous tissues following Hather (1988, 2000) and Paz (2001).

Micro: Starch Granule Recovery

In the laboratory, soil samples were initially sieved to remove the coarse fraction, then two 5 g samples were taken from each sedimentary unit. The lightest fraction of organics and minerals was removed using a low density solution of sodium polytungstate, and the starches and any other material with a specific gravity less than 1.5 were then removed in a high density solution. This sediment was resuspended in 500 ml of ultrapure water, from which a 50 ml aqueous sample was removed by pipette. The extract was left to air dry on a covered glass slide and mounted with glycerine.

The starch grains were recorded using a combination of size and morphological characteristics and a series of morpho-types defined (Barton 2005b). Maximum length and width were recorded using an eyepiece micrometer, and shape and other attributes were recorded at 200–1000x magnifications. Starch grains are three-dimensional objects, and to ensure that accurate shape identification is made, it is highly desirable to try and rotate the grains beneath the coverslip. The determination of size and/or shape alone is not sufficient to enable accurate identification in most cases, as overlap does exist between species. However, when combined with a more detailed assessment of the morphological characteristics of granules, such as surface features, colour, hilum position, hilum features, and the number and nature of granule facets, a higher level of taxonomic identification is possible (eg Loy et al 1992; Piperno and Holst 1998; Torrence 2006).

Confidence Levels in Plant Identification

The identification of plant taxa uses a scale of confidence developed by Paz (2001). The highest confidence is a binomial identification without any prefix. Such samples match closely with reference material and can be shown to fit within the regional flora and the correct geographic area. For parenchyma, the sample tissue fits all measurement ranges on the plant cells including size, shape and cell wall thickness. For starch, granules match very closely with reference material including size, shape and other internal and external features of the granule. A prefix of 'prob' indicates a close match with an illustration or image but is material not physically present within the reference set. The prefix 'cf' indicates that the sample matches an image previously identified by an archaeobotanist but the morphological fit of the sample is not exact. The prefix 'elim' is the lowest level of confidence when attempting a binomial classification. It indicates that the material may be the genus/species proposed, but the determination was derived without reference material or illustrations. Classification is based on the taxonomic description of a plant, its fit with its geographical range and its elimination from other material within the reference collection.

RESULTS

Owing to the manner of sediment investigation at Niah, results cannot be reported from a single trench or section as samples come from various sedimentary deposits and trenches across the site. As a consequence, individual finds will be discussed in relation to their radiocarbon ages or relative dating sequences.

Roots, Tubers and Rhizomes

The remains of tubers from the Dioscoreaceae family of true yams and Araceae rhizomes (which includes the group known as taro) have been recovered from deposits dating to the late Pleistocene and early Holocene in the West Mouth. Preservation of these plant organs is in the form of whole roots, charred parenchyma tissue from inside the tuber, and starch granules.

Dioscoreaceae: cf **Dioscorea hispida** *Dennst.*

Two fragments of charred yam parenchyma were recovered from context 1047, Area B, dated by AMS using the ABOX-SC pretreatment protocol at 21,130 ± 80 uncal (OxA-V-2077-8) and sealed by an upper layer dated to 17,560 ± 70 (OxA-V-2077-7) (Table 4.2). The tissues of these fragments were greatly distorted and diagnosis largely relied upon identification of the remains of vascular bundles (Figure 4.4). Comparison of the pitted-reticulated pattern of tracheary elements showed they resemble those found in *Dioscorea alata*, *D. hispida*, *vintakey* (Agta term for a Dioscoreaceae) and *lakapen* (Ivatan term for a Dioscoreaceae). The size and nature of the thickenings on the tracheary elements are highly diagnostic of *D. hispida* and overall the size of the elements falls within the middle of the *D. hispida* reference population. The measurements are all at the small end of the range for *D. alata* but do not fit any specific reference population measurement range. On this basis the determination can be placed as cf *Dioscorea hispida*.

Table 4.1 Starch granule counts for Hell Trench

Excavation Area	Context Number	Radiocarbon Age (BP)	Type 5 *Alocasia* sp.	Type 8 *Alocasia* sp.	Type 6 *Dioscorea* sp.	Type 9 *Caryota mitis*/ *Eugeissona* sp.
Hell	3112					
Hell	3128			2		
Hell	3129		1	5		
Hell	3130					1
Hell	3131	43,810 + 810/ − 660 (OxA-V-2057-27)				1
Hell	3132	42,600 ± 670 (Niah-310)				
Hell	3134	33,930 + 310/ − 290 (OxA-V-2059-11)				2
Hell	3136					1
Hell	3137				2	1
Hell	3138					1
Hell	3140	43,440 + 680/ − 630 (OxA-V-2057-29)				
Hell	3143	43,4400 + 680/ − 630 (OxA-V-2057-30)				1
Hell	3150					
Hell	3158	45,040 + 840/ − 760 (OxA-V-2057-31)	1	7	2	8
		Total	2	14	4	16

Dioscorea hispida is a vigorous climber forming a bulky tuber that is widely distributed and has been recorded in India, southern China through Southeast Asia to New Guinea (Burkill 1966: 831). Burkill (1966) describes the tuber as rare throughout the Malay Peninsula but common in areas where it was once under cultivation. While toxic, the tuber is shallow growing, easy to collect and produces tubers of prodigious size. Kuchikura (1987: 52) calculated that on average, the Semaq Beri of the Malay Peninsula recovered approximately 33 kg of *D. hispida* per person per hour, compared with 9 kg/person/hour for *D. pentaphylla* and 2.5 kg/person/hour for *D. gibbiflora*, *D. hamiltonii* and *D. orbiculata*.

Unidentified Root Tuber

Also recovered from this context (1047) was a fragment of root tuber. From the gross morphological features alone, the material looked like a small tuber. After dissection the internal tissue showed the structure of a root but did not match any of the reference material.

Dioscorea alata *L.*

Starch granules from a species of *Dioscorea* tuber (Barton 2005b) were recovered from sediments in Area A and Hell (Tables 4.1 and 4.2; Figure 4.5). A total of four individual granules were found in sediments dating to c 40,000 years BP and an undated but likely Holocene layer from Area A, Block A. Granules were all of the same type (Type 6), large ovates from a minimum of 20 microns to a maximum of 45 microns in maximum length with an eccentric hilum only visible under cross-polarised light. Such granules are characteristic of yam species and the features of these grains fall within the range known for the Southeast Asian species of *Dioscorea alata*. Accurate taxonomic identification of starch is dependent upon the size and scope of the reference material, so some caution is necessary, but the closest match was with the *D. alata* material.

D. alata is a climber rarely reaching a height above 20 feet with a deeply rooting tuber, though Burkill (1966: 827) argues that deliberate selection through cultivation has produced many shallower rooting varieties. In the wild this plant may be found in jungles and thin forests up to around 800 m elevation (Ochse and Bakhuizen 1980: 231) and is generally distributed throughout the lowlands of the Malay Peninsula (Burkill 1966: 827). The origins of this plant are unclear, but it is argued to be a true domesticate artificially selected from its two likely wild progenitors *D. persimilis* Prain & Burkill and *D. hamiltonii* Hook. f. (Burkill 1966: 826) at some time in prehistory.

Araceae

A fragment of charred rhizome was recovered from the basal layer of Section 12.1, Area A, context *1020,* dated by AMS radiocarbon to 19,650 ± 90 BP

Table 4.2 *Starch granule counts and charred macroplant remains for Area A, Block A*

			Starch Granule Count					Macroplant Remains											
Excavation Area	Context Number	Radiocarbon Age (BP)	Type 4 cf *C. merkusii* or *Alocasia*	Type 5 *Alocasia* sp.	Type 8 *Alocasia* sp.	Type 6 *Dioscorea* sp.	Type 9 *Caryota mitis/Eugeissona* sp.	cf *Colocasia* elim *esculenta*	cf *Dioscorea hispida*	elim. Araceae	Root Tuber	cf Fabaceae	prob Moraceae	cf Umbrelliferaceae	cf Urticaceae	Fruit	Large nut frag-1	Nut frag	Unident parenchyma
Area A Block A	1034			1		2	1												
Area A Block A	1033		1	1									F					F	
Area A Block A	1027		2																
Area A Block A	1025	27,960 ± 200 (OxA-11304)																	
Area A Block A	1023																		
Area A Block A	1086/1026		2	1						X						X	R	F	X
Area A Block B	1015	8630 ± 45 (OxA-11549)																	
Area A Block B	1016										X						R	F	X
Area A Block B	1018																		
Area A Block B	1019																		
Area A Block B	1020	19,650 ± 90 (OxA-11550)			1			X									R	F	
Area B	2075~2100	17,560 ± 70 (OxA-V-2077-7)																	
	2078~1047	21,130 ± 80 (OxA-V-2077-8)							X		X	R	F	X	R	R	C	C	X
	2079~1053	20,220 ± 80 (OxA-V-2077-9)																	
		Total	5	3	1	2	1												

X = fragment; R = rare (less than five fragments); F = few (less than twenty fragments); C = common (less than 100 fragments).

(OxA-11550) and calibrated at the 95.4% level to 23,850–23,020 cal BP (Bronk Ramsey 2000). The tissue sample contained rounded to elongate parenchyma cells and limited diagnostic features, but based on size, shape and cell wall thickness most closely resembled tissue from the taro *Colocasia esculenta* (Figure 4.6). Due to the lack of clear diagnostic elements the level of determination can only be given as cf *Colocasia* elim *esculenta*.

Colocasia esculenta or 'taro' is an aroid of the Araceae family that forms a large edible rhizome and is still heavily cultivated throughout Southeast Asia and the Pacific (Wilson and Siemonsma 1996: 69), often as a companion plant to wet paddy rice. Wild varieties are known to be very acrid (Yoshino 2002), irritating to the mouth, and most cultivated varieties, though more palatable, still require some processing before they can be eaten (Wilson and Siemonsma 1996: 69). Overall the main physical difference between wild and cultivated varieties is the presence of acridity in the wild forms and its general absence in cultivated varieties, a condition that Yoshino (2002: 113) believes may have arisen by mutation many times without human intervention.

A sample of elim Araceae derives from context 1026, from Section 12.1, Area A (Table 4.2) and is slightly younger than 27,960 ± 200 BP (OxA-11304). The best match with this sample is with that of Colocasia sp. reference material but cell size and wall thickness were outside the range for a taxonomic identification to be made with confidence (Figure 4.7).

Unidentified Vegetative Organ

This sample of elim Araceae was recovered from context 1016, Section 10.1, Area A (Table 4.2) sealed by the uppermost unit in this remnant section dated to 8630 ± 45 BP (OxA-11549), calibrated at the 95.4% level to 9690–9525 cal BP (Bronk Ramsey 2000). The sample seems to be parenchyma from vascular bundles, or perhaps from a wild type of vegetative organ with angular to rounded parenchyma cells. The tissue also contained several raphides in idioblastic cells. While there was no fit with the reference material, the presence of idioblasts filled with raphides is strongly suggestive of an aroid (Araceae). Such cells are particularly common throughout the rhizomes and leaves of *Colocasia esculenta* (Sunell and Healey 1979, 1985) and other wild *Colocasia* sp. are known for their high levels of acridity (Yoshino 2002: 96), which is related to the presence of numerous raphides and irritating proteases (Bradbury and Nixon 1998).

Alocasia longiloba *(Longiloba Complex) Hay and* **Cyrtosperma merkusii** *(Hassk.) Schott*

Starch granules from a large group of Alocasias, now termed the Longiloba Complex due to present difficulties in their taxonomy (Hay 1998), were found throughout the Niah Cave sediments and are identified at a very high level of

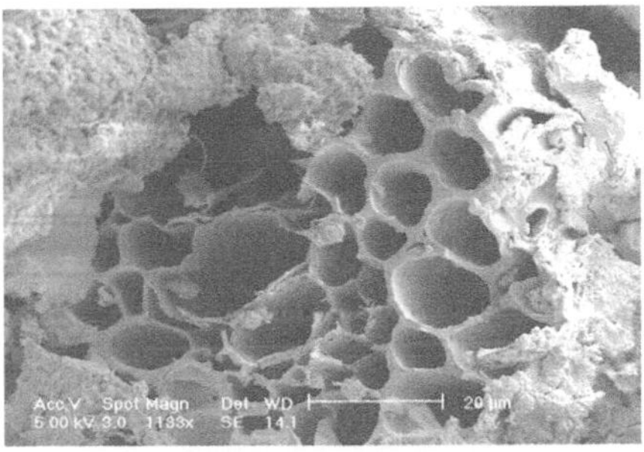

Figure 4.4 Cf *Dioscorea hispida*, parenchyma tissue, context 1047, Area B, Niah Cave.

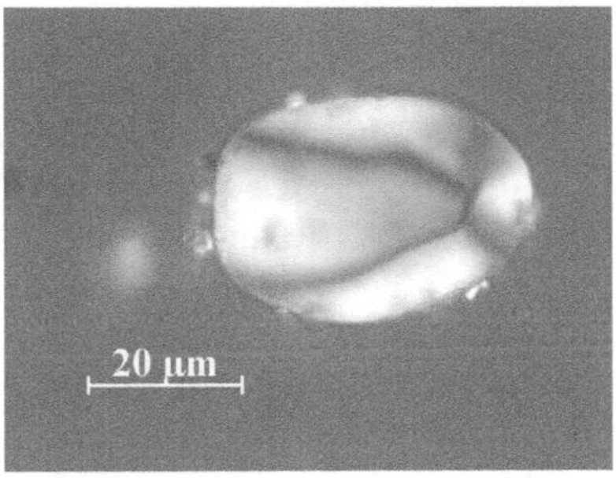

Figure 4.5 *Dioscorea* sp. starch granule, context 3137, Hell Trench, Niah Cave, c 40,000 BP.

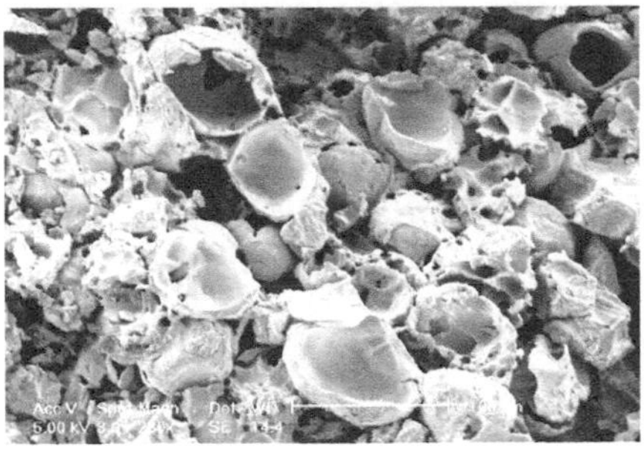

Figure 4.6 cf *Colocasia* elim *esculenta*, parenchyma tissue, context 1020, Area A, Niah Cave.

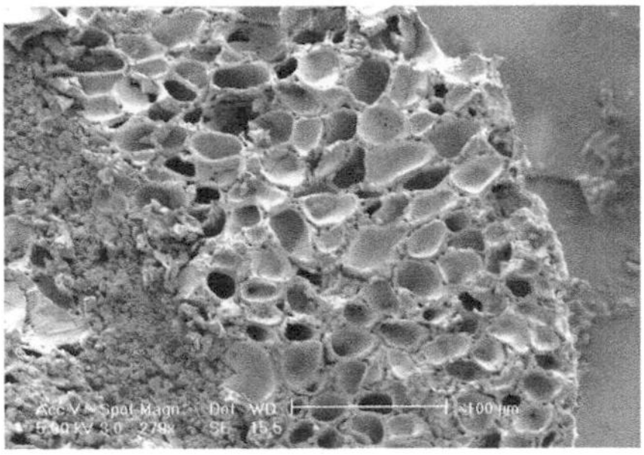

Figure 4.7 elim Araceae, parenchyma tissue, context 1026, Area A, Niah Cave.

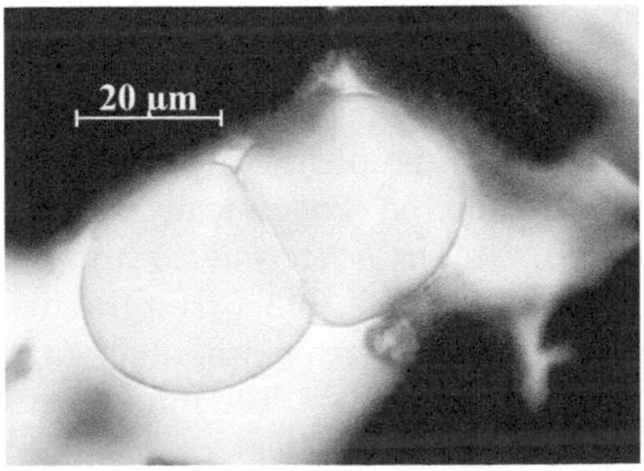

Figure 4.8 *Alocasia longiloba* Longiloba Complex, starch granules, context 3129, Hell Trench, c 40,000 BP.

Figure 4.9 Fragments of prob Moraceae fruit exocarp, scale in centimetres.

confidence. These granules are highly distinctive and, unlike other Alocasia species, are known to occur in the region. Further, Barton has voucher-quality material for this taxon (Tables 4.1 and 4.2; Figure 4.8). These granules have distinctive morphologies, particularly in relation to the growth form around the hilum, nature of compound granules, and the presence of banded pigment in some granules (Barton 2005b). One form of these granules, however, was not readily distinguishable from those of *Cyrtosperma merkusii*, an important food plant in Southeast Asia, which therefore must be included in this group for the present. Starch from these plants was found in the upper levels of Hell Trench dating to <40,000 years BP and levels in Area A, Block A pre- and postdating the radiocarbon-dated level of 27,960 ± 200 BP (OxA-11304) and in the basal layer of Block B, 19,650 ± 90 BP (OxA-11550), calibrated at the 95.4% level to 23,850–23,020 cal BP (Bronk Ramsey 2000).

Alocasia is a large genus of the family Araceae with some 70 species from Southeast Asia to the Pacific (Hay 1990: 27). They are generally found in various types of lowland rain forest and on limestone (Brown 2000: 176). Their growth form is similar to that of 'taro' though they are often larger plants with long stalks and sagittate (arrow-shaped) leaves. Plants identified as part of the Longiloba Complex in the Niah National Park were often found in partially shaded areas and areas of previously disturbed forest. Collectively these plants are generally referred to as 'birah' by the Iban because of their irritating properties, with several species used medicinally or as components of animal poisons (Burkill 1966: 106). It is not known whether they can be eaten, but it is possible. Because of their irritating compounds, alocasias have been marginalised in favour of other foods, but they may well have been far more important in the past as a source of starch and possibly for their medicinal or poisonous properties.

Other Plant Remains

Charred remains of fruit and nut fragments were recovered from all samples analysed by Paz (Table 4.2). Taxonomic identification is only at the level of the family but revealing nonetheless. Of particular note are the probable remains of Moraceae from Block A and from the Pleistocene context of 1047 in Area B (Figure 4.9). These tissue fragments resemble the thick fleshy exocarps of fruits of *Artocarpus* spp., which include the economically important breadfruit (*Artocarpus altilis*), jackfruit (*Artocarpus heterophyllus*) and some less well known but important Southeast Asian fruits. Other fruit fragments occur in an upper and lower level of Block A and were commonly in association with unidentified nut fragments from context 1047. The nut fragments are not yet firmly identified but they are not members of the species *Canarium* (Paz 2001).

The material excavated in 1977 solely consists of nut fragments recovered from the small soundings within the Hell Trench system (Zuraina 1982; Table 4.3). Material was recovered by hand from the sediments; no flotation was used (Zuraina 1982: 41). Recovered organics were identified by Paul Chai (forest botanist, Sarawak) as predominantly pieces of the toxic but edible nut *Pangium edule* R. and the edible *E. stipularis* B. and possibly an example of *Rattan* sp. A single fragment of *E. zwageri* was also recovered but

Table 4.3 Summary of macroplant remains (after Zuraina 1982)

Niah Caves Project Area	Square	Depth (inches)	*Elaeocarpus stipularis* B. or *Elaeocarpus subpuberis* M.	*Rattan* sp. or *other Arecaceae*	*Pangium edule* R.	*Eusideroxylon zwageri* T. et B.
Cemetery	77/DN4	25	X	X		
Hell	HQ/9	84			X	
Hell	HQ/1	87–90			X	
Hell	HQ/6	92				X
Hell	HQ/7	95			X	
Hell	HQ/1	103			X	
Hell	HE/22	105–108			X	
Hell	HQ/1	108			X	
Hell	HP/7	111–114			X	

this is not known to be an edible species. The Hell squares as defined by Harrisson and used by Zuraina (1982) have not yet been firmly tied into the NCP context system, but the charcoal sample collected from a depth of 84 to 90 inches in Square HQ/6 dated to 21,410 ± 760 BP (Gx 4834) is in accord with the current understanding of the antiquity of these sediments. Based on sediment descriptions and the depth profile, some of the *P. edule* remains recovered by Zuriana are likely to derive from sediments approximately 40,000 years old (see Barker et al 2002a: 153).

Barton (2005b) also recovered starch granules from sago palm (*Eugeissona* sp. or *Caryota* spp.) in Pleistocene levels and in an upper layer in Block A (Table 4.2). Such granules may indicate the presence of sago pith in deposits that are considered to have formed during human occupation of the site, as it is very difficult to see how palm starch could have entered the cave by another vector. Granules are similar in size and shape to the staple *Eugeissona utilis* used by the eastern Penan and also to *Caryota mitis* (Tomlinson 1961: 364) and *C. urens*. The large stately palm, *Caryota no*, is common in the forest in front of the cave, although samples are not yet in the starch reference collection at Leicester. The term 'sago' collectively refers to all starch flours recovered from the pith of a wide variety of tropical palms, which in Malaysia include *Arenga brevipes, A. pinnata, A. undulatifolia, Caryota mitis, C. no, Eugeissona utilis* (Puri 1997: 198) and possibly *Eugeissona insignis*. In Borneo, the most common palm used for sago is the clonal hill sago, *Eugeissona utilis*, which is still a starch staple for the eastern Penan of Sarawak (Brosius 1993). The labour required to fell a large palm and beat the interior fibres or pith to release the starch is high, but the energetic yields are considerable and far outweigh the effort of harvesting (Ulijaszek and Poraituk 1993).

POSSIBILITIES OF RAIN FOREST FORAGING

Macro and microplant evidence from the West Mouth at Niah indicates that during the late Pleistocene and into the early Holocene hunter-gatherers were using a variety of roots and tubers, fruit, nuts and sago pith extracted from the surrounding rain forest. The main tuberous/rhizomatous species identified were cf *Dioscorea hispida*, *D. alata*, cf *Colocasia* elim *esculenta*, *Alocasia longiloba* (Longiloba Complex), elim Araceae and some other aroid-type fragments. Fruit fragments include an identification of prob Moraceae (a family which includes *Artocarpus* spp.) nut exocarps of *Pangium edule* and a number of unidentified fruit and nut fragments. Starch grains from sago palm prob *Eugeissona utilis* or *Caryota* spp. were also recovered from late Pleistocene layers in Hell Trench and the upper level of the Block A sediments.

The ages of deposits from which identified material were collected are wide ranging, and cover some 35,000 years of prehistory at the site. The majority of macroplant remains derive from deposits dating to ⩾ 20,000 years BP. At this time climate was cooler, though not necessarily drier, vegetation more patchy and sea level at or near its lowest point of the last glacial cycle. Between 50,000 BP and 30,000 BP, little is known about the vegetation of northern Borneo and the surrounding region; however, sea levels were still low, around –80 m and dropping, placing Niah Cave at least 100 km from the Pleistocene shoreline. It seems likely that throughout this time foragers may have been living within and using a forested environment that was structurally quite different from that of later Holocene periods. During climatic amelioration after the LGM, the coastline of northern Borneo underwent gradual and sometimes rapid transformation as the rising seas swept across the low gradient slope of the Sunda Shelf (Hanebuth et al 2000).

During the Pleistocene period, the faunal evidence from Niah indicates a rather mixed foraging strategy that includes a wide range of terrestrial and arboreal mammals, reptiles, amphibians, fish and shellfish (Barker et al 2002a). Marine species of shellfish are notably absent throughout much of this period and preliminary analysis of the fish remains suggests that many are freshwater species (Piper pers comm 2004). In the late Pleistocene we are definitely dealing with a terrestrial foraging economy at Niah. And what is most telling in this regard is the use of plant species that are considered toxic or highly acrid: *Alocasia longiloba*, *Colocasia esculenta*, *Dioscorea hispida*, and *Pangium edule*.

THE USE OF TOXIC PLANTS

The family of edible aroids, the Araceae, which includes the Southeast Asian taros (*Colocasia esculenta*), giant taro (*Alocasia macrorrhizos*), giant swamp taro (*Cyrtosperma merkusii*), elephant foot yam (*Amorphophallus paeoniifolius*), and coco yam (*Xanthosoma sagittifolium* and *X. nigrum*) from South America, are well known for their acrid taste and the irritation and soreness of the skin if handled or eaten raw. These physiological effects are caused by needle-shaped crystals of calcium oxalate called 'raphides' and associated proteases which act as irritants entering the skin punctured by the raphides (Bradbury and Nixon 1998: 615). Acridity is often much higher in wild forms of the plant, such as *Colocasia* taro, rendering some variants inedible (Yoshino 2002: 96).

Toxins within the family of *Dioscorea* yams fall into three main categories: the alkaloids, tannins and saponins (Coursey 1967). The tannins and saponins are the least likely to present a problem for human consumption, though they do affect the taste of the tuber. Tannins are thought to give tubers an acrid taste (Coursey 1967: 208) and the saponins cause bitterness (Coursey 1967: 210). Saponins have also been identified as the active component in many fish poisons and are readily detectable in tubers, characteristically forming a soapy lather when agitated in water (Coursey 1967: 209). At least one of the alkaloids, dioscorine, which has been identified in tubers of *Dioscorea hispida*, are extremely toxic and cause paralysis of the nervous system and – if taken in sufficient quantities – death (Burkill 1966: 833). While some of the chemicals and cellular crystals in tubers and rhizomes cause discomfort and some are downright deadly, the toxins and other physiological effects are usually destroyed or alleviated by heat, applied either through boiling or roasting (Table 4.4). The toxin hydrocyanic acid, present within the seeds and leaves of *Pangium edule*, like the toxins within yams and aroids, is also destroyed by prolonged boiling (Burkill 1966: 1683). While it is certainly the case that the additional processing time required to prepare some of these plants will decrease their overall food value within the diet, it may not have caused such a deficit as to render them inaccessible to Pleistocene hunter-gatherers. In fact, current evidence seems to suggest that toxic plants were regular targets of rain forest foragers. For example, many species of Araceae and Dioscoreaceae are still used for their medicinal properties or as sources of hunting poisons (Burkill 1966; Christensen 2002: 109–14).

Recent archaeological evidence of preserved starch grains on stone tools and occupation deposits from the Melanesian islands of New Ireland and the Solomons has also shown that rhizomes of Araceae were a component of the diet of Pleistocene and Holocene foragers (Barton 1991; Barton and White 1993; Loy et al 1992). Evidence for the use of *Colocasia esculenta*,

Table 4.4 Treatment of toxic plants (after Burkill 1966)

Species	*Toxin/Complaint*	*Treatment*
Pangium edule	hydrocyanic acid	heat/boiling
Colocasia esculenta	acrid	heat/boiling
Dioscorea hispida	alkaloid dioscorine	heat/washing/boiling
Dioscorea gibbiflora	bitter	heat/boiling
Dioscorea orbiculata	acrid	heat/boiling
Dioscorea piscatorum	bitter/'fish poison'	heat/roasting
Dioscorea prainiana	bitter	heat
Dioscorea pyrifolia		heat/boiling/roasting

Alocasia sp. and *Cyrtosperma* sp. has been recovered from tool surfaces up to 28,000 years old in the Solomon Islands (Loy et al 1992) and evidence of *Alocasia* sp. from artefacts up to 14,000 years old in New Ireland (Barton 1991; Barton and White 1993). Recent research on stone tools found at Kuk Swamp in the New Guinea Highlands has recovered *Colocasia* taro from levels dating to 10,220–9910 cal BP (Denham et al 2003). Today the known species of wild *Colocasia* taros are generally regarded as inedible (Yoshino 2002); early users of this plant must have encountered similar phenotypes.

Dioscorea hispida is generally found in the lowlands, in jungle, and in more open localities such as thin forest and forest borders up to an elevation of around 850 m (Ochse and Bakhuizen 1980: 241). The plant forms large lobed tubers very near the surface, which are often covered in thick, stiff rootlets. There are several varieties recorded in use or cultivation, with white to yellow flesh, but all seem to be toxic. The tubers are soft skinned and fleshy, easy to dig, but poisonous and require careful processing by washing, boiling or steeping in saltwater to remove the toxin; even then the careless may still be poisoned (Burkill 1966: 832). Burkill (1966: 831) regarded this plant as the chief famine food of the tropical East, often reverted to when the rice harvest failed. While it appears to have been extensively cultivated at different times, there are no records of non-toxic varieties.

D. hispida is described as an important food for the Semang and nomadic Senoi of Peninsula Malaysia (Burkill 1966; Endicott 1984; Kuchikura 1993) and for the Batak of the Philippines (Eder 1978) and it is likely that this particular species of yam was very important in prehistory. For the Batak, collection of *D. hispida* accounted for three-quarters of the total wild yam calories, contributing 1739 cal/hr to the daily diet (Eder 1978: 61). Gathering of this yam by the Semaq Beri accounted for about 35% of the total yam yield, about 4.55 kg/hr, equivalent to about 3891 kcal/hr (Kuchikura 1993: 92), though patches of *D. hispida* were mostly considered to be feral rather than wild, arising from abandoned Malay gardens (Kuchikura 1993: 87). The Batek Dé recorded by Endicott (1984) appear less dependent upon wild yams; they ignored *D. hispida* most of the time because they considered it boring (Endicott 1984: 48). Even while ignoring what Endicott considered 'the most plentiful and prolific yam', the small amount of *D. hispida* taken in association with other wild yams still contributed 946 cal/hr (Endicott 1984: 47). The Semang's avoidance of the poisonous yam may be the result of recent historic trends, because Schebesta (1954) records the use of five yam species requiring pretreatment before eating including *D. hispida* and *D. prazeri* (*tulegen, hubi kapor*). Preparation of the former involved soaking for a long time in the river before being boiled and for the latter 'to make it edible they burn the leaves of the *gobn* tree and mix the ashes with the *tulgen*, which is first pounded to a pulp. This dish is edible and tastes like paste' (Schebesta 1973: 116).

The number of edible tubers and roots listed by Schebesta (1954) contains far more species than any lists produced subsequently. Included in the list are edible yams (members of the family Dioscoreaceae) and edible roots from rattans (members of the family Arecaceae). Schebesta's inclusion of 10 species of edible rattans is of great interest as rattans occur in wide diversity and profusion throughout the rain forests of the peninsula and Borneo. He also lists 19 types of *Dioscorea* tuber and the botanical names of 10 species (Schebesta 1954: 54) and at one camp alone counted 12 different types of edible root (Schebesta 1973: 117). While collecting ethnobotanical data in 1975, Endicott (1984) recorded 10 different species of *Dioscorea* tuber, but only names *gadong* or *Dioscorea hispida*, the very toxic yam that he describes as 'plentiful', but largely ignored by the Batek in favour of more interesting foods. Kuchikura (1987, 1993) lived with the seminomadic Semaq Beri between 1978 and 1979 and recorded a total of eight wild yams of which six were identified as true yams belonging to the Dioscoreaceae. Eder (1978) records the use of only two yams by the Philippine Batak, the endemic *Dioscorea luzonensis* and the familiar *Dioscorea hispida*. Plant lists such as those recorded by Schebesta (1954: 54–57) further indicate just how quickly diet breadth can change as the nature of resource distribution changes, whether that change is ecological and/or reflects the changing nature of social interactions between groups and the mechanisms and currency of exchange (Junker 2002: 144).

Ethnographic information of recent or even historic plant use should only be used as a guide to understanding the composition of prehistoric diet. Understanding past subsistence patterns in the absence of direct archaeological evidence may to a large extent require assessment of 'potential' foods. Such items may never have been staples or even food items used on a regular basis, but occasionally provided the necessary calories to maintain individuals above the starvation threshold. It may also be the case, as illustrated by O'Connell and Hawkes (1981), that a large number of food items may be dropped from the diet following the introduction of one or two highly ranked alternatives (Bettinger 1991). The opposite will also hold that if one or two highly ranked resources become unavailable, a large number of food items may be reintroduced to the diet on a more regular basis. It may be problematic to determine the total diet breadth of some foraging groups, particularly at the low end of the energetic scale, if groups have had access to high-ranking alternatives such as agricultural produce for an extended period of time, as is the case in the Southeast Asian tropics. Low-ranked but important resources may no longer be included in the diet, biasing later assessments of foraging potential that are dependent upon modern observations.

HUNTER-GATHERERS AND AGRICULTURE

The new archaeological evidence from Niah Cave demonstrates, without any doubt, that the earliest hunter-gatherers in Borneo were well adapted to exploiting plant and animal resources within an inland forested landscape from at least 45,000 years ago (Barker et al 2002a). The recovery of sago palm starch, the aroid, *Alocasia longiloba*, and a yam that is possibly *D. alata*, or a close relation, at Niah Cave demonstrates that people began using starchy plants of the families of Arecaceae (palms), Araceae (aroids) and Dioscoreaceae (yams) at least 40,000 years ago. The remains of animal bones and molluscs reflect reliance upon lowland game and freshwater resources, while the botanical evidence from around 20,000 BP reflects a relatively broad-spectrum subsistence base that included toxic plant foods and taro, *Colocasia esculenta*. These plants did entail some energetic costs in their processing to render them edible, and there remained a real risk of poisoning and even death, particularly with species such as *D. hispida* and *P. edule* (Burkill 1966: 1681), but overall, energetic returns could be high. The Pleistocene rain forest may have been more favourable to human occupation, and average energetic return rates from foraging may not have been as poor as recent ethnographic analysis might suggest. For example, settled groups in the New Guinea Highlands today characteristically return a negative energy balance from hunting and gathering (Sillitoe 2002), as their very low residential mobility places severe pressures on the frequency of high-ranked vegetation and game. Holocene rain forest is generally depauperate, preventing sedentary occupation without some form of food production. In a more productive Pleistocene landscape, groups maintaining relatively high residential mobility may have found subsistence based on wild resources a far easier task than groups living in the postglacial period.

It has long been speculated that hunter-gatherers in the early Holocene (eg Bayliss-Smith 1996; Harris 1989) and possibly Late Pleistocene (eg Spriggs 1996, 2000) were involved in some kind of plant manipulation of wild species, variously labelled as 'wild plant food production' or as systems of incipient 'horticulture' (Yen 1998: 164). With such a long 'people-plant interaction' (Harris 1989), it might be speculated that the emergence of such systems in northern Borneo could be very early. Establishing the structure of the environment, the range of plants that were in the diet, and the antiquity of their exploitation is a necessary first step. Vegetative propagation of tubers (yams) and rhizomes (taro) is relatively simple (Burkill 1966: 827; Hather 1996: 547) and these plant organs are suitable for long distance transport (Burkill 1951, 1966). However, knowing this and proving that people were deliberately manipulating the distribution of favourable plants in the Pleistocene or, as in the case of *Colocasia* taro, selecting for less acrid varieties, is another matter. As Yen (1998: 164) notes, 'we [still] have the problem of discriminating between introduced and endemic species, including the yams and taros that are common to the west and east of the region [Oceania], as indicative of possible human transfer'. From archaeobotanical remains, the beginning of rice and millet cultivation in Island Southeast Asia is around 4000–3850 cal BP (Paz 2004: 320), an increasingly short chronology against the backdrop of hunter-gatherer occupation in the region. Even without direct evidence for the deliberate manipulation of environment by Pleistocene foragers in Borneo, the later introduction of cereal-based systems of agriculture clearly did not occur within a cultural vacuum, and perhaps met with some initial resistance in the interior against an already established and successful subsistence system based on starchy tubers, nuts, fruits, sago palm and hunting.

ACKNOWLEDGMENTS

The research for this paper was funded by a grant from the AHRB and the British Academy for Southeast Asian Studies. This paper has benefited from the contributions of Tim Denham, Graeme Barker and two anonymous reviewers. Thanks also to Efrosyni Boutsikas for the German translation of Paul Schebesta's work. The original conference paper was presented at the 4th World Archaeology Congress in Washington, DC.

REFERENCES

Anshari, G, Kershaw, A P and van der Kaars, S (2001) 'A late Pleistocene and Holocene pollen and charcoal record from peat swamp forest, Lake Sentarum Wildlife Reserve, West Kalimantan, Indonesia', *Palaeogeography, Palaeoclimatology, Palaeoecology* 171: 213–28

Bailey, R C and Headland, T N (1991) 'Human foragers in tropical rainforest', *Human Ecology* 19: 261–85

Bailey, R C, Head, G, Jenike, M, Owen, B, Reichtman and Zechenter, E (1989) 'Hunting and gathering in tropical rainforest: is it possible?', *American Anthropologist* 91: 59–82

Barker, G, Badang, D, Beavitt, P, Bird, M, Daly, P, Doherty, C, Gilbertson, D, Glover, I, Hunt, C, Manser, J, McClaren, S, Paz, V, Pyatt, B, Reynolds, T, Rose, J, Rushworth, G and Stephens, M (2001) 'The Niah Cave project: the second (2001) season of fieldwork', *The Sarawak Museum Journal* 51(77): 37–120

Barker, G, Barton, H, Beavitt, P, Bird, M, Cole, F, Daly, P, Gilbertson, D, Hunt, C, Krigbaum, J, Lampert, C, Lewis, H, Manser, J, McClaren, S, Menotti, F, Paz, V, Piper, P, Pyatt, B, Rabett, R, Reynolds, T, Stephens, M, Thompson, J, Trickett, M and Whittaker, P (2002a) 'Prehistoric foragers and farmers in south-east Asia: renewed investigations at Niah Cave, Sarawak', *Proceedings of the Prehistoric Society* 68: 147–64

Barker, G, Barton, H, Beavitt, P, Bird, M, Cole, F, Daly, P, Gilbertson, D, Hunt, C, Krigbaum, J, Lampert, C, Lewis, H, Manser, J, McClaren, S, Menotti, F, Paz, V, Piper, P, Pyatt, B, Rabett, R, Reynolds, T, Stephens, M, Thompson, J, Trickett, M and Whittaker, P (2002b) 'The Niah Cave project: the third season of fieldwork', *The Sarawak Museum Journal* 62(78): 87–178

Barker, G, Barton, H, Bird, M, Cole, F, Daly, P, Dykes, A, Farr, L, Gilbertson, D, Higham, T, Hunt, C, Knight, S, Kurui, E, Lewis, H, Loyd-Smith, L, Manser, J, McLaren, S, Menotti, F, Piper, P, Pyatt, B, Rabett, R, Reynolds, T, Shimmin, J, Thompson, J and Trickett, M (2003) 'The Niah Cave project: the fourth season (2003) of fieldwork', *The Sarawak Museum Journal* 57(78): 45–119

Barton, H (1991) 'Raw material and tool function: a residue and use wear analysis of artefacts from a Melanesian rockshelter', unpublished BA Honours thesis, University of Sydney

Barton, H (2005a) 'Hunter-gatherer technology and mobility in Sundaland: a long-term perspective from Niah Cave, Sarawak', paper presented at the 70th annual meeting of the Society for American Archaeology, Salt Lake City, UT

Barton, H (2005b) 'The case for rain forest foragers: the starch record at Niah Cave, Sarawak', in Baker, G and Gilbertson, D (eds), *The Human Use of Caves in Peninsula and Island Southeast Asia*, special volume of *Asian Perspectives* 44: 56–72

Barton, H and White, J P (1993) 'Use of stone and shell artefacts at Balof 2, New Ireland, Papua New Guinea', *Asian Perspectives* 32(2): 169–81

Bayliss-Smith, T (1996) 'People-plant interactions in the New Guinea Highlands: agricultural hearthland or horticultural backwater', in Harris, D R (ed), *The Origins and Spread of Agriculture and Pastoralism in Eurasia*, pp 499–523, London: University College London Press

Bettinger, R L (1991) *Hunter-Gatherers: Archaeological and Evolutionary Theory*, New York: Plenum Press

Bird, M I, Hope, G and Taylor, D (2004) 'Populating PEP II: The dispersal of humans and agriculture through Austral-Asia and Oceania', *Quaternary International* 118–119: 145–63

Bradbury, J H and Nixon, R W (1998) 'The acridity of raphides from the edible aroids', *Journal of the Science of Food and Agriculture* 76: 608–16

Bronk Ramsey, C (2000) *OxCal Program v 3.5 Manual*, Oxford: Oxford Radiocarbon Accelerator Unit

Brosius, J P (1993) 'Contrasting subsistence strategies among Penan foragers, Sarawak (East Malaysia)', in Hladik, C M, Hladik, A, Linares, O F, Pagezy, H, Semple, A and Hadley, M (eds), *Tropical Forests, People and Food*, pp 515–22, Paris: UNESCO

Brown, D (2000) *Aroids, Plants of the Arum Family*, Portland: Timber Press

Burkill, I H (1951) 'The rise and decline of the greater yam in the service of man', *Advancement of Science* (London) 7(28): 443–48

Burkill, I H (1966) *A Dictionary of the Economic Products of the Malay Peninsula*, Singapore: Government Printer

Carey, I (1976) *Orang Asli: The Aboriginal Tribes of Peninsula Malaysia*, Kuala Lumpur: Oxford University Press

Chappell, J A, Omura, T E, McCulloch, M, Pandolfi, J, Ota, Y and Pillans, B (1996) 'Reconciliation of late Quaternary sea levels derived from coral terraces at Huon Peninsula with deep-sea oxygen isotope records', *Earth and Planetary Science Letters* 141: 227–37

Christensen, H (2002) *Ethnobotany of the Iban and the Kelabit*, Aarhus, Denmark: NEPCon and University of Aarhus and Kuching, Sarawak, Malaysia: Forest Department Sarawak

Colinvaux, P A and Bush M B (1991) 'The rain-forest ecosystem as a resource for hunting and gathering', *American Anthropologist* 93: 153–62

Coursey, D G (1967) *Yam: An Account of the Nature, Origins, Cultivation and Utilisation of the Useful Members of the Dioscoreaceae*, London: Longmans

Cranbrook, Earl of (2000) 'Northern Borneo environments of the past 40,000 years: archaeozoological evidence', *The Sarawak Museum Journal* 55(76): 61–110

Denham T P and H Barton (2006) 'The emergence of agriculture in New Guinea: continuity from pre-existing foraging practices', in Kennett D and Winterhalder B (eds), *Behavioral Ecology and the Transition to Agriculture*, pp 237–64, Berkeley: University of California Press

Denham, T P, Haberle, S G, Lentfer, C, Fullagar, R, Field, J, Therin, M, Porch, N and Winsborough, B (2003) 'Origins of agriculture at Kuk Swamp in the Highlands of New Guinea', *Science* 301: 189–93

Dentan, R N (1991) 'Potential food sources for foragers in Malaysian rainforest: sago, yams and lots of little things', *Bijdragen tot de taal-, land- en volkenkunde* (The Hague) 147: 420–44

Eder, J F (1978) 'The caloric returns to food collection: disruption and change among the Batak of the Philippine tropical forest', *Human Ecology* 6: 55–69

Ellen, R F (1988) 'Foraging, starch extraction and the sedentary lifestyle in the lowland rainforest of central Seram', in Woodburn, J, Ingold, T and Riches, D (eds), *History, Evolution and Social Change in Hunting and Gathering Societies*, pp 117–34, London: Berg

Endicott, K (1984) 'The economy of the Batek of Malaysia', *Research in Economic Anthropology* 6: 29–52

Englyst, H N, Kingman, S M and Cummings, J H (1992) 'Classification and measurement of nutritionally important starch fractions', *European Journal of Clinical Nutrition* 46: S33–S50

Flenley, J R (1998) 'Tropical forests under the climates of the last 30,000 years', *Climatic Change* 39: 177–97

Haberle, S G, Hope, G S and van der Kaars, S (2001) 'Biomass burning in Indonesia and Papua New Guinea: natural and human induced fire events in the fossil record', *Palaeogeography, Palaeoclimatology, Palaeoecology* 171: 259–68

Hanebuth, T J J and Stattegger, K (2003a) 'Depositional sequences on a late Pleistocene–Holocene tropical siliclastic shelf (Sunda Shelf, southeast Asia)', *Journal of Asian Earth Sciences* 23: 113–26

Hanebuth, T J J and Stattegger, K (2003b) 'The stratigraphic evolution of the Sunda Shelf during the past fifty thousand years', in Hasan Sidi, F, Nummedal, D, Imbert, P, Darman, H and Posamentier, H (eds), *Tropical Deltas of Southeast Asia – Sedimentology, Stratigraphy, and Petroleum Geology*, pp 189–200, Tulsa, OK: Society for Sedimentary Geology (SEPM)

Hanebuth, T, Stattegger, K and Grootes, P M (2000) 'Rapid flooding of the Sunda Shelf: a late-glacial sea-level record', *Science* 288: 1033–35

Harris, D R (1989) 'An evolutionary continuum of people-plant interaction', in Harris, D R and Hillman, G C (eds), *Foraging and Farming: The Evolution of Plant Exploitation*, pp 11–26, London: Unwin Hyman

Harrisson, B (1967) 'A classification of stone age burials from Niah Great Cave, Sarawak', *Sarawak Museum Journal* 50(30–31): 126–200

Harrisson, T (1957) 'The Great Cave of Niah: a preliminary report on Bornean prehistory', *Man* 57: 161–66

Harrisson, T (1958) 'The caves of Niah: a history of prehistory', *Sarawak Museum Journal* 8: 549–95

Harrisson, T (1959) 'Radio carbon C^{14} datings from Niah: a note', *Sarawak Museum Journal* 9: 136–42

Harrisson, T (1970) 'The prehistory of Borneo', *Asian Perspectives* 13: 17–45

Hather, J G (1988) 'The morphological and anatomical interpretation and identification of charred vegetative parenchymatous plant remains', unpublished PhD thesis, University College London

Hather, J G (1996) 'The origins of tropical vegeculture in Southeast Asia', in Harris, D R (ed), *The Origins and Spread of Agriculture and Pastoralism in Eurasia*, pp 538–51, London: University College London Press

Hather, J G (2000) *Archaeological Parenchyma*, London: Archetype Publications

Hay, A (1990) *Aroids of Papua New Guinea*, Madang, Papua New Guinea: Christensen Research Institute

Hay, A (1998) 'The genus *Alocasia* (Araceae-Colocasieae) in west Malaysia and Sulawesi', *Gardens Bulletin Singapore* 50: 221–334

Hazebroek, H P (2001) *National Parks of Sarawak*, Kota Kinabalu, Borneo: Natural History Publications

Headland, T N (1987) 'The wild yam question: how well could independent hunter-gatherers live in a tropical rain forest ecosystem?', *Human Ecology* 15: 463–91

Heaney, L R (1991) 'A synopsis of climatic and vegetational change in southeast Asia', *Climatic Change* 19: 53–61

Hladik, A, Leigh Jr, E G and Bourlière, F (1993) 'Food production and nutritional value of wild and semi-domesticated species – background', in Hladik, C M, Hladik, A, Linares, O F, Pagezy, H, Semple, A, and Hadley, M (eds), *Tropical Forests, People and Food*, pp 127–38, Paris: UNESCO

Hope, G, Kershaw, A P, van der Kaars, S, Xiangjun, S, Liew, P-M, Heusser, L E, Takahara, H, McGlone, M, Miyoshi, N and Moss, P T (2004) 'History of vegetation and habitat change in the Austral-Asian region', *Quaternary International* 118–119: 103–26

Horrocks, M, Irwin, G, Jones, M and Sutton, D (2004) 'Starch grains and xylem cells of sweet potato (*Ipomoea batatas*) and bracken (*Pteridium esculentum*) in archaeological deposits from northern New Zealand', *Journal of Archaeological Science* 31: 251–58

Junker, L (2002) 'Southeast Asia: introduction', in Morrison, K and Junker, L (eds), *Forager-Traders in South and Southeast Asia*, pp 131–66, Cambridge: Cambridge University Press

Kedit, P M (1982) 'An ecological survey of the Penan', *Sarawak Museum Journal* 30: 225–79

Kershaw, A P, Penny, D, van der Kaars, S, Anshari, G and Thanotherampillai, A (2001) 'Vegetation and climate in lowland Southeast Asia at the Last Glacial Maximum', in Metcalfe, I, Smith, J M B, Morwood, M and Davidson, I (eds), *Faunal and Floral Migrations and Evolution in SE Asia-Australasia*, pp 227–36, Rotterdam: Balkema

Kuchikura, Y (1987) *Subsistence Ecology among Semaq Beri Hunter-Gatherers of Peninsula Malaysia*. Hokkaido Behavioural Science Report, Series E, no 1, Sapporo, Japan: Hokkaido University

Kuchikura, Y (1993) 'Wild yams in the tropical rain forest: abundance and dependence among the Semaq Beri in Peninsula Malaysia', *Man and Culture in Oceania* 9: 81–102

Lambeck, K, Yokoyama, Y and Purcell, T (2002) 'Into and out of the Last Glacial Maximum: sea-level change during oxygen isotope stages 3 and 2', *Quaternary Science Reviews* 21: 343–60

Langub, J (1989) 'Some aspects of the Penan', *Sarawak Museum Journal* 50: 169–84

Loy, T, Wickler, S and Spriggs, M (1992) 'Direct evidence for human use of plants 28,000 years ago: starch residues on stone artefacts from the northern Solomon Islands', *Antiquity* 66: 898–912

MacKinnon, K, Hatta, G, Halim, H and Mangalik, A (1997) *The Ecology of Kalimantan*, Singapore: Oxford University Press

Maloney, B K (1985) 'Man's impact on the rainforest of west Malesia: the palynological record', *Journal of Biogeography* 12: 537–58

Maxwell, A L and Liu, K-B (2002) 'Late Quaternary pollen and associated records from the monsoonal areas of continental South and SE Asia', in Kershaw, A P, Tapper, N J, David, B, Bishop, P M and Penny, D (eds), *Bridging Wallace's Line*, pp 189–228. *Advances in Geoecology* 34, Cremlingen, Germany: Catena Verlag

Medway, Lord (1960) 'The Malay tapir in late Quaternary Borneo', *Sarawak Museum Journal* 9: 356–60

Ochse, J J and Bakhuizen, R C (1980) *Vegetables of the Dutch East Indies*, Amsterdam: A Asher and Co

O'Connell, J F and Hawkes, K (1981) 'Alyawara plant use and optimal foraging theory', in Winterhalder, B and Smith, E A (eds), *Hunter-Gatherer Foraging Strategies: Ethnographic and Archaeological Analyses*, pp 99–125, Chicago: University of Chicago Press

Payne, J and Francis, C M (1998) *A Field Guide to the Mammals of Borneo*, Kota Kinabalu, Sabah, Malaysia: Sabah Society

Paz, V (2001) 'Archaeobotany and cultural transformation: patterns of early plant utilisation in northern Wallacea', unpublished PhD thesis, University of Cambridge

Paz, V (2004) 'Crop domestication in Southeast Asia', in Goodman, R M (ed), *Encyclopedia of Plant and Crop Science*, pp 320–22, New York: Marcel Dekker

Penny, D (2001) 'A 40,000 year palynological record from north-east Thailand: implications for biogeography and palaeo-environmental reconstruction', *Palaeogeography, Palaeoclimatology, Palaeoecology* 171: 97–128

Peterson, J A, Hope, G S, Prentice, M and Hantoro, W (2002) 'Mountain environments in New Guinea and the Late Glacial Maximum "warm seas/cold mountains" enigma in the West Pacific Warm Pool region', in Kershaw, A P, Tapper, N J, David, B, Bishop, P M and Penny, D (eds), *Bridging Wallace's Line*, pp 173–87. Advances in Geoecology 34, Cremlingen, Germany: Catena Verlag

Piperno, D R and Holst, I (1998) 'The presence of starch granules on prehistoric tools from the humid neotropics: indications of early tuber use and agriculture in Panama', *Journal of Archaeological Science* 25: 765–76

Piperno, D R and Pearsall, D M (1998) *The Origins of Agriculture in the Lowland Neotropics*, London: Academic Press

Puri, R K (1997) 'Penan Benalui knowledge and use of tree palms', in Sorensen, K W and Morris, B (eds), *People and Plants in Kayan Mentarang*, pp 194–226, London: WWF-IP/UNESCO

Richards, P W (1996) *The Tropical Rain Forest*, Cambridge: Cambridge University Press

Schebesta, P (1954) *Die Negrito Asiens.* 12 Studia Instituti Anthropos, Vienna-Mödling: St-Gabriel-Verlag

Schebesta, P (1973) *Among the Forest Dwarfs of Malaya*, Kuala Lumpur: Oxford University Press

Sillitoe, P (2002) 'Always been farmer-foragers? Hunting and gathering in the Papua New Guinea Highlands', *Anthropological Forum* 12(1): 45–76

Spencer, J E (1996) *Shifting Cultivation in Southeastern Asia*, Berkeley: University of California Press

Spriggs, M (1996) 'Early agriculture and what went before in Island Melanesia: continuity or intrusion?', in Harris, D R (ed), *The Origins and Spread of Agriculture and Pastoralism in Eurasia*, pp 524–37, London: University College London Press

Spriggs, M (2000) 'Can hunter-gatherers live in tropical rain forests?', in Schweitzer P P, Biesele, M and Hitchcock, R K (eds), *Hunters and Gatherers in the Modern World*, pp 287–304, New York: Berghahn Books

Steinke, S, Keinast, M and Hanebuth, T (2003) 'On the significance of sea-level variations and shelf paleo-morphology in governing sedimentation in the southern China Sea during the last deglaciation', *Marine Geology* 201: 179–206

Sun, X, Li, X, Luo, Y and Chen, X (2000) 'The vegetation and climate at the last glaciation on the emerged continental shelf of the South China Sea', *Palaeogeography, Palaeoclimatology, Palaeoecology* 160: 301–16

Sunell, L A and Healey, P L (1979) 'Distribution of calcium oxalate crystals idioblasts in corms of taro (*Colocasia esculenta*)', *American Journal of Botany* 66(9): 1029–32

Sunell, L A and Healey, P L (1985) 'Distribution of calcium oxalate crystals idioblasts in leaves of taro (*Colocasia esculenta*)', *American Journal of Botany* 72(12): 1845–60

Taylor, D, Yen, O H, Sanderson, P G and Dodson, J (2001) 'Late Quaternary peat formation and vegetation dynamics in a lowland tropical swamp: Nee Soon, Singapore', *Palaeogeography, Palaeoclimatology, Palaeoecology* 171: 269–87

Terrell, J E (2002) 'Tropical agroforestry, coastal lagoons, and Holocene prehistory in greater near Oceania', in Shuji Y and Matthews, P J (eds), *Vegeculture in Eastern Asia and Oceania*, pp 195–216. International Area Studies Conference VII, JCAS Symposium Series 16, Osaka: National Museum of Ethnology

Tija, H D (1996) 'Sea-level changes in the tectonically stable Malay–Thai Peninsula', *Quaternary International* 31: 95–101

Tija, H D and Liew, K K (1996) 'Changes in tectonic stress field in northern Sunda Shelf basin', in Hall, R and Blundell, D (eds), *Tectonic Evolution of Southeast Asia*, pp 291–306. Special publication 106, London: Geological Society of London

Tomlinson, P B (1961) *Anatomy of the Monocotyledons*, Oxford: Clarendon Press

Torrence, R (2006) 'Description, classification, and identification', in Torrence R and Barton H (eds), *Ancient Starch Research*, pp 115–43. California: Left Coast Press

Ulijaszek, S J and Poraituk, P (1993) 'Making sago in Papua New Guinea: is it worth the effort?', in Hladik, C M, Hladik, A, Linares, O F, Pagezy, H, Semple, A and Hadley, M (eds), *Tropical Forests, People and Food*, pp 271–79, Paris: UNESCO

Urquhart, I A N (1954) 'Jungle Punans are human', *The Sarawak Gazette* June 30: 121–23

Voris, H K (2000) 'Maps of Pleistocene sea levels in Southeast Asia: shorelines, river systems and time durations', *Journal of Biogeography* 27: 1153–67

Wilson, J E and Siemonsma, J S (1996) '*Colocasia esculenta* (L.) Schott', in Flach, M and Rumawas, F (eds), *Plants Yielding Non-Seed Carbohydrates*, pp 69–72. Plant Resources of South-East Asia, no 9, Leiden: Backhuys Publishers and Bogor, Indonesia: PROSEA

Yen, D E (1998) 'Subsistence to commerce in Pacific agriculture: some four thousand years of plant exchange', in Pendergast, H D V, Etkin, N L, Harris, D R and Houghton, P J (eds), *Plants for Food and Medicine*, pp 161–83. Kew, England: Royal Botanic Gardens

Yoshino, H (2002) 'Morphological and genetic variation in cultivated and wild taro', in Shuji, Y and Matthews, P J (eds), *Vegeculture in Eastern Asia and Oceania*, pp 95–116. International Area Studies Conference VII, JCAS Symposium Series 16, Osaka: National Museum of Ethnology

Zuraina, M (1982) 'The West Mouth, Niah, in the Prehistory of Southeast Asia Sarawak', Special Monograph No 3, *Sarawak Museum Journal* 31(ns 52): 1–200

CHAPTER 5

Early to Mid-Holocene Plant Exploitation in New Guinea: Towards a Contingent Interpretation of Agriculture

Tim Denham

Multidisciplinary investigations at Kuk Swamp, upper Wahgi Valley, Papua New Guinea, have yielded evidence to support claims for the early and independent origins of agriculture in New Guinea (Denham et al 2003, 2004a, 2004b; Golson 1991a; Hope and Golson 1995). Much research on early agriculture in New Guinea, however, has been conducted without an explicit definition of agriculture vis-à-vis other subsistence activities and against which multidisciplinary lines of evidence can be measured (exceptions include Denham 2005a, 2006; Denham and Barton 2006; Golson 1997a: 44–46). Terms including 'agriculture', 'horticulture' and 'cultivation' have often been used interchangeably and without clear definition.

In this paper, I seek to overcome this 'semantic confusion' (after Harris 1996a: 3) and propose a working framework for the interpretation of early agricultural practices in New Guinea. In the first half, I present scenarios for the three early phases at Kuk and other highland sites that are based on results of recent multidisciplinary research. In the second half, I dissect these phase scenarios to evaluate alternative conceptions of 'agriculture' that have been applied in New Guinea and elsewhere. My starting points are the twin themes of morphogenetic change and environmental transformation, which have been widely applied individually, or in combination, as indicators of agriculture in New Guinea and other parts of the world. As an alternative, I present a more contingent and contextual framework for understanding and identifying agriculture in the past grounded on archaeological evidence of past cultivation practices.

PHASE SCENARIOS: KUK AND BEYOND

Of all the wetland archaeological sites in the Highlands, Kuk has been investigated in greatest detail, provides the most comprehensive evidence and is the 'type site' for the interpretation of past agricultural practices in New Guinea (Tables 5.1 and 5.2; Figure 5.1). Following extensive excavations at Kuk, Golson identified six phases of human manipulation of the wetland for agriculture (Golson 1977a, 1981, 1982, 1990, chapter 6; Golson

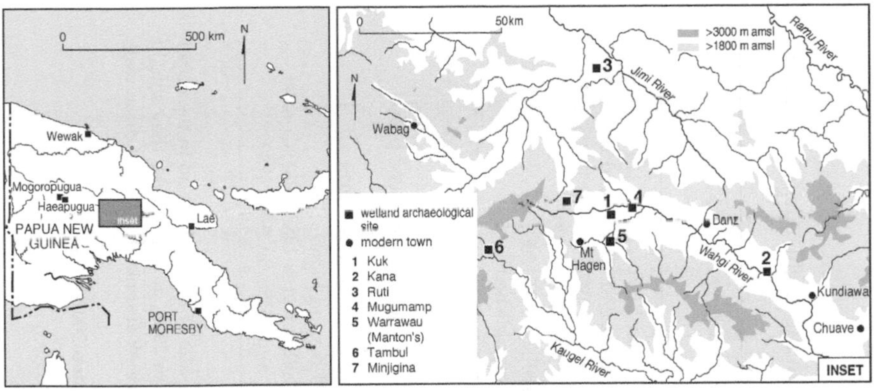

Figure 5.1 Map of New Guinea showing the location of wetland archaeological sites with evidence of early agriculture (after Denham et al 2004a: Figure 1).

and Hughes 1980). The phases mark periods of human activity on the wetland margin that are represented by in situ archaeological remains (Table 5.2). The three early phases at Kuk are of greatest significance because they ground interpretations of early agricultural development in New Guinea and predate the influence of Austronesian language speakers in the region (Denham 2004a; Denham et al 2003, 2004a, 2004b; Golson 1991a; Hope and Golson 1995).

Phase 1 (10,220–9910 cal BP)

Evidence of human activities on a wetland margin dating to the early Holocene (Phase 1) in New Guinea is unique to Kuk. Archaeological evidence of Phase 1 was previously characterised as comprising an artificial palaeosurface and an undeniably artificial palaeochannel that were chronologically and functionally associated (Golson 1977a: 613–14, 1981: 55–56, 1985: 308–09, 1991b; Golson and Hughes 1980; Hope and Golson 1985: 824). The Phase 1 evidence was interpreted to be similar to the overlying five phases and was similarly interpreted to represent 'wetland management for cultivation' (Hope and Golson 1995: 824). However, both the constitution and interpretation of Phase 1 require revision (Denham 2003a: 123–59, 329–32, 2004b, 2005a; Denham et al 2004a: 267–78, 293–94). At present, there is insufficient information to determine whether the palaeochannel was artificial and it is considered by this author to be a natural watercourse (see Denham et al 2004a: 269–74).

The palaeosurface consists of pits, runnels, stake and postholes with associated artefacts and heterogeneous feature fills (Figures 5.2b and 5.3a–b). These features are cut into black organic-rich clay, which provided the land surface at the time, and are sealed by grey-brown clay. The artificiality of the features is interpreted from their morphology, fills and associations. For example, the pits, and stake and postholes on the palaeosurface are feasibly

Table 5.1 Summary of wetland archaeological excavations and evidence for prehistoric agriculture in the Highlands (see Figure 5.1 for site locations; based on Denham and Barton 2006: Table 11.4)

Site Name[1]	*Elevation (m)*	*Location*	*Field Seasons*	*Phases Represented*	*Key References*
Tambul	2170	upper Kaugel Valley	1976	3	Golson 1997b
Mogoropugua	1890	Tari Basin	1980	5, 6	Golson 1982: 121; Ballard 1995: 193–95
Minjigina	1890	upper Wahgi Valley	1967	4	Lampert 1970; Powell 1970a: 172–74; Golson 1982: 121
Haeapugua	1650	Tari Basin	1991–92	3, 4, 5, 6	Ballard 1995, 2001
Kindeng	1600	upper Wahgi Valley	1968	n/a[2]	Unpublished
Warrawau (Manton's)	1590	upper Wahgi Valley	1966, 1977	2, 3, 5	Golson et al 1967; Lampert 1967; Powell 1970a: 142–46; Golson 1982: 121, 2002
Kuk	1560	upper Wahgi Valley	1972–77, 1998-99	1, 2, 3, 4, 5, 6	See Table 5.2 for detailed phase references
Mugumamp	1560	upper Wahgi Valley	1977	2, 5 (6)[3]	Harris and Hughes 1978
Kana	1480	middle Wahgi Valley	1993–94	(2)[3], 3, 4, 5	Muke and Mandui 2003
Ruti Flats	480	lower Jimi Valley	1983–85	(2)[3]	Gillieson et al 1985, 1987; Gorecki and Gillieson 1984, 1989

Notes

[1]Other wetland sites were inspected by archaeologists, although none was investigated in detail. For example, the site at Kotna (1580 m) in the upper Wahgi Valley was village land under drainage for coffee. The site was visited by Jack Golson and John Muke in 1988, at which time they sought permission to record features exposed in drain walls. Permission was refused, but while waiting they were able to look at some stretches of drain wall, in which ditches comparable to those of Phase 5 at Kuk were exposed (Jack Golson pers comm 2002).

[2]The archaeological finds at Kindeng have not been correlated to those at other wetland sites (Jack Golson pers comm 2001).

[3]Phases in brackets indicate provisional identifications (Denham 2003b).

Table 5.2 Archaeological phases at Kuk Swamp, Wahgi Valley (Denham et al 2003: Tables 1, S1 and S2; 2004b: Table 1). Another possible Phase 2 subphase predates R ash deposition at 3980–3630 cal BP, although it is not well characterised and is not included

Phase	*Age (cal BP)*	*Description*	*Stone artefacts*	*Wooden artefacts*[1]	*House sites*[2]	*Ditches*	*Cultivation features*[3]	*Artificial channels*[4]	*Key references*
6	260–100	rectilinear field systems	X	X	X	X		X	Golson 1977a, 1982
5	420–260	rectilinear field systems	X	X	X	X		X	Golson 1977a, 1982; Bayliss-Smith et al 2005
4	1940–1100	rectilinear field systems	X	X	?	X		X	Golson 1977a, 1982; Bayliss-Smith and Golson 1992a, 1992b, 1999
3	pre-3260–2800	late subphase: rectilinear/dendritic ditch networks	X			X	?	X	Golson 1977a, 1982; Denham 2003a, 2005b; Denham et al 2003, 2004b
	4350–3980	early subphase: rectilinear ditch networks	X			X		X	
2	6950–6440	mounded palaeosurface	X				X	?	Golson 1977a, 1982; Denham 2003a, 2003b; Denham et al 2003, 2004a, 2004b
1	10,220–9910	amorphous palaeosurface	X				X	?	Golson 1977a, 1991a; Golson and Hughes 1980; Denham 2003a, 2004b, 2005a; Denham et al 2003, 2004a, 2004b

Notes
[1]No wooden artefacts were collected from Phase 1–3 contexts (contra Powell 1982a: Table 2).
[2]E C Harris (1977) noted unexcavated house remains at a multioccupation house site and these could predate Phase 5.
[3]Occasional features interpreted to represent 'within plot' cultivation features have been recorded for late Phase 3.
[4]Palaeochannels have been differentiated from ditches at Kuk on the basis of scale, although the mode of formation of some palaeochannels is uncertain (Denham et al 2004a). Golson and Hughes have argued for the artificiality of all palaeochannels (Golson 1977a: 613–15; Golson and Hughes 1980: 298), whereas Denham has suggested that the Phase 1 and 2 palaeochannels are not anthropogenic (Denham 2003b: 163–64, 2004a: 47–53).

consistent with the digging of corms and tubers and the staking and supporting of plants within a cleared plot, respectively (after Powell et al 1975: 24–26). Corms and tubers were used in the vicinity based on starch grain residues of *Colocasia* taro and a yam (*Dioscorea* sp.) on a contemporaneous stone flake (K76/S29B) and a pestle or grinding/pounding stone (K75/S178), respectively (Fullagar et al 2006). The paucity of lithic artefacts is expected within a subsistence plot, since most earth-working tools would have been wooden (Golson 1977b), and those that are present were stone tools used to process plants prior to consumption. Several plants were feasibly supported using stakes, eg edible *pitpit* (cf *Setaria palmifolia*) and bananas, as done today in the vicinity (author's observations; Powell et al 1975: 4–11). *Musa* spp. phytoliths were present at Kuk prior to Phase 1; diagnostic seed morphotypes of species of the Musa section of bananas (formerly Eumusa), including *Musa acuminata* morphotypes, appear with the onset of grey-brown clay deposition at c 10,000 years ago (Denham et al 2003: 191–92).

The palaeoecology of the Kuk vicinity changed approximately 10,000 years ago (Denham et al 2004b: 847–48). Sediments, botanical remains and charcoal in the palaeochannel reflect disturbance using fire. High pulses of charcoal represent periodic, high-intensity fires within the catchment. The increasing frequency of fires is accompanied by decreasing frequencies of primary forest taxa and increasing frequencies of disturbance and secondary taxa. Disturbance of the primary forest occurred at a time when the climate was possibly warmer and wetter than at present (Brookfield 1989), ie forests were not increasingly sensitised to disturbance. Prolonged disturbance using fire and cumulative and sustained forest disturbance indicate clearance by people (after Haberle 1994). In the context of the Highlands, these signatures are more likely to be associated with extensive swidden cultivation of slash-and-burn or slash-and-mulch types (Haberle 2003) than house-garden horticulture (Harris 1995: 853, 1996b: 568). Similarly, the signal is not consistent with adventitious human exploitation of patches within the forest caused by landslides, lightening, tree throw and other natural phenomena.

The multidisciplinary evidence for the Phase 1 palaeosurface is consistent with a utilised plot. The nature of the contemporary palaeoecological signal suggests a swidden-type plot and represents the spatial extension of dryland practices onto the wetland margin. The patch of higher ground was probably only dry enough to utilise for a short period, perhaps only once, with occasional reuse, potentially to harvest wild and cultivated plants surviving within abandoned plots, eg *Colocasia* taro and bananas. In contrast to previous portrayals (Golson 1977a, 1991b, 2000; Golson and Hughes 1980), Phase 1 need not signify specialised wetland management. The dryland practices were merely occurring on the wetland edge where the morphology of the constituent features was preserved, whereas these traces were destroyed by subsequent erosion, gardening and other practices on adjacent dryland slopes.

For the next 3000 years, during grey-brown clay deposition, there was no specialised use of the wetland margin at Kuk. A mosaic of open and disturbed, anthropic habitats was superimposed on a gradual succession from wetter (*Typha* reed and *Pandanus* swamp forest) to drier (Compositae, ferns and forest regrowth) conditions locally, within an increasingly deforested dryland landscape (Denham et al 2004b: 847–48). Intact chains of phytoliths and occasional high frequencies within grey-brown clay indicate that bananas were growing in the vicinity, although probably not in the wetland given the aversion of cultivated *Musa* spp. plants to water-saturated soil conditions.

Although the antiquity of the archaeological and palaeoecological records is unique to Kuk, the practices they record were probably not localised to the site or its catchment. Practices represented by Phase 1 could have easily diffused and were probably occurring elsewhere in the upper Wahgi Valley and the Highlands. The evidence of forest clearance supports such an interpretation for the mid-Holocene at three sites in the upper Wahgi Valley: Draepi-Minjigina, Lake Ambra and Warrawau (Denham et al 2004b: 845–47; Powell 1982b: 218). Although the timing of initial forest clearance is unknown for all three sites, the Kuk evidence suggests it may have occurred in the early rather than the mid-Holocene. Additionally, a palaeoecological record for the Baliem Valley indicates forest clearance was initiated before 7800 cal BP (Haberle et al 1991). Other widely dispersed sites from highland and lowland locations on New Guinea indicate major forest disturbance in the late Pleistocene and early Holocene (Denham 2004b: Table 2; Haberle 1994: Figure 8.2).

Phase 2 (6950–6440 cal BP)

Multiple sites in the Highlands contain purported evidence of similar character to Kuk Phase 2 (Table 5.1), although only those in the upper Wahgi Valley are considered directly comparable (Denham 2003b; Golson 1982: 121). At Kuk, Phase 2 was previously characterised as comprising artificial palaeosurfaces that articulated with at least four sequentially operating and artificially constructed palaeochannels (Golson 1977a: 614–16). The Phase 2 evidence was interpreted to represent agricultural activities on the wetland margin associated with the cultivation of indigenous crops (Golson 1977a: 617). The previous constitution and interpretation of Phase 2 at Kuk and other sites require revision based on recent research (Denham 2003a, 2003b; Denham et al 2003a, 2004b; cf Denham et al 2004a: 278–88).

Phase 2 at Kuk can be divided into two subphases (Denham 2003a: 160–215). The early palaeosurface dates to 6950–6440 cal BP and, as Golson interpreted, consists of integrated and discrete areas (1977a: 616–17). The late palaeosurface predates the deposition of a tephra, called R ash, at 3980–3630 cal BP. As with the Phase 1 evidence, the chronological and functional associations between palaeosurfaces and palaeochannels for both Phase 2 subphases are questionable (Denham 2003b). There is doubt as to whether three early Phase 2 palaeochannels are artificial (Denham 2003b: 163–64; see authors' debate in Denham et al 2004a: 279–82) and, consequently, I consider them to

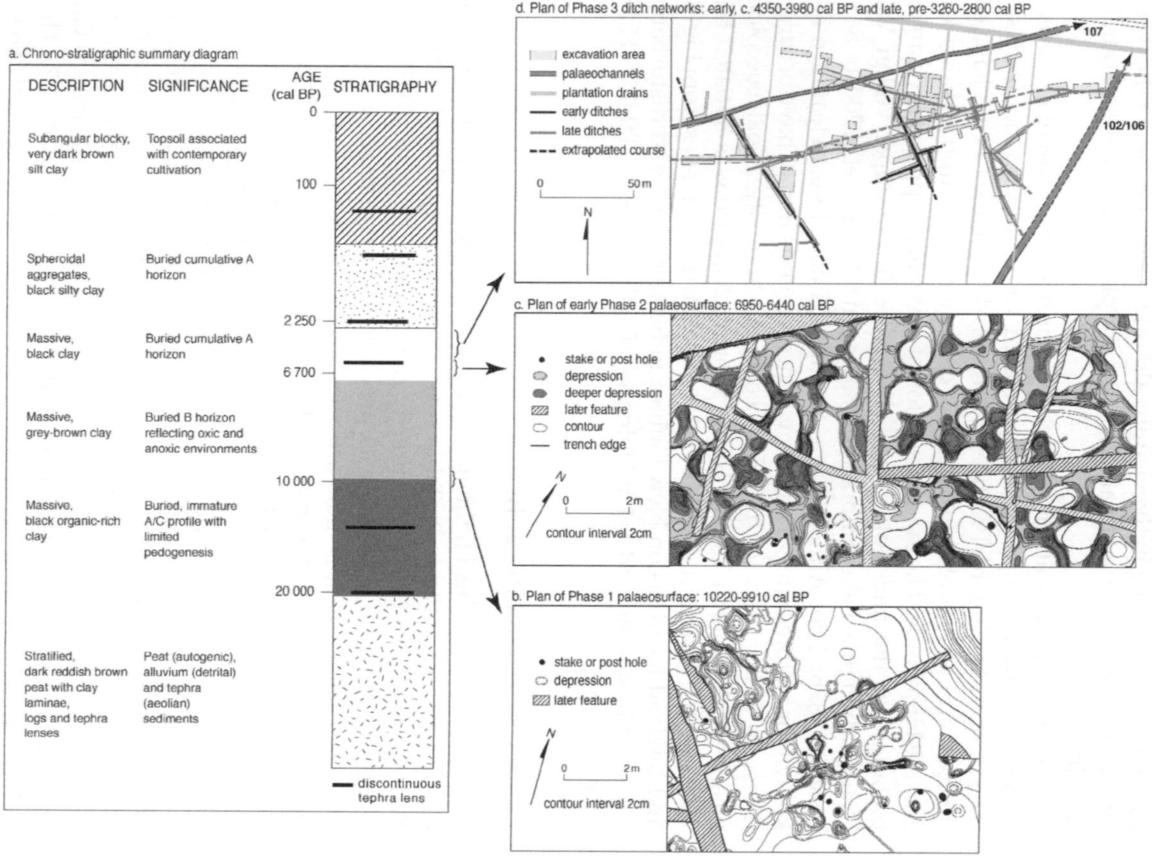

Figure 5.2 Archaeostratigraphic summary of phases 1, 2 and 3 at Kuk (amended version of Denham et al 2003: Figure 2). Late Phase 2 is not represented.

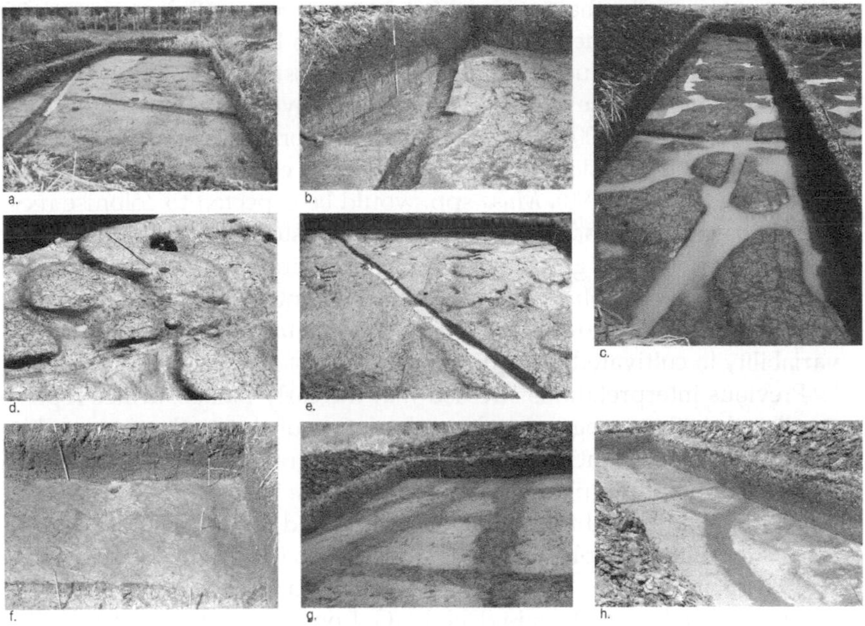

Figure 5.3 Photographs of archaeological features indicative of early agricultural practices at Kuk (all photographs courtesy of Jack Golson): (a) overview of Phase 1 palaeosurface (after Denham 2004b: Figure 8); (b) close-up image of Phase 1 features; (c) overview of integrated, early Phase 2 palaeosurface; (d) close-up image of early Phase 2 'mound' base; (e) overview of discrete, early Phase 2 palaeosurface (after Denham 2003b: Plate I); (f) close-up image of a pit with post hole on the discrete, early Phase 2 palaeosurface; (g) Late Phase 3 ditch junction of triangular type (see cross-cutting ditch in the centre of the image); (h) Late Phase 3 ditch junction of dendritic type.

be natural watercourses. The associations between palaeosurfaces and the ditch at Mugumamp and the palaeochannel at Warrawau are similarly problematic (Denham 2003b).

The early palaeosurface consists of integrated and discrete areas (Figures 5.2c and 5.3c–f). An explanation for the diversity of palaeosurface morphologies focuses on the nature of mound cultivation. Relocating mounds within plots during successive plantings could produce the discrete palaeosurface type, whereas the reestablishment of mounds in the same place would preserve mound bases. These two types of mound construction, 'shifting' and 'nonshifting' mounds, respectively, are still practiced around Mount Hagen today (Powell et al 1975: 11, 12). Although the morphology of the discrete palaeosurface type in some areas is consistent with denudation and truncation of more integrated forms, in other areas the scattered features may well represent less intensive or adventitious practices between cultivated plots.

Artefacts collected from Phase 2 contexts are sparse, as expected for an agricultural site at which wooden tools were probably used (after Golson 1977b). The lithic artefacts exhibit evidence of plant processing including

Colocasia taro and other unidentified plant residues (Fullagar et al 2006). Additionally, one palaeosurface feature exhibits highly anomalous, elevated banana phytolith frequencies, including diagnostic seed phytoliths of Musa section bananas. These attain 9% of the total phytolith assemblage (Denham et al 2003: Figure 3, 2004b: 849), although proportions attain 15% during the period of open grassland, ie from 6950–6440 cal BP to 4840–4440 cal BP (Denham et al 2003: 190). *Musa* spp. would be expected to colonise recently created patches from adjacent disturbed or forested habitats, rather than be a major component of a grassland assemblage maintained by periodic burning. The locally elevated *Musa* spp. frequencies are interpreted to represent deliberate planting, with lower values in some feature fills representing spatial variability in cultivated plant densities.

Previous interpretations stressed that the early palaeosurface represents small-scale microtopographical manipulation of the wetland to enable the multicropping of plants with different edaphic requirements (Golson 1977a: 617). Water-tolerant plants, eg taro (*Colocasia esculenta*), were plausibly planted in damp conditions along the edges and in the bases of the runnels, and water-intolerant plants, eg sugarcanes (*Saccharum* spp.), bananas (*Musa* spp.), yams (*Dioscorea* spp.), *Setaria palmifolia* and mixed vegetables, were planted and staked on the raised 'beds' (Golson 1977a: 616, 1981: 57–58; Powell et al 1975: 42). Archaeobotanical remains from feature fills, associated contexts and stone tool residues provide evidence consistent with this type of cultivation.

At the time of early Phase 2, the ecology of the Kuk vicinity underwent a dramatic change (Denham et al 2004b: 845–48). Mounded cultivation occurred within an open, grassed landscape at Kuk maintained by periodic burning. Primary and secondary forest taxa decreased steadily in phytolith and pollen frequency taxa distributions during grey-brown clay deposition, ie before early Phase 2, except for the invasion of *Pandanus* swamp forest and Arecaceae/Zingiberaceae. However, at the beginning of Phase 2, the forest signal dropped significantly with dramatic increases in grass taxa frequencies. Associated charcoal frequencies suggest the formation and management of extensive grasslands using fire within the catchment and on the wetland margin with only isolated stands of forest surviving. Such degraded environmental conditions are characteristic of previously cultivated areas across the Highlands today (Gillison 1972). Although the formation and maintenance of these grasslands was human-induced, the susceptibility of the environment to burning may have increased around this time with the intensification of El Niño/Southern Oscillation (ENSO) and greater interannual climatic variability. ENSO is considered to have intensified between 5000 and 3500 years ago (Haberle 1993: 308–09, 1998: 5), but these are estimates and some seasonal or interannual climatic variability may have occurred earlier (Moy et al 2002). The formation of grasslands may represent a continuation of previous land-use practices, but with greater impacts due to a climatically sensitised environment.

Information on the late subphase palaeosurface, ie predating R ash deposition, is limited and focuses on multidisciplinary investigations of the fills of one feature (504) dated to c 4840–4440 cal BP, which was formerly classified under Phase 3 (Denham et al 2003), but has been reclassified as part of ongoing investigations. Although its mode of formation is uncertain, the feature may be a rill that formed through concentrated surface wash following clearance in the vicinity, or an artificial curvilinear runnel excavated to facilitate soil moisture control within cultivated plots on the wetland margin (Bourke 2001: 231–32; Brookfield and Hart 1971: 101). The admixture of materials within the fill of Feature 504, including dense macrocharcoal, highly admixed fills and extremely high *Musa* banana phytolith frequencies (attaining 15% of total phytoliths) are suggestive of clearance using fire, cultivation and planting in the vicinity. Clearance and physical disturbance of the ground surface was necessary to break up dense grass root mats. Grasslands dominated the palaeoecology of the area at this time.

Palaeosurface forms similar to but later than the early integrated Phase 2 type at Kuk were documented at Warrawau (Golson 1982: 121; 2002) and Mugumamp (Harris and Hughes 1978). At all three sites, there are problems with determining whether the palaeochannels (Kuk, Warrawau) or ditch (Mugumamp) are artificial and whether they are chronologically or functionally associated with palaeosurfaces (Denham 2003b). Similarly aged, more equivocal evidence of former agricultural activities was documented at Kana near Minj (Muke and Mandui 2003) and the Ruti Flats (Gillieson et al 1985). The Phase 2 palaeosurfaces at Kuk, Warrawau and Mugumamp were not contemporaneous and may represent continual, spatially variable or episodic use of wetlands in the upper Wahgi Valley throughout a 3000-year period (Table 5.3).

Significantly, the palaeosurfaces at the three upper Wahgi Valleys sites can all be correlated to palaeoecological records of human clearance. The correlation at Kuk is most direct as samples were taken from the archaeological site, including palaeosurface feature fills. At Kuk, a dramatic degradation to grassland is synchronous with the palaeosurface. At Warrawau, the correlations are relatively secure given that the palaeoecological record from the M1 core was taken directly from a palaeochannel adjacent to the palaeosurface (Powell 1970a: 155–59). Although the contemporaneity of palaeochannel and palaeosurface at Warrawau is uncertain, the palaeosurface formed in a highly disturbed environment. At Mugumamp, the correlation is less secure, as the nearest palaeoecological site on the valley floor is at Kuk (Figure 5.1). The palaeoecological records from Kuk and Warrawau, and a third at Draepi-Minjigina, indicate significant disturbance of primary forest by the mid-Holocene (Denham et al 2004b: 845–47). Degradation of grasslands occurred at Kuk and Draepi-Minjigina with a mosaic of disturbance communities forming at Warrawau. Although the initiation of forest clearance at the latter two sites is not securely known, it predates or is potentially contemporary with the palaeosurfaces at the archaeological sites.

Table 5.3 Relative intersite chronology for early and mid-Holocene remains at wetland agricultural sites in the Highlands of New Guinea

Date (cal BP)	*Deposit*	*Earliest innovation*	*Kuk*[1]	*Warrawau*[2, 4]	*Mugumamp*[2]	*Kana*[3, 4]	*Ruti*[3]	*Tambul*[4]	*Haeapugua*[4]
2650–1950	Y ash								
2750–2150				X			X		X
pre-3260–2800			late P3			X			
			mid-late P3						
(3980–3630)[5]	(R ash?)[5]								
c 4350–3980			early P3						
(3980–3630)[5]	(R ash?)[5]								
		ditches						X	
					X		X?		
c 4840–4440			late P2			X?			
				X					
6440–5990	'R+W'								
6950–6440		mounds	early P2						
	(grey clay deposition)								
10,220–9910			P1						

Notes
[1]Early and late phase dates at Kuk are based on tephrochronology and radiocarbon dates (see Denham 2003a).
[2]The relative dates of mounded palaeosurfaces at Mugumamp and Warrawau are based on tephrochonology, ie the lie of R ash in palaeosurface features, and on radiocarbon dates for a palaeochannel at the latter (Denham 2003b).
[3]The relative dates for Kana and Ruti are based on tephrochronology, ie the lie of R ash, but the presence of Phase 2 at both sites is considered provisional (Denham 2003b).
[4]The dating of ditches at Warrawau, Kana, Tambul and Haeapugua is based on radiocarbon dates (Denham 2005b).
[5]Radiocarbon dates for R ash deposition postdate those for the earliest Phase 3 ditch, but their stratigraphic relationship is uncertain.

Taken together, the archaeological and palaeoecological records indicate that Phase 2 cultivation and clearance were not restricted to Kuk, but occurred throughout the upper Wahgi Valley over a 3000-year period. Given the extent of forest clearance and anthropic grasslands (Flenley 1979: 122; Powell 1982b: 224), cultivation was potentially widely practiced in intermontane valleys across the Highlands by 4500 cal BP.

Phase 3 (4350–3980 cal BP)

The earliest construction of ditches in New Guinea is represented at Kuk by Phase 3 (Denham 2003a: 216–62, 2005b; Denham et al 2003: 190–91, 2004a: 288–93; Golson 1977a: 619–20, 1977c: 49–50, 1981: 58), with similar finds reported elsewhere in the Highlands (Figure 5.1; Tables 5.1 and 5.3). At Kuk, Phase 3 ditches have been allocated to subphases according to tephrochronology, stratigraphy (distinctive fills and boundary type), articulations and alignments.

Three early ditch networks (c 4350–3980 cal BP) were identified at Kuk (Figure 5.2d). They share rectilinear alignments, dominant orientation and are each associated with a deposit called 'new grey clay'. The innovation of ditch construction suggests an increased water management problem at the site, although alternative potential causes should not be overlooked, eg boundary markers to demarcate plots or territories. One early ditch complex forms a rectangular enclosure measuring 29 m × 24 m that defines a drained plot.

The palaeoecology of early Phase 3 is similar, but distinct from that of Phase 2 (Denham 2003a: 320–22). There is limited recovery of forest species within the catchment, even though grasslands still predominate. There is also a dramatic decrease in the significance of *Musa* banana phytoliths, which continues in later subphases and throughout black clay deposition. The increased water control problems at the site, reflected in the construction of ditches, may have discouraged the cultivation of bananas.

Based on general stratigraphic interpretations, numerous ditches are assigned to a 'mid-late' subphase. Mid-late ditches may represent semicontinuous and spatially variable cultivation of the wetland margin between the better characterised early and late subphases. The alignments of these ditches form a marked rectilinear network in the southeastern corner of the wetland. The interpolation of these alignments forms a network of rectangular enclosures, as undertaken for one early Phase 3 complex (Denham 2003a: 237–39) and Phase 4 ditches (Bayliss-Smith and Golson 1992a, 1992b, 1999).

Numerous late ditches (predate 3260–2800 cal BP) were identified in modern plantation drain walls. Two articulated complexes and numerous other late ditches were traced in the excavations across the southeast portion of the wetland (Figure 5.2d). Both complexes shared rectilinear, triangular and dendritic components that are artificial in design (Figures 5.3g–h).

The subphases in which the rectilinear ditches at Kuk occur can be correlated to those at other sites in the Highlands (Denham 2005b; Tables 5.1

and 5.3). Agricultural practices predating 4000 cal BP are suggested by straight ditches arranged into rectilinear complexes at Kuk and the recovery from a ditch at Tambul of a spade associated with its maintenance (Steensberg in Golson 1997b: 161–62). The intervening drained land, in the absence of domestic herd animals, was intended for cultivation. Ditch-digging may have enabled the expansion of settlement into higher elevations at this time, as witnessed at Tambul (Golson 1997b) and Sirunki (Walker and Flenley 1979: 340).

The straighter arrangements of early Phase 3 ditches represent a break with older, more amorphous and curvilinear forms. At Kuk, the transition between older forms (R ash-marked Phase 2 subphase) to later rectilinear forms (early Phase 3 subphase) was rapid, although the reasons for changes in spatial form and design are unknown.

A later expansion of agricultural activities through the Wahgi Valley and beyond is witnessed at c 2750–2150 cal BP based on the synchronous dates from Warrawau (Golson et al 1967; Lampert 1967), Kana (Muke and Mandui 2003) and Haeapugua (Ballard 1995). At Kuk, the large number of ditches assigned to a mid-late subphase probably reflects spatially variable and semicontinuous cultivation of the wetland margin between the early and late subphases, ie cultivation of individual plots for a few years with intermittent fallowing. Such a model conforms to the use and abandonment of different parts of the wetland margin at Lake Haeapugua documented during archaeological and ethnoarchaeological investigations by Ballard (1995, 2001).

The distinction between Phases 3 and 4 at Kuk may largely be an artefact of investigation. Phase 3 ditch networks were not abandoned with the adoption of tillage and deposition of 'crumby black' on the wetland edge, a stratigraphic unit that most plausibly represents in situ pedogenesis. Given that multiple subphases are also present in Phase 4 ditch networks, the wetland may have been utilised periodically and spatially variably throughout Phases 3 and 4.

The early subphase ditch complexes predate the dispersal of Austronesian language speakers and the Lapita cultural complex in the Bismarck Archipelago. A revised estimate for the date of arrival of Austronesian speakers in the Bismarck Archipelago is approximately c 3300–3200 cal BP (Spriggs 2001: 240), although some favour an earlier date, eg c 3550–3450 cal BP (Kirch cited in Spriggs 2001: 240). The diffusion of ditch-digging and associated implements is unlikely to have occurred ahead of direct colonisation by Austronesian language speakers and would not accord with Bellwood's propositions of demic diffusion and replacement of previous populations by agriculturalists throughout most of Indo-Malaysia (Bellwood 1997: 201–67). No antecedents of rectilinear ditches have yet been discovered in the regions west of New Guinea. Additionally, there is no well-documented Austronesian influence on mainland New Guinea contemporary with any Phase 3 ditch network.

QUESTIONS OF AGRICULTURE

Although recent multidisciplinary research at Kuk addressed archaeological, archaeobotanical and palaeoecological lacunae with previous research (Spriggs 1996: 528–29, 1997: 62), thereby solidifying formerly speculative chronologies of agricultural development in New Guinea (eg Harris 1995, 1996b: 567–69; Hope and Golson 1995), questions of agricultural definition remain. Below, the relevance of three different conceptions of agriculture to the New Guinean context is considered: domestication, environmental change and practices (Denham 2006).

Potential Markers of Agriculture: Archaeobotanical Finds

Based on ethnographic and historical accounts, Harris has proposed terminology and a continuum of human-environment interactions for subsistence (Harris 1989, 1996a: 3–5). The continuum is based on two entwined processes: the manipulation of biotic resources leading to eventual domestication and the transformation of natural and artificial ecosystems (Harris 1989). Wild-plant food production includes two categories of cultivation varying from small to large-scale clearance of vegetation, minimal to systematic tillage, and minor to moderate dependence on domesticated plants. For Harris, a system in which wild-food plant production is dominant can be differentiated from agriculture because the latter is 'based largely or exclusively on the cultivation of domesticated plants' (Harris 1996a: 4). As well as the propagation and cultivation of genotypic and phenotypic variants, ie domesticated crops, agriculture involves the establishment of agroecosystems. Agriculture is also associated with 'such activities as soil preparation, the maintenance of soil fertility, weeding, seed selection and storage, and the exclusion of potential predators attracted by the enlarged food-storage organs of domesticated plants' (Harris 1989: 21–22).

Using a combination of techniques, archaeobotanical evidence for over 30 edible plants was documented for late Pleistocene, early Holocene and mid-Holocene contexts at Kuk (Denham 2005a: Table 2). There is a major hiatus in the microfossil record during the late Pleistocene, potentially extending for the period of post-Last Glacial Maximum (post-LGM) climatic amelioration to 10,000 cal BP. The archaeobotanical inventory includes major starchy staples (bananas of Musa section, *Colocasia* taro and yam) and minor sources of starch (probable *Setaria palmifolia*), which may have been more significant contributors to diet in the past. Unfortunately there is little archaeobotanical data from early to mid-Holocene contexts at other sites in the Highlands to complement the Kuk findings, although the collection of plant remains from Manim 2 offers great potential (Christensen 1975). Other plants used for medicine, ritual, ornamentation and construction have not been inventoried, but would greatly enhance the potential value of botanical resources in the Kuk vicinity (see Harris 1998: 90;

Powell 1976). Several significant findings emerge from this archaeobotanical research.

First, phytolith and pollen taxa frequencies of most edible plants do not correlate with the archaeological phases, and many are present in Pleistocene sediments and throughout grey-brown clay deposition between Phases 1 and 2 (Denham 2003a: 322–28). In terms of archaeological interpretation, if Phase 1 represents an extension of dryland practices onto the wetland margin, then a correlation between taxa frequencies and archaeological phases would not be expected. The distribution of archaeobotanical remains merely confirms that these plants were in the landscape, were potentially available to cultivators, and were gradually brought into cultivation. Such a scenario is consistent with the use of many of these plants in their wild forms today, ie for gathering and consumption, for transplanting to stock gardens, or within relatively continuous cultivation (Powell 1970b, 1976; Powell et al 1975: 15–32). The likely intercropping of these plants within mixed gardens, as occurs in the Highlands today, would not necessarily produce frequency peaks that could be differentiated from distributions away from gardens. Moreover, the visibility of some plants would be expected to be low because they are harvested before seeding (Powell 1970b: 199), as opposed to after seed production in cereal-based harvesting.

Second, frequencies of banana phytoliths are extremely high in some contexts given low rates of *Musa* phytolith production relative to grasses (Denham et al 2003: 191–92). The frequencies suggest high populations of bananas locally either within the wetland during drier periods or on adjacent slopes. High *Musa* spp. frequencies prior to Phase 2 are unusual and conform to an interpretation of mixed cultivation within swidden plots. The higher frequencies in Phase 2 features (both subphases) are highly anomalous, because they occurred in a grassed landscape maintained by fire, and are suggestive of deliberate and intensive planting, probably within mixed and intercropped plots. In contemporary New Guinea, bananas are the dominant staple in drier areas, including some grasslands (Gagné 1982: 236). Although phyto-liths of fire-tolerant *Ensete* bananas are present from Phase 2 onwards, most phytoliths appear to be derived from *Musa* spp.

Third, diagnostic seed phytoliths from Musa section bananas, including morphotypes consistent with *Musa acuminata* ssp. *banksii*, from early Holocene contexts are extremely significant. Lebot (1999: 621–22) and De Langhe and de Maret (1999: 378, 380) have argued that most contemporary cultivated varieties of bananas are ultimately derived from Musa section, AA diploids and more specifically from two subspecies, *Musa acuminata* ssp. *banksii* and *errans*. The former subspecies underwent the initially long process of domestication (De Langhe and de Maret 1999: 380) in the New Guinea region (Lebot 1999: 622). The phytolith evidence from Kuk provides circumstantial corroboration of Lebot's genetic interpretations by indicating the antiquity of one significant subspecies in New Guinea. The potential effects of human selection on banana phytolith morphotypes have not been identified from the relatively few diagnostic seed phytoliths available.

Fourth, *Colocasia* taro starch grain residues on artefacts from Phase 1, grey-brown clay and Phase 2 contexts indicate the presence and use of this plant at Kuk from the early Holocene (Fullagar et al 2006; cf Yen 1990: 263). Yen interpreted *Colocasia* taro to be of lowland origin (1995: 835) based on the greater phenotypic variation in wild varieties at lower elevations (Yen 1980, 1990: 77). The presence of *Colocasia* taro at Kuk in the early Holocene is suggestive of deliberate movement, as previously hypothesised by Golson (1991b: 88–89) and Yen (1991: Table 2). As with bananas, the antiquity of *Colocasia* taro at Kuk circumstantially corroborates the potential for independent domestication in New Guinea from wild types (Lebot 1999: 623–24; Matthews 1991; Yen 1998: 163). Although documented at a lowland lake of similar antiquity (Lake Wanum, Haberle 1995) and tentatively in Island Melanesia with much greater antiquity (Kilu Cave on Buka, Loy et al 1992), the *Colocasia* taro at Kuk is the first documented occurrence at a formerly cultivated site in the New Guinea region.

Fifth, starch grain residues from lithic artefacts collected from Phase 1 and grey-brown clay contexts have been identified as yam (*Dioscorea* sp.) (Fullagar et al 2006). The presence of yam is not necessarily surprising as some yam species grow wild in the Kuk vicinity (Powell et al 1975: 35), although knowledge of yam in New Guinea is poorly advanced (Martin and Rhodes 1977; Yen 1991: 81). Taken together with *Colocasia* taro and bananas of Musa section, the yam residues indicate a great antiquity for the exploitation of a significant suite of Pacific cultigens in highland New Guinea. Significantly, the earliest occurrence of two (*Colocasia* taro and yam) and, potentially, three (Musa section bananas) of these starch-rich plants is roughly synchronous with the earliest evidence of plant exploitation at Kuk, ie Phase 1. Golson, drawing on the work of Doug Yen, has long argued that these staples were brought into the Highlands by the expansion of agricultural populations from the lowlands (Golson chapter 6). Colleagues and I consider that these plants could equally have diffused to the Highlands naturally following post-LGM climatic amelioration, or were dispersed by people engaged in more extensive foraging practices during this period (Denham and Barton 2006; Denham et al 2004b: 851–52; Haberle 1993: 299–305).

From a morphogenetic perspective, the archaeobotanical evidence in New Guinea does not provide a clear signature of plant domestication. In Southwest Asia, the domestication of cereal grains is marked by morphogenetic changes that enhanced productivity and ease of processing (eg Zohary 1999). These changes are thought to have occurred rapidly, perhaps even within decades for some species (Hillman and Davies 1990, 1999). However, based on a consideration of tropical vegeculture in the humid tropics, many plants need not develop characteristic signatures similar to those documented for cereal grains (Harris 1996b: 568; Piperno and Pearsall 1998: 8; Smith 2001: 16–17; Yen 1985), or as rapidly (Pearsall 1995), although some do

(Piperno 1998; Piperno et al 2000). Insufficient research has been undertaken on phytolith and starch grain morphologies of cultivated plants in New Guinea to investigate whether diagnostic wild and domesticated forms are present. Carol Lentfer and Luc Vrydaghs's ongoing work on *Musa* banana phytoliths are exceptions. This work is needed to infer the morphological effects of human selection and to corroborate increasingly refined genetic interpretations.

Several factors call into question the value of using morphogenetic markers to differentiate subsistence practices in the Pacific context. Morphogenetic change need not be a major genetic threshold (Jones et al 1996) and may commence once selective pressures exerted by humans commence (Hather 1996), which can occur under various types of plant exploitation. However, there are other reasons to call into question the validity of using domestication marked by morphogenetic change as an indicator of agriculture. Some traditional food plants in the Highlands are still cultivated in agricultural systems from wild or semidomesticated forms. Cultivation need not be marked by significant morphological changes to enable clear differentiation between wild or cultivated varieties. Other plants with a long history of cultivation in New Guinea and elsewhere are not, in a classical sense, fully domesticated (Yen 1985). Although morphogenetic changes have occurred in cultivated bananas (Lebot 1999: 621–22; Lebot et al 1993, 1994), the domestication of species of Callimusa section (De Langhe and de Maret 1999: 378) and several species of Musa section bananas, is an ongoing process after millennia.

Given these conceptual shortcomings, alternative signatures of early agriculture in New Guinea are required.

Social Dependence and Landscape Change: Palaeoecological Evidence

A reliance on the concept of domestication has given too much significance to one epiphenomenon of human practices in the past. A focus on domestication has sometimes shifted the emphasis from studying 'what people were doing in the past' to a fixation on defining these practices solely in terms of their effects on plant and animal morphogenetics, which in itself is problematic (Jones and Brown chapter 3). Effects are often used to trace more fundamental causes, and morphogenetic transformations in animal and plant species under human selection are significant processes. However, debates in archaeology have divorced agriculture and domestication from subsistence and other realms of people living-in-the-past (Thomas 1999: 15–16). As archaeologists, we should prioritise subsistence activities in the past over botanical and zoological attributes created by those activities. Morphogenetic transformations do not necessarily provide a reliable guide to the degree of involvement or dependence of people on different modes of subsistence and are even more inappropriate in the tropics where the signatures of vegeculture may be different from those of cereal-based practices.

Spriggs (1996: 525–26) has modified Harris's schema for application to the Pacific context, a modification accepted in part by Harris (1995: 849–50). Drawing on Yen's (1985) observation that many Pacific cultigens are not fully domesticated today, Spriggs abandons Harris's reliance on domestication as a threshold demarcating cultivation and agriculture because it 'no longer seems relevant' (Spriggs 1996: 525). Spriggs replaces 'domestication' with 'when dependence on agriculture began, defined here in terms of the creation of agro-ecosystems that limit subsistence choice because of environmental transformation or labour demands' (Spriggs 1996: 525).

In revising Harris's schema, Spriggs is refocusing the investigation of agriculture away from plant and animal morphogenetics and onto 'human behaviour and organization' (Spriggs 1996: 525). This shift reflects the need to recognise that the investigation of agriculture in the past is primarily a study of people's subsistence in the past. Spriggs's shift reflects a general awareness that agriculture needs to be considered in more social terms (Hather 1996: 548; Ingold 1996: 21; Thomas 1999: 7–33).

According to Harris's schema, swidden cultivation would only be considered agriculture when accompanied by a reliance on domesticated plants. Given the reported incomplete domestication of species such as *Colocasia* taro, some yam species and cultivated bananas in New Guinea (Yen 1985), many contemporary practices would not be considered agricultural. Swidden cultivation accompanied by extensive clearance and systematic tillage, but only a weak-to-moderate reliance on domesticated plants, would be considered a system in which wild-food plant production is dominant (Harris 1996a: 4). The problem with Harris's schema is the focus upon plant domestication to the near exclusion of other spheres of human involvement, dependence and impacts on the environment. Such distinctions are not necessarily relevant to all historical contexts. The predominance and hence reliance on wild food resources characterised early forms of shifting cultivation in Southwest Asia (Hillman 1981: 189), neotropics (Piperno and Pearsall 1998: 8) and lowland Britain (Thomas 1999: 24–25). These same societies are generally accepted as being agricultural or 'Neolithic'.

According to Spriggs, large-scale swidden cultivation is 'agriculture' if humans were dependent upon these activities for subsistence and had limited alternative subsistence choices due to 'environmental transformation *or* labour demands' (Spriggs 1996: 525, my emphasis). These criteria apply to intensive swidden cultivation irrespective of the relative composition of wild and domesticated resources. Spriggs's scenario better accords with multidisciplinary evidence and interpretations for New Guinea and the Pacific. At present, however, the social aspects accompanying subsistence practices in the early to mid-Holocene in the Highlands are archaeologically invisible and, therefore, unknown (Harris 1995: 849–50; Yen 1998: 164). Consequently, palaeoecological signals of cultivated landscapes (Kennedy and Clarke 2004), domesticated environments (Yen 1989), domesticated landscapes (Terrell et al 2003) and social landscapes (Bayliss-Smith and Golson 1999) have been used as guides to the nature and extent of different subsistence practices in the past and the likely levels of human dependence upon them.

Complementary charcoal particle, phytolith and pollen analyses document disturbance using fire of the mixed-montane forest in the Kuk catchment from the early Holocene (Denham et al 2003: 189–90, 191; 2004b: 845–48). There is a relative decline in primary forest species with increased levels of secondary and disturbance taxa, which are interpreted to represent clearance and the creation of a vegetation mosaic. These palaeoenvironmental analyses confirm earlier interpretations that grey-brown clay represented accelerated deposition along the wetland margin following forest clearance and subsequent erosion in the catchment (Golson and Hughes 1980: 296–98; Hughes et al 1991). At c 6950–6500 cal BP, or Phase 2 on the wetland, the vegetation of the catchment degraded to open grassland. Grasslands were subsequently maintained in the vicinity for thousands of years by fire, with limited recovery of secondary forest and disturbance taxa.

As previously discussed, the record of vegetation change is not unique to Kuk. The early to mid-Holocene clearances in the Highlands were initiated at a time of wetter and slightly warmer climates (Brookfield 1989); hence they are unlikely to be climatically driven. Haberle (2003; Denham et al 2004b: 845–48) has interpreted similar cumulative changes at other sites in the Highlands to represent the emergence of agricultural landscapes from the early Holocene. The clearances of primary forest in highland New Guinea are certainly comparable in nature and antiquity to agricultural clearances inferred for other parts of the world (eg Hillman 1996; Piperno and Pearsall 1998).

The early Holocene clearances at Kuk and elsewhere in the Highlands limited subsistence choice because the resource base within the changing environment was substantially reduced (Golson 1982: 126–28). Fire had probably long been used to hunt and to enhance the availability of edible and other useful species (Groube 1989). Through time, and following the use of fire to aid the clearing of plots for cultivation, the cumulative and unforeseen consequences were certainly deleterious, ie the replacement of a mosaic of habitats (early Holocene) by anthropic grasslands (mid-Holocene). Grasslands are a resource-poor environment in the Highlands, limit subsistence choice and require different subsistence practices (Yen 1989: 62).

As with using morphogenetic markers of animals and plants, landscape transformation is just another effect, or proxy record (Jones and Colledge 2001: 395), used to track a more fundamental cause, ie agriculture. As is the case with morphogenetic markers, there is no necessary correspondence between landscape transformation and agriculture. Just as non-agricultural practices exert selective pressures that may account for morphogenetic transformations in some plant species, eg *Canarium* sp. (Yen 1996), tubers and corms (Hather 1996), so too extensive alterations of vegetation communities in Australia are not associated with agricultural activities, eg the Atherton Tablelands (Kershaw et al 1997). Conversely, practices considered to be agricultural in lowland Britain (Thomas 1999: 7–33) and Southwest Asia

(Hillman 1996) did not produce synchronous palaeoecological signals of forest clearance. Thus agriculture need not produce large-scale environmental change and the latter need not indicate the former. Given these problems of inferring agriculture from its morphogenetic and palaeoecological signals, on what basis can a framework for the interpretation of early agriculture in New Guinea be made?

Towards a Contingent Interpretation of Agriculture: Wetland Archaeological Evidence

Following from the above, the characteristics of agriculture in the humid tropics are different from those for semiarid regions, such as Southwest Asia. These differences centre on the types of plant, mode of reproduction, lack of clear morphogenetic indicators of domestication, resource distribution, continued reliance on wild food sources and cultivation practices in the tropics. Consequently, agriculture in the tropics needs to be studied 'in its own terms' (Piperno and Pearsall 1998: 1–38).

In casting a broad comparative net, some of the New Guinea evidence is less specific than for other regions, particularly in terms of evidence for plant domestication, eg for cereals in Southwest Asia (Hillman and Davies 1999; Zohary 1999) and maize and other crops in the neotropics (Piperno 1998, 2001; Piperno and Pearsall 1998; Piperno et al 2000). However, in a review of the Neolithic in lowland Britain, Thomas characterises agricultural activities as being broad spectrum, with a heavy reliance on wild species (1999: 24–25), and 'the evidence for domesticated plants can best be seen as representing rather small-scale, garden horticulture, carried out on a sporadic basis' (Thomas 1999: 25). Associated evidence of forest clearance is not ubiquitous, as large tracts of woodland remained uncleared (Thomas 1999: 31). For Thomas, previous portrayals of agricultural adoption being rapid and ubiquitous with subsequent demographic growth and expansion leading to environmental degradation are misapplications of the evidence. In contrast, he sees 'the seemingly contradictory character of the evidence as an indication of a high degree of variability, both between and within regions' (Thomas 1999: 25). The lowland Britain scenario of spatial variability and chronological discontinuity conforms to the New Guinean evidence of early agriculture. Furthermore, the identification of agriculture even in regions of assumed certainty can be questioned. Hodder has drawn on asynchronies in plant domestication and environmental transformation to claim: 'It is increasingly difficult to identify any point within a 4000 year period at which agriculture 'began' in the Near East' (1999: 175).

If we are to avoid an absolute foundation of agriculture based on *a priori* notions of plant domestication and landscape change, upon what basis is 'agriculture' to be defined and its 'origins' identified? The twin concepts of plant and landscape domestication have dominated thinking and illuminated much about agriculture in the Pacific and beyond (Harris 1989; Yen 1989; Terrell et al 2003). The links, or degrees of correspondence, among agriculture and morphogenetic change or environmental transformation, are dependent upon the types of practices in given geographical and historical situations. In attempting to interpret past subsistence we need to incorporate the specificity and contingency of practices to develop a more contextual understanding (Smith 1998: 208, 2001: 16).

Practice, Specificity and Contingency

The key to understanding early agriculture in the Highlands of New Guinea lies in uncovering the 'indeterminate relations between large-scale processes and individual lives' (Hodder 1999: 175). These indeterminate relations can be characterised by what people were doing in the past, either individually or communally, ie their practices. The evidence of past practices links large-scale processes to meaningful activity (Bourdieu 1990; Gardner 2001). Archaeological features derived from past subsistence practices are preserved in the wetlands, and this evidence grounds the interpretation of associated archaeobotanical and palaeoecological information. Direct, physical remains of past practices, such as features and associated artefacts and materials, enable multiple lines of contextual evidence to be woven into robust and specific interpretations. The archaeological remains enable specificity in interpreting what people were doing in the past, their use of plants and the cumulative effects of these practices on the environment. Such specificity is necessarily contingent on given social and environmental contexts. Incorporating contingency into definitions of early agriculture in New Guinea requires consideration of how agriculture in contemporary New Guinean societies is conceived and practised, ie we need to have an idea of which contemporary practices are considered agricultural in this region.

Contemporary Practices as a Guide

Contemporary agricultural practices in New Guinea are diverse, well studied (eg Brookfield and Hart 1971: 94–124) and recently inventoried as part of the Mapping of Agricultural Systems in Papua New Guinea Project (MASP; Allen and Ballard 2001; Bourke 2001; Hide et al 2002). Agricultural practices include various forms of wetland cultivation (Ballard 1995, 2001; Bayliss-Smith 1985; Powell and Harrison 1982), intensive forms of dryland cultivation (Brookfield and Brown 1963; Powell et al 1975; Waddell 1972) and shifting cultivation (Clarke 1971; Rappaport 1984). Other practices are more ambiguous and include a reliance to varying degrees on managed stands of *Musa* banana, sago (*Metroxylon sagu*) and nut-bearing trees supplemented by fishing (Dwyer and Minnegal 1991; Roscoe 2002; Terrell 2002). These diverse practices, agricultural and ambiguous, coexist in New Guinea.

Given the diversity of plant exploitation in contemporary New Guinea, we should allow for similar diversity in the past. However this diversity does not form a unilinear continuum based on energetic thresholds and degrees of dependence on domesticated species. Although the classification of contemporary practices could be undertaken with respect to such a schema, such taxonomies are predicated on unilinear, evolutionary conceptions of the past that embody the dualistic fallacies of 'the primal exemplar', ie foragers, and 'the pre-conceived end-point', ie intensive societies (Minnegal and Dwyer 2001: 282). As with other evolutionary continua applied to human societies, this approach to understanding subsistence practices is teleological and constitutes a 'progressionist trap' (Jones and Meehan 1989: 131). Despite claims to the contrary (Harris 1996a: 4), such unilinear schemes implicitly place ourselves as the modern and complex endpoint, with those societies practising less-intensive practices representing earlier and simpler stages of development. Harris's subsistence schema adopts criteria that may be highly relevant to much of Eurasia, ie the significance of domestication as marked by degrees of morphogenetic change, but domestication is too inflexible a marker to classify the present-day diversity of New Guinean practices. To avoid such ethnocentric interpretations of the past, although unavoidable at one level (Gadamer 1975: 153–341, 1976; Hodder 1999: 80–104), it is necessary to attempt to see the New Guinean past with reference to the New Guinean present. This does not represent a retreat to ethnographic analogues, but opens the past to different readings that at least acknowledge the diversity of contemporary practices in the region.

In summary, following Hather (1996), Ingold (1996) and Spriggs (1996), agriculture needs to be defined in social terms of dependence and involvement. Given the acknowledged problems in measuring such a social basis in the distant past (Harris 1998: 88; Yen 1998: 164), early agriculture in New Guinea needs to be differentiated from other forms of plant exploitation using an evidential triptych through which its effects are potentially visible (Figure 5.4). The effects of agriculture are evident in plant use (which may or may not lead to domestication), environmental transformation, and in the archaeological remains of former cultivation practices. The nature and diversity of contemporary practices serve as heuristic guides for translating these lines of evidence into specific and contingent interpretations of early agriculture – contingent because different conditions and varieties of practice produce different signals, and specific because any interpretation needs to be grounded in the physical evidence of those past practices, ie archaeological remains of cultivation.

Interpreting Evidence of Past Practices

The interpretation of past practices at Kuk is based upon the archaeological remains along the wetland margin. These remains provide direct evidence of former activities that ground the interpretation of associated archaeobotanical and palaeoecological finds. In contrast to previous readings (Golson 1977a, 1982, 1991b, 2000; Golson and Hughes 1980), the

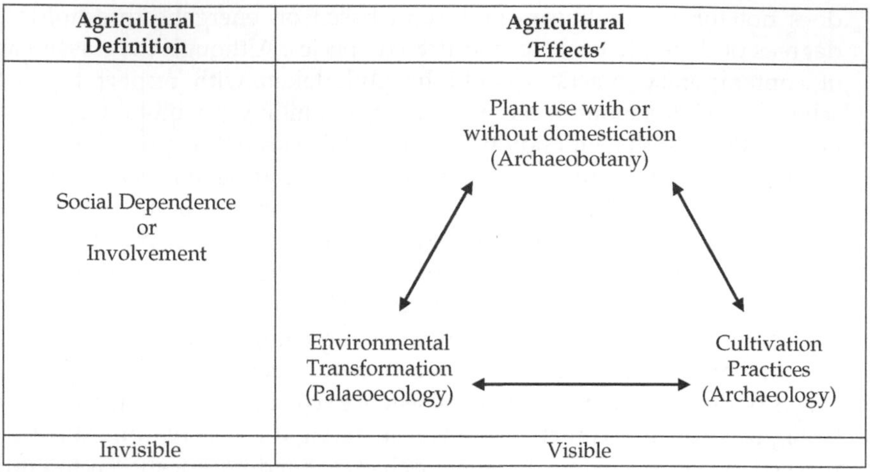

Figure 5.4 Evidential 'effects' through which early agriculture can be identified in New Guinea.

archaeological evidence at Kuk is interpreted conservatively here. The emphasis is on removing equivocal evidence, such as that for Phase 1 and early Phase 2 palaeochannels. Only through such a critical approach can robust evidence for these early phases provide a basis upon which an argument for agriculture can rest.

The Phase 1 palaeosurface at Kuk is plausibly consistent with the remains of a swidden-type plot cleared on a wetland margin for plant cultivation, although this interpretation requires further ethnoarchaeological corroboration, eg through the excavation of contemporary swidden plots in New Guinea. Several feature types within a plot on higher ground adjacent to a palaeochannel are suggestive of plant exploitation, with the remains of utilised plants being found in associated sediments (banana) and on stone tools (*Colocasia* taro and yam). These plants are major starchy staples in contemporary agricultural practices in New Guinea (Powell 1976). This plot exists within a catchment subject to forest disturbance and clearance using fire, with resultant soil degradation and erosion. Such a signal of forest disturbance during a period of potentially warmer and wetter climates is interpreted here to be more indicative of swidden cultivation than house-garden horticulture. Greater detail is still required, particularly in determining the range of plants used, but a more coherent picture of early Holocene activities and subsequent impacts is emerging.

The archaeobotanical, palaeoecological and archaeological findings exhibit some continuities between the early Holocene and recently documented practices. These continuities are visible in the plants used and present, the nature of forest clearance and its effects, and the types of features potentially associated with cultivation. Some of these signatures continue during grey-brown clay deposition and probably represent off-site processes. The lower interpretative resolution for Phase 1 is expected as emergent agriculture would likely be poorly distinguished from previous practices. Archaeological remains of past practices enable specificity in the interpretation of associated plant use and environmental change. They are broadly consistent with shifting cultivation, which is generally considered in New Guinea to be a form of agriculture (Allen and Ballard 2001; Bourke 2001; Brookfield and Hart 1971: 94–124).

Phase 2, like Phase 1, may represent the extension of dryland practices onto the wetland margin, or the development of specialised mounded cultivation along the wetland edge (Golson 1977a: 617–18). The regularity of the integrated, early Phase 2 palaeosurface is suggestive of mounding and is associated with more definitive evidence of agriculture including the deliberate planting of bananas in a grassed landscape maintained by fire and the processing of *Colocasia* taro. Deliberate planting would exert selective pressures on the genetic stock of *Musa* spp., including those of Musa section, and constitutes part of the domestication process.

Phase 3 represents the earliest unequivocal management of a wetland environment through the construction of ditch networks to facilitate drainage. In the absence of domesticated animals, it is inconceivable that these ditch networks were constructed for anything but plant cultivation. This interpretation is consistent with the form of the Phase 3 ditch networks, more recent ditch networks and contemporary ditched agricultural practices. The form of these alignments changed over the duration of Phase 3, approximately 1500 years, as various parts of the wetland were brought in and taken out of cultivation.

INTERPRETING AGRICULTURE: A CONCLUSION

The practice-oriented and contingent interpretation of agriculture advocated here is not a prescriptive, *a priori* definition. Rather, I present a general interpretative framework, which has variable, contingent and specific application in different parts of the world. Within this interpretative framework, different types of evidence are used to infer past agricultural practices in different regions. Morphogenetic changes and environmental transformations are not rejected as significant potential effects of agricultural practices, but they are viewed contextually within a contingent interpretation of agriculture based on direct evidence of past cultivation practices derived from archaeological excavation. Where possible, a consideration of recent and contemporary practices are used as guides, but only in the sense that they open up the past to a range of interpretations based on the archaeobotanical, archaeological and palaeoecological evidence.

In terms of archaeobotany, there is evidence for the presence, movement, use and planting of food plants considered to have been 'domesticated' in New Guinea, including most significantly *Colocasia* taro, *Musa acuminata* ssp. *banksii* and a yam. In terms of landscape transformation and a dependence on agroecosystems, clearances of vegetation were cumulative during much of the early Holocene at Kuk with the resultant need of people to obtain their subsistence in a highly degraded, grassland environment by the mid-Holocene. In terms of past practices, archaeological sites provide direct evidence of former cultivation on the wetland margins. All these lines of evidence seem to conform to ideas of agriculture widely applied to contemporary practices in New Guinea and the Pacific region. The nature of the earliest activities requires further clarification and the emergence of agriculture was certainly not inevitable, but similar ambiguities cloud the classification of contemporary and past subsistence practices in New Guinea and elsewhere in the world.

ACKNOWLEDGMENTS

I am indebted to Jack Golson for his continual support during my research into early agriculture and plant exploitation in Papua New Guinea and for permission to use the photographs from the 1970s excavations at Kuk. I thank Soren Blau, Jack Golson, Simon Haberle, José Iriarte, Robin Torrence and Luc Vrydaghs for comments on earlier drafts of this paper. I would also like to thank *Antiquity, Archaeology in Oceania, Records of the Australian Museum* and *Science* for permission to reproduce in amended form previously published images.

REFERENCES

Allen, B and Ballard, C (2001) 'Beyond intensification: reconsidering agricultural transformations', *Asia Pacific Viewpoint* 42(2–3): 157–62

Ballard, C (1995) 'The death of a great land: ritual, history and subsistence revolution in the Southern Highlands of Papua New Guinea', unpublished PhD thesis, Australian National University

Ballard, C (2001) 'Wetland drainage and agricultural transformations in the Southern Highlands of Papua New Guinea', *Asia Pacific Viewpoint* 42(2–3): 287–304

Bayliss-Smith, T P (1985) 'Pre-Ipomoean agriculture in the New Guinea Highlands above 2000 metres: some experimental data on taro cultivation', in Farrington, I (ed), *Prehistoric Intensive Agriculture in the Tropics*, pp 285–320. International Series 232, part I, Oxford: British Archaeological Reports

Bayliss-Smith, T P and Golson, J (1992a) 'Wetland agriculture in New Guinea Highlands prehistory', in Coles, B (ed), *The Wetland Revolution in Prehistory*, pp 15–27, Exeter: Prehistoric Society and Wetland Archaeological Research Project

Bayliss-Smith, T P and Golson, J (1992b) 'A Colocasian revolution in the New Guinea Highlands? Insights from Phase 4 at Kuk', *Archaeology in Oceania* 27: 1–21

Bayliss-Smith, T P and Golson, J (1999) 'The meaning of ditches: deconstructing the social landscapes of drainage in New Guinea, Kuk, Phase 4', in Gosden, C and Hather, J (eds), *The Prehistory of Food: Appetites for Change*, pp 199–231, London: Routledge

Bayliss-Smith, T P, Golson, J, Hughes, P, Blong, R and Ambrose, W (2005) 'Archaeological evidence for the Ipomoean revolution at Kuk Swamp, Papua New Guinea', in Ballard, C, Brown, P, Bourke, R M and Harwood, T (eds), *The Sweet Potato in Oceania: A Reappraisal*, pp 109–120. Oceania Monograph, Sydney: University of Sydney and Pittsburgh, PA: Department of Anthropology, University of Pittsburgh

Bellwood, P (1997) *Prehistory of the Indo-Malaysian Archipelago*, Honolulu: University of Hawaii Press

Bourdieu, P (1990) *The Logic of Practice*, Cambridge: Polity Press

Bourke, M (2001) 'Intensification of agricultural systems in Papua New Guinea', *Asia Pacific Viewpoint* 42(2–3): 219–36

Brookfield, H C (1989) 'Frost and drought through time and space, part III: what were conditions like when the high valleys were first settled?' *Mountain Research and Development* 9: 306–21

Brookfield, H C and Brown, P (1963) *Struggle for Land*, Melbourne: Oxford University Press

Brookfield, H C and Hart, D (1971) *Melanesia*, London: Methuen

Christensen, O A (1975) 'Hunters and horticulturalists: a preliminary report of the 1972–4 excavations in the Manim Valley, Papua New Guinea', *Mankind* 10(1): 24–36

Clarke, W C (1971) *Place and People: An Ecology of a New Guinea Community*, Berkeley, CA: University of California Press

De Langhe, E and de Maret, P (1999) 'Tracking the banana: its significance in early agriculture', in Gosden, C and Hather, J (eds), *The Prehistory of Food: Appetites for Change*, pp 377–96, London: Routledge

Denham, T P (2003a) 'The Kuk Morass: multi-disciplinary evidence of early to mid Holocene plant exploitation at Kuk Swamp, Wahgi Valley, Papua New Guinea', unpublished PhD thesis, Australian National University

Denham, T P (2003b) 'Archaeological evidence for mid-Holocene agriculture in the interior of Papua New Guinea: a critical review', *Archaeology in Oceania* 38(3): 159–76

Denham, T P (2004a) 'The roots of agriculture and arboriculture in New Guinea: looking beyond Austronesian expansion, Neolithic packages, and indigenous origins', *World Archaeology* 36(4): 610–20

Denham, T P (2004b) 'Early agriculture in the Highlands of New Guinea: an assessment of Phase 1 at Kuk Swamp', in Attenbrow, V and Fullagar, R (eds), *A Pacific Odyssey: Archaeology and Anthropology in the Western Pacific: Papers in Honour of Jim Specht*, pp 47–57. Records of the Australian Museum, Supplement 29, Sydney: Australian Museum

Denham, T P (2005a) 'Envisaging early agriculture in the Highlands of New Guinea: landscapes, plants and practices', *World Archaeology* 37(2): 290–306

Denham, T P (2005b) 'Agricultural origins and the emergence of rectilinear ditch networks in the Highlands New Guinea', in Pawley, A, Attenborough, R, Golson J and Hide, R (eds), *Papuan Pasts: Cultural, Linguistic and Biological Histories of Papuan-Speaking Peoples*, pp 329–62. Pacific Linguistics 572, Canberra: Research School of Pacific and Asian Studies (RSPAS), Australian National University

Denham, T P (2006) 'The origins of agriculture in New Guinea: evidence, interpretation and reflection', in Lilley, I (ed), *Blackwell Guide to Archaeology in Oceania: Australia and the Pacific Islands*, pp 160–88, Oxford: Blackwell

Denham, T P and Barton, H (2006) 'The emergence of agriculture in New Guinea: continuity from pre-existing foraging practices', in Kennett, D J and Winterhalder, B (eds), *Behavioral Ecology and the Transition to Agriculture*, pp 237–64, Berkeley: University of California Press

Denham, T P, Haberle, S G, Lentfer, C, Fullagar, R, Field, J, Therin, M, Porch, N and Winsborough, B (2003) 'Origins of agriculture at Kuk Swamp in the Highlands of New Guinea', *Science* 301: 189–193

Denham, T P, Golson, J and Hughes, P J (2004a) 'Reading early agriculture at Kuk (Phases 1–3), Wahgi Valley, Papua New Guinea: the wetland archaeological features', *Proceedings of the Prehistoric Society* 70: 259–98

Denham, T P, Haberle, S G and Lentfer, C (2004b) 'New evidence and revised interpretations of early agriculture in highland New Guinea', *Antiquity* 78: 839–57

Dwyer, P D and Minnegal, M (1991) 'Hunting in lowland, tropical rain forest: towards a model of non-agricultural subsistence', *Human Ecology* 19: 187–212

Flenley, J (1979) *The Equatorial Rainforest: A Geological History*, London: Butterworths

Fullagar, R, Field, J, Denham T P and Lentfer, C (2006) 'Early and mid Holocene processing of taro (*Colocasia esculenta*), yam (*Dioscorea* sp.) and other plants at Kuk Swamp in the Highlands of Papua New Guinea', *Journal of Archaeological Science* 33: 595–614

Gadamer, H-G (1975) *Truth and Method*, Barden, G (ed) and Cumming, J (trans), New York: Seabury Press

Gagné, W C (1982) 'Staple crops in subsistence agriculture', in Gressitt, J L (ed), *Biogeography and Ecology of New Guinea*, vol 1, pp 229–59, The Hague: Junk

Gardner, D (2001) 'Intensification, social production and the inscrutable ways of culture', *Asia Pacific Viewpoint* 42(2–3): 193–207

Gillieson, D, Gorecki, P and Hope, G (1985) 'Prehistoric agricultural systems in a lowland swamp, Papua New Guinea', *Archaeology in Oceania* 20: 32–37

Gillieson, D, Gorecki, P, Head, J and Hope, G (1987) 'Soil erosion and agricultural history in the central Highlands of Papua New Guinea', in Gardiner, V (ed), *International Geomorphology*, part II, pp 507–22, London: John Wiley and Sons

Gillison, A N (1972) 'The tractable grasslands of Papua New Guinea', in Ward, M W (ed), *Change and Development in Rural Melanesia*, pp 161–72, Port Moresby, Papua New Guinea: University of Papua New Guinea and Canberra: Australian National University

Golson, J (1977a) 'No room at the top: agricultural intensification in the New Guinea Highlands', in Allen, J, Golson J and Jones, R (eds), *Sunda and Sahul: Prehistoric Studies in Southeast Asia, Melanesia and Australia*, pp 601–38, London: Academic Press

Golson, J (1977b) 'Simple tools and complex technology: agriculture and agricultural implements in the New Guinea Highlands', in Wright, R V S (ed), *Stone Tools as Cultural Markers: Change, Evolution and Complexity*, pp 154–61, Canberra: Australian Institute of Aboriginal Studies

Golson, J (1977c) 'The making of the New Guinea Highlands', in Winslow, J H (ed), *The Melanesian Environment*, pp 45–56, Canberra: Australian National University Press

Golson, J (1981) 'New Guinea agricultural history: a case study', in Denoon, D and Snowden, C (eds), *A Time to Plant and a Time to Uproot*, pp 55–64, Port Moresby, Papua New Guinea: Institute of Papua New Guinea Studies

Golson, J (1982) 'The Ipomoean revolution revisited: society and sweet potato in the upper Wahgi Valley', in Strathern, A (ed), *Inequality in New Guinea Highlands Societies*, pp 109–36, Cambridge: Cambridge University Press

Golson, J (1985) 'Agricultural origins in Southeast Asia: a view from the east', in Misra, V N and Bellwood, P (eds), *Recent Advances in Indo-Pacific Prehistory*, pp 307–14, New Delhi: Oxford and IBH Publishing Co

Golson, J (1990) 'Kuk and the development of agriculture in New Guinea: retrospection and introspection', in Yen D E and Mummery, J M J (eds), *Pacific Production Systems: Approaches to Economic Prehistory*, pp 139–47, Canberra: Research School of Pacific and Asian Studies, Australian National University

Golson, J (1991a) 'Bulmer Phase II: early agriculture in the New Guinea Highlands', in Pawley, A (ed), *Man and a Half: Essays in Pacific Anthropology and Ethnobiology in Honour of Ralph Bulmer*, pp 484–91, Auckland: Polynesian Society

Golson, J (1991b) 'The New Guinea Highlands on the eve of agriculture', *Bulletin of the Indo-Pacific Prehistory Association* 11: 82–91

Golson, J (1997a) 'From horticulture to agriculture in the New Guinea Highlands', in Kirch, P V and Hunt, T L (eds), *Historical Ecology in the Pacific Islands: Prehistoric Environmental and Landscape Change*, pp 39–50, New Haven, CT: Yale University Press

Golson, J (1997b) 'The Tambul spade', in Levine, H and Ploeg, A (eds), *Work in Progress: Essays in New Guinea Highlands Ethnography in Honour of Paula Brown Glick*, pp 142–71, Frankfurt: Peter Lang

Golson, J (2000) 'A stone bowl fragment from the early Middle Holocene of the upper Wahgi Valley, Western Highlands Province, Papua New Guinea', in Anderson, A and Murray, T (eds), *Australian Archaeologist: Collected Papers in Honour of Jim Allen*, pp 231–48, Canberra: Coombs Academic Publishing, Australian National University

Golson, J (2002) 'Gourds in New Guinea, Asia and the Pacific', in Bedford, S, Sand, C and Burley, D (eds), *Fifty Years in the Field: Essays in Honour and Celebration of Richard Shutler Jr.'s Archaeological Career*, pp 69–78. New Zealand Archaeological Journal Monograph 25, Auckland: Auckland Museum

Golson, J and Hughes, P J (1980) 'The appearance of plant and animal domestication in New Guinea', *Journal de la Société des Océanistes* 36: 294–303

Golson, J, Lampert, R J, Wheeler, J M and Ambrose, W R (1967) 'A note on carbon dates for horticulture in the New Guinea Highlands', *Journal of the Polynesian Society* 76(3): 369–71

Gorecki, P and Gillieson, D S (1984) 'The highland fringes as a key zone for prehistoric developments in Papua New Guinea: a progress report', *Bulletin of the Indo-Pacific Prehistory Association* 5: 93–103

Gorecki, P and Gillieson, D S (1989) *A Crack in the Spine: Prehistory and Ecology of the Jimi-Yuat River, Papua New Guinea*, Townsville, Australia: Division of Anthropology and Archaeology, James Cook University of North Queensland

Groube, L (1989) 'The taming of the rainforests: a model for Late Pleistocene forest exploitation in New Guinea', in Harris, D R and Hillman, G C (eds), *Foraging and Farming: The Evolution of Plant Exploitation*, pp 292–304, London: Unwin Hyman

Haberle, S G (1993) 'Late Quaternary environmental history of the Tari Basin, Papua New Guinea', unpublished PhD thesis, Australian National University

Haberle, S G (1994) 'Anthropogenic indicators in pollen diagrams: problems and prospects for late Quaternary palynology in New Guinea', in Hather, J G (ed), *Tropical Archaeobotany: Applications and New Developments*, pp 172–201, London: Routledge

Haberle, S G (1995) 'Identification of cultivated *Pandanus* and *Colocasia* in pollen records and the implications for the study of early agriculture in New Guinea', *Vegetation History and Archaeobotany* 4: 195–210

Haberle, S G (1998) 'Late Quaternary change in the Tari Basin, Papua New Guinea', *Palaeogeography, Palaeoclimatology, Palaeoecology* 137: 1–24

Haberle, S G (2003) 'The emergence of an agricultural landscape in the Highlands of New Guinea', *Archaeology in Oceania* 38(3): 149–58

Haberle, S G, Hope, G S and de Fretes, Y (1991) 'Environmental change in the Baliem Valley, Montane Irian Jaya, Republic of Indonesia', *Journal of Biogeography* 18: 25–40

Harris, D R (1989) 'An evolutionary continuum of people-plant interaction', in Harris, D R and Hillman, G C (eds), *Foraging and Farming: The Evolution of Plant Exploitation*, pp 11–26, London: Unwin Hyman

Harris, D R (1995) 'Early agriculture in New Guinea and the Torres Strait divide', *Antiquity* 69 (265): 848–54

Harris, D R (1996a) 'Introduction: themes and concepts in the study of early agriculture', in Harris, D R (ed), *The Origins and Spread of Agriculture and Pastoralism in Eurasia*, pp 1–9, London: University College London Press

Harris, D R (1996b) 'The origins and spread of agriculture and pastoralism in Eurasia: an overview', in Harris, D R (ed), *The Origins and Spread of Agriculture and Pastoralism in Eurasia*, pp 552–73, London: University College London Press

Harris, D R (1998) 'Introduction: the multi-disciplinary study of cross-cultural plant exchange', in Pendergast, H D V, Etkin, N L, Harris, D R and Houghton, P J (eds), *Plants for Food and Medicine*, pp 85–91, Kew: Royal Botanic Gardens

Harris, E C and Hughes, P J (1978) 'An early agricultural system at Mugumamp Ridge, Western Highlands Province, Papua New Guinea', *Mankind* 11(4): 437–45

Hather, J G (1996) 'The origins of tropical vegeculture: Zingiberaceae, Araceae and Dioscoreaceae in Southeast Asia', in Harris, D R (ed), *The Origins and Spread of Agriculture and Pastoralism in Eurasia*, pp 538–50, London: University College London Press

Hide, R L, Bourke, R M, Allen, B J, Fritsch, D, Grau, R, Hobsbawn, P and Lyon, S (2002) *Western Highlands Province*. Agricultural Systems of Papua New Guinea Working Paper 10, Canberra: Research School of Pacific and Asian Studies, Australian National University

Hillman, G C (1981) 'Crop husbandry: evidence from macroscopic remains', in Simmons, I and Tooley, M (ed), *The Environment in British Prehistory*, pp 183–91, London: Duckworth

Hillman, G C (1996) 'Late Pleistocene changes in wild plant-foods available to hunter-gatherers of the northern Fertile Crescent: possible preludes to cereal cultivation', in Harris, D R (ed), *The Origins and Spread of Agriculture and Pastoralism in Eurasia*, pp 159–203, London: University College London Press

Hillman, G C and Davies, M S (1990) 'Measured domestication rates in wild wheats and barley under primitive cultivation, and their archaeological implications', *Journal of World Prehistory* 4: 157–222

Hillman, G C and Davies, M S (1999) 'Domestication rate in wild wheats and barley under primitive cultivation', in Anderson, P C (ed), *Prehistory of Agriculture: New Experimental and Ethnographic Approaches*, pp 70–102. Institute of Archaeology Monograph 40, Los Angeles: University of California at Los Angeles

Hodder, I (1999) *The Archaeological Process: An Introduction*, Oxford: Blackwell

Hope, G S and Golson, J (1995) 'Late Quaternary change in the mountains of New Guinea', *Antiquity* 69 (265): 818–30

Hughes, P J, Sullivan, M E and Yok, D (1991) 'Human induced erosion in a Highlands catchment in Papua New Guinea: the prehistoric and contemporary records', *Zeitschrift für Geomorphologie* 83: 227–39

Ingold, T (1996) 'Growing plants and raising animals: an anthropological perspective on domestication', in Harris, D R (ed), *The Origins and Spread of Agriculture and Pastoralism in Eurasia*, pp 12–24, London: University College London Press

Jones, M K and Colledge, S (2001) 'Archaeobotany and the transition to agriculture', in Brothwell, D R and Pollard, A M (eds), *Handbook of Archaeological Sciences*, pp 393–401, Chichester: John Wiley and Sons

Jones, M K, Brown, T and Allaby, R (1996) 'Tracking early crops and early farmers: the potential of biomolecular archaeology', in Harris, D R (ed), *The Origins and Spread of Agriculture and Pastoralism in Eurasia*, pp 93–100, London: University College London Press

Jones, R and Meehan, B (1989) 'Plant foods of the Gidjingal: ethnographic and archaeological perspectives from northern Australia on tuber and seed exploitation', in Harris, D R and Hillman, G C (eds), *Foraging and Farming: The Evolution of Plant Exploitation*, pp 120–35, London: Unwin Hyman

Kennedy, J and Clarke, W C (2004) *Cultivated Landscapes of the Southwest Pacific*. Resource Management in the Asia-Pacific (RMAP) Working Paper 50, Canberra: RMAP, Australian National University

Kershaw, A P, Bush, M B, Hope, G S, Weiss, K-F, Goldammer, J G and Sanford, R (1997) 'The contribution of humans to past biomass burning in the tropics', in Clark, J, Cachier, H, Goldammer, J G and Stocks, B (eds), *Sediment Records of Biomass Burning and Global Change*, pp 413–42, Berlin: Springer Verlag

Lampert, R J (1967) 'Horticulture in the New Guinea Highlands – C14 dating', *Antiquity* 41: 307–09

Lampert, R J (1970) 'Archaeological report of the Minjigina site', appendix 5 in Powell, J M, 'The impact of man on the vegetation of the Mount Hagen region, New Guinea', unpublished PhD thesis, Australian National University

Lebot, V (1999) 'Biomolecular evidence for plant domestication in Sahul', *Genetic Resources and Crop Evolution* 46: 619–28

Lebot, V, Meilleur, B A, Manshardt, R M and Meilleur, B (1993) 'Genetic relationships among cultivated bananas and plantains from Asia and the Pacific', *Euphytica: International Journal of Plant Breeding* 67: 163–75

Lebot, V, Meilleur, B A and Manshardt, R M (1994) 'Genetic diversity in eastern Polynesian Eumusa bananas', *Pacific Science* 48: 16–31

Loy, T, Spriggs, M and Wickler, S (1992) 'Direct evidence for human use of plants 28,000 years ago: starch residues on stone artefacts from northern Solomon Islands', *Antiquity* 66: 898–912

Martin, F W and Rhodes, A M (1977) 'Intra-specific classification of *Dioscorea alata*', *Tropical Agriculture* (Trinidad) 54: 1–13

Matthews, P (1991) 'A possible wild type taro: *Colocasia esculenta* var. *aquatilis*', *Bulletin of the Indo-Pacific Prehistory Association* 11: 69–81

Minnegal, M and Dwyer, P D (2001) 'Intensification, complexity and evolution: insights from the Strickland-Bosavi region', *Asia-Pacific Viewpoint* 42(2–3): 269–86

Moy, C M, Seltzer, G O, Rodbell, D T and Anderson, D M (2002) 'Variability of El Niño/ Southern Oscillation activity at millennial timescales during the Holocene epoch', *Nature* 420(6912): 162–65

Muke, J and Mandui, H (2003) 'In the shadows of Kuk: evidence for prehistoric agriculture at Kana, Wahgi Valley, Papua New Guinea', *Archaeology in Oceania* 38(3): 177–85

Pearsall, D M (1995) '"Doing" paleoethnobotany in the tropical lowlands: adaptation and innovation in methodology', in Stahl, P W (ed), *Archaeology in the Lowland American Tropics: Current Analytical Methods and Recent Applications*, pp 113–129, Cambridge: Cambridge University Press

Piperno, D R (1998) 'Paleoethnobotany in the neotropics from microfossils: new insights into ancient plant use and agricultural origins in the tropical forest', *Journal of World Prehistory* 12(4): 393–449

Piperno, D R (2001) 'On maize and the sunflower', *Science* 292(5525): 2260

Piperno, D R and Pearsall, D M (1998) *The Origins of Agriculture in the Lowland Neotropics*, San Diego: Academic Press

Piperno, D R, Ranere, J A, Holst, I and Hansell, P (2000) 'Starch grains reveal early crop horticulture in the Panamanian tropical forest', *Nature* 407: 894–97

Powell, J M (1970a) 'The impact of man on the vegetation of the Mount Hagen region, New Guinea', unpublished PhD thesis, Australian National University

Powell, J M (1970b) 'The history of agriculture in the New Guinea Highlands', *Search* 1(5): 199–200

Powell, J M (1976) 'Ethnobotany', in Paijmans, K (ed), *New Guinea Vegetation*, pp 106–83, Canberra: Commonwealth Scientific and Industrial Research Organisation and Australian National University Presses

Powell, J M (1982a) 'Plant resources and palaeobotanical evidence for plant use in the Papua New Guinea Highlands', *Archaeology in Oceania* 17: 28–37

Powell, J M (1982b) 'The history of plant use and man's impact on the vegetation', in Gressitt, J L (ed), *Biogeography and Ecology of New Guinea*, vol 1, pp 207–27, The Hague: Junk

Powell, J M and Harrison, S (1982) *Haiyapugua: Aspects of Huli Subsistence and Swamp Cultivation*. Department of Geography Occasional Paper no 1 (ns), Port Moresby, Papua New Guinea: University of Papua New Guinea

Powell, J M, Kulunga, A, Moge, R, Pono, C, Zimike, F and Golson, J (1975) *Agricultural Traditions in the Mount Hagen Area*. Department of Geography Occasional Paper no 12, Port Moresby, Papua New Guinea: University of Papua New Guinea

Rappaport, R (1984) *Pigs for the Ancestors: Ritual in the Ecology of a New Guinea People*, New Haven, CT: Yale University Press

Roscoe, P (2002) 'The hunters and gatherers of New Guinea', *Current Anthropology* 43(1): 153–62

Smith, B D (1998) *The Emergence of Agriculture*, New York: Scientific American Library

Smith, B D (2001) 'Low-level food production', *Journal of Archaeological Research* 9: 1–43

Spriggs, M (1996) 'Early agriculture and what went before in Island Melanesia: continuity or intrusion?', in Harris, D R (ed), *The Origins and Spread of Agriculture and Pastoralism in Eurasia*, pp 524–37, London: University College London Press

Spriggs, M (1997) *The Island Melanesians*, Oxford: Blackwell

Spriggs, M (2001) 'Who cares what time it is? The importance of chronology in Pacific archaeology', in Anderson, A, Lilley, I and O'Connor, S (eds), *Histories of Old Ages: Essays in Honour of Rhys Jones*, pp 237–49, Canberra: Pandanus Books, Australian National University

Terrell, J E (2002) 'Tropical agroforestry, coastal lagoons and Holocene prehistory in Greater Near Oceania', in Shuji, Y and Matthews, P J (eds), *Proceedings of the International Area Studies Conference VII: Vegeculture in Eastern Asia and Oceania*, pp 195–216, Osaka: National Museum of Ethnology

Terrell, J E, Hart, J P, Barut, S, Cellinese, N, Curet, A, Denham, T P, Haines, H, Kusimba, C M, Latinis, K, Oka, R, Palka, J, Pohl, M E D, Pope, K O, Staller, J E and Williams, P R (2003)

'Domesticated landscapes: the subsistence ecology of plant and animal domestication', *Journal of Archaeological Method and Theory* 10(4): 323–68

Thomas, J (1999) *Understanding the Neolithic*, London: Routledge

Waddell, E (1972) *The Mound Builders: Agricultural Practices, Environment and Society in the Central Highlands of New Guinea*, Seattle, WA: University of Washington Press

Walker, D and Flenley, J R (1979) 'Late Quaternary vegetational history of the Enga Province of upland Papua New Guinea', *Philosophical Transactions of the Royal Society of London* 286: 265–344

Yen, D E (1980) 'The southeast Asian foundations of Oceanic agriculture', *Journal de la Société des Océanistes* 66–7: 140–46

Yen, D E (1985) 'Wild plants and domestication in Pacific islands', in Misra, V N and Bellwood, P (eds), *Recent Advances in Indo-Pacific Prehistory*, pp 315–26, New Delhi: Oxford and IBH Publishing Co

Yen, D E (1989) 'The domestication of environment', in Harris, D R and Hillman, G C (eds), *Foraging and Farming: The Evolution of Plant Exploitation*, pp 55–75, London: Unwin Hyman

Yen, D E (1990) 'Environment, agriculture and the colonisation of the Pacific', in Yen, D E and Mummery, J M J (eds), *Pacific Production Systems: Approaches to Economic Prehistory*, pp 258–77, Canberra: Research School of Pacific and Asian Studies, Australian National University

Yen, D E (1991) 'Polynesian cultigens and cultivars: the questions of origin', in Cox, P A and Banack, S A (eds), *Islands, Plants and Polynesians: An Introduction to Pacific Ethnobotany*, pp 67–95, Portland, OR: Dioscorides Press

Yen, D E (1995) 'The development of Sahul agriculture with Australia as bystander', *Antiquity* 69 (265): 831–47

Yen, D E (1996) 'Melanesian arboriculture: historical perspectives with emphasis on the genus *Canarium*', in Evans, B R, Bourke, R M and Ferrar, P (eds), *South Pacific Indigenous Nuts*, pp 36–44, Canberra: ACIAR

Yen, D E (1998) 'Subsistence to commerce in Pacific agriculture: some four thousand years of plant exchange', in Pendergast, H D V, Etkin, N L, Harris, D R and Houghton, P J (eds), *Plants for Food and Medicine*, pp 161–83, Kew: Royal Botanic Gardens

Zohary, D (1999) 'Domestication of the Neolithic Near Eastern crop assemblage', in Anderson, P C (ed), *Prehistory of Agriculture: New Experimental and Ethnographic Approaches*, pp 42–50. Institute of Archaeology Monograph 40, Los Angeles: University of California at Los Angeles

CHAPTER 6

Unravelling the Story of Early Plant Exploitation in Highland Papua New Guinea

Jack Golson

This paper provides a historical context for, and some post factum comments on, Tim Denham's comprehensive review in this volume of the two stages of archaeological and associated investigations focused on Kuk Swamp in the upper Wahgi Valley of highland Papua New Guinea. As a result of these multidisciplinary investigations, claims to the antiquity and independence of New Guinea agricultural origins have been made, strengthened and refined. The first stage, involving colleagues and myself, took place in the 1970s and early 1980s, the second, involving Denham, his associates and myself, began in the late 1990s.

The paper does not deal with differences in the interpretation of specific aspects of the archaeological evidence; these are the subject of continuing dialogue elsewhere (eg Denham et al 2004a). Its purpose is to discuss the thinking about agricultural history in Southeast Asia and the South Pacific in the context of which the Kuk project began in 1972 and the subsequent changes in interpretation which it underwent, and to which it contributed, as advances in knowledge took place over a range of associated fields of study. Towards the end I venture some thoughts about conclusions reached in the second stage of work looked at from the standpoint of the first.

A minor subtheme in my discussion is the use of the term 'agriculture', which is a major concern in Denham's contribution. I call it 'minor' because I discuss it very much in historical, not theoretical, context. I had my archaeological education in England in the late 1940s, when it was conventional to talk of the Neolithic revolution (cf Clark 1972: viii) and the break this marked between a hunter-gatherer stage of human history and an agricultural 'food producer' stage (Higgs and Jarman 1972: 12). I had my first experience in agricultural archaeology in the 1950s in New Zealand, where Polynesian settlers had brought agricultural plants and practices from the tropical Pacific and adapted them to high latitudes (cf Yen 1961). In the mid-1960s I found myself engaged in another exercise in agricultural archaeology, this time at high elevation in the central ranges of New Guinea, which led on to the investigations at Kuk. At the time, the major crops of traditional New Guinea agriculture were thought to have been imported, as in the New Zealand case, from overseas. The following account of the first stage of work at Kuk will explain how this belief in external derivation operated initially to justify the use of the term 'agriculture' for cultivation in New Guinea as a whole and how later the factor of elevation served the same purpose to describe the adaptation of tropical crops for cultivation in the central highlands.

At a late stage (Golson 1997: 45) I made a terminological distinction between horticulture and agriculture in central highlands cultivation practice on the basis of the evidence from Kuk, though the distinction itself had been made explicit many years before (in Golson 1982b: 120). It involved a perceived difference between the first three and the second three phases of the six-phase sequence of cultivation in the Kuk Swamp. The earlier phases were seen as horticultural, with their structural evidence of water control and cultivation interpreted as making provision for the growth in the same planting of a variety of crops of different edaphic and hydrologic requirements. The later phases were called agricultural, with ditches defining standardised patterns of square or rectangular plots thought to be for the intensive cultivation of a single crop type.

THE KUK PROJECT STAGE 1: BACKGROUND AND CONTEXT

The investigations on which the history of New Guinea agriculture has been constructed took place at 1550 m elevation. Little has so far been contributed to that history by investigations at lower elevations. It is important to look at agriculture in the highlands within the context of New Guinea as a whole in order to understand how its history relates to that of agriculture in the South Pacific and beyond. The following sketch (amplifying Golson 1991a: 82–83) is based on unpublished research by Michael Bourke (nd) on the altitudinal ranges of economic plants in Papua New Guinea.

Bourke's data are a detailed illustration of a recognised zonation in New Guinea food procurement: the lowlands, up to 600 m; an intermediate zone, 600 m to 1200 m; the highlands, 1200 m to 1800 m; and a high-elevation zone, 1800 m to 2700 m. Bourke points out that within the intermediate zone fall the upper altitudinal limits of a number of important tree species, such as sago (*Metroxylon sagu*), breadfruit (*Artocarpus altilis*), coconut (*Cocos nucifera*), galip nut (*Canarium indicum*), okari nut (*Terminalia kaernbachii*) and *Pangium edule*, while the large number of species that grow up to 1750–1900 m and define the upper limit of the highland zone includes most of the yams (*Dioscorea* spp.) and most of the bananas (*Musa* spp.). Taro (*Colocasia esculenta*) grows higher, as does sweet potato (*Ipomoea batatas*), a tropical American plant of post-Magellan introduction into Southeast Asia and the western Pacific which came to dominate the agriculture of highland New Guinea.

Ethnobotanical Evidence

Fifty years ago the general belief among scholars was that agriculture based on the cereal rice (*Oryza sativa*) as staple had replaced in Southeast Asia an older system characterised by the normally vegetative reproduction of a series of tuberous plants, fruit and nut trees. This older system, it was assumed, is the one still practised in New Guinea and the South Pacific islands (Golson 1985: 307 and references cited therein).

In an address at the Twelfth Congress of the Pacific Science Association in Canberra in 1972, on the eve of the Kuk project with which he was shortly to become associated, Douglas Yen (1973: 68) noted how the topic of plant evidence for the movement of people in Pacific prehistory had become 'a minor tradition' at successive congresses of the association. He pointed specifically to the Honolulu congress of 1961 as initiating a more coordinated and interdisciplinary phase of research.

In one of the volumes produced from the earlier congress, Jacques Barrau (1963: 4, 6) said of the subsistence plants of Oceania that most of them were 'traceable to what the great Vavilov called the "Indo-Malaysian centre of origin of cultivated plants"', citing taro (*Colocasia esculenta*), yam (*Dioscorea alata*) and banana (*Musa sapientum*). He pointed out (1963: 6), however, that the definition of Vavilov's Indo-Malaysian centre needed to be revised, since New Guinea appeared to have been the place of origin of a number of Pacific cultigens, such as the sago palm (*Metroxylon sagu*), the *fehi* banana (*Musa troglodytarum*), sugarcane (*Saccharum officinarum*) and the related species *Saccharum edule*. This suggested New Guinea as an important Melanesian subdivision, or extension, of Vavilov's Indo-Malaysian centre.

For Yen (1995: 831), 'the evidence for the domestication of unique New Guinea plant species has been the spur for the hypothesis that western Melanesia is a site of agricultural origin, parental to the Oceanic subsistence systems'. He cautiously suggested this in his first review paper in the field (Yen 1971: 4). He did so more confidently in his 1972 Canberra address (1973: 70, 73) because there was now archaeological evidence for New Guinea settlement in the Pleistocene, allowing the necessary time for the suggested developments to take place.

A quarter of a century and a string of publications later (Yen 1980, 1982, 1985a, 1985b, 1989, 1990, 1991), Yen (1995: 833) reviewed the evidence from new work in phytogeography and ethnobotany and a range of botanical sources, from taxonomy to recent biomolecular techniques, that confirmed the initial tentative hypothesis about western Melanesia as a site of agricultural origins. At the same time the old theory of diffusion out of Asia was not to be discarded, since Southeast Asian plants had had a role in the formation of agriculture in New Guinea subsequent to local domestications in the region (Yen 1995: 831). This was seen to be the case with the three major starch staples of New Guinea field agriculture, bananas of the Eumusa section, yam (*Dioscorea* spp.) and taro (*Colocasia esculenta*). The clearest case was made for Eumusa bananas, which Yen (1995: 838) recorded as having their origin on the Southeast Asian mainland, citing Simmonds (1962), who at the same time pointed out that New Guinea was the region of greatest known diversity for the group, evidence often taken as indicating a centre of origin.

As regards yam and taro, the results of limited isozymic and DNA analyses allowed for both local domestication and introduction from Asia (Yen 1995: 841). In both cases the possibility of dual origins arose from the fact that the two genera were present in northern Australia as traditional Aboriginal food plants (Yen 1995: 835, 836). Of the five species of yam in Australia, three were shared with New Guinea and two thought to be endemics, which suggested a long history in the region (Yen 1995: 836).

Archaeological Evidence

Throughout the work that I have referenced above, Yen was alive to archaeology as a source of chronology for developments to which his other types of data bore witness, but could not date. This is perfectly illustrated by Loy et al (1992: 909–10) who, with an appropriate warning against possible misidentification, reported starch grains of both *Colocasia* and *Alocasia* on stone tools from the northern Solomons dating around 28,000 BP. They also made reference to two recent unpublished studies about the observation of starch grains of *Colocasia* type on stone tools going back to the terminal Pleistocene at other sites in Island Melanesia. A few years later Haberle (1995: 207) confirmed the presence of *Colocasia esculenta* pollen at 9000 BP at the lowland site of Lake Wanum in the Markham Valley of Papua New Guinea.

Yen was aware that direct evidence of plant remains, as in the north Solomons case, was relatively scarce (eg 1995: 833), but saw the relevance of a range of indirect evidence, including that of field systems. The cultivation systems stratified in the swamp at Kuk provided the extended time scale required for his developing thesis of the independent origins of New Guinea agriculture, while this thesis in turn informed Golson's interpretation of the evidence from the site (Golson 1991b: 487–88; cf Golson 1977: 614, 618, 1982a: 297–300, 1989: 682, 1991a: 82–83, 89; Golson and Hughes 1980: 299–300).

In the 1970s archaeology was fairly new on the New Guinea scene, having only arrived with the work of Susan Bulmer in the Papua New Guinea Highlands in 1959–60 (Bulmer 1966; Bulmer and Bulmer 1964). By the mid-1970s there was emerging a consensus among the few archaeologists in the field that what they called agriculture (or horticulture) had made an appearance in the Highlands, the only area for which relevant evidence was available, by around 6000 BP (Golson and Hughes 1980: 294). The evidence in question included palynological indications of change in forest composition by that date, attributed to clearance in the course of shifting cultivation, and arguments from the presence of bones of the non-indigenous pig, an animal nevertheless very much at home in traditional agricultural economies of the Pacific. All this was compatible with agriculture resulting from the migration of Austronesian speakers out of Southeast Asia bringing pigs as husbanded animals across Wallace's Line – the boundary between the Oriental and Australasian faunal regions – together with the root crops to feed them. Their arrival at the threshold of the Pacific was at the time estimated in collaborative overviews of the archaeological and linguistic evidence to fall around

6000 years ago (Yen 1980: 143, citing Shutler and Marck 1975: 104 and Pawley and Green 1975: 52, who, however, made a more cautious suggestion of 'no later than 3,000 BC').

A discordant note was struck by Susan Bulmer's report (1975: 18, 19, 36) of a pig incisor from the basal layer of a Simbu rockshelter dating, at 10,350 ± 140 BP, to the terminal Pleistocene. Golson and Hughes (1980: 300) were unwilling to discount this as evidence on the grounds that the body of excavated data from early New Guinea sites was not large enough to allow them to do so. Indeed, they found possible support for pigs in the presence in Phase 1 (9000 BP) and Phase 2 (starting at 6000 BP) contexts at Kuk of oval hollows resembling those commonly made by contemporary pigs to lie in when at rest (1980: 299; Hope and Golson 1995: 824, Figure 4). As a result they were ready to entertain a hypothesis (1980: 301) of the introduction of Southeast Asian pigs into New Guinea as regularly hand-fed animals as early as 10,000 years ago, accompanied by at least some of the cultivated Southeast Asian plants that kept them firmly attached to people. These plants would have been grown in the gardens inferred from the archaeological evidence to be present on the Kuk Swamp margin from Phase 1. In an earlier draft of the Golson and Hughes article, the cultivated plants in question were explicitly spelt out as including taro and particular varieties of yam (Golson 1977: 613).

In terms of the argument with which their paper began (1980: 294), Golson and Hughes were describing a situation where major plant staples and a domesticated animal were of exotic origin in New Guinea. If this were so, their appearance would provide evidence of the beginnings of agriculture in the island and its earlier presence in the region from which the plants and animal had been introduced. Yen (1982: 293) summed the position up more dramatically: 'If the evidence for taro and pig at early time levels in New Guinea (or, indeed, Melanesia) can be affirmed ..., where does it leave the alternate hypothesis for endogenous agricultural development? Exploded, probably.'

In fact, this line of argument, though not its influence, was running out of steam. In 1980, four years or so after the original case had been made, Golson (1982a: 298–99) found that there was little new evidence to bring to bear. On the matter of the antiquity of pig, Bulmer (1982: 188) announced the identification of a pig incisor from a terminal Pleistocene level of Yuku Rockshelter, one of her highland sites excavated some years before, but after this there was nothing relevant to report for several years. The so-called pig hollows at Kuk could have alternative explanations, for example as marks of cultivation for New Guinea plants, as already suggested by Yen (1980: 143–44). Was it indeed possible then that the Southeast Asian elements in New Guinea husbandry came after all with the arrival of Austronesian speakers? The multi-institution Lapita Homeland Project launched by Jim Allen in the Bismarck Archipelago in 1985 was designed to throw light on this and other matters (Allen 1991: 3), given the identification of Lapita with Austronesian speakers by some scholars on the basis of work in Vanuatu, New Caledonia, Fiji and Western Polynesia (Shutler and Marck

1975: 95; Pawley and Green 1975: 49–50 give a more qualified view). When the relevant results from the Homeland Project were assessed, the earliest dates for Lapita sites in the region did not go back beyond about 3500 BP (Gosden et al 1989: 576), falling well short of the 5000–6000-year estimates cited above from Pawley and Green (1975) and Shutler and Marck (1975) that allowed developments in the New Guinea interior, including pigs, to be tied into the Austronesian story (cf Yen 1980: 143).

There is an ironic twist to the tale. Direct AMS dating of six pig teeth from two highland New Guinea rockshelters, Kafiavana (White 1972) and Nombe (Mountain 1991), have given results within the last few hundred years (Hedges et al 1995: 428), much younger than the mid-Holocene age expected for three of them and the late Pleistocene age stratigraphically indicated for a fourth (a find at Nombe in 1979–80, Mountain pers comm). The standard view now is that the first pigs in the New Guinea region, and the first pottery, were those that appeared on Lapita sites in the Bismarck Archipelago after 3500 BP, brought by Austronesian-speaking newcomers. This suspends judgement on a series of claims for earlier pigs and pottery at sites in coastal West Sepik, on the lower Ramu and in the Bismarck Range and for early pig in New Ireland. Denham (2004: 615) indicates the relevant literature for the sites in question, including critical sources, to which Spriggs (1996a: 329, 335) should be added.

Palaeoenvironmental Evidence

Palynological investigations in the Central Highlands of Papua New Guinea, begun in the 1960s by Donald Walker and his students at the Australian National University, have been described as providing striking pollen evidence of early forest clearance, 'speaking on a tropical or even a world scale' (Flenley 1979: 122). The most relevant work for present purposes was done by Jocelyn Powell, whose central interest was the human impact on the vegetation, first as one of Walker's graduate students, later as a member of the Kuk research team. Reviewing eight pollen diagrams from sites in the upper Wahgi Valley, she reported (1982: 218) that forest reduction attributable to human influence had occurred before 5300–5000 BP and assumed that a system of shifting cultivation with long-term fallow was in practice. Unfortunately there was no pollen evidence to say when the forest disturbance attributed to shifting cultivation had begun. This was because the vegetation record of five of the eight pollen diagrams started at or later than 5300–5000 BP, while in the other three there was no record because of the absence of suitable sediments from that time back to c 25,000–20,000 BP.

At Kuk, however, the early Holocene part of this hiatus saw the deposition, between 9000 and 6000 BP, of a distinctive grey clay (Denham's 'grey-brown clay', chapter 5), part of a fan deposit in the swamp, thickest at its southern margin where the sediment-bearing waters enter the basin and thinning to the north. This clay represented a striking acceleration in the rate of sedimentation in the swamp or, expressed another way, of soil loss from the southern catchment, compared with what had gone before (Golson and Hughes 1980: 296–97; cf Hughes et al 1991: 233–35 for recalculations on the basis of more complete information). The soil loss was attributed to the establishment of a regime of shifting cultivation in the catchment and the repeated forest clearing that this involved. It was an explanation that fitted well with two notions about grey clay. The first was that grey clay deposition started as the Phase 1 palaeochannel, whether artificial or natural, became unable to cope with the organic detritus and sediment load resulting from sustained forest clearance. The second notion was that grey clay deposition ceased when a Phase 2 palaeochannel, whether artificial or natural, became available (Golson 1977: 487; Golson and Hughes 1980: 298). Subsequently, Golson (1991a: 84–88, 1991b: 486–88) supported this interpretation of Phase 1 and grey clay in terms of shifting agriculture by arguing that they represented a new type of activity in the highland region. The markers for this activity showed continuity with subsequent evidence for cultivation in the Kuk basin, but discontinuities with the late Pleistocene past of the Highlands (Golson 1991a: 84), including that of the Kuk Swamp itself (Golson 1991a: 88).

In the course of the discussion Golson (1991a: 85, 1991b: 487) drew on new evidence from Kelela Swamp at 1420 m elevation on the floor of the Baliem Valley in West Papua (Haberle et al 1991: 31–37). In the light of the fact that there was no early Holocene vegetation record for the upper Wahgi to register the forest disturbances thought to have produced the grey clay, Kelela Swamp was an apt case for comparison. Here a pollen core produced an all but continuous vegetation history from beyond 7000 BP to the present that reflected the characteristic impact of shifting cultivation on forest composition. The beginning of clearance was signalled by clays and silts at the base of the core that marked increased sediment input into the local drainage and initiated swamp formation.

Considerations of Elevation

The new activity seen as represented by Phase 1 and grey clay at Kuk, and interpreted to be shifting agriculture, made an appearance about 9000 BP, around the time that temperatures were beginning to reach their modern levels in the course of late glacial climatic amelioration. Under such modern conditions the main starchy food species of traditional highland New Guinea agriculture had a mean usual upper limit to their productive growth of around 2000 m, with two notable exceptions: sweet potato (*Ipomoea batatas*), which as an introduction of the past few hundred years was irrelevant to the argument; and taro (*Colocasia esculenta*), whose present altitudinal limit may have owed much to the efforts of millennia of highland gardeners to increase it (Golson 1991a: 82–83, based on data in Bourke nd). In these circumstances it was difficult to think that any of the potential species in question, with the possible exception of taro and that of the now unimportant tuber *Pueraria lobata* (Watson 1964), would have been viable in the upper Wahgi Valley much before the date at which cultivation was archaeologically registered at Kuk at around 1550–1600 m. Before that, it was presumed, the plants were being grown at lower elevations and were moved upslope in step with the ameliorating climate.

The origins of New Guinea agriculture in the lowlands had been an early conclusion from the Kuk work, based on the 9000 BP date for the grey clay and what lay immediately beneath, both by Golson and Hughes (1980: 301) on straightforward climatic grounds and by Yen (1982: 291–92) in a more complex argument. This was a time when, because of Bulmer's 10,000 BP date for pig, the idea was canvassed (Golson and Hughes 1980) of New Guinea agriculture being a fairly straightforward transfer of an animal and cultigens from Southeast Asia. We have seen how this proposition began to weaken as the archaeological record failed to produce decisive support for early, or early enough, pig in New Guinea. The question then was asked as to what indeed the plants might be that were grown in the claimed early cultivations at Kuk. There was an attempt to determine this directly by the (for Kuk) pioneering use of phytoliths, with the aim of classifying those of *Musa* as belonging to the indigenous Australimusa section or, as was considered at the time, the exotic Eumusa section (Wilson 1985: 94). The results were promising, though for the time being limited: Eumusa phytoliths could not be differentiated from those of *Musa ingens*, the wild New Guinea banana of the Ingentimusa section, and though Australimusa phytoliths could be differentiated from both, only two were found in the archaeological samples, the earlier one in association with Phase 2 (Wilson 1985: 94, Table 3).

Interestingly, Wilson (1985: 96, Figure 3) saw his phytolith analysis as supporting the interpretation of grey clay as the result of erosion due to forest clearance in the catchment. The phytolith evidence showed two changes over the early Holocene period of grey clay deposition: there was a drop in total phytolith density due to a faster rate of sediment deposition and a 'dramatic drop in percentage of grass phytoliths'.

It was about the same time that Doug Yen, now on the staff of the Australian National University, began to interest himself in the implications of the 'distribution of wild species, used by the Aborigines, of the major cultivated genera of root crops in Papua New Guinea and Oceania', comprising *Dioscorea* yams, *Colocasia*, *Alocasia* and *Amorphophallus* taros and the sweet potato genus *Ipomoea* (Yen 1985b: 494, Table 2; cf Yen 1982: 284, 1985a: 317–18, 1989: 59–60, 68–69). This geographical sharing of genera, and indeed species, resulted from the contact of two different floras and faunas brought about by the collisions of the Australian–New Guinea fragment of Gondwanaland with the Laurasian Plate at its margin in eastern Sulawesi during the Tertiary Period (Yen 1990: 259, on the basis of Whitmore 1981). In the later development of agriculture in New Guinea '[d]omestication, as the key to such development, was enacted largely on the extension of the Indo-Malayan or Malesian flora, and its results were the principal reason for the attribution of completely Southeast Asian origins in the past' (Yen 1990: 259).

As a result, Yen went on to argue elsewhere (1991: 563) that the exotic origin of the various species of domesticated yams and genera of taro, 'a contention ... as old as ethnobotany in the Pacific', was under question, with (Yen 1990: 260) the possibility of 'Laurasian origin and New Guinea domestication' an alternative to human introduction in the Holocene for 'the taro-yam Oceanic complex'. In the case of *Colocasia* taro, a new chromosome study adding karyotyping to simple enumeration suggested two separate domestications as a possibility, one Southeast Asian or Indian, the other Oceanic or Sahulian (Yen 1990: 260, 1991: 563). The situation with *Dioscorea* was much less clear, partly because the wild yams of New Guinea had not been sufficiently studied (Yen 1991: 563).

Shortly afterwards, as we have seen, *Colocasia* and *Alocasia* starch grains were reported on stone tools some 28,000 BP in date from a site in the northern Solomons (Loy et al 1992: 910), starch of *Colocasia* type from the terminal Pleistocene at other sites in Island Melanesia (Loy et al 1992: 910) and *Colocasia* pollen in a 9000 BP lowland context at Lake Wanum in the Markham Valley of Papua New Guinea (Haberle 1995: 207). *Colocasia* taro was thus present and potentially available to be, as had always been suggested, the most appropriate candidate in the swamp-margin cultivations reconstructed for Kuk. The argument was that it would have arrived there from lower elevations around 9000 BP, when on the one hand the attainment of modern temperatures made this a viable proposition and on the other there was evidence in the form of grey clay deposition for the appearance of the cultivation regime with which the plant was associated. This suggested a systematic process for which the description 'shifting agriculture' was appropriate (Hope and Golson 1995: 824–25).

THE KUK PROJECT STAGE 2: AGRICULTURAL ORIGINS AND DEVELOPMENTS

Origins

The multidisciplinary investigations at Kuk recently directed by Tim Denham (chapter 5) have supported claims for the antiquity and independence of New Guinea agricultural origins by a combination of established archaeological and palaeoecological procedures systematically applied and new and improved techniques in stratigraphy and archaeobotany (Denham et al 2003, 2004b). They have led to two main conclusions (Denham et al 2004b: 850–53).

(1) Agriculture was practised in the upper Wahgi Valley in the mid-Holocene, by at least Phase 2 of the Kuk sequence, initiated around 6000 BP, on the basis of clear evidence of systematic planting (Denham et al 2003: 190, 2004b: 845; chapter 5), but probably much earlier (Denham et al 2004b: 852). However, the archaeological evidence of plant exploitation practices in early Holocene Phase 1, dated around 9000 BP, is not thought sufficient to qualify and is badly in need of confirmation at other sites (Denham et al 2003: 190, 2004b: 843, 852; chapter 5).

(2) Agriculture may have emerged in highland New Guinea, not the lowlands, as argued by others. The new archaeobotanical evidence from Kuk shows a range of edible plants to have been present in the late Pleistocene, which, together with hunting, could have allowed permanent occupation of the island interior (Denham et al 2003: Table S3, 2004b: 850, 852, Table 2). Increasingly interventionist strategies of plant exploitation resulted from the climatic fluctuations of the late Pleistocene and the environmental effects of those strategies in the early to mid-Holocene (Denham et al 2004b: 850, 852). The identification of starch grains of *Colocasia esculenta* and *Dioscorea* sp. on tools of Kuk Phase 1 and of Eumusa phytoliths in the base of the grey (-brown) clay sealing the Phase 1 palaeochannel not only attests their appearance in the early Holocene, as a result, it is suggested, of natural dispersal processes from lower elevations, but also their incorporation in existing plant exploitation practices at Kuk (Denham chapter 5; Denham et al 2004b: 849–52, but note the absence there of reference to *Dioscorea* sp., which had not at the time been reported). Agriculture developed out of such practices in the context of 'an increasing focus' on major sources of starch such as these plants represented (Denham et al 2004b: 852–3).

The above is a new version, in the light of the results from the recent work at Kuk, of an argument for agricultural origins in highland New Guinea put forward by Simon Haberle (1993). This position was discussed by Hope and Golson (1995: 827), with Golson adopting the contrary view. One of the relevant points made by Hope was that the palynological evidence for the prevalence of upper montane forest and hence cool and misty conditions in the highland valleys of the late Pleistocene meant that none of the possible crop plants of a potential cultivation regime would have flourished, except for *Pandanus*. Nothing is radically changed by the list of 21 plants identified in late Pleistocene levels at Kuk, for there are no positive major starch foods among them to sustain year-round settlement from local resources (Denham et al 2003: Table S3, 2004b: Table 2). When those major starch foods make an appearance, in the form of *Colocasia* taro, *Dioscorea* yam and Eumusa banana, that of the first two and potentially that of the third as well are, as Denham (chapter 5) says, 'roughly synchronous with the earliest evidence of plant exploitation at Kuk, ie Phase 1'.

That early evidence of plant exploitation, comprising the presumed marks of cultivation belonging to Phase 1 and the grey (-brown) clay layer that seals them in, is itself, I have argued above, synchronous with the attainment of levels of temperature that would make the plants in question viable at Kuk. Moreover, the palaeoenvironmental analyses carried out in connection with the Stage 2 work at Kuk have confirmed the grey (-brown) clay as representing 'accelerated deposition along the wetland margin following forest clearance and subsequent erosion in the catchment' (Denham chapter 5), an early conclusion from Stage 1 of the Kuk work (eg Golson and Hughes 1980: 296–97) already supported by Wilson's (1985: 97) phytolith analysis. Such forest clearance was part of Phase 1 activity at Kuk and the associated palaeochannel was initially able to dispose of the erosional products from the catchment for which it was responsible. This seems to indicate a complex of attributes – plants, forest clearance and cultivation plot – appearing at Kuk together when climatic conditions became suitable and arriving there from lower elevations. Food plants spread with the system of shifting agriculture in which they were embedded.

I conclude this section with a quotation from Yen (1995: 842–43):

> The highland discoveries have always implied a derivative origin, largely through the genetic factor that the field crops like taro, and at lower altitudes yams, are lowland plants. Thus it follows that their adaptation to drier and seasonally cooler higher-altitude climates reflects an earlier domestication in the more tropical environments.

Developments

In his paper in this volume, Denham proceeds from a review of the results of the multidisciplinary work on the early stages of the Kuk sequence to look at different concepts of agriculture in their light. Considerations of cultivated plants and cultivation practices as they are treated in his discussion are not directly addressed here, but have been covered from a different standpoint in the preceding section. In what follows I take his section on 'Social Dependence and Landscape Change: Palaeoecological Evidence', where he considers Matthew Spriggs' definition of agriculture 'in terms of the creation of agro-ecosystems that limit subsistence choice because of environmental transformation or labour demands' (Spriggs 1996b: 525) and looks at the palaeoecological evidence as 'a guide to the nature and extent of different subsistence practices in the past and the likely levels of human dependence on them' (Denham chapter 5). These are issues that were raised by work on the later phases of the Kuk sequence in the 1970s and formed an important part of the discussions that resulted from it.

The new pollen records from the Stage 2 investigations at Kuk have filled the early Holocene gap in the vegetation history of the upper Wahgi Valley and link up with Jocelyn Powell's (1982) pollen diagrams from the 1960s and 1970s relating to the mid (c 5000 BP) and later Holocene to illustrate the effects of long-term disturbance of the forest under shifting cultivation. By the mid-Holocene a diversified early Holocene landscape of primary and secondary forest, regrowth and grassland had been replaced by one of regrowth and grassland (Denham et al 2004b: 850–1). This transformation and its effects were precisely what had provided the explanatory framework for Golson's discussions of later agricultural history in the upper Wahgi (Golson 1977: 604–09, 1982a: 300–05, 1982b: 123–30; Golson and Gardner 1990: 398–403; Hope and Golson 1995: 827–28).

Shifting cultivation depends on forest regeneration to renew the nutrient store and rehabilitate soil structure on land going into fallow after limited cropping. Hence interruptions or impediments to the cycle of regeneration could lead to the deflection of the vegetation succession to degraded regrowth and grassland. In these circumstances new strategies were needed in the sphere of agricultural practice. Parallel with this, there were ultimately deleterious effects on the plant and animal resources of the bush, the loss of which made increased demands on agricultural production (Golson 1982a: 127–28). This was an argument of which much was made in connection with the pig, whose replacement of bush animals as a source of protein was seen as constituting such a charge on garden produce that its possession became a symbol of wealth and authority. However, instead of this taking place in the mid-Holocene, as proposed by Golson writing when pigs were acceptable by at least 6000 BP, it has to be delayed until an unknown date after Lapita appeared with pigs in the Bismarck Archipelago after 3500 BP (Hope and Golson 1995: 828 n).

There are some indications that the forests of highland New Guinea may be slow to recover after disturbance, as a result, among other things, of lowered temperatures and diminished amounts of photosynthetically active radiation, which decrease stature, biomass and productivity (Grubb 1977: 102–03). There were, moreover, limiting conditions to the practice of highland cultivation that increased the risk of disturbance. Cultivation operated in a favourable but laterally and vertically restricted zone (Brookfield 1964: 22–23), between a ceiling on the productive growth of the pre-Ipomoean staples of 2000 m or so and a virtual lower limit ranging between some 1250 and 1550 m, defined by a variety of negative factors including steep terrain, persistent cloud, excessive rainfall or drought (and today at least, malaria) (Brookfield 1964: 31–34). Given that the old field staples taro and yam are not very tolerant of poorer soils and given the reasons for slower forest regeneration and thus soil rehabilitation, it seems likely that larger areas of land per person would have been required for the system of shifting agriculture to work. Once it had reached the geographical limits specified above, the process would have turned in on itself, leading to ever shorter fallow periods and accelerating environmental change.

In the publications that I have cited here, Golson made attempts to put the various strands of evidence together and offer explanations of their interactions. A particular point that made an appearance in each narrative concerned a widespread change in the depositional regime in the Kuk Swamp, whereby sediments in the form of dark clay gave way to others in the form of soil aggregates, marking the end of Phase 3 about 2500 BP (Golson 1977: 621, 1982a: 304, 1982b: 121–22; Golson and Gardner 1990: 399–400). This change was interpreted as signalling the appearance of soil tillage in the dryland agricultural sector, from where some of its products were washed into the swamp. Soil tillage itself, which has no place in shifting agriculture under forest or woody regrowth, but breaks up and aerates the grassland sod, was thought to register the appearance of permanent grasslands in the local environment at Kuk, as the culmination of the ecological transformation set in train by the operations of shifting cultivation millennia before.

The fuller palaeoenvironmental evidence now available from Kuk is interpreted as indicating that the transformation from forest to grassland had been completed on the upper Wahgi Valley floor by the mid-Holocene, some 6000–5500 BP (Denham et al 2004b: 851), rather than Golson's 2500 BP.

As a result, Denham (chapter 5) suggests that the stratigraphic change in the Kuk Swamp at the latter date is more plausibly seen as in situ pedogenesis than the onset of soil tillage. What this difference of opinion means in the context of the present discussion, however, is that the problem of dealing with the spread of grassland may have confronted the inhabitants of the upper Wahgi much earlier than hitherto suggested.

CONCLUDING REMARKS

In this essay I have been concerned with the complicated story of the interaction between different lines of investigation relevant to the origins and early history of plant cultivation in New Guinea, as each of them was pursued and influenced the others in the latter part of the last century. The story has run the gamut from one of the wholesale introduction of an agricultural complex (including domesticated pig) as proposed at an early stage of investigation to the current view of independent agricultural origins (without the pig). A claim of independent origins raises questions as to the definition of agriculture employed (as in Denham chapter 5), and thus of its recognition in the available evidence in order to decide on when, where and why. These are questions beyond answer at present, given that we have no record for lowland New Guinea to set against that of the highland interior, where the major cultivated plants were overwhelmingly of lowland origin.

From the recent work directed by Denham at Kuk, three of those plants – *Colocasia* taro, *Dioscorea* yam and Eumusa banana – made an appearance in the upper Wahgi Valley of montane Papua New Guinea in the very early Holocene, following the late glacial climatic amelioration and the attainment of modern temperature levels. For Denham and his colleagues this was the result of the natural spread of the plants from lower elevations and their incorporation into existing highland subsistence systems that had hitherto lacked starch-rich sources of food with the capacity to perform as staples on a year-round basis. The development of that capacity through management procedures appropriate to elevations close to the margin of production provided highland societies with an agricultural base. As Golson and Hughes read the evidence, the three plants made their appearance not only at a critical juncture for their viability in montane conditions, but in association with structures compatible with their cultivation on the swamp margin, artefacts for their processing and a sealing layer of inwashed clay indicative of forest clearance for cultivation on dry land in the swamp catchment. For them this is a new combination of features in the subsistence domain and marks the introduction from lower elevations of the structured system of cultivation known as shifting agriculture.

Commenting on these different points of view, Denham and colleagues (2004a: 294) say that '[d]ifferent readings of the archaeological evidence, as well as of wider spatial and chronological contexts, lead to discordant interpretations of the locus and nature of the earliest agricultural practices'.

The resolution of the disagreements about the archaeology of Kuk Phase 1 is highly important for the resolution of the larger question, as is further investigation of Phase 1 at Kuk and confirmation of similar practices elsewhere.

In either case, however, whether the early Holocene adoption of starch-rich plants into existing highland subsistence systems or their introduction from lower elevations as part of a new system, the result was the same, an immediate commitment to their cultivation that qualifies as agricultural. This started from the fact that their possession seems essential to permanent occupation of the Highlands. It continued because of the procedures of shifting cultivation by which the plants were managed and the particular circumstances of highland geography in which their cultivation took place, which began an irreversible ecological transformation that locked highland societies into the agricultural condition.

ACKNOWLEDGMENTS

I thank Mike Bourke, Tim Denham and Philip Hughes, who read the first draft of this paper and contributed to improvements of style and substance in the published version. My debt to Doug Yen is apparent throughout. Full acknowledgment will be made in an appropriate place to colleagues in the Kuk enterprise, to agencies who provided specialist services, to funding authorities who supported it, to the Papua New Guinea authorities who authorised it and to members of the Kuk community who were partners in carrying it out.

REFERENCES

Allen, J (1991) 'Introduction', in Allen, J and Gosden, C (eds), *Report of the Lapita Homeland Project*, pp 1–8. Occasional Papers in Prehistory 20, Canberra: Department of Prehistory, Research School of Pacific Studies, Australian National University

Barrau, J (1963) 'Introduction', in Barrau, J (ed), *Plants and the Migrations of Pacific Peoples: A Symposium*, pp 1–6, Honolulu: Bishop Museum Press

Bourke, M (nd) 'Altitudinal limits of 220 economic crop species in Papua New Guinea', unpublished manuscript on file in the Department of Archaeology and Natural History, Research School of Pacific and Asian Studies, Australian National University, Canberra

Brookfield, H C (1964) 'The ecology of highland settlement: some suggestions', in Watson, J B (ed), *New Guinea: The Central Highlands*, pp 20–38. Special Publication of *American Anthropologist* vol 66, no 4, part 2, Washington, DC: American Anthropological Association

Bulmer, S (1966) 'The prehistory of the Australian New Guinea Highlands: a discussion of archaeological field survey and excavations 1959–1960', unpublished MA thesis, University of Auckland

Bulmer, S (1975) 'Settlement and economy in prehistoric Papua New Guinea: a review of the archaeological evidence', *Journal de la Société des Océanistes* 31: 7–75

Bulmer, S (1982) 'Human ecology and cultural variation in prehistoric New Guinea', in Gressitt, J L (ed), *Biogeography and Ecology of New Guinea*, vol 1, pp 169–206. Monographiae Biologicae 42, The Hague: Junk

Bulmer, S and Bulmer, R (1964) 'The prehistory of the Australian New Guinea Highlands', in Watson, J B (ed), *New Guinea: The Central Highlands*, pp 39–76. Special Publication of *American Anthropologist* vol 66, no 4, part 2, Washington, DC: American Anthropological Association

Clark, G (1972) 'Foreword', in Higgs, E S (ed), *Papers in Economic Prehistory: Studies by Members and Associates of the British Academy Major Research Project in the Early History of Agriculture*, pp vii–x, Cambridge: Cambridge University Press

Denham, T P (2004) 'The roots of agriculture and arboriculture in New Guinea: looking beyond Austronesian expansion, Neolithic packages and indigenous origins', *World Archaeology* 36: 610–20

Denham, T, Haberle, S, Lentfer, C, Fullagar, R, Field, J, Therin, M, Porch, N and Winsborough, B (2003) 'Origins of agriculture at Kuk Swamp in the Highlands of New Guinea', *Science* 301: 189–93

Denham, T P, Golson, J and Hughes, P J (2004a) 'Reading early agriculture at Kuk (Phases 1–3), Wahgi Valley, Papua New Guinea: the wetland archaeological features', *Proceedings of the Prehistoric Society* 70: 259–98

Denham, T, Haberle, S and Lentfer, C (2004b) 'New evidence and revised interpretations of early agriculture in highland New Guinea', *Antiquity* 78: 839–57

Flenley, J R (1979) *The Equatorial Rain Forest: A Geological History*, London: Butterworths

Golson, J (1977) 'No room at the top: agricultural intensification in the New Guinea Highlands', in Allen, J, Golson, J and Jones, R (eds), *Sunda and Sahul: Prehistoric Studies in Southeast Asia, Melanesia and Australia*, pp 601–38, London: Academic Press

Golson, J (1982a) 'Kuk and the history of agriculture in the New Guinea Highlands', in May, R and Nelson, H (eds), *Melanesia: Beyond Diversity*, pp 297–307, Canberra: Research School of Pacific Studies, Australian National University

Golson, J (1982b) 'The Ipomoean revolution revisited: society and the sweet potato in the upper Wahgi Valley', in Strathern, A (ed), *Inequality in New Guinea Highlands Societies*, pp 109–36, Cambridge: Cambridge University Press

Golson, J (1985) 'Agricultural origins in Southeast Asia: a view from the east', in Misra, V N and Bellwood, P (eds), *Recent Advances in Indo-Pacific Prehistory*, pp 307–14, New Delhi: Oxford and IBH Publishing Co

Golson, J (1989) 'The origins and development of New Guinea agriculture', in Harris, D R and Hillman, G (eds), *Foraging and Farming: The Evolution of Plant Exploitation*, pp 678–87, London: Unwin Hyman

Golson, J (1991a) 'The New Guinea Highlands on the eve of agriculture', *Bulletin of the Indo-Pacific Prehistory Association* 11: 82–91

Golson, J (1991b) 'Bulmer Phase II: early agriculture in the New Guinea Highlands', in Pawley, A (ed), *Man and a Half: Essays in Pacific Anthropology and Ethnobiology in Honour of Ralph Bulmer*, pp 484–91, Auckland: Polynesian Society

Golson, J (1997) 'From horticulture to agriculture in the New Guinea Highlands: a case study of people and their environments', in Kirch, P V and Hunt, T L (eds), *Historical Ecology in the Pacific Islands: Prehistoric Environmental and Landscape Change*, pp 39–50, New Haven, CT: Yale University Press

Golson, J and Gardner, D S (1990) 'Agriculture and sociopolitical organization in New Guinea Highlands prehistory', *Annual Review of Anthropology* 19: 395–417

Golson, J and Hughes, P J (1980) 'The appearance of plant and animal domestication in New uinea', *Journal de la Société des Océanistes* 36: 294–303

Gosden, C, Allen, J, Ambrose, W, Anson, D, Golson, J, Green, R, Kirch, P V, Lilley, I, Specht, J and Spriggs, M (1989) 'Lapita sites of the Bismarck Archipelago', *Antiquity* 63: 561–86

Grubb, P J (1977) 'Control of forest growth and distribution on wet tropical mountains', *Annual Review of Ecology and Systematics* 8: 83–107

Haberle, S G (1993) 'Late Quaternary environmental history of the Tari Basin, Papua New Guinea', unpublished PhD thesis, Australian National University

Haberle, S G (1995) 'Identification of cultivated *Pandanus* and *Colocasia* in pollen records and the implications for the study of early agriculture in New Guinea', *Vegetation History and Archaeobotany* 4: 195–210

Haberle, S G, Hope, G S and DeFretes, Y (1991) 'Environmental change in the Baliem Valley, montane Irian Jaya, Indonesia', *Journal of Biogeography* 18: 25–40

Hedges, R E M, Housley, R A, Bronk Ramsey, C and van Klinken, G J (1995) 'Radiocarbon dates from the Oxford AMS system: *Archaeometry* datelist 20', *Archaeometry* 37: 417–30

Higgs, E S and Jarman, M R (1972) 'The origins of animal and plant husbandry', in Higgs, E S (ed), *Papers in Economic Prehistory: Studies by Members and Associates of the British Academy Major Research Project in the Early History of Agriculture*, pp 3–13, Cambridge: Cambridge University Press

Hope, G and Golson, J (1995) 'Late Quaternary change in the mountains of New Guinea', *Antiquity* 69(265): 818–30

Hughes, P J, Sullivan, M E and Yok, D (1991) 'Human-induced erosion in a highlands catchment in Papua New Guinea: the prehistoric and contemporary records', *Zeitschrift für Geomorphologie* 83: 227–39

Loy, T H, Spriggs, M and Wickler, S (1992) 'Direct evidence for human use of plants 28,000 years ago: starch residues on stone artefacts from the northern Solomon Islands', *Antiquity* 66: 898–912

Mountain, M-J (1991) 'Highland New Guinea hunter-gatherers from the Pleistocene: Nombe Rockshelter, Simbu', unpublished PhD thesis, Australian National University

Pawley, A and Green, R C (1975) 'Dating the dispersal of the Oceanic languages', *Oceanic Linguistics* 12: 1–67

Powell, J M (1982) 'The history of plant use and man's impact on the vegetation', in Gressitt, J L (ed), *Biogeography and Ecology of New Guinea*, vol 1, pp 207–27. Monographiae Biologicae 42, The Hague: Junk

Shutler Jr, R and Marck, J C (1975) 'On the dispersal of the Austronesian horticulturalists', *Archaeology and Physical Anthropology in Oceania* 10: 81–113

Simmonds, N W (1962) *The Evolution of Bananas*, London: Longman

Spriggs, M (1996a) 'What is Southeast Asian about Lapita?', in Akazawa, T and Szathmáry, E J E (eds), *Prehistoric Mongoloid Dispersals*, pp 324–48, Oxford: Oxford University Press

Spriggs, M (1996b) 'Early agriculture and what went before in Island Melanesia: continuity or intrusion?', in Harris, D R (ed), *The Origins and Spread of Agriculture and Pastoralism in Eurasia*, pp 524–37, London: University College London Press

Watson, J B (1964) 'A previously unreported root crop from the New Guinea Highlands', *Ethnology* 3: 1–5

White, J P (1972) *Ol Tumbuna: Archaeological Excavations in the Eastern Central Highlands, Papua New Guinea*. Terra Australis 2, Canberra: Department of Prehistory, Research School of Pacific Studies, Australian National University

Whitmore, T C (ed) (1981) *Wallace's Line and Plate Tectonics*, Oxford: Clarendon Press

Wilson, S M (1985) 'Phytolith analysis at Kuk, an early agricultural site in Papua New Guinea', *Archaeology in Oceania* 20: 90–97

Yen, D E (1961) 'The adaptation of the sweet potato by the New Zealand Maori', *Journal of the Polynesian Society* 70: 338–48

Yen, D E (1971) 'The development of agriculture in Oceania', in Green, R C and Kelly, M (eds), *Studies in Oceanic Culture History*, vol. 2, pp 1–12. Pacific Anthropological Records 12, Honolulu: Department of Anthropology, Bishop Museum

Yen, D E (1973) 'The origins of Oceanic agriculture', *Archaeology and Physical Anthropology in Oceania* 8: 68–85

Yen, D E (1980) 'The Southeast Asian foundations of Oceanic agriculture', *Journal de la Société des Océanistes* 36: 140–46

Yen, D E (1982) 'The history of cultivated plants', in May, R and Nelson, H (eds), *Melanesia: Beyond Diversity*, pp 281–95, Canberra: Research School of Pacific Studies, Australian National University

Yen, D E (1985a) 'Wild plants and domestication in Pacific islands', in Misra, V N and Bellwood, P (eds), *Recent Advances in Indo-Pacific Prehistory*, pp 315–26, New Delhi: Oxford and IBH Publishing Co

Yen, D E (1985b) 'The genetic effects of agricultural intensification', in Farrington, I S (ed), *Prehistoric Intensive Agriculture in the Tropics*, pp 491–99. International Series 232, part II, Oxford: British Archaeological Reports

Yen, D E (1989) 'The domestication of environment', in Harris D R and Hillman, G C (eds), *Foraging and Farming: The Evolution of Plant Exploitation*, pp 55–75, London: Unwin Hyman

Yen, D E (1990) 'Environment, agriculture and the colonisation of the Pacific', in Yen D E and Mummery, J M J (eds), *Pacific Production Systems: Approaches to Economic Prehistory*, pp 258–77. Occasional Papers in Prehistory 18, Canberra: Department of Prehistory, Research School of Pacific Studies, Australian National University

Yen, D E (1991) 'Domestication: the lessons from New Guinea', in Pawley, A (ed), *Man and a Half: Essays in Pacific Anthropology and Ethnobiology in Honour of Ralph Bulmer*, pp 558–69, Auckland: The Polynesian Society

Yen, D E (1995) 'The development of Sahul agriculture with Australia as bystander', *Antiquity* 69(265): 831–47

CHAPTER 7

The Meaning of Ditches: Interpreting the Archaeological Record from New Guinea Using Insights from Ethnography

Tim Bayliss-Smith

INTRODUCTION

On 1 June 1860 Charles Darwin wrote a letter to the geologist Charles Lyell in which he remarked that explaining the origin of species was different from explanation in the physical sciences. When tracing remote origins, the researcher could not meet the same standard of proof as in the physical sciences. 'On [that] standard of proof, *natural* science would never progress; for without the making of theories, I am convinced there would be no observation' (Burchardt et al 1993: 233). As well as the 'making of theories' that Darwin himself had just achieved in *The Origin of Species*, he was perhaps thinking also of Lyell's book *Principles of Geology* (1830–33) which had been his inspiration throughout the voyage of the *Beagle*. Lyell had provided an integrated theory of landforms that framed all of Darwin's observations of mountain building, fluvial erosion and atoll formation. But Darwin's remark also foreshadows some twentieth-century debates in the philosophy of science, about the problems of distinguishing between 'facts' and 'theories' if each is dependent upon the other. It also reminds us that *evolutionary* theories about agriculture – theories that were originally inspired by the Darwinian revolution in biology – continue to influence archaeological enquiry and may constrain our 'observations' by providing an inappropriate set of subsistence categories.

Evolutionary Theories as Procrustean Beds

Melanesia provides a good example of a region in which a discourse of prehistory based on assumptions that derive from other regions ('theory') may not generate useful insights from fieldwork ('observation'). Jim Specht (2003), for example, has argued that the categories in which subsistence practices in New Guinea are discussed reflect strongly certain evolutionary assumptions. These subsistence categories, he suggests, fit so poorly New Guinea practices of hunting, gathering, fishing, forest management and plant cultivation as to constitute a kind of 'Procrustean bed', either stretching or mutilating the meaning of what really exists.

Procrustes – 'he who stretches' – was the innkeeper in ancient Greek legend who only kept one bed to fit all his guests. If the guest was too short for the bed Procrustes stretched him on a rack; if he was too long Procrustes cut his feet off. If our archaeological categories are like Procrustean beds, then as Specht (2003: 209) suggests, the analysis becomes 'problematic and futile' (Figure 7.1).

It is indeed difficult to position many communities in Melanesia along the conventional continuum that places hunter-gatherers at one end and intensive agriculturalists and full domestication at the other (Gosden 1995; Latinis 2000). The subsistence strategies in this region usually combine elements of both hunter-gathering and agriculture alongside fishing, arboriculture (cultivation of trees), wild plant management and complex patterns of exchange (Harris 1996; Kennedy and Clarke 2004). Many of the plants and animals used for subsistence are not fully domesticated. The domesticated plants include root crops (yams, taro), bananas and trees, and they are

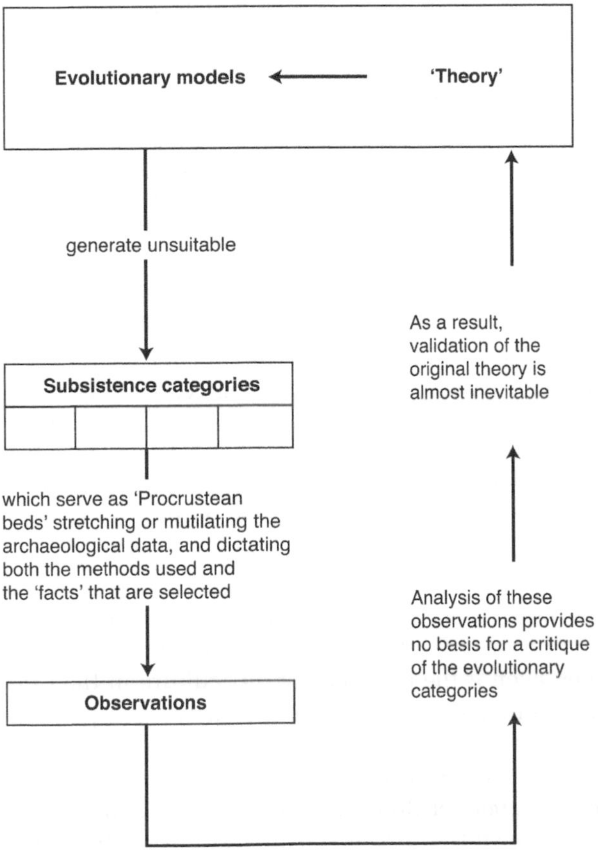

Figure 7.1 Specht's 'futile circle' of explanation in New Guinea, in which models based on outmoded theories of cultural evolution generate unsuitable subsistence categories, so that a critical analysis becomes difficult and validation of the original 'explanation' is almost inevitable.

grown in 'gardens'; their cultivation could be regarded as horticulture rather than agriculture (Leach 1997, 1999). The term 'agroforestry' has sometimes been used to describe these cultivation practices, which are found throughout Oceania (Clarke and Thaman 1993; Terrell 2002). Agroforestry implies an integration of crops with trees, rejecting Old World assumptions that fields and forests are separate domains and that shifting cultivation is something primitive that 'evolved' into proper agriculture.

Agricultural Change at Kuk: A Case Study

The interactions between theory and observation and the possibility of Procrustean categories will be critically examined in this paper. However, the focus will be on assumptions about what has driven agricultural change in the Highlands of Papua New Guinea, rather than assumptions about the actual subsistence categories. The case study for this exercise is the sequence of changes observed at the Kuk site, where separate 'phases' of wetland drainage have been described (eg Golson 1977, 1997). In recent millennia each 'phase' has consisted of a widespread network of small linear features (field ditches) that articulate with larger drains (major disposal channels). These features are preserved in the peat and clay deposits at Kuk Swamp in the upper Wahgi Valley, about 1590 m above sea level.

Jack Golson has divided the ditch systems at Kuk into six phases of drainage spanning the past 9000 years (Golson 1977, 1982). I focus in this paper on the systems that span the last two millennia: Phase 4 from about 2000–1200 years ago, Phase 5 from about the thirteenth century AD until 1665, and Phase 6 from about AD 1700–1900 (Table 7.1). Phases 4, 5 and 6 are all accepted as representing systems of wetland agriculture, but precisely what crops were being grown remains uncertain (Figure 7.2). There is now direct evidence that taro (*Colocasia esculenta*), pandanus nuts (*Pandanus* spp.) and bananas (*Musa* spp.) were in cultivation at Kuk from Phase 2 onwards, starting about 6500 years ago (Denham et al 2003). At present we must infer the presence of other crops, notably yam (*Dioscorea* spp.), sugarcane (*Saccharum officinarum*), winged bean (*Psophocarpus tetragonolobus*) and various green vegetables. The sweet potato (*Ipomoea batatas*) arrived in the Highlands more recently, being adopted at Kuk about 300 years ago with the onset of Phase 6 (Bayliss-Smith et al 2005).

However, even if the existence of agriculture in the upper Wahgi Valley from Phase 4 onwards is undisputed, exactly what phases 4, 5 and 6 represent is not clear. There are certain morphological differences between the ditches in phases 4, 5 and 6. Should these differences be read as evidence for labour *intensification*, or do they reflect technical *innovation*? Were phases of wetland use driven by subsistence crisis, or were they motivated by wealth incentives? Can we make causal connections between phases of ditch digging and periods of environmental degradation or climate change? Are there links to symbolic capital, the emergence of big men, or gender inequality?

Merely describing the sequence of changes at Kuk has proved to be a substantial task requiring a range of specialist techniques. Interpreting the

Table 7.1 Summary of phases 4, 5 and 6 at Kuk (after Bayliss-Smith et al 2005)

Phase	*Chronology*	*Main Crops*	*Major Disposal Channel*	*Minor Field Drains*	*Fields*
	date of start, date of finish		*width at grey clay level x depth below grey clay*	*mean depth below black clay x mean width at black clay level (block locations)*	*mean length x mean breadth (block locations)*
4	c 2000 –1200 BP	taro?	Neringa's Baret: ~ 2.4 m wide × > 0.8 m deep	41 × 46 cm (in A9, A10 and A11); 34 × 54 cm (in C9)	13 × 9 m (in A9, A10 and A11)
5	After twelfth century AD – 1665/1666, with three subphases (early, middle and late)	taro, yam and banana (?)	Wai's Baret: early subphase: minimum 2.8 m wide × 1.6 m deep	57 × 96 cm (in A9); 49 × 87 cm (in C9)	14 × 7 m (in A10 and A11)
			late subphase: 3.3 m wide × 1.1 m deep		19 × 9 m (in A10 and A11)
6	c AD 1700–1900	sweet potato (?)	Wai's Baret: minimum 2.0 m wide × 0.95 m deep	42 × 109 cm (in A9)	20 × 10 m (in A10 and A11); 11 × 7 m (in A9)

changes is even more daunting, since the evidence available relates only to the wetland sphere of agriculture and consists of little more than ditches, some crop residues and occasional tools. It thus provides a very insecure basis for reconstructing the societies that produced these artefacts. In this situation there is indeed a danger that a 'Spechtian futile circle' will be constructed, with preconceived evolutionary ideas dictating the conceptual categories which then become 'Procrustean beds' in subsequent analysis. Observations appear to match the theories because the archaeological evidence is too sparse to offer any real challenge to the circular argument.

AVOIDING PROCRUSTES: (1) USE OF ETHNOGRAPHIC ANALOGIES

Six Ethnographic Models

Ethnography is a form of in-depth qualitative research that seeks to understand the beliefs, practices, and artefacts of individuals and groups in the context of their normal everyday life and surroundings. Using such insights about beliefs, practices and artefacts today, we can seek to escape from the Procrustean impasse outlined above through a cautious but explicit use of

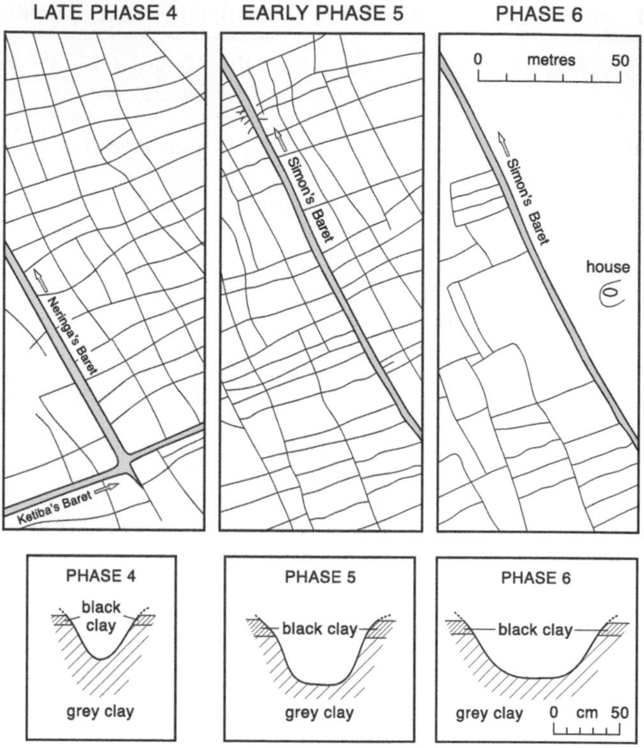

Figure 7.2 Reconstructed networks of field ditches and major disposal channels within the same area of blocks A10/A11 at Kuk (Phase 4 after Bayliss-Smith and Golson 1999; Phases 5 and 6 after Bayliss-Smith et al 2005), and the average size and shape of field ditches in the three phases, based on measurements of numerous cross-sections in block C9 (Phases 4 and 5) and block A9 (Phase 6) (Bayliss-Smith nd). A complete change in the layout of the ditch network between Phases 4 and 5 is very clear, whereas Phase 6 appears to be based on the same basic network that had been established in early Phase 5. The change in depth and shape of the field ditches may indicate a change in crop, probably from taro in Phase 4 to yams/bananas/sugarcane in Phase 5, and possibly to sweet potato in Phase 6 (Bayliss-Smith et al 2005).

ethnographic analogy (Figure 7.3). Hidden analogies permeate all archaeological interpretations, of course, enabling one feature to be called a drainage ditch and another feature an agricultural field because of their resemblance to ditches and fields in the present-day landscape. What is needed for a better interpretation of agricultural change is for ethnographic analogies to be found that will help us to identify *social process* as well as *material artefact*. Appropriate analogies provide the only secure route towards constructing a range of alternative models that might explain the meaning of evidence.

If ditches, a few crop residues and certain tools are all that survive from a site such as Kuk, then to explain their shape and form, their appearance in the record (the onset of a 'phase') and then their sudden disappearance (the end of a 'phase'), we will need to widen the range of observations to

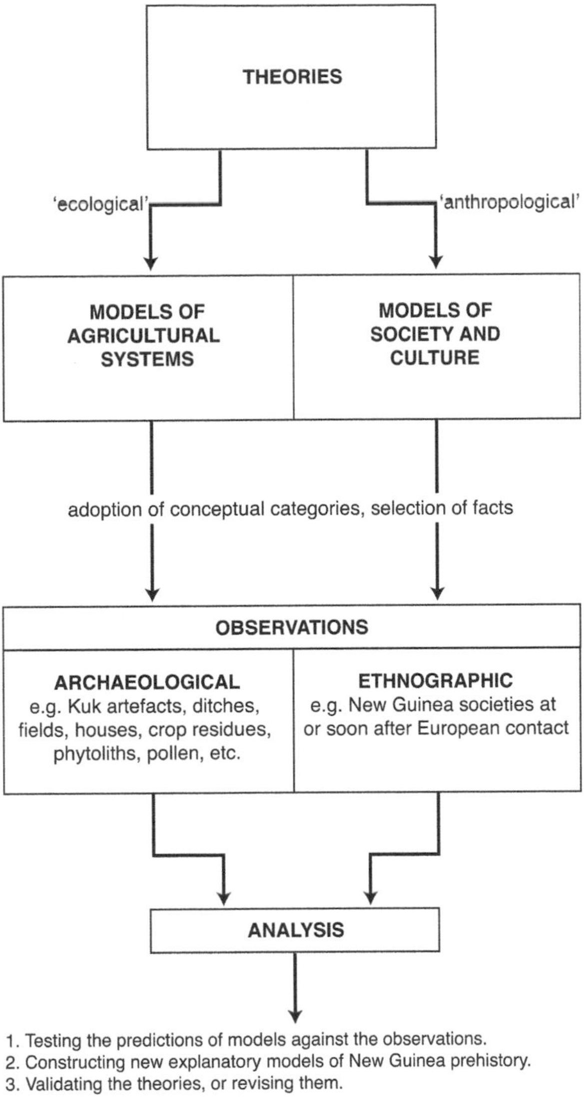

Figure 7.3 A more virtuous circle of explanation, in which both ecological and anthropological theories structure the fieldwork observations. A more rigorous critique of theory becomes possible through an integrated analysis of both archaeological and ethnographic data.

include neighbouring areas in which wetland agriculture is still practised. Through ethnographic research in these neighbouring societies, can we uncover the rationale behind the decisions that result in ditches being dug and wetlands reclaimed? What social processes underlie these individual or collective efforts, and if patterns do emerge from ethnography, how valid is it to transfer these to other times and other places? We need to be explicit about where our explanatory models come from, and we need to combine ecological insights about how wetland systems function with sociological insights about the circumstances in which investments (or perhaps experiments) in wetland cultivation take place. Above all, we need to be very explicit about how our explanatory models have been constructed and how we can improve them.

If we review the full range of studies available in the field of agrarian ethnography, we find that the various explanations for agricultural change fall into six groups. Generalising from these explanations, we can construct six alternative theories that might account for the digging of ditches and the cultivation of wetlands (Table 7.2):

- Ditches as a reflection of economic rationality;
- Ditches as the outcome of social inequality;
- Ditches as a response to population crisis;
- Ditches as an investment in future security;
- Ditches as a symbol of property in the landscape;
- Ditches as an adaptation to climate change.

These six explanations do not constitute a complete list of all possibilities, and they may not be mutually exclusive. We can regard them as working hypotheses – in other words models in need of testing against 'data' (both archaeological evidence and ethnographic analogue) from an appropriate place or context, before we accept that one or more might contribute to an explanation of wetland drainage at a given site.

Ditches as a Reflection of Economic Rationality

The first category of explanatory models emphasises the economic rationality of cultivators faced with soils and sites of varying potential. In Europe 200 years ago, at a time of widespread 'agricultural improvement', this assumption seemed to be mere common sense to political economists such as Malthus, Ricardo and von Thünen. In their models, both the type and the intensity of land use reflected the economic rationality of farmers, who reacted to spatial variations in the production functions for different crops grown on different sites. Land use 'intensity' is the ratio of inputs to land area, with labour the primary input in any preindustrial context. 'Production function' is the term used to describe the relationship between input and output (Found 1971: 12–13).

An ethnographic example of relevance to New Guinea is the contrast between the production functions of two alternative land uses in rural Indonesia, described by Clifford Geertz (1963) as swidden (forest fallow shifting cultivation) and *sawah* (wet rice cultivation). *Sawah*, like many other forms of wetland cultivation, generates a lower level of average product per capita, but has a much greater capacity for increase in total production, absorbing vast amounts of labour and showing much less tendency than swidden for ecological collapse. If swidden does collapse, then the

Table 7.2 The rationale for digging ditches: six working hypotheses

Model	*Origins*	*The Local Scenario*	*The Ultimate Explanation*
Ditches as a reflection of economic rationality	Ricardo, Von Thünen and other classical economists modelling the decision-making of farmers.	Digging ditches represents a rational land-use decision in situations where the production functions of dryland alternatives favour wetland land use, perhaps because drainage has become less burdensome, because surplus production is in greater demand, or because of dryland soil degradation.	Farmers are economically rational and will attempt to optimise their land use. Their decisions are based on what they perceive to be the economic rent of different options for both type and intensity of land use on different soils and sites.
Ditches as the outcome of social inequality	Karl Marx, via historians such as Karl Wittfogel and archaeologists such as Gordon Childe.	Coercive relations between leaders and followers make possible the digging of ditches. Such relations derive from the growing social inequalities that emerge in hydraulic societies.	The physical infrastructure (including ditches) is determined by the ideological superstructure. This infrastructure depends on surplus value being extracted from the labour of the disempowered, such as low-class men, slaves and women.
Ditches as a response to population crisis	Ester Boserup and many processual archaeologists.	Wetland reclamation (requiring ditches) is a reflection of diminishing returns to labour in the dryland sphere of agriculture following population growth. Under this pressure fallow periods become shorter and soils deteriorate, making necessary the shift to wetland land use.	Intensification (eg ditch-digging) is an unwilling response to population pressure, as producers will always minimise their expenditure of effort in the satisfaction of their subsistence needs. Dryland soils offer higher returns to labour and so are used up first.
Ditches as an investment in future security	Geographers such as Harold Brookfield and Piers Blaikie studying decision-making in agrarian societies.	Digging of ditches reflects a decision to invest labour in the present so as to achieve improved productivity in the future, as insurance against the possibility of crisis or shortage.	'Landesque capital' formation (eg land drainage) is designed to increase production in the future, either for subsistence or for exchange purposes, as part of a social strategy of risk avoidance.

(*Continued*)

Table 7.2 *(Continued)*

Model	*Origins*	*The Local Scenario*	*The Ultimate Explanation*
Ditches as a symbol of property	Anthropologists such as Pierre Bourdieu; political scientists such as Elinor Ostrom.	Ditches are a form of symbolic capital and mark the boundaries of private property. Their absence implies that wetlands are a communal or open-access resource.	Those who undertake land improvements safeguard their investment by establishing new and more exclusive property rights and by building a fund of symbolic capital that they can exchange for labour.
Ditches as an adaptation to climate change	Catastrophists, chaos theorists and some palaeoclimatologists	Wetland land use (hence ditch digging) represents land-use expansion during periods of drought, eg El Niño. Wetland abandonment occurs following wars, epidemics or natural hazards.	In preindustrial societies neither the technology nor the social institutions are sufficient to shield farmers from extreme events. Land use adjusts through an opportunistic strategy of expansion or contraction.

adoption of *sawah* will involve an intensification of land use, as wet rice cannot be grown without large and frequent investments of labour.

In a market economy land use decisions are not just based on production functions, they are also strongly influenced by prices. The cost of the inputs and outputs determines the economic rent of each piece of land, and a 'rational' farmer will choose the land-use system that will optimise this variable (Found 1971: 20). Wetland drainage may suddenly become the best option when either the value of outputs improves (eg because of new crops or expanded markets) or when the cost of inputs falls. In the fens of eastern England, lower costs following new technologies for drainage, as well as new institutions for raising capital, transformed the economic rent of marshland in the seventeenth century, leading to widespread drainage. The beneficiaries were mainly outsiders, 'venturers' with enough money, political clout and tech-nical know-how to privatise the fenland commons (Darby 1983). A similar process of wetland reclamation occurred in the upper Wahgi Valley between the 1950s and the 1980s, when mechanised technology, cheap labour and new crops (coffee, tea) made profitable land drainage on a large scale, with Australian planters as the main beneficiaries.

Can a similar model be applied to New Guinea wetlands in the past? Intensification as rational land management takes on new meanings when the context is a premodern and nonmonetised society. It would, however, be wrong to assume that in prehistory the production functions of different types of land were fixed and unchanging. Innovations in crops and techniques, changes in the regional scale of gift exchange, and changes in labour availability can all be modelled as reasons for the production functions of wetland, grassland and forest to change over time. Moreover unsustainable land-use practices can transform the resource base itself, as grasslands with degraded soils replace forests, for example.

We can demonstrate such changes over the past 300–400 years in the Highlands, following the 'Ipomoean revolution' that occurred with the adoption of the sweet potato. Both the ethnographic and the palaeoenvironmental evidence suggest that wetlands were abandoned, dry grassland soils began to be cultivated, and high elevation areas were colonised, following the transformation in root crop yields and labour requirements on different sites (Golson 1982).

Is it possible that earlier abrupt changes in land use, for example drainage phases 4, 5 and 6 at Kuk, can be explained in similar ways? Phase 4, for example, looks like an agricultural system based on wetland cultivation of *Colocasia* taro. It could be that the adoption of this new crop 2000 years ago generated a 'Colocasian revolution' for those populations in highland valleys that had access to drained wetlands (Bayliss-Smith and Golson 1992a, 1992b). In fact more recent evidence has demonstrated that far from being newly introduced, taro was already present at Kuk 10,000 years ago and was being cultivated by 6500 cal BP (Denham et al 2003). The new evidence makes the onset of Phase 4 look more like a rational land use decision following the degradation of dryland soils, as Golson (1977) originally proposed. According to this scenario, the rising 'costs' of dryland cultivation and its diminishing returns made taro cultivation in the wetlands a more attractive option, thus causing the shift in land use that we now identify as Phase 4.

In summary, the 'rational land use' model would see phases of prehistoric drainage in New Guinea as intrinsically no different from the drainage of the English fenland from Roman times onwards. In both cases phases of drainage mark periods when external factors (land degradation, trade, new technology, labour availability) encourage the reclamation of a type of land once thought marginal, but where intensive cultivation now becomes rational.

Ditches as the Outcome of Social Inequality

As a reaction against the rather narrow assumptions of classical economics, other writers emphasise the wider political economy of intensive land use, and see social inequality as both the cause and the consequence of land reclamation. In Marx's labour theory of value, wealth is extracted by capitalist entrepreneurs from the labour of the working class, whose wages are an unequal reward for their inputs of skill and effort. It is this transfer of surplus value that enables land managers to profit from exploiting potentially valuable resources such as wetlands. The same idea can be applied to societies where more overtly coercive relations between land managers and the work force exist, as in the hydraulic civilisations of the past where irrigation, enslavement and despotism were, according to Karl Wittfogel (1957), mutually reinforcing processes. Gender relations of a semi-coercive form, in other words women's labour under patriarchal control, is a further example of social inequality permitting, and in turn reinforcing, a process of agricultural intensification.

In a New Guinea context the social inequality model would see big man ambitions as the key to intensive land management. Big man ambitions can only be satisfied through the support of a faction of male followers, who in turn are dependent on the productive efforts of a female labour force, mainly wives and daughters. In more extreme cases, as in the Eastern Highlands, these gender relations were indeed coercive (Donaldson 1982). The extraction of surplus value from subordinate men, achieved through their labour and that of the women they controlled, raises crop production to a level beyond that required for local needs, and this process enables big men to enlarge their factions (Sahlins 1963; Terrell 1986). In Melanesia the motive for mobilising surplus production was redistribution on occasions of ritual exchange, rather than the accumulation of surplus for personal consumption or reinvestment, as occurs in market economies. However, the end result, status enhancement, was the same.

Before the sweet potato, it has been argued, surplus production of crops such as taro was most readily achieved through land reclamation in places like Kuk Swamp (Golson and Gardner 1990; Modjeska 1982). These mid-elevation wetlands thus became 'hotspots' in a process of growing social differentiation (Bayliss-Smith and Golson 1992a). Those farmers able to exploit the new social relations of production recognised that drainage generated surplus, surplus created wealth, and wealth enhanced status.

Ditches as a Response to Population Crisis

A third category of models derives from Ester Boserup's (1965) revival of the notion that, in agriculture, necessity is the mother of invention, or rather that necessity is the spur for innovations that allow more intensive land use. Contrary to the assumptions of both classical and Marxist economics, Boserup did not believe that producers were motivated primarily by the urge towards optimisation and surplus, at least not in agrarian societies with a primarily subsistence orientation. The economist A V Chayanov made similar claims for the peasants in precommunist Russia (Sahlins 1974). In such societies a self-exploitation of household labour always accompanies any attempt to gain surplus. Farmers will manage their land in a way that satisfies basic needs while minimising their household labour input, so as to avoid self-exploitation and provide time for leisure and social activities. Therefore, once subsistence has been achieved plus the modest surplus that satisfies cultural norms, additional effort in cultivation (intensification) is avoided if possible.

Intensification of labour input only becomes necessary if population growth and consequent land shortage obliges farmers to shorten their fallow periods. In this scenario soils deteriorate, yields will begin to decline, and the farmer will experience diminishing returns unless he or she takes remedial action. If this process is modelled using the conventions of agricultural economics, then the point when the marginal productivity of labour falls to zero is where remedial action becomes urgent (Brookfield 1972).

One possible remedy is migration of the surplus population out of the area, which removes the pressure and restores the balance. Alternatively the crisis can be resolved by the adoption of innovations, such as soil tillage, composting, mounding, terracing, drainage or irrigation. Such innovations generally involve more work, but they also give a boost to subsistence production, so in the short term at least the pressure is relieved. In summary, population growth is seen as the ultimate cause of land reclamation, because the innovations that generate more production from marginal land, for example digging ditches so that wetlands can be reclaimed, are laborious and difficult, and so are avoided until the onset of crisis.

In the New Guinea Highlands during the Holocene, the absence of malaria and the other diseases of the lowlands plus the evidence of soil erosion and vegetation change all point to the possibility of Boserupian scenarios. Whether or not it resulted from population growth, the slashing and burning of forests for swidden cultivation caused soil degradation in the mid-elevation valleys and an expansion in the area of grasslands as early as 7500 cal BP (Haberle 1998). Swidden cultivation would always have been the preferred option, since it is a low labour input system that uses the forest fallows to restore soil nutrients and eliminate pests such as taro beetle (Bayliss-Smith 1985; Bayliss-Smith and Golson 1992b). When forests were gone, the restoration of soil fertility had to be achieved in some other way.

In New Guinea it is tempting to see subsistence crisis as the outcome of deforestation, particularly before the arrival of the sweet potato 300–400 years ago. The new crop allowed new areas to be cultivated in the dryland sphere, particularly grasslands and areas at higher elevations, and it also allowed much shorter fallows. At Kuk the Ipomoean revolution ultimately led to the abandonment of wetland cultivation (Bayliss-Smith et al 2005). Pre-Ipomoean phases of drainage could therefore be viewed as indirect evidence for recurrent subsistence crises in the dryland sphere, with reclamation as the only option in highland valleys where there was 'no room at the top' and hence no scope for out-migration (Golson 1977).

Ditches as an Investment in Future Security

One possible shortcoming of all of the preceding models is their underlying assumption that wetlands are difficult but rewarding places. As a result, therefore, reclamation by drainage will be delayed or avoided, but if circumstances do allow it to happen then the reward will be an enhanced level of production. Questioning these assumptions, Blaikie and Brookfield (1987) point to the high level of initial investment of labour that is needed when any kind of landesque capital is formed (eg terraces, drains, irrigation channels, tree plantations). They define 'landesque capital' as an improvement in the agrarian landscape that lasts for longer than a single cropping cycle, and they suggest it is unlikely that the farmer will see a significant return on such investments in the short or even the medium term. In other words, it seems questionable if there is any economic rationale for making such investments, if 'rational decision' is defined as an action that is expected to achieve greater economic benefits than the costs incurred over a 'reasonable' period of time, perhaps ten or fifteen years.

We therefore need to expand our concept of 'benefit' if we are to make sense of such investments in agricultural improvement. One way forward is to consider not just the average returns available from the new system of land use, but also the reliability of yield. If the perceived problem is fluctuating returns from unimproved land, and hence the risk of bad harvests, or food shortage, or even famine, then investments in an improved system make sense however labour-intensive (hence 'uneconomic') they may be. Susan Bulmer (1999) provides an example from New Zealand, seeing Maori land management practices as a response to the difficulties of adapting tropical crops to more temperate conditions. She suggests that the activities that created ditched swamps, terraced slopes, earthen mounds, mulched basins and composting were principally motivated by the need to reduce risk. Rather than reflecting pressure from population or elite consumption, for the Maori their investments in landesque capital 'produced systems … aimed at insuring minimum crop loss through storms, droughts, frosts and other periodic changes' (Bulmer 1999: 326).

Wetland drainage, therefore, can make available land that is less liable to drought, weeds, pests, or a collapse in fertility. Drainage should be understood as one out of many strategies for risk avoidance in agrarian societies almost wholly dependent on local resources and with limited capacity for food storage. In modern societies farmers may invest some of their income in insurance to cover such risks. In a premodern context landesque capital formation, together with the establishment of exchange relationships with neighbours, are comparable strategies.

Ditches as a Symbol of Property in the Landscape

Natural hazards such as pests, droughts or floods can be mitigated by certain farming strategies, but to avoid social risks (war, theft, desertion of the workforce, loss of land tenure), an investment in symbolic capital may be more effective. Defined by Pierre Bourdieu (1977: 179) as 'the form of prestige and renown attached to a family name', symbolic capital is maintained by gifts, is legitimated by rituals, and is displayed by transformations in the cultural landscape. Among the farmers of Kabylia in Algeria, Bourdieu showed how, by accumulating a 'fund' of honour and prestige, a leader gains access to credit and can mobilise a work group. In this way leaders can convert symbolic capital into more economic forms (ploughed and planted fields, irrigation channels, terraces, ditches), some of which may leave behind archaeological traces.

Liz Watson (1999) analyses the dynamics of this process in southern Ethiopia, where Konso leaders exist in a privileged but precarious position, trying to maintain their control over terraced land and the symbolic capital that derives from their sacred and ritual roles. Their inherited status plus the flow of gifts that they make to subordinate men, made possible by surplus production from their fields, enables these leaders to obtain access to labour. The work of men and women is crucial for building stone terraces, digging irrigation channels and cultivating fields, and this human capital is the kind that is most difficult to accumulate. However, without symbolic capital no Konso leader could assemble the necessary labour force, nor mobilise the feelings of collective pride in their terraced landscape that the Konso people express. At the same time, without intensified labour and landesque capital, the flow of gifts that maintains a leader's prestige would dry up and the system of terraced agriculture would collapse.

Symbolic capital transformed into visible forms, for example terraces and ditches, can also help to legitimate land rights. A successful drainage project is itself a lasting reminder of the efforts and skills of those responsible, as well as a means towards long-term benefits in surplus production or subsistence security. To enjoy such benefits, the special and lasting claims of the leader or the group responsible need to be remembered by others, and here the visibility of the new landmark in the cultural landscape carries powerful messages, signifying new claims to long-term rights. The major disposal channels at Kuk, for example, certainly had vital hydraulic functions, but perhaps they also need to be seen as signifiers of property and prestige.

All land boundaries in Melanesia tend to be marked in meaningful ways. For example, at the clan boundary between Fungai and Bomagai-Angoiang territory in montane New Guinea, Bill Clarke (1971: 10–11) describes an archway of poles bedecked with ferns and other foliage and set in a corridor of painted sticks, *Cordyline fruticosa* and other ritually important plants. Such landmarks do not merely constitute land claims:

> Boundaries and 'marks' also have psychic and social meanings. They enclose a province that is at least partially protected from the inimical world that lies beyond a man's own place. The 'marks' along the trail prevent the entry of disease-causing spirits, who are frightened or repelled by the power resident in the passageway; similarly, human enemies from other clans are likely to fall sick if they pass through the ritual archways (Clarke 1971: 11).

The boundaries of swiddens are also marked with fences that are often more symbolic than functional and are decorated with special plants. In Marovo, western Solomon Islands, for example, *Cordyline terminalis* and *Codiaeum variegatum* are planted along the margins of gardens (Hviding and Bayliss-Smith 2000: 85). These brightly-coloured plants signify ownership and are imbued with magical properties that deter trespass and the theft of crops.

Swiddens represent relatively small and short-lived investments of labour. Larger investments that produce landesque capital have the potential to create new forms of property over a longer period, privatising the commons to some degree. Even in Marovo, where forests are abundant and communal rights to use them are widely shared, any trees that are deliberately planted by individuals become their private property. Agricultural improvements are subject to the same transformation in property rights. In Marovo terraced fields for taro cultivation (called *ruta*) were laboriously constructed with stone walls and were served by a network of irrigation channels. *Ruta* became a form of private property, unlike the land used for swiddens: 'as an old man from the bush people of south Vangunu … expressed it: "taro is big work to keep. Growing taro is like feeding a child. It is hard work to keep *ruta*"' (Hviding and Bayliss-Smith 2000: 120).

In the late nineteenth century changes in the political economy of Marovo and the migration of population to the coast destroyed the rationale for *ruta* cultivation, but even today these sites remain visible, despite a hundred years of forest regrowth. Like skull houses and sacred shrines, they are landmarks of symbolic capital within the Marovo cultural landscape.

Cross-cultural studies of land management show that if farming is to be sustainable and effective, whether by individuals or by groups, then mutually agreed boundaries, both social and spatial, need to be maintained (Ostrom 1990). Perhaps drainage works need to be seen in the same light, as a transformation to the cultural landscape in which the ditches are not just functional in the hydraulic sense but also serve as markers proclaiming the existence of symbolic capital and new property rights in the wetland.

At Kuk, once they had been dug, the major disposal channels would have drained the swamp effectively for several decades. In Phase 5, for example, Wai's Baret was initially dug to a depth of 1.6 m, was at least 2.8 m wide, and it stretched for 2.4 km in a straight line to its outfall (Bayliss-Smith et al 2005). In Phase 4 the digging of Ketiba's Baret was almost as substantial a project, requiring about 1000 man-days of work, the equivalent to about two months labour by a gang of sixteen men (Bayliss-Smith and Golson 1999: 222). The digging of the channels could have been men's work, women's work, or the work of mixed groups, but today it is usually men who are most involved in ditch construction. We can also infer from ethnography (eg Heider 1970 in the Baliem Valley) that such channels, as well as playing a key role in drainage, also helped to form symbolic capital.

In all phases at Kuk there are networks of smaller ditches that articulate with the major disposal channels, and here again their hydraulic function may only be part of their rationale. For example in Phase 4 these small ditches form a tight grid, defining square plots measuring on average 13 m × 9 m. Each plot has the appearance of a garden cultivated by one family or individual. Their size, shape and hydrology are difficult to explain in purely functional terms (Bayliss-Smith and Golson 1999: 220). Through these channels and ditches, the wetland wilderness seems to have been transformed into a privatised domain of symbolic capital accumulation. The rights of those who invested effort in these drainage works were safeguarded by the durability of the new landmarks that had been created.

Ditches as an Adaptation to Climate Change

All the preceding models have been based on a rather rigid dichotomy between 'the wet' and 'the dry', but in many cases the margins of a wetland fluctuate, both seasonally and in response to occasional drought. For example in the upper Wahgi Valley, 'most of the drylands become wetlands during the wet season, whereas swamp lands become wetlands, and even drylands on their margins, during dry spells' (Gorecki 1985: 325). Episodes of ditch-digging may represent an opportunistic response to new dry land becoming available, rather than being an actual reclamation of the wetland.

According to this view, the archaeological evidence of drainage ditches in present-day wetlands may represent a phase of land use expansion during drier periods, rather than innovation (creation of a new agricultural system suited to the wet) or intensification (increase of labour input to cope with problems of waterlogging, but using existing crops and skills). The surviving evidence may in fact be an index of climate change rather than social change. In a New Guinea context, the El Niño/Southern Oscillation (ENSO) phenomenon is the main cause of climatic variability, and it is therefore important to establish whether droughts associated with ENSO were more or less prevalent in the past than they are today.

In the New Guinea Highlands, the mid-1600s saw very intense El Niño activity (Cobb et al 2003), but Phase 5 probably began much earlier, around 750–800 years ago (Golson pers comm). If so, then the interval between phases 4 and 5 may correspond with the Little Climatic Optimum, which was a time of warmer and more stable climate (Haberle and David 2004: 176). In contrast, about AD 1300 a period of abrupt climate and sea level change in the Pacific islands began, with far-reaching effects on settlement and agriculture (Nunn 2000; Nunn and Britton 2001). In the Sigatoka Valley in Fiji, for example, episodic droughts and floods associated with ENSO events may have intensified the competition for resources and led to increased warfare in the period AD 1500–1700 (Field 2004). If the New Guinea Highlands also began to experience an increased incidence of ENSO droughts about 700 years ago, then the greater food security offered by wetland sites may have been an additional reason for undertaking large-scale drainage.

Phase 4 might also be susceptible to this kind of explanation. New data from northern New Guinea on oxygen isotopes in corals enable sea surface temperatures to be reconstructed in remarkable detail. After the warm, wet and stable climate of the early Holocene, there was a transition to slightly cooler and more fluctuating conditions. The period 2500–1700 years ago in particular was marked by a high frequency of extreme and prolonged ENSO events (Gagan et al 2004; McGregor and Gagan 2004). This period of severe drought matches the pollen evidence for the spread of grasslands in the upper Wahgi Valley (Haberle 1998), also matches the peaks in concentrations of charcoal in sediments (Gagan et al 2004). This period also saw the beginning of Phase 4 at Kuk, about 2000 years ago.

The role of drought in initiating phases 4 and 5 of wetland land use cannot be proven, but catastrophes of a different kind certainly played some role in wetland abandonment after each of these two phases. The fall of Olgaboli tephra about 1190 BP signalled the end of Phase 4 and a contraction of land use (swamp abandonment) can be inferred, perhaps because of social disruption or because the fertility of dryland agriculture received such a boost. Similarly the end of Phase 5 occurred with the fall of Tibito tephra, probably in 1665 or 1666, and this catastrophic event, 'the time of darkness' in highlands legend, again seems to have triggered a contraction of land use and the end of Phase 5 (Bayliss-Smith et al 2005). In this case, however, the hiatus in wetland use was short-lived.

Further progress might be possible using ethnographic observations of wetland land use during ENSO events. For example, Rebecca Robinson (1999, 2001) has studied the effects of the 1997–98 ENSO on the region of Lake Kopiago in the Southern Highlands Province. While dryland land use in the Kopiago basin was being devastated by drought and frost, Lake Kopiago itself dried up and almost disappeared, providing opportunities for an expansion of sweet potato cultivation across the former lake bed. New gardens were established which enabled the lakeside communities to cope well with what might otherwise have been a major crisis. When normal rainfall resumed in 1998, the drowning of these gardens had the potential to produce an archaeological signature, and in future this evidence could easily be confused with a drainage episode.

Before extrapolating such observations into the past, we must be sure that the analogy is accurate. By the late twentieth century sweet potato production at Kopiago was driven by the subsistence needs of an expanding population, and by an escalating prestige economy based on pig exchange, and hence the growing demand for pig fodder. It is therefore uncertain how far we can extrapolate to pre-Ipomoean times the rationale that lay behind the apparent expansion in 'wetland' land use in 1997–98 and the subsequent contraction in cultivated area after the drought ended. The possible effects of climate change at various time scales nonetheless deserve more explicit attention, and this 'catastrophe' model can be expanded to incorporate other forms of social disturbance, such as epidemics, wars and natural disasters. In societies unable to control such forces, land use must be opportunistic, expanding and contracting into different environments as circumstances permit, or even dictate.

Wetland reclamation may therefore be an index of opportunistic land use. Apparent episodes of drainage may in fact be the result of a shift in the dryland/wetland margin, so that dryland systems expand into former swamps. As with irrigation in deserts, where the evidence from marginal land survives it may be because such incursions into the margins of cultivation are rare events, and subsequent history has not erased their archaeological signature (Farrington 1985).

AVOIDING PROCRUSTES: (2) REGIONAL SCENARIOS

These six models provide alternative meanings for ditches, and they constitute a set of working hypotheses, possibly an incomplete set. The range of possible explanations could be enlarged still further using ethnographic analogy, including for example the role of chiefs in organising intensive land use as proposed for certain islands in the Pacific (Kirch 1994; Spriggs 1986). Establishing alternative models from ethnographic analogy is one necessary task, but ultimately the challenge is to test them in particular cases, and this process is never straightforward in archaeology. The data we can make available, our 'observations' in other words, are not just influenced by our theories; they are also intrinsically sparse. What we can actually 'observe' in the past is really a pathetically partial record of the activities of past societies. The archaeological record usually covers such a narrow range of material that the information we can generate from local sites is more or less consistent with every possible explanation, none of which can be completely rejected.

Fortunately the predictions of each theory are not confined to a set of changes observable at one local site, such as Kuk. By thinking through the wider implications of each theory, we can construct larger-scale models (scenarios) in which the regional as well as the local consequences are specified. These effects can include environmental as well as social changes, and some may be traceable in the archaeological record. In this way new data will be generated from new sites, and the archaeological database will be widened beyond the local assemblage, for example the ditch systems and crop residues that a wetland site like Kuk provides. The enlarged database provides a more rigorous test for the various models that we can derive from theory.

An example of this procedure is provided by the attempt to test Model 2 (wetland drainage as outcome of social inequality) against the archaeological evidence available for Kuk Phase 4, between c 2000 and 1200 years ago (Bayliss-Smith and Golson 1992a, 1992b, 1999). The working hypothesis derived from Model 2 states that digging ditches in Phase 4 was organised by leaders (big men) motivated by the desire to produce surplus root crops as a means to wealth. In such a case the wider effects would be analogous to the response of highlands societies to new wealth following the sweet potato and other recent agricultural innovations. In particular we can note the implications of an expanding surplus for regional exchange.

Ethnographic analogies are available from New Guinea highland societies in the period c 1930–70, when surplus production was boosted by new crops and improved sweet potato varieties, better tools, new imported valuables, and expanded opportunities for exchange following the Pax Australiana. In this period, analogous perhaps to Phase 4 of drainage, the effect of the boost in surplus production (sweet potato in the twentieth century, taro in Phase 4) was an expansion in the geographical scale of exchange systems and an escalation in demand for scarce valuables.

Thus we can hypothesise that in Phase 4 the wetlands became 'hotspots' boosting the regional exchange systems that big men organised and at the same time encouraging greater social stratification including different gender roles (Modjeska 1982). The same process, in a more exaggerated form, accompanied the changes introduced by Europeans in the first 50 years after first contact. For such an analogy to be plausible, we must accept that generalised reciprocity and latent big man ambitions were likely to have been as important social principles in the Highlands in the past as they are now. If so, then a predicted outcome of expanded food surplus will be an intensified production of prestige artefacts and a wider spatial distribution of their exchange.

Testing this model depends upon archaeological evidence surviving, and stone axes are ideal for this purpose. Axes made of high-grade stone are among the artefacts that were traded, and both their manufacture and their ultimate deposition are potentially 'visible' in an archaeological sense. An expansion of stone quarrying activity during Phase 4 therefore becomes a new working hypothesis, and one that is consistent with the wider implications of the 'social inequality' model. Tuman in the upper Wahgi Valley was the prime source of high-grade stone in the central Highlands. John Burton (1984: 228, 1989) investigated Kamapuk Rockshelter and interpreted the lithic evidence there as showing the onset, between 2500 and 1500 years ago, of large-scale stone axe quarrying at Tuman, designed to serve regional exchange systems and not just the local demand for stone. Expanding the site model to explore its wider regional implications provided, in this case, a way of validating – not, of course, 'proving' – the original theory about the underlying rationale for Phase 4 of wetland drainage (Bayliss-Smith and Golson 1992a).

CONCLUSION

This example shows how some of the 'Procrustean beds' of existing concepts and categories, themselves often embedded in evolutionary assumptions, can be broken up and replaced with better-grounded explanations. However, even if some models are shown to be more truthful than others, 'truth' in archaeology is always partial and provisional. New data can always overturn old conclusions, but it is never valid to reject altogether any explanation for past behaviour if that explanation has a firm grounding in ethnographic analogy. I have argued that the six alternative models for wetland drainage outlined above are not equally attractive as explanations for phases 4, 5 and 6 at Kuk Swamp, but ranking these alternatives is almost impossible because we cannot devise any rigorous way of testing them. In other areas of science the approach of 'multiple hypothesis testing' is routine and has proven to be a sound basis for constructing better understandings of phenomena. In relation to agrarian change in prehistory this whole approach founders because, as Darwin suggested, without the making of theories there would be no observations. Our facts reflect our ideas, which in turn reflect our ethnographic analogies, especially those analogies we do not consciously apply as explanations but allow to creep in through the language and theory of knowledge, the epistemology that we employ.

If we accept this conclusion and recognise that in archaeology we must allow for the permanent coexistence of alternative explanations, then at first sight the outlook appears grim – Procrustes triumphant, perhaps? Does this version of how we should perceive the past not open the floodgates to almost anything, allowing us to construct 'the-past-as-wished-for' in a post-modern style of explanation in which all narratives have equal merit?

The answer is no, because in reality the limits to alternative explanation based on ethnoarchaeology are quickly reached. An example from central Burma demonstrates these limits. At Sri Ksetra (fifth century AD) the circular pattern of concentric channels dug around the city can be seen as a symbolic expression of Buddhist beliefs. Water was made to flow in a clockwise and convergent way, like the rivers and seas of the cosmos, or like the Buddhist Wheel of the Law (Stargardt 2003). In this case the archaeologist's explanation derives from historical and ethnographic accounts of Buddhist sacred geography, but such insights cannot be transposed to New Guinea. For instance, we cannot construct a 'sacred geography' model to account for the strikingly geometric chequerboard pattern of drainage channels found at Kuk. In New Guinea wetland drainage as a reflection of sacred geography receives no support at all from ethnographic analogy and cannot be invoked as a seventh alternative model.

Of course we need to keep open the possibility of new explanations emerging, based on new data or new ethnographic insights, and resulting in new models quite different (and perhaps better) than those so far imagined. And where the ethnographic analogies available are scanty or unconvincing, then archaeology must fall back on a more deductive approach. However, the poor track record of deductive models, for example those based on assumptions of cultural evolution or geographical diffusion, suggests that this 'non-analogic' alternative will be much less satisfactory.

In this paper, six different models of wetland drainage have been identified, all more or less plausible and having some grounding in ethnographic reality. The challenge is to explore the implications of alternative theories for new archaeological observations and so evaluate the models against new data. We also need to enlarge our range of possible explanations by utilising new ethnographic insights. These two strategies provide a way forward, reducing the threat of 'Procrustean beds' and enabling us to construct a more truthful prehistory of agriculture.

ACKNOWLEDGMENTS

I should like to thank Tim Denham, Jack Golson, Philip Hughes, Jean Kennedy, Janice Stargardt and Liz Watson for their helpful comments on an earlier draft of this paper.

REFERENCES

Bayliss-Smith, T P (1985) 'Pre-Ipomoean agriculture in the New Guinea Highlands above 2000 metres: some experimental data on taro cultivation', in Farrington, I S (ed), *Prehistoric Intensive Agriculture in the Tropics*, pp 285–320. International Series 232, part I, Oxford: British Archaeological Reports

Bayliss-Smith, T P and Golson, J (1992a) 'A Colocasian revolution in the New Guinea Highlands? Insights from Phase 4 at Kuk', *Archaeology in Oceania* 17: 1–21

Bayliss-Smith, T P and Golson, J (1992b) 'Wetland agriculture in New Guinea Highlands prehistory', in Coles, B (ed), *The Wetland Revolution in Prehistory*, pp 15–27, Exeter: Prehistoric Society and Wetland Archaeology Research Project

Bayliss-Smith, T P and Golson, J (1999) 'The meaning of ditches: deconstructing the social landscapes of New Guinea, Kuk, Phase 4', in Gosden, C and Hather, J (eds), *The Prehistory of Food: Appetites for Change*, pp 199–231, London: Routledge

Bayliss-Smith, T P, Golson, J, Hughes, P, Blong, R and Ambrose, W (2005) 'Archaeological evidence for the Ipomoean revolution at Kuk Swamp, Papua New Guinea', in Ballard, C, Brown, P, Bourke, R M and Harwood, T (eds), *The Sweet Potato in Oceania: A Reappraisal*, pp 109–20. Oceania Monograph no 56, Sydney: University of Sydney and Pittsburgh: Department of Anthropology, University of Pittsburgh

Blaikie, P and Brookfield, H C (1987) *Land Degradation and Society*, London: Routledge

Boserup, E (1965) *The Conditions of Agricultural Growth: The Economics of Agrarian Change under Population Pressure*, London: Allen and Unwin

Bourdieu, P (1977) *Outline of a Theory of Practice*, R Nice (trans), Cambridge: Cambridge University Press

Brookfield, H C (1972) 'Intensification and disintensification in Pacific agriculture: a theoretical approach', *Pacific Viewpoint* 13: 30–48

Bulmer, S (1999) 'Comment on intensification in the Pacific', *Current Anthropology* 40: 325–26

Burchardt, F, Porter, D M, Browne, J and Richmond, M (eds) (1993) *The Correspondence of Charles Darwin, 1860*, vol 8, Cambridge: Cambridge University Press

Burton, J (1984) 'Axe makers of the Wahgi: pre-colonial industrialists of the Papua New Guinea Highlands', unpublished PhD dissertation, Australian National University

Burton, J (1989) 'Repeng and the salt-makers: 'ecological trade' and stone axe production in the Papua New Guinea Highlands', *Man* (ns) 24: 255–72

Clarke, W C (1971) *People and Place: An Ecology of a New Guinea Community*, Berkeley, CA: University of California Press

Clarke, W C and Thaman, R R (1993) *Agroforestry in the Pacific Islands: Systems for Sustainability*, Tokyo: United Nations University Press

Cobb, R M, Charles, C D, Cheng, H and Edwards, R L (2003) 'El Niño/Southern Oscillation and tropical Pacific climate during the last millennium', *Nature* 424: 271–76

Darby, H C (1983) *The Changing Fenland*, Cambridge: Cambridge University Press

Denham, T P, Haberle, S G, Lentfer, C, Fullagar, R, Field, J, Therin, M, Porch, N and Winsborough, B (2003) 'Origins of agriculture at Kuk Swamp in the Highlands of New Guinea', *Science* 301: 189–93

Donaldson, M (1982) 'Contradictions, mediation and hegemony in pre-capitalist New Guinea: warfare, production and sexual antagonism in the Eastern Highlands', in May, R J and Nelson, H (eds), *Melanesia: Beyond Diversity*, vol 2, pp 435–60, Canberra: Research School of Pacific Studies, Australian National University

Farrington, I S (1985) 'The wet, the dry and the steep: archaeological imperatives and the study of agricultural intensification', in Farrington, I S (ed), *Prehistoric Intensive Agriculture in the Tropics*, pp 1–9. International Series 232, part I, Oxford: British Archaeological Reports

Field, J S (2004) 'Environmental and climatic considerations: a hypothesis for conflict and the emergence of social complexity in Fijian prehistory', *Journal of Anthropological Archaeology* 23: 79–99

Found, W C (1971) *A Theoretical Approach to Rural Land-Use Patterns*, London: Edward Arnold

Gagan, M K, Hendy, E J, Haberle, S G and Hantoro, W S (2004) 'Post-glacial evolution of the Indo-Pacific Warm Pool and El Niño–Southern Oscillation', *Quaternary International* 118–119: 127–43

Geertz, C (1963) *Agricultural Involution: The Process of Ecological Change in Indonesia*, Berkeley, CA: University of California Press

Golson, J (1977) 'No room at the top: agricultural intensification in the New Guinea Highlands', in Allen, J, Golson, J and Jones, R (eds), *Sunda and Sahul: Prehistoric Studies in South-East Asia, Melanesia and Australia*, pp 601–38, London: Academic Press

Golson, J (1982) 'The Ipomoean revolution revisited: society and the sweet potato in the upper Wahgi Valley', in Strathern, A (ed), *Inequality in New Guinea Highlands Societies*, pp 109–36, Cambridge: Cambridge University Press

Golson, J (1997) 'From horticulture to agriculture in the New Guinea Highlands: a case study of people and their environment', in Kirch, P V and Hunt, T L (eds), *Historical Ecology in the Pacific Islands: Prehistoric Environmental and Landscape Change*, pp 39–50, New Haven, CT: Yale University Press

Golson, J and Gardner, D (1990) 'Agriculture and sociopolitical organisation in New Guinea Highlands prehistory', *Annual Reviews in Anthropology* 19: 395–417

Gorecki, P P (1985) 'The conquest of a new 'wet and dry' territory: its mechanism and its archaeological consequences', in Farrington, I S (ed), *Prehistoric Intensive Agriculture in the Tropics*, pp 321–45. International Series 232, part I, Oxford: British Archaeological Reports

Gosden, C (1995) 'Arboriculture and agriculture in coastal Papua New Guinea', *Antiquity* 69: 807–17

Haberle, S G (1998) 'Late Quaternary vegetation change in the Tari Basin, Papua New Guinea', *Palaeogeography, Palaeoclimatology, Palaeoecology* 137: 1–24

Haberle, S G and B David (2004) 'Climates of change: human dimensions of Holocene environmental change in low latitudes of the PEPII transect', *Quaternary International* 118–119: 165–79

Harris, D R (1996) 'Domesticatory relationships of people, plants and animals', in Ellen, R and Kukui, K (eds), *Redefining Nature: Ecology, Culture and Domestication*, pp 437–63, Oxford: Berg

Heider, K G (1970) *The Dugum Dani: A Papuan Culture in the Highlands of West New Guinea*, New York: Wenner-Gren Foundation for Anthropological Research

Hviding, E and Bayliss-Smith, T P (2000) *Islands of Rainforest: Agroforestry, Logging, and Ecotourism in Solomon Islands*, Aldershot, England: Ashgate

Kennedy, J and Clarke, W C (2004) *Cultivated Landscapes of the Southwest Pacific*. Resource Management in Asia-Pacific Working Paper 50, Canberra: Research School of Pacific and Asian Studies, Australian National University

Kirch, P V (1994) *The Wet and the Dry: Irrigation and Agricultural Intensification in Polynesia*, Chicago: University of Chicago Press

Latinis, K D (2000) 'The development of subsistence system models for Island Southeast Asia and Near Oceania: the nature and role of arboriculture and arboreal-based economies', *World Archaeology* 32: 41–67

Leach, H M (1997) 'The terminology of agricultural origins and food production systems: a horticultural perspective', *Antiquity* 71: 135–48

Leach, H M (1999) 'Intensification in the Pacific: a critique of the archaeological criteria and their application', *Current Anthropology* 40: 311–39

McGregor, H V and Gagan, M K (2004) 'Western Pacific coral $\delta^{18}O$ records of anomalous Holocene variability in the El Niño–Southern Oscillation', *Geophysical Research Letters* 31: L11204

Modjeska, N, (1982) 'Production and inequality: perspectives from central New Guinea', in Strathern, A (ed), *Inequality in New Guinea Highlands Societies*, pp 50–108, Cambridge: Cambridge University Press

Nunn, P D (2000) 'Environmental catastrophe in the Pacific islands about AD 1300', *Geoarchaeology* 15: 715–40

Nunn, P D and Britton, J M R (2001) 'Human-environment relationships in the Pacific islands around AD 1300', *Environment and History* 7: 3–22

Ostrom, E (1990) *Governing the Commons: The Evolution of Institutions for Collective Action*, Cambridge: Cambridge University Press

Robinson, R P (1999) 'Big wet, big dry: the role of extreme periodic environmental stress on the development of the Kopiago agricultural system, Southern Highlands Province, Papua New Guinea', unpublished MA thesis, Australian National University

Robinson, R P (2001) 'Subsistence at Lake Kopiago, Southern Highlands Province, during and following the 1997–98 drought', in Bourke, R M, Allen, M G and Salisbury, J G (eds), *Food Security for Papua New Guinea*, pp 190–200. ACIAR Proceedings 99, Canberra: Australian Centre for International Agricultural Research

Sahlins, M D (1963) 'Poor man, rich man, big man, chief: political types in Melanesia and Polynesia', *Comparative Studies in Society and History* 5: 285–303

Sahlins, M D (1974) *Stone Age Economics*, London: Tavistock

Specht, J (2003) 'On New Guinea hunters and gatherers', *Current Anthropology* 44: 209

Spriggs, M (1986) 'Landscape, land use and political transformation in southern Melanesia', in Kirch, P V (ed), *Island Societies: Archaeological Approaches to Evolution and Transformation*, pp 6–19, Cambridge: Cambridge University Press

Stargardt, J (2003) 'City of the wheel, city of the ancestors: spatial symbolism in a Pyu royal city', *Indo-asiatische Zeitschrift* (Berlin) 6–7: 144–67

Terrell, J (1986) *Prehistory in the Pacific Islands: A Study of Variation in Language, Customs and Human Biology*, Cambridge: Cambridge University Press

Terrell, J (2002) 'Tropical agroforestry, coastal lagoons, and Holocene prehistory in Greater Near Oceania', in Shuji Y and Matthews, P J (eds), *Vegeculture in Eastern Asia and Oceania*, pp 195–216, Osaka: National Museum of Ethnology

Watson, E E (1999) 'Ground truths: land and power in Konso, Ethiopia', unpublished PhD thesis, University of Cambridge

Wittfogel, K P (1957) *Oriental Despotism: A Comparative Study of Total Power*, New Haven, CT: Yale University Press

CHAPTER 8

Perspectives on Traditional Agriculture from Rapa Nui

Geertrui Louwagie and Roger Langohr

INTRODUCTION

Land evaluation is a method to determine the suitability of existing and potential land uses. It provides a holistic framework to investigate the complexity of any land use, including agriculture, by incorporating a range of biophysical (eg climate, landscape, soil) and socioeconomic (eg farming strategies, perceptions) components and by examining the linkages between them. Land evaluation thus has great potential as a framework for characterising and assessing the performance (success or failure) of land use and, more broadly, of farming systems in the past.

In this contribution, diverse components of the traditional Rapanui agricultural system, both biophysical and socioeconomic, are characterised. These components suggest the degree of complexity needed in our thinking to understand much earlier practices and perceptions. Characterisations have been undertaken using a questionnaire-based survey, supplemented by biophysical data, archaeological evidence and ethnographic information (see Louwagie 2004). Traditional Rapanui agriculture, which is the focus here, refers to early twentieth century practices; these practices have in turn been used to shed light on prehistoric farming dating from the earliest sweet potato cultivation on the island at c AD 1400–1650 (Orliac and Orliac 1999; Yen 1988) to the decline in sociopolitical complexity of Rapanui society by the end of the seventeenth century. The fully elaborated land evaluation model and ensuing prehistoric Rapanui land suitability classification, and associated studies, are reported elsewhere (Louwagie 2004; Louwagie and Langohr 2005; Louwagie et al 2006).

PALAEOLAND EVALUATION

Land evaluation is the process of assessing land performance for specific purposes (FAO 1976). In this paper we look at agriculture and take crop production or yield as the measure of land performance. Indeed, either yield optimisation or risk spread, avoiding yield fluctuations, often forms the rationale of a farming system. The concept of land is holistic (Sombroek 1997); land comprises the physical environment, including climate, landforms and topography, soil, hydrology, vegetation and the effects of past and present human activity. Land evaluation also encompasses socioeconomic resources, such as farm management strategies, availability of manpower, market possibilities or perceptions. Socioeconomic components can express what drives a society and how it responds to external stimuli; these drivers and responses influence land management decisions.

There are several dominant land evaluation models. The FAO framework (1976) was adopted and modified for application to the past because it allows the integration of biophysical and socioeconomic information, is crop-specific and can accommodate data scarcity, a characteristic of palaeoland evaluation. Land evaluation in an archaeological, or prehistoric, context is referred to as palaeoland evaluation. Four sets of factors are fundamental to palaeoland evaluation; they each influence crop production, a fundamental component of any agricultural system:

- Climate: Palaeoclimatic data are obtained through proxies from palaeoecology, sedimentology and associated dating methods;
- Landscape and soil: Combined use of geomorphology and pedology provides data on palaeolandscapes and soils;
- Crop ecology: Crop palaeoecology, such as crop development stages and crop yield, is inferred through archaeobotany, experimental archaeology, ethnopedology and ethnobotany;
- Agricultural practices: Information on past agricultural systems and agropedological regions is gathered from archaeology, archaeopedology and ethnopedology.

Problems in undertaking palaeoland evaluation are numerous. For example, and of most relevance to investigations of prehistoric agriculture, inferences regarding crop or plant productivity in the past are largely hypothetical. Reference data are mostly derived from present-day cultivars or wild plants, which are likely to have been very different in the past and in many cases no longer exist. Similarly, crop productivity experiments designed to replicate agricultural practices in the past, such as those encountered in archaeological contexts, are not conducted under the same environmental and soil management conditions as those that existed in the past (Louwagie and Langohr 2005).

RAPA NUI FARMING IN CONTEXT

Rapa Nui is a remote island located in the eastern Pacific approximately 3700 km west of Santiago de Chile, Chile. Its triangular shape is formed by three extinct volcanoes: Poike to the east, Rano Kau to the south and Maunga Terevaka to the north (Figure 8.1). The island's land area is approximately 170 km^2 and elevations range from 0–506 m at Maunga Terevaka.

Polynesians colonised Rapa Nui by AD 690, although there is some debate concerning the timing of human arrival (Flenley and Bahn 2003: 75–77). At initial settlement, the vegetation was composed of large trees with shrub undergrowth; *Jubaea chilensis*, a now extinct palm tree, was the dominant canopy species. This vegetation was mostly cleared by the end of

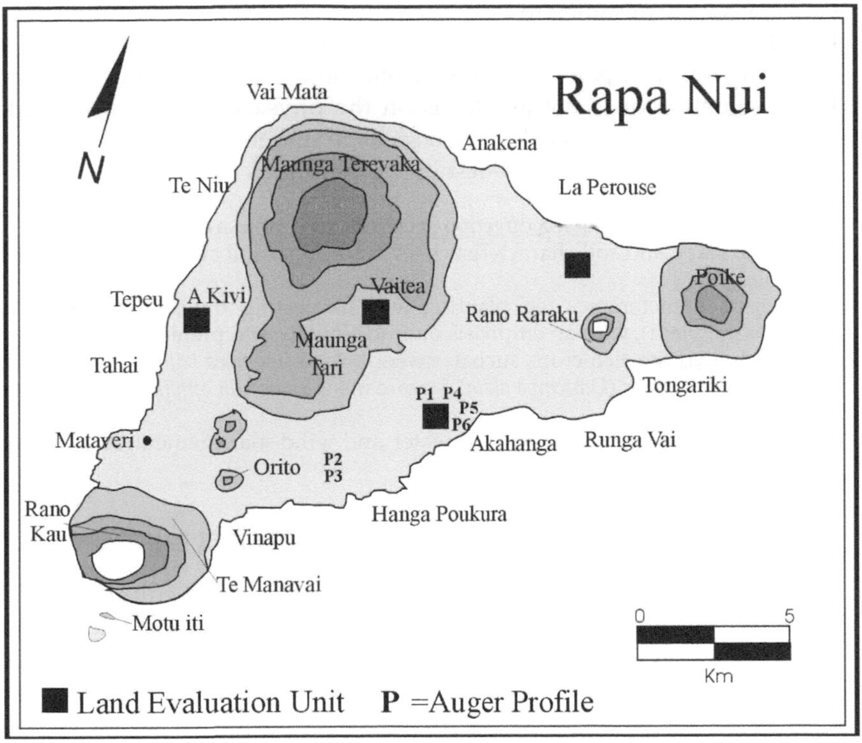

Figure 8.1 Map of Rapa Nui with location of investigated land evaluation units (squares), including auger profiles (P) for traditional soil capability classification, contour intervals of 100 m (after Louwagie et al 2006).

the 16th (Flenley 1996) or mid-seventeenth century (Mann et al 2003; Orliac and Orliac 1999). Rapanui culture is marked by the famous ceremonial centres composed of *ahu* (ie platform) and *moai* (ie statue), which were built from AD 1000–1600 (Stevenson 2002). Sociopolitical complexity of Rapanui society declined by c AD 1680 (Louwagie et al 2006; Renfrew and Bahn 1996: 251). Throughout prehistory and up to the beginning of the twentieth century, agriculture was central to Polynesian occupation of Rapa Nui (Stevenson et al 1999), and gave rise to a rich agricultural tradition, described in the early twentieth century by Englert (1974), McMillan Brown (1996) and Métraux (1957).

Given the peculiar cultural characteristics and complexity of Rapanui society, its subsequent decline towards the end of the seventeenth century and the central role of agriculture in the economy throughout its prehistory, land evaluation was applied to investigate agriculture in the past. However, in the absence of sufficiently detailed data on prehistoric agriculture (AD 1400–1700), traditional agriculture (beginning of the twentieth century) was characterised because it could 'bridge the gap' and shed light on earlier practices.

During the current study, interviews targeted people or their relatives who were farming on the island at the beginning of the twentieth century (see Louwagie 2004 for full discussion of method and associated problems).

Questions were designed to assess people's perceptions of the four main factors that affect land performance, thereby incorporating some fundamental socioeconomic components into the palaeoland evaluation. As stated above, land evaluation does not just focus on the physical resources available to people within a given locale; it also seeks to understand the ways in which people perceive and use those resources. Questions focussed on:

- climatic conditions (eg effective precipitation, wind activity);
- landscape and soil characteristics (eg traditional soil classification, ethnopedology);
- crop characteristics (eg planting and harvesting date, crop development stages, yield), with an emphasis on traditionally and prehistorically important carbohydrate-rich crops such as sweet potato (*Ipomoea batatas*), taro (*Colocasia esculenta*), yam (*Dioscorea alata*), sugarcane (*Saccharum officinarum*) and banana (*Musa* sp.);
- soil management practices (eg water and wind management, fertiliser application).

PERCEPTIONS OF TRADITIONAL AGRICULTURE

Each of the four main components of the Rapanui agricultural system is discussed. This characterisation contributes to developing relevant scenarios for land evaluation (such as assessing land performance for crop production) in traditional, and by extension, prehistoric, farming contexts.

Climate

Rapa Nui has a subtropical climate with buffering oceanic character (FAO 1985) (Figure 8.2). The Rapanui climate has been variable through time and across space (Norero 1998). Goosse ran simulations for the Rapanui climate in 2004 that suggest that the present-day climatic conditions are similar to those occurring during AD 1400–1700 (on the ECBILT-CLIO-model, see Goosse et al 2004). Climates are conducive to crop growth for most of the year. However, limiting factors for crop growth include: considerable loss of precipitation to surface runoff or deep percolation, resulting in a net (effective) precipitation of only 30–50% (Hauser 1998 cites estimates of 46–62%); and exposure to wind, which has a mechanical impact and increases evapotranspiration, particularly on bare soil surfaces.

Landscape and Soil

Soils on Rapa Nui are variable, reflecting two important factors: (a) parent material ranging from basic to siliceous rocks and from solid lava flows to pyroclastic materials; and (b) age of parent materials ranging from 3 million years at Poike to 2000 years in the vicinity of Maunga Terevaka (Charola 1997; Fischer and Love 1993). Extensive areas have very shallow soils and extensive rock outcrops. The weathering of tephra produced andic soil properties (USDA 1999: 80) and limited available phosphorus and potassium to plants (Louwagie 2004: 258). Andic soil properties are associated with short-range

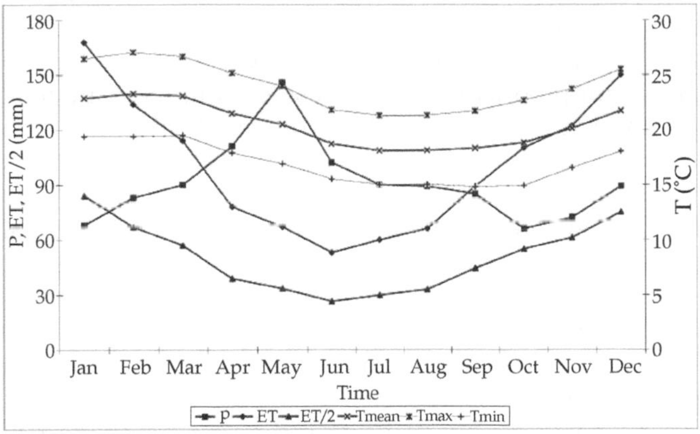

Figure 8.2 Precipitation (P), evapotranspiration (ET), half of the evapotranspiration (ET/2), mean (Tmean), minimum (Tmin) and maximum (Tmax) temperature at Mataveri (after FAO 1985; Louwagie et al 2006). Monthly averages from series of at least 30 years (P) or ten years (other parameters).

ordered minerals, reflected in high silicon, aluminium and/or iron contents (on extraction with ammonium oxalate). Andic horizons generally have a low bulk density, mostly silt loam or finer textures, a high organic matter content, variable charge surfaces and high phosphate-retention capacity. Farmed land is preferentially located at the base of slopes to take advantage of deeper soil, additional water supply from slope runoff and nutrient inputs from upslope.

Traditional soil classification on Rapa Nui refers to vaguely defined, rather subjective and experiential soil characteristics. The main traditional soil types were sampled (see location of auger profiles in Figure 8.1) to compare traditional observations with chemically determined soil characteristics, such as pH, organic carbon and nitrogen contents and base cation concentrations (Louwagie 2004). Each traditional soil class is characterised below.

Oone hatu ('uri) connotes good (black) soil and arable land and is considered the optimal soil type for crop growth. It is called a 'spring soil', the season when it is most appropriate for planting; it becomes hard during drier weather in the summer. This traditional soil type has relatively high available phosphorus content that increases with depth. The relative yield is assessed at 100%.

Oone (hatu) mea connotes (good) red soil. This soil type has andic soil properties: poorly (or short-range) ordered soil minerals, such as ferrihydrite and allophane, are reflected in relatively high iron (Childs 1985) and silicon (Parfitt 1990) levels (on extraction with ammonium oxalate), respectively. Chemical analysis of soils with such properties often show underestimated clay contents (due to a lack of soil dispersion) and overestimated cation exchange capacities when the soil pH is below 7 (laboratory measurement at pH of 7). Characteristics of agricultural significance are high phosphorus retention and relatively low potassium.

Oone reherehe connotes poor soil – very loose and porous with a low bulk density. High porosity hampers moisture retention and people tend to grow shallow-rooting crops. It has andic soil properties and relatively high amounts of silicon (on extraction with ammonium oxalate) compared to the *Oone hatu 'uri* type. The low available phosphorus content contributes to low agricultural productivity. It is considered to have moderate capability for crop growth; the relative yield is assessed at 80%.

Oone mararía connotes unproductive soil or barren land. Soils are shallow and contain common small stones; parent rock occurs at 0.40–0.50 m depth. There is low available phosphorus content and the soil is considered moderate to marginal for crop growth; the relative yield is assessed at 50–60%.

Oone rautú is land that is exhausted of nutrients due to ongoing planting (over the past 50 years) and needs regeneration, eventually requiring fertiliser. This soil class is considered the most unsuitable for crop growth; its relative yield is estimated at 20%.

Oone pakahía is described as being rather hard at the top, but deep and humid and hence optimal for deep root growth. This soil type is considered very suitable for yam, which requires a deep soil, and sugarcane, which requires considerable moisture.

Oone ranga connotes flooded, swampy soil. It has a water table perched on rock and is suitable for banana.

Oone riku connotes abundant growth. It actually refers to fallow land, allowing the soil nutrients to regenerate and preparing the soil for a new cycle of crop production.

Crop Ecology

Crop-specific information was gathered during interviews for five traditionally (Englert 1974: 213; McMillan Brown 1996; Métraux 1957) and prehistorically (Orliac and Orliac 1999) important carbohydrate-rich crops: sweet potato, taro, yam, sugarcane and banana. Reference is made to traditional crop varieties as opposed to modern varieties that were recently (from 1960 onwards) introduced to the island. Most of the ten interviewees suggest that crop production is strongly determined by, and hence dependent upon, soil type, lunar phases, weather conditions (particularly at planting time), applied techniques (crop rotation, fertilisation, fallow for soil nutrient regeneration, etc) and the farmer's knowledge and effort.

Several aspects of crop ecology are presented for each carbohydrate-rich crop: list of traditional varieties (Table 8.1), landscape and soil suitability, cultivation methods, crop cycle and storage. Crop cycle, or development stages are defined according to Sys et al (1991: 245):

(1) Initial stage: germination and early growth with ground cover under 10%;
(2) Crop development stage: from end of initial stage to effective full, ie 70–80%, ground cover;
(3) Mid-season stage: from effective full ground cover to start of maturation, ie when leaves discolour or shed; and
(4) Late season or maturation stage: from end of mid-season to full maturity or harvest.

Table 8.1 *Named traditional varieties of carbohydrate-rich crops; (see Louwagie 2004; additional sources used for verification: 1 = Kondratov (1965); 2 = Barthel 1978; 3 = Hotus Chávez 2000; 4 = UNDP 1998; 5 = McMillan Brown 1996)*

Sweet potato *Kumara*	Taro *Taro*	Yam *Uhi*	Sugar cane *Toa*	Banana *Maika*
ari(n)ga rikiriki[1,2,3]	*hamea*	*kunekune*[3]	*ruma*[1]*/kuma*[4]	*ve'i*
ha 'u pu (tea	*(harahara) hiva*[3]	*pukehe 'u/apuka 'he/apokahe'*	*mari kuru*[1]	*hihi*[2,3]
tea/'uri 'uri)[1,2,3]	*(teatea/'uri'uri)/hara hara hiva*[4]	*u/ puke heu*	*peteriko*[4]	*hiva*[3,4]
pene[3]	*harahara rapanui*[2,3]*/harahara*[5]	*keri tangata*	*viti viti*[4]*/*	*hore hore tapatea*
pita[1,3,4]	*hore hore tapatea*[5]	*keri vahine/navihe*	*vitiviti*[1,3]	*koratea*[3]*/koro tea*[2,4]
puku[3]	*kave heke*	*marikuru*		*naho'o/naoho*[3]*/nahoo*[2]
re(n)ga moe tahi	*ketu anga mea*[2,3,4]	*meamea*		*pahu*[3]
(teatea/'uri 'uri)[1,2,3,4]	*ngaatu*[2,5]*/nga'atu*[3]*/ma'atu*	*mingo mingo*		*perorá*[3,4]
teatea/tua tea[2]*/tu'a tea*[3]	*ngeti ('uri/tea)*[2,3]*/ngete*[4,5]	*(he) neke*		*pia*[2,3]
te tuá	*mangó*[2,3,4]	*take take*		*pukapuka*[2,3]*/puka*
tipataipaka[3]	*mataheke*[3]	*teatea*		*puka*[4]
uka te tuá[3]	*paíki*	*'uri'uri*		*purau'ino*[3]*/purau ino*[4]
uka tire	*teatea*[2,3,4]	*ravei*[3]		*ri'o*[3,4]
ure omo	*vaihi*[3]*/vai hi*[4]	*vae manu*		*ta'amoa/tamoa*[3]
(teatea/'uri'uri)[1,2,3,4]	*vaihoiti*[3]*/vai ho iti*[4]			*ve'i*
ure vai[1,2,3]				

Sweet Potato *(Ipomoea batatas; Kumara)*

Traditional soil types suitable for sweet potato are: *Oone hatu*, *Oone pakahía* and *Oone reherehe*, the latter particularly for the varieties *Kumara tipataipaka* and *Kumara ure omo*. One interviewee stated basal slopes were appropriate for sweet potato, as this crop needs sufficient soil moisture and loose soil; another indicated the advantage of using rock fragments to improve soil moisture retention, whereas a third suggested slightly hard soil for optimal production.

Propagation of sweet potato occurs by planting up to five stem cuttings of about 0.5–0.6 m length to form one plant cluster. These cuttings are taken from mature sweet potato plants and are collected up to one week before planting, preferably during a waning or waxing moon. Planting distance varies from 0.4–1 m in the row and from 0.5–2 m between rows. About one month after planting, prior to complete vegetative development, a heap of earth (*puke*) is put on top of the sweet potato plants.

The main sweet potato planting periods are January to April and August/September. Planting preferentially occurs with a new moon, although some considered a full moon favourable as well. The most commonly suggested crop cycle lengths for sweet potato were 120 and 180 days. Answers on the length of the different crop development stages were in general poor, as farmers do not pay attention to ground cover or leaf development. The end of the initial stage is referred to as the moment when, with only 10% ground cover, the sweet potato plants are covered with soil (*puke*). However, it was also mentioned that leaves only started growing once sweet potato plants were covered with this *puke*, or soil heap. With regard to the transition from the mid-season to the late season stage, some interviewees mentioned that sweet potato leaves do not shed prior to harvest, although others referred to the time of flowering as the start of the late season stage. Responses were variable regarding flowering: one interviewee believed only one variety, *Kumara aringa rikiriki* flowered prior to harvest; others suggested that only some varieties give flowers, some seldom flower, and in one case, the plant only flowers after one year. Actual rooting depth was commonly estimated from 0.20–0.30 m.

Young leaves of sweet potato, taro and *ti* (*Cordyline terminalis*; *ha'a*) are eaten. Once the sweet potato tubers are harvested, the dried leaves can be used as fuel. Storage practices for sweet potato are limited, as production and harvest is focused on direct consumption. Sweet potato is thus kept growing until needed. Only when larger quantities are needed, such as for festivities or ceremonies, did people store sweet potato tubers. The largest sweet potato tubers are commonly stored underground for a maximum of one month; once the tubers start shooting, they are dug and eaten. In order to prevent animals from eating the stored tubers, they are dried in the sun for a few days prior to burial. Alternatively, sweet potato tubers can be stored after sun drying, but need to be rehydrated one day prior to consumption. Interviewees claimed that dried sweet potato can be stored for up to eight months, whereas for only 20–30 days untreated. Similarly, sweet potato baked in an earth oven (*umu* or *umu pae*) enables storage for up to a month.

Taro *(Colocasia esculenta; Taro)*

Traditional soil types preferred for taro cultivation are *Oone hatu* and *Oone pakahía*; topographically, a plateau position protected from wind was considered ideal. In most other Polynesian islands taro is raised using some form of irrigation (McMillan Brown 1996; Yen 1988). On Rapa Nui, surface rocks were used to conserve soil moisture and protect the tubers from direct sunshine, ie lithic mulch.

Taro is propagated using stem cuttings (*uru taro*); two to three stem cuttings are put together at 0.1–0.2 m distance to form one taro plant cluster. For *Taro hiva* only one stem cutting is planted and only the mother plant is harvested. For other taro varieties, eg *Taro vaihi*, only the 'children' or new tubers are harvested for human consumption, while the mother plant is used as animal fodder. Planting distance varies from 0.4–1.0 m in the row and from 0.4–2.0 m between rows. Cultivars producing several tubers ('children'), eg *Taro vaihi*, are planted at greater distances than varieties producing only a 'mother' tuber, eg *Taro hiva*. As opposed to sweet potato, taro is not covered with soil after planting, although taro tubers grow deeper in looser soil. Taro needs high humidity at planting. *Colocasia* taro likes shade and is often grown under bananas or intercropped with yam; it is rarely planted as the first crop in rotation (Raemaekers 2001: 221–228).

Taro is commonly planted between December and March, and occasionally in June/July. Planting preferentially occurs with a new moon and occasionally with a full moon. Summer with its relatively dry conditions is favourable for harvest, as vegetative development decreases. *Taro vaihi* is the only variety that never dries out completely; this variety is commonly grown in *manavai* that provide sufficient humidity and compost.

Crop cycle lengths for taro varied from 240 to 360 days. Answers on the length of the different crop development stages were poor and too diverse to rely upon, as people did not know how much time ground cover or leaf development took to occur. Leaf shooting was identified as part of the initial stage. Regarding the start of the late season stage, interviewees noted that taro leaves do not shed prior to harvest. For those varieties that flower, however, interviewees regarded flowering as an indication of the late season stage. Rooting depth was commonly estimated from 0.15–0.40 m.

Storage practices for taro are comparable to those for sweet potato. Taro can be left to continue growing in the ground for one year after maturation, although it was noted that *Taro vaihi* needs to be harvested when ready for consumption. Nevertheless, taro harvesting can be timed to provide food for social events, ie wedding and engagement ceremonies.

Yam *(Dioscorea alata; Uhi)*

As opposed to sweet potato and taro, yam produces large tubers and favours deep soils. Interviewees said that yam needs soils that are either slightly hard and retain moisture at depth, eg *Oone hatu*, or a rather soft soil type, intermediate between *Oone hatu* and *Oone reherehe. Oone pakahía*, rather hard at the top, but deep and humid, fulfils this latter requirement. Topographically, both plateau and basal slope positions are favoured.

The cultivation of yam requires more manual work than other root crops. Formerly, people piled soil (*puke*) up to 1.0–1.5 m high to plant yam. Alternatively, a stone was put at about one metre underground to control root growth and prevent the tuber of deep-rooting varieties (eg *Uhi vae manu*) from growing too deep. According to a Rapanui legend, people cultivating yam put a flat stone underground so that roots would feel pain when they touched the stone and consequently would stop growing. *Uhi keri tangata, Uhi (he) neke* and *Uhi ravei* are quite tall, grow to depth and were traditionally harvested by men; *Uhi puke heu, Uhi keri vahine* and *Uhi vae manu* are relatively small, grow superficially and were harvested by women. In recent historic times, yam was commonly harvested using an iron bar. *Uhi puke heu* was commonly grown in *manavai*. Some interviewees reported that yam, like taro, is planted in groups of two or three 0.1–0.2 m from one another. Planting distance in and between rows varies from one to three metres.

Traditionally, yam is seen as a very sensitive crop, comparable to watermelon cultivated today. This notion probably corresponds to yam being much less tolerant of low fertility than other tuber crops, such as sweet potato (Norman et al 1995: 305–318). Prohibitions associated with yam cultivation were numerous: people who had eaten meat or fish were not allowed in the vicinity to prevent the yam from becoming inedible; neither women nor animals were allowed near yam gardens; and finally, yam should not be cut with a knife formerly used to cut meat (*enfermedad de uhi*).

The main yam planting periods mentioned were October and December to February. One person mentioned that planting preferentially occurs with a new moon, and occasionally with a full moon. The most commonly suggested crop cycle length for yam was 360 days. Answers on the length of the different crop development stages were in general poor and too diverse to rely upon, as people do not pay attention to the timing of ground cover and leaf development. Leaf-shooting was occasionally identified as part of the initial stage. Interviewees observed no indications of other crop development stages. Yam was not observed to flower. Rooting depth was commonly estimated between one and two metres.

Storage practices for yam are comparable to those for sweet potato and taro. However, according to one interviewee, yam cannot be kept underground due to the risk of rot, which is consistent with observations elsewhere (Raemaekers 2001).

Sugarcane *(Saccharum officinarum; Toa)*

Traditional soil types suitable for sugarcane are *Oone hatu* and *Oone pakahía*. Topographically, valleys protected from wind by surrounding hills were considered suitable for sugarcane growth.

Sugarcane is a perennial crop that is vegetatively propagated. Planting distance varies from 0.5–2.0 m. Sugarcane is preferentially grown in *manavai,* which maintain humidity and fertility. Once the sugarcane reaches a height of two metres, the stalks are tied together (*kuku*) and pulled up (*hakatere*), so that they can grow taller and upright. Sugarcane stalks can reach up to three metres, after which they fall down (*papau*) and reach full ripeness. New shoots announce a new crop cycle. Observations of flowering were highly variable: flowers within eight to 10 months, does not flower, or only flowers after two or several years.

Planting occurs preferentially between November and April, and occasionally in October. New and occasionally full moons are favoured for planting. The most commonly proposed crop-cycle length was 360 days. Answers on the length of the different crop development stages were unreliable. Actual rooting depth is commonly estimated at 0.15–0.40 m.

Sugarcane is harvested for direct consumption and can be kept for one month after harvest, although up to only three days if baked in an earth oven (*umu pae*). A site protected from humidity and heat, eg underground, and covered at the surface with banana leaves can be used for storage.

Banana *(Musa* sp.; *Maika)*

Traditional soil types suitable for banana are: *Oone hatu, Oone pakahía* and *Oone ranga*; however, one interviewee contradicted the latter soil type by saying banana should not be planted in wet soil. The crop can also be grown during the fallow period within the crop rotation cycle (*Oone riku*) and favours earthworm (*koreha henua*) activity.

Banana is a perennial crop that reproduces through vegetative propagation. Like sugarcane, banana is preferentially grown in *manavai,* which maintain humidity and fertility through the piling up of fallen leaves. Banana production commonly lasts for five to 10 cycles and even reaches up to 50 cycles. According to a Rapanui legend, the first banana stems, called *huri ariki,* would bend their bunches in the same direction; this characteristic testifies to the magical properties of banana. The banana bunch develops from the (female) flowers; one banana stem produces one bunch. Once the bananas are ripe, the mother is cut in order to permit young shoots, or daughters, to grow, announcing a new crop cycle. Planting distance in and between rows was variably reported at 1–4 m; plants are occasionally planted closer to one another to provide shade.

Planting preferentially occurs from December to February, occasionally in October/November or March, and on the first or third day of a full moon or with a waxing moon. These lunar phases are considered to ensure adequate rain. The most commonly given crop-cycle length is 360 days. The mid-season stage is understood as the period when the bunch emerges. Flowering indicates the start of the mid-season stage. Data on duration of crop development stages are scarce. Rooting depth was commonly estimated at 0.2–0.6 m.

Banana ripening can occur either in the sun or underground (*rua maika*). In order to ripen banana underground, people used to dig a hole in the ground, burn it, remove the ash, put lots of dry grass or dried banana leaves and up to 1000 bananas on top, and finally cover with fresh soil or/and banana leaves. The bananas were thus left for two to five days and eaten green (Englert 1974: 211). This practice is considered to render the banana flesh very sweet.

Banana production is aimed at direct consumption. Once ripe, bananas can be sun dried for prolonged storage up to eight months, but need to be rehydrated one day prior to consumption. The drying practice involves exposing bananas to the sun on dry banana leaves, and subsequently covering them with dry banana leaves and branches. After such treatment, it was claimed by one interviewee that bananas could be stored for years.

Agricultural Management

Of the numerous agricultural practices recorded for Rapa Nui (Englert 1974; McMillan Brown 1996; Métraux 1957; Stevenson et al 1999; Wozniak 2001; Yen 1988), several are briefly discussed below. Emphasis is given to the environmental and social contexts of agriculture, with special consideration of how the Polynesian worldview influenced production and how Polynesian agricultural practices were adapted to the island's distinctive environment.

Crop Lunar Calendar

On Rapa Nui, planting traditionally occurred according to lunar phases (see Handy and Handy 1991: 130 for similar observations on Hawaiian practices). The crop lunar calendar (Figure 8.3) gives an overview of the lunar phases and months that are particularly favourable for sweet potato, taro, yam, sugarcane and banana planting. The different lunar phases – full moon, waxing moon, new moon and waning moon – are indicated. The crop lunar calendar is illustrated for sweet potato (indicated in grey), with an average crop-cycle length of five months.

The period (four to seven days) around new and full moons coincides with a change in sea currents and wind direction and is considered to provide more rain. It also represents a specific fertility period for animals and crops. A new moon is considered favourable for tuber crops, a waxing moon favourable for leaf-holding crops, eg lettuce. A full moon combined with dry conditions (from November to January) is considered appropriate for planting stem-holding crops, such as banana; too much humidity causes the stem to rot. According to one interviewee, new and waxing moons are favoured for planting yam, banana and sugarcane.

The size and shape of the moon are also considered important. A large moon that opens up towards Anakena, on the north side of the island, appearing for three to four days, is supposed to bring rain and is considered favourable for planting. A small moon that opens up towards the sky is considered unsuitable for planting.

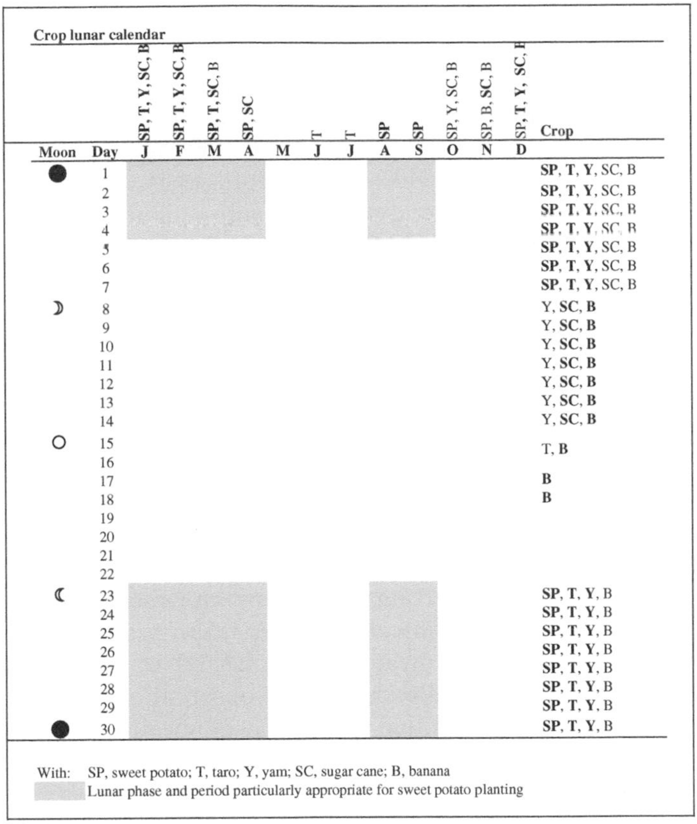

Figure 8.3 Crop lunar calendar based on Chávez Haoa's (1998) calendar for fishing and agriculture and adapted according to interview data.

Water and Wind Management

The main climatic constraints on crop production on Rapa Nui in the present and the past are water availability and wind strength. Winter was considered to be ill-suited for planting due to intense wind. From mid-August (*hora iti,* literally 'small summer') onwards, conditions are favourable for crop growth. The unique Rapanui climate required adaptations of traditional Polynesian practices, such as the construction of *manavai, pu(pu)* and lithic mulching, which were all intended to increase soil water retention.

Manavai are stone circles, or walls, preferably situated in low shady spots that retained soil moisture and served as windbreaks. They were used to grow banana, sugarcane and paper mulberry (*Broussonetia papyrifera* L.; *Mahute*), as well as taro and sweet potato. *Manavai* accumulated organic matter (falling leaves, fruit, etc) that provided compost. Métraux (1957: 64) mentions structures, either natural hollows or artificial constructions, which were frequently located where vegetable waste accumulated near abandoned dwellings to form a thick layer of fertile humus. The natural hollows were cavities of lava tunnels, in which the so-called sunken gardens emitted an odour of rotting vegetation (Métraux 1957: 63). The latter were observed during our 2001 fieldwork in the most recent lava flows along the coast close to Tepeu (Figure 8.1). Like *manavai*, eucalyptus (*Eucalyptus* sp.), cypress (*Cupressus* sp.), fig trees (*Ficus* sp.) and *Miro tahiti* (*Melia azedarach*) act as windbreaks, or alternatively, provide shelter for animals against rain and sun.

Pu(pu) are stone circles with central depressions of half a metre diameter used for planting. The depression is subsequently covered with rock fragments to retain moisture (through reduced evaporation), prevent rainwater from deep percolation and provide shade. *Pu* were traditionally used to plant sweet potato or taro, and today are used for pineapple.

Métraux (1957: 64) mentions other management practices to overcome water shortage: (1) covering the vegetable beds with a thin carpet of grass to keep them moist (ie organic mulch); (2) lithic mulch to protect taro tubers from the sun and retain sufficient moisture; and (3) furrows drawn across the slopes of the volcanoes to temporarily hold rainwater. Lithic mulch consists of lithic material (stones) at the surface to retain moisture in the ground, primarily during drier months; this adaptive strategy was possibly developed in the AD 1300s to deal with Rapa Nui's weather and climate (Stevenson et al 1999). Caves also retain moisture and are used to grow crops such as banana and taro.

The only sources of freshwater on Rapa Nui were the Rano Raraku crater lake and caves. Rainwater was collected in holes carved in flat stones. Yen (1988) refers to Beechey's 1825 observation of irrigation on the northern part of the island that was archaeologically verified by Ferdon in 1955; he notes an example recorded by Englert at Vaipu in the island's interior. Stevenson et al (1999), however, found no evidence of irrigation at La Pérouse (northeast coast), Maunga Tari (centre of the island, southwest of Vaitea) and Te Niu (northwest coast).

Planting and Crop Rotation

In the early twentieth century, different crops and species were planted in the same parcel, but cultivated in lines or rows separated from one another. Crops were grouped within a parcel according to their characteristics. Banana, sugarcane and *Taro vaihi* can grow for many years and were therefore planted together, with sweet potato along the side. Row cultivation enabled appreciable quantities to be harvested separately and for horses and oxen to be used to weed fields. Between 1940 and 1960, oxen were used to till the land; horses and tractors have been used only since the 1960s.

Crop rotation prevented soil exhaustion. The soil was cultivated for five years and then left fallow for one to three years. Moreover, a particular crop could only be grown for a maximum of two years on the same spot. Deeper rooting crops, such as sweet potato, taro and yam, were alternated with shallower rooting crops, such as banana, sugarcane and gourd (*Lagenaria* sp.) (or maize [*Zea mays*] during the time of the Williamson and Balfour Company).

In traditional agroforestry practices on Pacific islands, the sequence of tuber crops in the rotation cycle is significant; yam preceded taro and sweet potato because yams are more nutrient-demanding (Thaman et al 2003); and, yams followed graminaceous plants, such as sugarcane (Raemaekers 2001). If exhausted from the cultivation of deep rooting plants, the soil was left fallow for five years or was planted with shallow rooting plants, trees or bananas (except nutrient-demanding trees such as eucalyptus and cypress). Banana is considered to regenerate soil nutrients (*Oone riku*, *Oone rautú*).

Most interviewees did not mention any form of direct fertilisation and even said that there was no need to fertilise soil in the past because people only used soil that was suitable for crop growth. During the first half of the twentieth century, when the Williamson and Balfour Company was in operation, cow and sheep excrement was used. Some mentioned ash from earth ovens and dried or rotten (*pudrido*) grass as traditional fertilisers. *Miritonu*, a kind of seaweed, was used from 1968–73 onwards. It was harvested in winter and consequently dried and pulverised prior to application. However, it is now thought to be extinct on the island.

Social and Gender Relations

In the past, men and women organised themselves in clans to cultivate the land. Interviewees stated that men planted large crops, such as banana, sugarcane and taro, whereas women planted small crops, such as sweet potato (occasionally in *manavai*). Others noted that men preferred to plant and women to harvest, except for varieties such as *Uhi keri tangata* that are physically hard to harvest. More recently, maize planting for the Williamson and Balfour Company was considered to be a task for men. According to two male interviewees, agriculture was a man's activity because it was physically demanding; women should stay at home to take care of the children and/or household; however, this scenario probably reflects European, colonial norms and does not reflect pre-European social practices.

During menstruation and for three days afterwards, women were not allowed to sow, plant or do anything in the fields. In particular, they should not touch yam (and watermelon nowadays) because the crop would be damaged (*taketake* or *mingomingo*) or yield poorly. Additionally, some clans and families, such as the king's family (*ariki*), were not allowed to approach any crop, due to their negative influence on crop growth and food production.

THE VALUE OF LAND EVALUATION TO UNDERSTANDING PREHISTORIC AGRICULTURE

The adoption of a palaeoland evaluation framework enables an integrated interpretation of interrelated biophysical (climate, landscape and soil, crop ecology) and socioeconomic (farming strategies, perceptions) aspects of agriculture in the past. The intention of this paper has been to describe palaeoland evaluation as a method, to demonstrate its value to the study of prehistoric agriculture, and to show the range of knowledge needed to characterise several components of the traditional agricultural system on Rapa Nui. Particular emphasis was placed on factors that determine crop production, an essential component of any farming system and a measure of its success or failure. Incorporating the multifaceted nature of human-environment relationships is central to characterising agriculture in different locales, whether in the recent (early twentieth century) or the more distant past (AD 1400–1700) (see Louwagie 2004 for a full discussion). Only by doing so will we begin to understand the complexity of any agricultural system and how and why people used resources differently in various regions of the world.

ACKNOWLEDGMENTS

The authors express their sincere thanks to the Flemish Fund for Scientific Research (FWO-Vlaanderen) for supporting the project that resulted in the interviews. The Museo Antropológico Padre Sebastian Englert with its director, Francisco Torres, acted as counterpart. We also wish to thank the members of the Consejo Provincial de Monumentos Nacionales Rapa Nui for their interest in the interview project, the Consejo de Ancianos, the governor of the province at the time of the mission, Enrique Pakarati Ika, the Corporación Nacional Forestal (CONAF), the Servicio Agrícola y Ganadero (SAG) and the Sociedad Agrícola y de Servicios Isla de Pascua Ltda (SASIPA). We are also grateful to Sonia Haoa and all the interviewees who participated in the ethnoarchaeological study for their cooperation and interest: Luis Avaka Paoa, Juan Chávez Haoa, Luisa Fati, Nicolas Haoa, Sr, Juan Haoa Hereveri, Maria Auxilia Hereveri Pakomio, Juan Hey, Alberto Hotus Chávez, Maria Kristina Manutomatoma Pakarati, Omar Castillo, Evi Pakarati Tepano, Rodrigo Paoa Atamu and 'Kipi'.

REFERENCES

Barthel, T (1978) *The Eighth Land: The Polynesian Discovery and Settlement of Easter Island*, Honolulu, HI: University Press of Hawaii

Charola, A E (1997) *Death of a Moai: Easter Island Statues, Their Nature, Deterioration and Conservation*. Easter Island Foundation Occasional Paper 4, Los Osos, CA: Easter Island Foundation

Chávez Haoa, J (1998) 'Lunar calendar for fishery and agriculture', unpublished manuscript, Rapa Nui (Easter Island), Chile

Childs, C W (1985) *Towards Understanding Soil Mineralogy, II: Notes on Ferrihydrite*. Laboratory Report CM7, Lower Hutt, New Zealand: New Zealand Soil Bureau

Englert, P S (1974) *La Tierra de Hotu Matu'a: Historia y Etnología de la Isla de Pascua*, Santiago de Chile, Chile: Editorial Universitaria

FAO (1976) *A Framework for Land Evaluation*. Soils Bulletin 32, Rome: Food and Agriculture Organization of the United Nations (FAO)

FAO (1985) *Agroclimatological Data: Latin America and the Caribbean*. Plant Production and Protection Series 24, Rome: Food and Agriculture Organization of the United Nations (FAO)

Fischer, S R and Love, C M (1993) 'Rapanui: the geological parameters', in Fischer, S R (ed), *Easter Island Studies: Contributions to the History of Rapanui in Memory of William T. Mulloy*, pp 1–6. Oxbow Monograph 32, Oxford: Oxbow

Flenley, J R (1996) 'Further evidence of vegetational change on Easter Island', *South Pacific Study* (Kagoshima, Japan) 16(2): 135–41

Flenley, J and Bahn, P (2003) *The Enigmas of Easter Island: Island on the Edge*, Oxford: Oxford University Press

Goosse, H, Masson-Delmotte, V, Renssen, H, Delmotte, M, Fichefet, T, Morgan, V, van Ommen, T, Khim, B K and Stenni, B (2004) 'A late medieval warm period in the Southern Ocean as delayed response to external forcing?', *Geophysical Research Letters* 31(6): L06203

Handy, E S C and Handy, G E (1991) *Native Planters in Old Hawaii: Their Life, Lore, and Environment*. Bernice P Bishop Museum Bulletin 233, Honolulu, HI: Bishop Museum Press

Hauser, Y A (1998) 'Capítulo 8 – localización y caracterización morfológica, geológica y geotécnica de eventuales sitios para la captación, almacenamiento y distribución de agua de riego en Isla de Pascua', in Ingeniería Agrícola Limitada (ed), 'Diagnóstico para el Desarrollo Integral de Isla de Pascua: Proyecto Piloto de Riego en Cultivos Hortofrutícolas, V Región', vol I: 'Diagnóstico de Situacion Actual', unpublished report, Departamento de Estudios, Comisión Nacional de Riego, Santiago, Chile

Hotus Chávez, A (2000) *Diccionario Etimológico Rapanui-Español*. Comisión para la Estructuración de la Lengua Rapanui, Valparaíso, Chile: Puntángeles Universidad de Playa Ancha Editorial

Kondratov, A M (1965) 'Appendix D. The hieroglyphic signs and different lists in the manuscripts from Easter Island', in Heyerdahl, T and Ferdon, Jr, E N (eds), *Reports of the Norwegian Archaeological Expedition to Easter Island and the East Pacific*, vol 2, pp 387–416, Chicago: Rand McNally and Co

Louwagie, G (2004) 'Palaeo-environment reconstruction and evaluation based on land characteristics on archaeological sites' [in English with Dutch abstract], unpublished PhD thesis, Ghent University

Louwagie, G and Langohr, R (2005) 'Relevance of archaeological experiments in crop productivity for evaluating past land-use systems', in Meyer, M and Wesselkamp, G (eds), *Zu den Wurzeln europäischer Kulturlandschaft*, pp 57–66. Materialhefte zur Archäologie in Baden-Württemberg 73, Stuttgart: Konrad Theiss

Louwagie, G, Stevenson, C M and Langohr, R (2006) 'The impact of moderate to marginal land suitability on prehistoric agricultural production and models of adaptive strategies for Easter Island (Rapa Nui, Chile)', *Journal of Anthropological Archaeology* 25(3): 290–317

Mann, D, Chase, J, Edwards, J, Beck, W, Reanier, R and Mass, M (2003) 'Prehistoric destruction of the primeval soils and vegetation of Rapa Nui (Isla de Pascua, Easter Island)', in Loret, J and Tanacredi, J (eds), *Easter Island*, pp 133–53, New York: Kluwer Academic/Plenum Publishers

McMillan Brown, J ([1924] 1996) *The Riddle of the Pacific*, Kempton, IL: Adventures Unlimited Press

Métraux, A (1957) *A Stone-Age Civilization of the Pacific*, London: Andre Deutsch

Norero, S A (1998) 'Capítulo 4 – Aspectos climáticos y ecofisiológicos de Isla de Pascua', in Ingeniería Agrícola Limitada (ed), 'Diagnóstico para el Desarrollo Integral de Isla de Pascua: Proyecto Piloto de Riego en Cultivos Hortofrutícolas, V Región', vol I: 'Diagnóstico de Situacion Actual', unpublished report, Departamento de Estudios, Comisión Nacional de Riego, Santiago, Chile

Norman, M J T, Pearson, C and Searle, P G E (1995) *The Ecology of Tropical Food Crops*, Cambridge: Cambridge University Press

Orliac, C and Orliac, M (1999) 'Evolution du couvert végétal à l'île de Pâques du 15ième au 19ième siècle', in Vargas Casanova, P (ed), *Easter Island and East Polynesian Prehistory*, pp 195–200, Santiago, Chile: Instituto de Estudios de Isla de Pascua, Facultad de Arquitectura y Urbanismo, Universidad de Chile

Parfitt, R L (1990) 'Allophane in New Zealand – a review', *Australian Journal of Soil Research* 28: 343–60

Raemaekers, R H (ed) (2001) *Crop Production in Tropical Africa*, Brussels: Directorate General for International Co-operation (DGIC), Ministry of Foreign Affairs, External Trade and International Co-operation

Renfrew, C and Bahn, P (1996) *Archaeology: Theory, Methods and Practice*, London: Thames and Hudson

Sombroek, W G (1997) 'Land resources evaluation and the role of land-related indicators', in *Land Quality Indicators and Their Use in Sustainable Agriculture and Rural Development*. Land and Water Bulletin 5, Rome: Food and Agriculture Organization of the United Nations (FAO)

Stevenson, C (2002) 'Territorial divisions on Easter Island in the sixteenth century: evidence from the distribution of ceremonial architecture', in Ladefoged, T N and Graves, M W (eds), *Pacific Landscapes, Archaeological Approaches*, pp 211–229, Los Osos, CA: Bearsville Press

Stevenson, C, Wozniak J and Haoa, S (1999) 'Prehistoric agricultural production on Easter Island (Rapa Nui), Chile', *Antiquity* 73: 801–12

Sys, C, Van Ranst, E and Debaveye, J (1991) 'Land evaluation, part I: principles in land evaluation and crop production calculations', unpublished lecture notes, International Centre for Post-Graduate Soil Scientists, State University of Gent [Ghent]

Thaman, R, Elevitch, C and Wilkinson, K (2003) 'Traditional Pacific island agroforestry systems', *The Overstory: Agroforestry ejournal* 49, Holualoa, HI: Agroforestry Net (www.agroforestry.net/overstory/overstory 49.html)

UNDP (United Nations Development Program) (1998) 'Recuperación de Flora Nativa Arbustiva y Comestible de Isla de Pascua (CHI/97/G01)', unpublished final report, Corporación Mata Nui A Hotu A Matu'a O Kahu Kahu O'Hera

USDA (US Department of Agriculture, Natural Resources Conservation Service) (1999) *Soil Taxonomy: A Basic System of Soil Classification for Making and Interpreting Soil Surveys.* Agriculture Handbook 436, Washington, DC: US Government Printing Office

Wozniak, J A (2001) 'Landscapes of food production on Easter Island: successful subsistence strategies', in Stevenson, C, Lee, G and Morin, F J (eds), *Pacific 2000: Proceedings of the Fifth International Conference on Easter Island and the Pacific*, pp 91–101, Los Osos, CA: Easter Island Foundation

Yen, D E (1988) 'Easter Island agriculture in prehistory: the possibilities of reconstruction', in Cristino, C, Vargas Casanova, P, Izaurieta, R and Budd, R (eds), *Proceedings of the First International Congress on Easter Island and East Polynesia, Hanga Roa, Easter Island, 1984*, vol I: *Archaeology*, pp 59–81, Santiago, Chile: Instituto de Estudios Isla de Pascua, Facultad de Arquitectura y Urbanismo, Universidad de Chile

CHAPTER 9

New Perspectives on Plant Domestication and the Development of Agriculture in the New World

José Iriarte

INTRODUCTION

The aim of this volume is to present diverse conceptual, archaeological and ethnoarchaeological perspectives on early plant domestication and the development and spread of agriculture in several non-Eurasian regions of the world. This chapter provides an overview of the major advances in these issues during the last decade in Central and South America (see Smith 1992 and Fritz chapter 10 for overviews of eastern North America). It is not my intention to provide a detailed description of all the new evidence and ideas that have accumulated over the past 10 years, but rather to draw attention to the most important developments. Comparison of our current understanding of the emergence of food production in the Americas with traditional Eurasian models[1] will serve as a point of departure to reflect on the current state of our knowledge of agriculture in the Americas (Vrydaghs and Denham chapter 1; see also Ammerman and Biagi 2003 and Price 2000 for alternative views to traditional Eurasian models). I chose Eurasia because this region is one of the best studied and consequently Eurasian models have had an enormous impact on concepts about how agriculture arose, including in the Americas.

Despite the fact that the neotropics provide more than 50% of crop plants cultivated in the Americas (Clement 1999; Piperno and Pearsall 1998) and that indigenous societies of the tropical forest 'domesticated a larger assemblage of root and tuber crops than anyone else on earth' (Harlan 1995: 190; see also Hawkes 1989), it is only during the last two decades that intensive research employing adequate techniques for archaeobotanical recovery has been carried out in this region. Historically, the initial work on the transition to agriculture in the Americas was biased toward those geographical areas exhibiting good preservation of desiccated and charred macrobotanical remains, such as the western desert coast of South America (Towle 1961) and the dry highland caves of central Mexico and Peru (Flannery 1986; Lynch 1980; MacNeish 1992). The results of these investigations suggested that plant domestication probably began independently only in two regions, Mexico and Peru. When and where the numerous indigenous American root and tuber crops had been originally brought under cultivation and domesticated, and what role the neotropics and areas beyond Mexico and Peru played in the development and spread of food production, were little understood. Nonetheless, decades earlier, Sauer (1952) had postulated the importance of seasonal tropical forests and the role of underground plant organs in early agriculture. Later, Lathrap (1970) proposed the priority of lowland vegeculture, with manioc as the founder crop that motivated the development and spread of food production throughout South and Central America. Lathrap also envisioned that house gardens played crucial roles as experimental plots in the early stages of neotropical crop domestication (Lathrap 1977). Sauer and Lathrap's views were influential and widely discussed, but empirical evidence to support them was rare. For one thing, few archaeological projects had been carried out in the lowland tropical forest. Only a small number of these investigations saw the systematic application of archaeobotanical recovery techniques and, in any case, macrobotanical remains were usually poorly preserved in early sites of interest. Over the last two decades, the investigation of plant domestication and the dispersals of cultigens in the Americas has been revitalised by (a) major technical methodological breakthroughs and their incorporation as standard components of archaeological excavation and data analysis; (b) new archaeological projects in previously unexplored regions; and (c) the elaboration of new models to conceptualise and form hypotheses concerning early plant domestication.

The systematic application of microfossil botanical analyses is revolutionizing our knowledge of early food production in the Americas. Until recently, we were not able to document root and tuber crop domestication unless sites investigated were located in arid environments (eg the Peruvian coast and the Tehuacán Valley), where conditions permitted the survival of these soft underground plant organs. Moreover, these arid regions were outside the original areas of domestication of many of the major crops for which they yielded evidence (eg manioc, sweet potato, squashes, cotton, chile peppers). These data lacunae have been overcome with the refinement of microfossil botanical techniques, in particular phytolith and starch grain studies (Pearsall et al 2004; Perry et al 2007; Piperno and Holst 1998; Piperno et al 2000; Piperno 2006a). Starch grain research, as well as studies of charred remains of parenchyma tissue from underground plant organs, are now providing robust information on root crop domestication and spread (and on various important seed crops) in tropical regions of the Americas and additionally in the Old World tropics (eg Denham et al 2003; Pearsall et al 2004; Piperno et al 2000; Piperno 2006a; Scheel-Ybert 2001). Many of these studies retrieved and studied starch grain residues from stone tools (eg grinding stones, chipped stone) that were used to process plants. Starch residue has also illuminated the functions of these tools and forced revisions to traditional interpretations of stone tools that were based on less direct forms of evidence, such as ethnographic analogy (eg Perry 2004).

Molecular research is a second major addition to the arsenal of researchers in the Americas and elsewhere who seek to understand crop plant domestication. Molecular data are helping to fingerprint the wild progenitors of domesticated plants and pinpoint their present day and, by inference, ancient geographical distributions. Molecular data have informed us that some crops were probably domesticated only once. Among these plants are maize (*Zea mays* L.) (Doebley 1990; Matsuoka et al 2002), manioc (*Manihot esculenta*) (Olsen and Schaal 1999, 2001), potatoes (*Solanum tuberosum*) (Spooner et al 2005), and cotton (*Gossypium barbadense*) (Westengen et al 2005) (Figure 9.2). In contrast, other major crop plants probably were domesticated more than once in different regions of the Americas. They include squash species such as *Cucurbita pepo* (Sanjur et al 2002), common beans (*Phaseolus vulgaris*) (Sonnante et al 1994), and sunflower (*Helianthus annuus*) (Harter et al 2004). The increasing incorporation of the molecular studies of plant domesticates' wild ancestors and their geographical distribution with more traditional archaeological studies is just beginning and almost certainly will reveal many fascinating details of plant domestication in the Americas (Armelagos and Harper 2005; Emschwiller 2006; Smith 2001a).

A third major area of data analysis pertinent to the origins of agriculture that has considerably informed our understanding of plant domestication is palaeoecological research. This has proven invaluable for identifying the natural environments and plant associations in which the first farming arose. Palaeoecological analysis has also been very useful in documenting early slash-and-burn agriculture, which unlike irrigation canals, raised fields, and agricultural terraces, does not leave visible imprints on landscapes (see Pearsall chapter 11). Direct AMS dating of macrobotanical remains and more recently phytoliths and pollen retrieved both from archaeological and palaeoecological contexts is providing more precise chronologies of plant domestication and environmental history (eg Kaplan and Lynch 1999; Long et al 1989; Piperno and Flannery 2001; Piperno and Stothert 2003; Smith 2005).

As a result of these advances, our understanding of prehistoric plant domestication has both increased and changed significantly. The new data are revising our perceptions of the geographical origins of agriculture and early dispersals, chronology, and the diversity of indigenous food-producing economies in the Americas. Each of these major areas of knowledge will be discussed below.

CENTRES AND NON-CENTRES

In contrast to traditional Eurasian models, but similar to the picture that is emerging in New Guinea and Africa (Vrydaghs and Denham chapter 1), the available data suggest that the domestication of plants in the neotropics was characterised by a diffuse spatial pattern (Figures 9.1 and 9.2)[2]. Endorsing previous views (Bray 2000; Harlan 1995; Piperno and Pearsall 1998), current archaeological evidence shows that there was no single centre of domestication or agricultural origins in the Americas. On the contrary, independent origins in Central and South America are now apparent and multiple origins within lowland South America are likely (Piperno 2006b). It is also becoming clear that crop plants in the same genus, such as squashes and gourds of *Cucurbita*, were sometimes brought under domestication early on in several different regions, such as highland Mexico (Smith 1997a), tropical southern Mexico (Sanjur et al 2002), southwestern Ecuador (Piperno and Stothert 2003), and other areas of South America (Sanjur et al 2002).

In addition to the high-altitude Andes and eastern North America, the Central and South American tropical forests are now well established as major and independent regions of plant domestication. The Andean mountain chain gave birth to a diversity of tubers including potato (*Solanum tuberosum*), oca (*Oxalis tuberosa*) and ulluco (*Ullucus tuberosus*), as well as pseudocereals such as quinoa (*Chenopodium quinoa*) (Bruno 2006; MacNeish 1992; Pearsall 1989, 1992; Smith 2001b). In eastern North America, a suite of four indigenous seed crops were cultivated and/or domesticated beginning during the mid-Holocene (Smith 1992; Fritz chapter 10). Completing the picture, current archaeological, botanical, and molecular evidence suggest

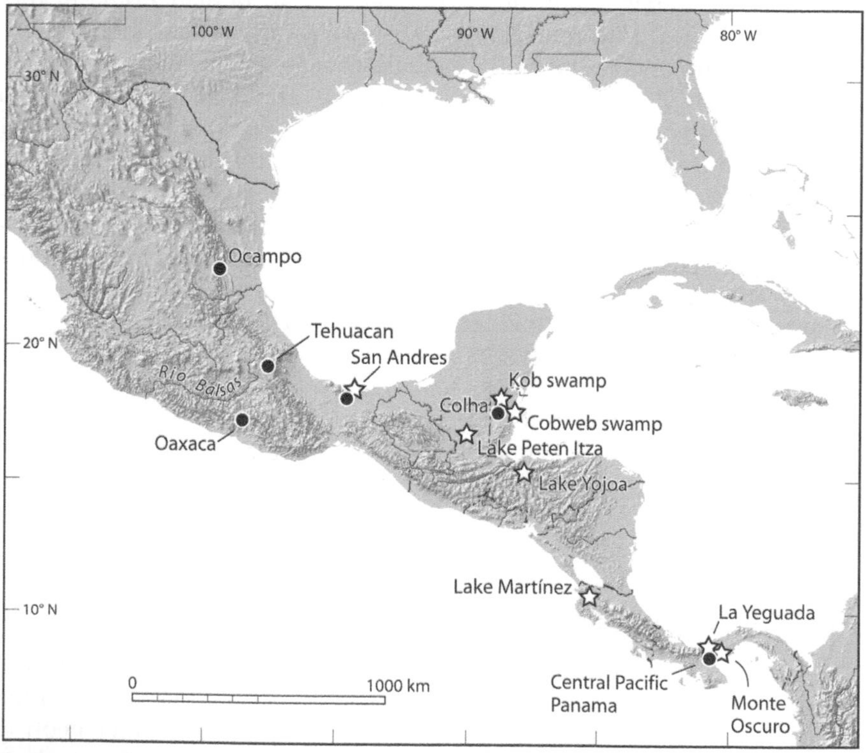

Figure 9.1 Map showing the location of archaeological (circles) and palaeoecological sites (stars) that have yielded early evidence of domesticated plants and food-producing practices in Central America.

independent origins of plant cultivation and domestication in at least four other areas: southwestern Ecuador/northern Peru, southwestern Amazonia, northern South America (Colombia/Venezuela/the Guianas/northern Brazil), and southwestern Mexico (Piperno and Pearsall 1998; Piperno 2006b) (see Figures 9.1–3, Table 9.1).

The new picture emerging from the neotropics is fundamentally different from traditional Eurasian models. The latter point to discrete centres of early agricultural development where only a few founder seed crops, usually displaying fairly easily identified morphogenetic changes, were brought under domestication. These crops subsequently spread as a package over most areas of Europe and they continued under cultivation without interruption

Figure 9.2 Map showing the location of South American archaeological (circles) and palaeoecological (stars) sites that have yielded early evidence of domesticated plants and food-producing practices along with the some of the proposed geographical areas for the domestication of cultivars.

up to the present. In contrast, the available data from the neotropics suggest a mosaic-like pattern diffuse in space, with multiple areas of early, independent agriculture involving different plants. Some of these plants would continue to be important dietary items in their native regions while others would be supplanted later in prehistory by different crops, some of which were introduced from elsewhere.

A MATTER OF TIME

The domestication of plants in the New World has been shown to have occurred as early as in other parts of the world, such as Southwest Asia, where plant food production developed shortly after the Pleistocene ended (Figure 9.3, Table 9.1). There were some concerns, brought about by the direct AMS dating of a few crop plants, including maize, recovered from the Tehuacán Valley in Mexico (eg Long et al 1989) and beans from Guitarrero Cave in Peru (Kaplan 1994), that the chronology of agricultural origins in the Americas would have to be considerably revised to reflect an age no older than 5000 BP[3] (Fritz 1994a). However, evidence accumulated during the past 15 years indicates otherwise. This evidence is currently strongest in a region from lower Central America to northwestern South America comprising Panama, Ecuador, and Colombia. Here, a suite of microbotanical studies (pollen, phytoliths and starch grains) indicate that human manipulation of neotropical plant species – including squashes and gourds (*C. moschata, C. ecuadorensis*), arrowroot (*Maranta arundinacea*), manioc (*Manihot esculenta*), leren (*Calathea allouia*), yams (*Dioscorea* spp.), and maize (*Zea mays*) – resulted in their domestication during the early Holocene (10,000–7000 BP) (Aceituno and Castillo 2005; Dickau et al 2007; Olivier 2001; Piperno 2006b; Piperno and Pearsall 1998; Pohl et al 2007). Macrobotanical analyses at a number of different South American sites have also documented the early use and human manipulation of palms (see Morcote-Ríos and Bernal 2001 for a review) and other tree fruits (eg Resende and Prous 1991; Roosevelt et al 1996).

Contextual evidence for early agriculture has also been forthcoming from palaeoecological records that were retrieved and studied from many of the same regions where archaeological research yielded data for early crop plants. These environmental records indicate that beginning between 7000 BP and 5000 BP, depending on the region, crops were grown under slash-and-burn techniques of field preparation (Athens and Ward 1999; Mora 2003; Piperno 2006b; Piperno and Pearsall 1998). Similarly, the use and manipulation of wetlands were earlier and more persistent activities than previously thought (eg Iriarte et al 2004; Pohl et al 1996; Pope et al 2001).

Recent evidence from the Zaña Valley, north coast of Peru, shows that irrigation agriculture probably started at 6500 BP and was well established by 5400 BP (Dillehay et al 2005). Previous work in this region (Dillehay et al 1989, 1997) has documented changes in settlement patterns by local populations to exploit fertile alluvial lands, the appearance of the earliest forms of public architecture and evidence of early cultigens by at least the 7th millennium BP. Molecular and other studies point to this region, called the southwestern Ecuador/northwestern Peru zone of domestication by Piperno and Pearsall (1998), as the potential domestication cradle for cotton, lima beans, *Canavalia* beans, and *Cucurbita ecuadorensis* (eg Westengen et al 2005; see Piperno and Pearsall 1998: 163–166 for an overview). In southwestern Ecuador, phytolith studies at sites belonging to the early Holocene Las Vegas cultural tradition have yielded evidence for the domestication of a native species of squash, *Cucurbita ecuadorensis* between 10,000 and 9300 BP (Piperno and Stothert 2003) and the arrival of maize to the region between around 7000 and 6600 BP (Piperno and Pearsall 1998: 183–189). In the same region of Ecuador,

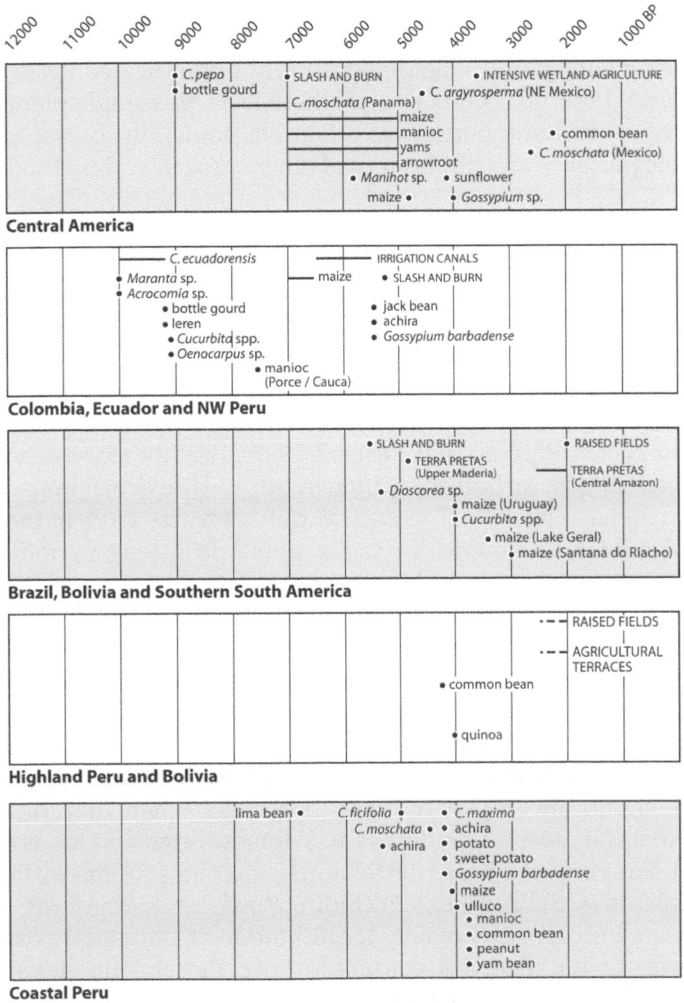

Figure 9.3 Chronology for the earliest appearance of domesticated plants and food-producing practices in different regions of Central and South America.

multiproxy macro- and microbotanical analyses of sediments and artefacts from the early ceramic Valdivia cultural occupations, including numerous grinding stones from intact Valdivia 3 house floors at the site of Real Alto, indicate the presence of maize, achira (*Canna edulis*), jack beans (*Canavalia plagiosperma*), and cotton (*Gossypium barbadense*) between 5500 and 4300 BP (Pearsall 2003; Pearsall et al 2004).

MAIZE, MAIZE AND MAIZE

Unlike Eurasian agriculture, which was traditionally focused on relatively few cereals, a number of different plants are known to have been brought under domestication in the Americas, including a variety of root crops (manioc, sweet potato, yams), beans and pulses, squashes, vegetables, tree crops and condiments (Figure 9.3 and Table 9.1)[3]. The American crop plant complex includes only one demonstrably domesticated grass grain: corn (*Zea mays*) (Piperno 2006b). Despite these facts, this single plant, maize, has long received disproportionate attention, somehow overshadowing the probable presence and importance of other crops in ancient food production systems.

Maize has what will probably be demonstrated to have been a long and complex history of domestication. Despite great advances made in understanding the ancestry of maize and the underlying genetic basis for the conversion of wild Balsas teosinte into domesticated maize (eg Doebley 1990; Jaenicke-Després et al 2003; Matsuoka et al 2002), plus the demonstrated presence of maize in Costa Rica, Panama, Colombia, and Ecuador between 7000 and 4700 BP (Arford and Horn 2004; Bush et al 1989; Dickau et al 2007; Piperno 2006b; Piperno and Pearsall 1998), the full story of maize evolution, including its earliest cultivation in tropical Mexico, remains to be documented archaeologically. The great variety of maize races in the Andes and in lowland South America probably points to a long period of manipulation, experimentation and adaptation to local conditions (Bonavia and Grobman 1989; Brieger et al 1958).

There is a tendency to perceive that none of the numerous other American crops were capable of being staple crops equivalent to maize during Preceramic times. Roots and tubers, beans, squash and tree fruits are usually considered accompaniments to the diet rather than staple items of early food-producing systems (eg Bellwood 2005; Browman et al 2005). The dietary significance of these crops, especially when they first appear in archaeological records as cultivars and domesticated species, is often downplayed. For example, Fritz (1994b) argues that most of the earliest domesticated plants in the Americas, including squashes and gourds of *Cucurbita* spp., were initially used mainly for utilitarian or industrial purposes (eg as containers or net floats). According to Fritz, these plants fit easily into the subsistence practices and highly mobile settlement cycles of groups that essentially remained foragers for a long time after they initially developed or obtained domesticated plants. Once again, it is not until maize becomes a staple crop that 'true' agriculture is perceived to have developed.

There are several problems with these views as they relate to a maize-centric perspective and also as they frame perspectives of early farming in general. First, with regard to plants such as *Cucurbita*, it is much more likely that various squash species were domesticated at early times in both Central and South America because early farmers valued their edible, nutritious, oil and protein-rich seeds, as had the pre-agrarian populations that first incorporated them into their diets. Furthermore, plants that were better suited to be containers and net floats, such as bottle gourd (*Lagenaria siceraria*) and *Crescentia* spp. (the tree gourds), were available. Recent recoveries and direct dating of bottle gourd remains make it clear that the plant was well dispersed throughout Central and South America by the early Holocene (Erickson et al 2005). The paucity of scalloped phytoliths from *Cucurbita* at many sites studied from the lowland tropical forest dating to 7000 BP and after also indicates that by that time squashes probably have soft rinds and edible, non-bitter flesh (Piperno 2006c; Piperno et al 2002).

These approaches also fail to take into account other important factors. They commonly assume, for example, that before sedentary villages were established, human populations throughout Central and South America that were food producers were primarily organised as highly mobile hunters and gatherers and cultivated plants were no more than minor dietary supplements (Browman et al 2005; Fritz 1994b). However, many of the sites in question exhibit dense middens starting by at least 7000 BP; settlement patterns have been reconstructed as occupations of small but numerous groups of farmers. Associated palaeoecological and settlement survey data indicate that slash-and-burn cultivation was being practiced and that these societies were organised as food producers, not foragers (see Piperno and Pearsall 1998). The early Holocene coastal Vegas sites and others may have been occupied year-round (Stothert et al 2003). It can be a difficult task to calculate how many calories were coming from cultivated/domesticated versus wild food products, but the presence of a variety of cultivars at these sites plus the strong evidence for slash-and-burn cultivation argue for societies that were deriving many calories from plants that they grew. It would be a misapprehension to call them hunters and gatherers.

Many root and tuber crops domesticated in the Americas (manioc, sweet potatoes, and achira) often exceed or are equivalent to maize in caloric production (Piperno and Pearsall 1998: 110–128; Raymond 1981: 811–15). The idea that root and tuber crops constituted a large part of the caloric intake of early food-producing economies in the Americas is now finding empirical support through recent starch grain studies of early and middle Holocene-age plant grinding stones. Mixed economies dating to the Preceramic period between 10,000 and 7000 BP that included arrowroot, leren, yams, manioc, squashes and gourds, and a variety of tree crops are well documented in southern Central and South America. In the mid- and high-altitude Andes potatoes, ulluco, oca, and quinoa must have played important roles in Preceramic economies (eg Burger and van der Merwe 1990). New research is needed to document the different crop histories in such an important region of the Americas. As in part reviewed above, maize has been documented at a number of sites ranging from Costa Rica to southeastern Uruguay between 7000 and 4000 BP (Arford and Horn 2004; Dickau et al 2007; Iriarte et al 2004; Pearsall et al 2004; Perry et al 2006; Piperno and Pearsall 1998; Piperno et al 2000). After decades of neglect, pre-maize early food-producing economies are now being investigated throughout the Americas and are beginning to yield results very different from traditional maize-centric formulations.

In summary, recent research has revealed a wide array of early indigenous food-producing economies that developed across the Americas. Some of these appear to have incorporated a combination of wild, managed, cultivated and domesticated plants. Where good palaeoecological and settlement data are available, it is apparent that human populations were actively transforming and maintaining their landscapes for the purpose of growing their food. Regardless of what we choose to call these societies for analytical purposes, characterising them as highly mobile hunters and gatherers or as groups of foragers who augmented their diets with small portions of cultivated foodstuffs (eg Browman et al 2005) is likely inappropriate. Pre-Hispanic cultures that were initially conceptualised as 'complex hunter-gatherers' are now understood to have practiced food production using a variety of domesticated crops. Some examples are the Hopewell of Midwestern United States (eg Asch and Asch 1985) and the mound builders of Uruguay (Iriarte et al 2004). Furthermore a variety of domesticated plants are now firmly documented at inland Late Preceramic sites along the coast of Peru (Raymond 1981; Sandweiss and Moseley 2001; Shady et al 2001). Similarly, recent starch grain analyses from Preceramic sites on Puerto Rico and Vieques in the Caribbean provide the first direct evidence for the presence of maize, beans, manioc, sweet potatoes, yams, and *Xanthosoma* and *Canna* tubers between 3300 and 2700 BP (Pagán Jiménez et al 2005). As in the regions mentioned above, these new results prompt a revaluation of the Archaic economies in this Caribbean region.

The coast of southcentral Brazil is another example. Charcoal analysis from six shell mounds along the southern coast of the state of Rio de Janeiro, Brazil, dated between c 5500 BP and 1400 BP, documented for the first time the use of yams (*Dioscorea* sp.) in addition to palm and fruit trees in this region (Scheel-Ybert 2001). This research corroborates previous investigations from other preceramic Archaic occupations (4200–3000 BP) from the southern coast of Brazil, where the study of dental wear patterns from 46 adult crania from the Corondo site (Rio de Janeiro) documented high rates of caries, suggesting these Archaic populations had a 'high-consumption' of starchy plants (Turner and Machado 1983). At the Santana do Riacho rockshelter in central Brazil, desiccated maize cobs, squash, several

Table 9.1 Early plant remains and food-producing practices in major regions of Central and South America.

Plant Taxon	*Common Name*	*Radicarbon Age (BP)*	*Site*	*Region*	*Reference*
Central America					
Lagenaria siceraria	bottle gourd	9000 (AMS)	Guila Naquitz	Oaxaca	Erickson et al 2005
Cucurbita pepo	squash	8910–8990 (AMS)	Guila Naquitz	Oaxaca	Smith 1997a: 933
Dioscorea sp.	yam	7000–5000	Aguadulce	Central Pacific Panama	Piperno et al 2000
Manihot esculenta	manioc	7000–5000	Aguadulce	Central Pacific Panama	Piperno et al 2000
Maranta arundinacea	arrowroot	7000–5000	Aguadulce	Central Pacific Panama	Piperno et al 2000
Zea mays	corn	7000–5000	Aguadulce	Central Pacific Panama	Dickau et al 2007; Piperno et al 2000
Zea mays	corn	6200	San Andrés	Tabasco	Pope et al 2001: 1372
Zea mays	corn	4760	Lake Martínez	Guanacaste, Costa Rica	Arford and Horn 2004
Zea mays	corn	5420	Guila Naquitz	Oaxaca	Piperno and Flannery 2001
Cucurbita argyrosperma	squash	4450 (AMS)	Tamaulipas	Ocampo	Smith 1997b: 373
Cucurbita moschata	squash	2620 (AMS)	Tamaulipas	Ocampo	Smith 1997b: 373
Cucurbita moschata	squash	8000–7000	Aguadulce	Central Pacific Panama	Piperno 2006b
Helianthus annuus	sunflower	4130 (AMS)	San Andrés	Tabasco	Lentz et al 2001
Gossypium sp.	cotton	4000	San Andrés	Tabasco	Pope et al 2001: 1372
Phaseolus vulgaris	common bean	2285	Coxcatlán	Tehuacán Valley	Kaplan and Lynch 1999
<u>Agricultural practices</u>					
Slash and burn		7000	La Yeguada	Central Pacific Panama	Piperno and Pearsall 1998
Intensive wetland agriculture		3500	Cobweb Swamp	Belize	Pohl et al 1996

(*Continued*)

Table 9.1 (*Continued*)

Plant Taxon	*Common Name*	*Radicarbon Age (BP)*	*Site*	*Region*	*Reference*
Colombia, Ecuador and Northwest Peru					
Acrocomia sp.	corozo palm	10,050–9539	San Isidro	Upper Cauca Valley	Piperno and Pearsall 1998: 199–203
Maranta sp.	arrowroot	10,050–9539	San Isidro	Upper Cauca Valley	Piperno and Pearsall 1998: 199–203
Cucurbita ecuadorensis	squash	9320–10,130	Vegas	Southwestern Ecuador	Piperno and Stothert 2003: 1055
Calathea alluoia	leren	9320	Vegas	Southwestern Ecuador	Piperno and Stothert 2003: 1055
Lagenaria siceraria	bottle gourd	9320	Vegas	Southwestern Ecuador	Piperno and Stothert 2003: 1055
Zea mays	corn	7000–6700	Las Vegas	Southwestern Ecuador	Piperno and Pearsall 1998: 187
Zea mays	corn	5300	Lake Ayacuchi	Ecuadorian Amazonia	Bush et al 1989
Cucurbita spp.	squash	9160	Peña Roja	Upper Caquetá Valley	Piperno and Pearsall 1998: 203–206
Manihot esculenta	manioc	7500	Middle Cauca and Porce valleys	Colombia	Aceituno and Castillo 2005
Canavalia plagiosperma	jack bean	5500	Real Alto	Southwestern Ecuador	Pearsall et al 2004
Canna edulis	achira	5500	Real Alto	Southwestern Ecuador	Pearsall et al 2004
Calathea alluoia	leren	9160	Peña Roja	Middle Caquetá Valley	Piperno and Pearsall 1998: 203–206
Lagenaria siceraria	bottle gourd	9160	Peña Roja	Middle Caquetá Valley	Piperno and Pearsall 1998: 203–206
Oenocarpus sp.	bataua	9160	Peña Roja	Middle Caquetá Valley	Piperno and Pearsall 1998: 203–206
Agricultural practices					
Slash-and-burn		5300	Ayauchi	Ecuadorian Amazonia	Bush et al 1989

Table 9.1 *(Continued)*

Plant Taxon	*Common Name*	*Radicarbon Age (BP)*	*Site*	*Region*	*Reference*
Irrigation canals		6500–5500	Zaña Valley	NW Peru	Dillehay et al 2005
Brazil, Bolivia, and southern South America					
Dioscorea sp.	yam	5500–1400	Sambaqui do Forte	Southern Brazil	Scheel-Ybert 2001
Zea mays	corn	4000	Los Ajos	SE Uruguay	Iriarte et al 2004
Cucurbita spp.	squashes	4000	Los Ajos	SE Uruguay	Iriarte et al 2004
Zea mays	corn	3350	Lake Geral	Eastern Amazonia	Bush et al 2000
Zea mays	corn	3000	Santana do Riacho	Central Brazil	Prous 1999
Agricultural practices					
Terra preta		2500–2000	Açatuba	Central Amazon	Neves et al 2003
Slash-and-burn		5500	Lake Geral	Eastern Amazonia	Bush et al 2000
Terra preta		4780	RO-PV-48 Cachoeirinha	Upper Madeira River	Miller 1992
Raised fields		2000	Llanos de Mojos	Eastern Bolivia	Erickson 1995; Walker 2004
Higland Peru and Bolivia					
Chenopodium sp.	quinoa	5000–4000	Panaulauca Cave		Pearsall 1989; Smith 2001b
Phaseolus vulgaris	common bean	4337	Guitarrero Cave		Kaplan and Lynch 1999
Coastal Peru					
Phaseolus lunatus	lima bean	6920	Chilca	Chillon Valley	Kaplan and Lynch 1999
Cucurbita ficifolia	squash	5000	Paloma	Chillon Valley	(D. Piperno pers comm 2005)
Cucurbita moschata	squash	4500	Ventanilla	Chillon Valley	Piperno 2006b
Ipomoea batatas	sweet potato	4250	Huaynuma	Casma Valley	Ugent et al 1981
Canna edulis	achira	4250	Huaynuma	Casma Valley	Ugent et al 1984
Solanum tuberosum	potato	4250	Huaynuma	Casma Valley	Ugent et al 1982
Cucurbita maxima	squash	4250	Huaynuma	Casma Valley	Ugent et al 1984

(Continued)

Table 9.1 *(Continued)*

Plant Taxon	*Common Name*	*Radicarbon Age (BP)*	*Site*	*Region*	*Reference*
Zea mays	corn	4100	Los Gavilanes	Huarmey Valley	Bonavia 1982
Ullucus tuberosus	ulluco	4000–3350	Late Preceramic sites	Ancón-Chillón Valley	Martins-Farias 1976
Manihot esculenta	manioc	3800		Casma Valley	Ugent et al 1986
Pachyrrhyzus sp.	yam bean	3750	Los Gavilanes	Huarmey Valley	Bonavia and Grobman 1989

palms and other tree-fruits have been minimally dated to 3000 BP (Prous 1999: 346; Resende and Prous 1991).

Additional work in little-explored regions of Central America, South America and the Caribbean, together with the systematic application of botanical recovery techniques, in particular microbotanical analyses, will surely provide more examples where our current understanding of the pre-Archaic economy will have to be revised.

LOOKING AHEAD: WHAT DOES THE FUTURE HOLD?

The neotropics and the Americas in general have become cornerstones for understanding the regional and global processes associated with the transition to agriculture. Despite these advances, basic data are still lacking for major areas of South America. In particular, there are unanswered questions about the chronology of the transition and the specific mechanisms (ie indigenous adoption versus colonization) that led to the adoption of domesticates and the spread of agriculture.

More systematic research is needed in the deciduous tropical forest regions of South and Central America. Seasonal tropical forest habitats appear to have been important foci of early human settlement, the natural habitats of many wild progenitors of important crop plants and origin places for many domesticated species (Piperno and Pearsall 1998: 50–72).

In contrast to their evergreen forest counterparts, deciduous tropical forests produce considerably more underground plant organs rich in edible starch, generally have more fertile soils for agriculture and are easier to clear using slash-and-burn technologies. Though these regions have begun to be explored in northwestern South America, there are considerable expanses of seasonal tropical forest lying in the southern fringes of the Amazon that are still largely uncharted. A major case in point is the deciduous tropical forests and savannas occupying the present day Brazilian states of Rondônia and Matto Grosso where, according to recent molecular studies, the wild ancestor of manioc (*Manihot esculenta* ssp. *flabellifolia*) grows (Olsen and Schaal 1999, 2001). This region also appears to have what is currently the oldest *terra preta* (black earth) soils, which are dated to around 4800 BP (Miller 1992). Southern Amazonia may prove crucial for elucidating persistent questions of manioc domestication and early dispersals. Until now, the oldest direct evidence for manioc in South America came from coastal Peru, where it dated to around 3800 BP (Ugent et al 1986), but recently, Aceituno and Castillo (2005) have reported the presence of manioc starch grains from edge-ground cobbles in the Porce and Cauca valleys of Colombia dating to 7500 BP. These data are consistent with the age of manioc appearance in southern Central America at c 7000 to 6000 BP (Piperno et al 2000).

Similarly, the temperate mountain valleys of the eastern slopes of the southcentral Andes, which exhibit a wide array of closely juxtaposed environmental zones and high plant diversity, may have supported early plant manipulation and domestication. This macroregion located between the eastern Andes, lowland Bolivia, and western Brazil has been proposed as the cradle for several crops, including common beans, jack beans, peanuts and some varieties of chile peppers (Piperno and Pearsall 1998: 164).

To the south in the La Plata Basin, little is known about early and mid-Holocene cultural developments. The expansion and colonization of the region by Tupi-Guarani tropical forest farmers during the late Holocene is often cited as the mechanism responsible for the arrival of the first farmers (Schmitz 1991). However, new multidisciplinary data from southeastern Uruguay indicates that this is not the case, as cultigens such as maize, squash and probably *Phaseolus* beans were introduced and became integrated into local food economies by c 4000 BP, much earlier than previously thought (Iriarte et al 2004). This study also showed that questions relating to native food production systems and the early dispersal of cultigens should not be restricted to the practice of shifting agriculture in tropical/subtropical seasonal forest. The role that flood-recessional agriculture and wetlands played in early economies was likely to have been considerable. The new data from Uruguay reinforce similar evidence from other regions of the Americas and show that the use and manipulation of wetlands was an earlier, more important and more frequent activity than previously thought (Pohl et al 1996; Pope et al 2001). A more nuanced analysis of the southern pampas of South America also demonstrates the potential of grasslands and wetlands – historically viewed as marginal areas – for pre-Hispanic cultural development (Stahl 2004).

While most of the recent advances in the study of agriculture in the Americas have to do with the early domestication and dispersal of crop plants, research on later agricultural landscapes has also been revitalised and will benefit from the new methodological approaches that are being increasingly applied by archaeobotanists. During the mid and late Holocene, South America experienced major transformations of its agricultural landscapes (Denevan 2001). Large complexes of raised-field systems were built in the seasonally flooded lowland savannas (eg Erickson 1995; Falchetti 2000; Rostain 1994; Spencer et al 1994; Walker 2004) and also in perennially wet highland habitats (eg Kolata 1993; Wilson et al 2002), allowing farmers to carry out agriculture in the rainy season with flood-intolerant crops.

Raised fields have also been recently documented along the wetlands of the rainier and cooler forests of the southcentral Andes dating to around AD 1280 (Dillehay 2007; Dillehay and Pino, pers comm 2005).

Extensive areas of *terra preta* soils (estimated to have occupied 6000 to 18,000 km^2) that are associated with large, dense archaeological sites occur in the central and southern Amazon Basin during the mid and late Holocene (Denevan 2001; Lehmann et al 2003; Miller 1992; Neves et al 2003). The creation of these black earth soils allegedly allowed native Amazonian farmers to intensively cultivate the heavily leached and infertile upland tropical soils. Research by soil scientists on *terra preta* suggests that the present-day Amazonian landscape has been subject to millennia of human management (Lehmann et al 2003). These discoveries are an effective critique of environmental determinism in Amazonia, which postulates that the limited agricultural potential of Amazonian soils was a limiting factor for cultural development in general (Meggers 1971). While the *terra preta* sites have long been a subject of investigation and debate, there is no empirical botanical evidence of the intense agricultural activity that is thought to have helped create them in the central Amazon and little data is available from elsewhere (Herrera et al 1992; Mora et al 1991). Though researchers have hypothesised about the intensive agricultural practices associated with *terra preta*, the primary botanical evidence for these assumptions has not been forthcoming (Lehmann et al 2003). More systematic application of microbotanical analysis in tandem with soils studies will help determine more precisely the formation processes and trace the agricultural history of these landscapes.

In future, palaeoethnobotanical studies will be formulated to take into account the great diversity of food-producing practices. The systematic application of the various compatible and complementary palaeoethnobotanical techniques, both macro and microfossil, in combination with the analysis of palaeoecological records oriented to resolve anthropological questions, will increasingly provide important insights into food-producing practices of the past. Information gained through the application of these new techniques will also lead to new theoretical understandings of agriculture in the Americas.

ACKNOWLEDGMENTS

I would like to thank Tim Denham, Tom Dillehay, Dolores Piperno and Luc Vrydaghs for providing insightful comments and editorial advice to various drafts of this manuscript. Seán Goddard patiently prepared the figures presented here. All statements made herein, however, are my own responsibility.

NOTES

1. In this chapter, when I refer to traditional Eurasian models I denote the 'consensus views' that arose from the large multidisciplinary projects that were carried out in the region until the mid-1980s. See chapters in Price (2000) and Ammerman and Biagi (2003) for the new synthesis that is emerging on the origins and dispersal of crops in Europe.

2. Figures 9.1–3 and Table 9.1 provide a synopsis of the earliest appearance of domesticated plants along with some of the proposed areas for the origin of crops of economic importance based on the latest molecular evidence. See Piperno and Pearsall (1998:164–165: Figures 3.18, 3.19) for a map detailing other proposed macroregions of domestication. The information summarised here allows major patterns to be pinpointed and traced through time. For example, only some of the more than 20 coastal Peruvian sites, which contain more than 20 crops, are shown (see Pearsall 1992 and Piperno and Pearsall 1998 for more details). The data in these figures are primarily extracted from Pearsall (1992), Piperno and Pearsall (1998), Piperno (2006b, 2006c), and Dillehay et al (2004). Proposed regions for the domestication of crops have been taken from the following authors: manioc (Olsen and Schaal 1999), potato (Spooner et al 2005), cotton (Westengen et al 2005), and quinoa (Wilson 1988). See Doebley (1990: 7, Figure 1) and Sanjur et al (2002: 537, Figure 1) for detailed maps of proposed areas of domestication of maize and squash, respectively.
3. All dates are reported in uncalibrated radiocarbon years BP.

REFERENCES

Aceituno Bocanegra, F J and Castillo Espitia, N (2005) 'Mobility strategies in Colombia's middle mountain range between the early and middle Holocene', *Before Farming* 2: 1–17

Ammerman, A J and Biagi, P (eds) (2003) *The Widening Harvest*, Boston: Archaeological Institute of America

Arford, M R and Horn, S (2004) 'Pollen evidence of the earliest maize agriculture in Costa Rica', *Journal of Latin American Geography* 3: 108–15

Armelagos, G J and Harper, K N (2005) 'Genomics at the origins of agriculture, part one', *Evolutionary Anthropology* 14: 68–77

Asch, D and Asch, N (1985) 'Prehistoric plant cultivation in west-central Illinois', in Ford, R I (ed), *Prehistoric Food Production in North America*, pp 149–204. Anthropological Papers no 75, Ann Arbor, MI: Museum of Anthropology, University of Michigan

Athens, S and Ward, J B (1999) 'The Late Quaternary of the western Amazon: climate, vegetation and humans', *Antiquity* 73: 287–301

Bellwood, P (2005) *First Farmers*, Oxford: Blackwell

Bonavia, D (1982) *Los Gavilanes*, Lima: Corporacion Financiera de Desarrollo SA Cofide

Bonavia, D and Grobman, A (1989) 'Andean maize: its origins and domestication', in Harris, D R and Hillman, G C (eds), *Foraging and Farming: The Evolution of Plant Exploitation*, pp 456–70, London: Unwin Hyman

Bray, W (2000) 'Ancient food for thought', *Nature* 408: 145–46

Brieger, F G, Gurdel, T A, Paterniani, E, Blumenschein, A and Alleoni, M R (1958) *Races of Maize in Brazil and other Eastern South American Countries*. National Academy of Science Publication 593, Washington, DC: National Academy of Science

Browman, D L, Fritz, G J and Watson, P J (2005) 'Origins of food-producing economies in the Americas', in Scarre, C (ed), *The Human Past*, pp 306–43, London: Thames and Hudson

Bruno, M (2006) 'A morphological approach to documenting the domestication of *Chenopodium* in the Andes', in Zeder, M, Emschwiller, E and Smith, B (eds), *Documenting Domestication: New Genetic and Archaeological Paradigms*, pp 32–45, Berkeley, CA: University of California Press

Burger, R L, and van der Merwe, N J (1990) 'Maize and the origin of highland Chavín civilization: an isotopic perspective', *American Anthropologist* 92(1): 85–95

Bush, M B, Piperno, D R and Colinvaux, P A (1989) 'A 6000 year history of Amazonian maize cultivation', *Nature* 340: 303–5

Bush, M B, Miller, M C, De Oliveira, P E and Colinvaux, P (2000) 'Two histories of environmental change and human disturbance in eastern lowland Amazonia', *The Holocene* 10: 543–53

Clement, C R (1999) '1492 and the loss of Amazonian crop genetic resources, I: the relation between domestication and human population decline', *Economic Botany* 53: 188–202

Denevan, W M (2001) *Cultivated Landscapes of Native Amazonia and the Andes*, New York: Oxford University Press

Denham, T P, Haberle, S G, Lentfer, C, Fullagar, R, Field, J, Therin, M, Porch, N and Winsborough, B (2003) 'Origins of agriculture at Kuk Swamp in the Highlands of New Guinea', *Science* 301: 189–93

Dickau, R, Ranere, A J and Cooke, R (2007) 'Starch grain evidence for the preceramic dispersal of maize and root crops into tropical dry and humid forests of Panama,' *Proceedings of the National Academy of Sciences* (Washington, DC) 104(9): 3651–3656

Dillehay, T D (2007) *Monuments, Empires, and Resistance: The Araucanian Polity and Ritual Narratives*, Cambridge: Cambridge University Press

Dillehay, T D, Netherly, P J and Rossen, J (1989) 'Middle Preceramic public and residential sites on the forested slope of the western Andes, northern Peru', *American Antiquity* 54: 733–59

Dillehay, T D, Rossen, J and Netherly, P J (1997) 'The Nanchoc Tradition: the beginnings of Andean civilization', *American Scientist* 85: 46–55

Dillehay, T D, Bonavia, D and Kaulicke, P (2004) 'The first settlers', in Silverman, H (ed), *Andean Archaeology*, pp 16–34, Oxford: Blackwell Publishing

Dillehay, T D, Eling, H H and Rossen, J (2005) 'Preceramic irrigation canals in the Peruvian Andes', *Proceedings of the National Academy of Science* (Washington, DC) 102: 17241–44

Doebley, J (1990) 'Molecular evidence and the evolution of maize', *Economic Botany* 44: 6–27

Emschwiller, E (2006) 'Genetic data and plant domestication', in Zeder, M, Emschwiller, E and Smith, B (eds), *Documenting Domestication: New Genetic and Archaeological Paradigms*, pp 99–122, Berkeley, CA: University of California Press

Erickson, D L (1995) 'Archaeological perspectives on ancient landscapes of the Llanos de Mojos in the Bolivian Amazon', in Stahl, P (ed), *Archaeology in the American Tropics: Current Analytical Methods and Applications*, pp 66–95, Cambridge: Cambridge University Press

Erickson, D L, Smith, B D, Clarke, A C, Sandweiss, D H and Tuross, N (2005) 'An Asian origin for a 10,000-year-old domesticated plant in the Americas' *Proceedings of the National Academy of Science* (Washington, DC) 102: 18315–20

Falchetti, A M (2000) 'Los Zenues de las llanuras del Caribe colombiano: organizacion regional y manejo del medio ambiente' in Duran, A and Bracco, R (eds), *Arqueología de las Tierras Bajas*, pp 83–97, Montevideo, Uruguay: Ministerio de Educación y Cultura, Comisión Nacional de Arqueología

Flannery, K V (1986) *Guilá Naquitz*, Orlando, FL: Academic Press

Fritz, G (1994a) 'Are the first American farmers getting younger?', *Current Anthropology* 35: 305–09

Fritz, G (1994b) 'Multiple pathways to farming in precontact eastern North America', *Journal of World Prehistory* 4: 387–435

Harlan, J R (1995) *The Living Fields: Our Agricultural Heritage*, Cambridge: Cambridge University Press

Harter, A V, Gardner, K A, Falush, D, Lentz, D L, Bye, R A and Rieseberg, L H (2004) 'Origin of extant domesticated sunflowers in eastern North America', *Nature* 430: 201–05

Hawkes, J G (1989) 'The domestication of roots and tubers in the American tropics', in Harris, D R and Hillman, G C (eds), *Foraging and Farming: The Evolution of Plant Exploitation*, pp 481–503, London: Unwin Hyman

Herrera, L F, Cavelier, I, Rodriguez, C and Mora, S (1992) 'The technical transformation of an agricultural system in the Colombian Amazon', *World Archaeology* 28: 98–113

Iriarte, J, Holst, I, Marozzi, O, Listopad, C, Alonso, E, Rinderknecht, A and Montaña, J (2004) 'Evidence for cultivar adoption and emerging complexity during the mid-Holocene in the La Plata Basin', *Nature* 432: 614–17

Jaenicke-Després, J, Buckler, E S, Smith, B D, Thomas, M, Gilbert, P, Cooper, A, Doebley, J and Pääbo, P (2003) 'Early allelic selection in maize as revealed by ancient DNA', *Science* 302: 1206–08

Kaplan, L (1994) 'Accelerator mass spectrometry dates and the antiquity of *Phaseolus* cultivation', *Annual Report of the Bean Improvement Cooperative* 37: 131–32

Kaplan, L and Lynch, T (1999) '*Phaseolus* (Fabaceae) in archaeology: new AMS radiocarbon dates and their significance for pre-Columbian agriculture', *Economic Botany* 53: 261–72

Kolata, A (1993) *The Tiwanaku*, Cambridge: Blackwell

Lathrap, D W (1970) *The Upper Amazon*, New York: Praeger

Lathrap, D W (1977) 'Our father the cayman, our mother the gourd: Spinden revisited, or a unitary model for the emergence of agriculture in the New World', in Reed, C A (ed), *Origins of Agriculture*, pp 713–51, The Hague: Mouton

Lehmann, J, Kern, D, Glaser, B and Woods, W (eds) (2003) *Amazonian Dark Earths*, Amsterdam: Kluwer

Lentz, D L, Pohl, M E D, Pope, K O and Wyatt, A R (2001) 'Prehistoric sunflower (*Helianthus annuus* L.) domestication in Mexico', *Economic Botany* 55: 370–77

Long, A, Benz, B F, Donahue, J, Jull, A and Toolin, L (1989) 'First direct AMS dates on early maize from Tehuacan, Mexico', *Radiocarbon* 31: 1035–40

Lynch, T (1980) *Guitarrero Cave: Early Man in the Andes*, New York: Academic Press

MacNeish, R S (1992) *The Origins of Agriculture and Settled Life*, Norman, OK: University of Oklahoma Press

Martins-Farias, R (1976) 'New archaeological techniques for the study of ancient root crops in Peru', unpublished PhD dissertation, University of Birmingham (UK)

Matsuoka, Y, Vigouroux, Y, Goodman, M M, Sanchez, J, Buckler, E and Doebley, J (2002) 'A single domestication for maize shown by multilocus microsatellite genotyping', *Proceedings of the National Academy of Science* (Washington, DC) 99: 6080–84

Meggers, B J (1971) *Amazonia: Man and Culture in a Counterfeit Paradise*, Chicago: Aldine-Atherton

Miller, E (1992) *Arqueología nos Empreendimentos Hidrelétricos da Eletronorte: Resultados Preliminares*, Brasília: Centrais Elétricas do Norte do Brasil – Eletronorte

Mora, S C (2003) *Early Inhabitants of the Amazonian Tropical Rain Forest*. University of Pittsburgh Latin American Archaeology Reports no 3, Pittsburgh, PA: University of Pittsburgh

Mora, S C, Herrera, L F, Cavalier, I and Rodriguez, C (1991) *Cultivars, Anthropic Soils, and Stability*. University of Pittsburgh Latin American Archaeology Report no 2, Pittsburgh, PA: University of Pittsburgh

Morcote-Ríos, G and Bernal, R G (2001) 'Remains of palms (Palmae) at archaeological sites in the New World: a review', *The Botanical Review* 67(3): 309–50

Neves, E G, Petersen, J B, Bartone, R N and Da Silva, C A (2003) 'Historical and sociocultural origins of Amazonian dark earth', in Lehmann, J, Kern, D, Glaser, B and Woods, W (eds), *Amazonian Dark Earths*, pp 29–50, Amsterdam: Kluwer

Olivier, J R (2001) 'The archaeology of forest foraging and agricultural production in Amazonia', in McEwan, C, Barreto, C and Neves, E (eds), *Unknown Amazon*, pp 50–85, London: British Museum Press

Olsen, K M and Schaal, B A (1999) 'Evidence on the origin of cassava: phylogeography of *Manihot esculenta*', *Proceedings of the National Academy of Science* (Washington, DC) 96: 5586–91

Olsen, K M and Schaal, B A (2001) 'Microsatellite variation in cassava (*Manihot esculenta, Euphorbiaceae*) and its wild relatives: further evidence for a southern Amazonian origin of domestication', *American Journal of Botany* 88: 131–42

Pagán Jiménez, J R, Rodríguez López, M A, Chanlatte Baik, L A, and Nargares Storde, Y (2005) 'La temprana introducción y uso de algunas plantas domésticas, silvestres, y cultivos en Las Antillas precolombinas: una primera revaloración desde la perspectiva del 'arcaico' de Vieques y Puerto Rico', *Diálogo Antropológico* 3: 7–33

Pearsall, D M (1989) 'Adaptations of prehistoric hunter-gatherers to the high Andes: the changing role of plant resources', in Harris, D R and Hillman, G C (eds), *Foraging and Farming: The Evolution of Plant Exploitation*, pp 318–33, London: Unwin Hyman

Pearsall, D M (1992) 'The origins of plant cultivation in the Americas', in Cowan, W C and Watson, P J (eds), *The Origins of Agriculture: An International Perspective*, pp 173–205, Washington, DC: Smithsonian Institution Press

Pearsall, D M (2003) 'Plant food resources of the Ecuadorian Formative: an overview and comparison to the Central Andes', in Raymond, J S and Burger, R L (eds), *Archaeology of Formative Ecuador*, pp 213–58, Washington, DC: Dumbarton Oaks

Pearsall, D M, Chandler-Ezell, K and Zeidler, J A (2004) 'Maize in ancient Ecuador: results of residue analysis of stone tools from the Real Alto site', *Journal of Archaeological Science* 31: 423–42

Perry, L (2004) 'Starch analyses reveal the relationship between tool type and function: an example from the Orinoco Valley of Venezuela', *Journal of Archaeological Science* 31: 1069–81

Perry, L, Sandweiss, D H, Piperno, D R, Rademaker, K, Malpass, M A, Umire, A and de la Vera, P (2006) 'Early maize agriculture and interzonal interaction in southern Peru', *Nature* 440 (7080): 76–79

Perry, L, Dickau, R, Zarrillo, S, Holst, I, Pearsall, D M, Piperno, D R, Berman, M J, Cooke, R G, Rademaker, K, Ranere, A J, Raymond, J S, Sandweiss, D H, Scaramelli, F, Tarble, K and Zeidler, J A (2007) 'Starch fossils and the domestication and dispersal of chili peppers (*Capsicum* spp. L.) in the Americas', *Science* 315(5814): 986–88

Piperno, D R (2006a) 'Identifying manioc (*Manihot esculenta* Crantz) and other crops in preColumbian tropical America through starch grain analysis: a case study from Central Panama', in Zeder, M, Emschwiller, E and Smith, B (eds), *Documenting Domestication: New Genetic and Archaeological Paradigms*, pp 46–67, Berkeley, CA: University of California Press

Piperno, D R (2006b) 'The origins of plant cultivation and domestication in the neotropics', in Kennett, D J and Winterhalder, B (eds), *Behavioral Ecology and the Transition to Agriculture*, pp 137–166, Los Angeles, CA: University of California Press

Piperno, D R (2006c) *Phytoliths: A Comprehensive Guide for Archaeologists and Paleoecologists*, Lanham, MD: Altamira Press

Piperno, D R and Flannery, K V (2001) 'The earliest archaeological maize (*Zea mays* L.) from highland Mexico: new accelerator mass spectrometry dates and their implications', *Proceedings of the National Academy of Science* (Washington, DC) 98: 2101–03

Piperno, D R and Holst, I (1998) 'The presence of starch grains on prehistoric stone tools from the humid neotropics: indications of early tuber use and agriculture in Panama', *Journal of Archaeological Science* 25: 765–76

Piperno, D R and Pearsall, D M (1998) *The Origins of Agriculture in the Lowland Neotropics*, San Diego, CA: Academic Press

Piperno, D R and Stothert, K E (2003) 'Phytolith evidence for early Holocene *Cucurbita* domestication in southwest Ecuador', *Science* 299: 1054–57

Piperno, D R, Ranere, J A, Holst, I and Hansell, P (2000) 'Starch grains reveal early root crop horticulture in the Panamanian tropical forest', *Nature* 407: 894–97

Piperno, D R, Holst, I, Wessel-Beaver, L and Andres, T C (2002) 'Evidence for the control of phytolith formation in *Cucurbita* fruits by the hard rind (*Hr*) genetic locus: archaeological and ecological implications', *Proceedings of the National Academy of Sciences* (Washington, DC) 99: 10923–28

Pohl, M D, Piperno, D R, Pope, K O and Jones, J G (2007) 'Pre-Columbian maize dispersals in the neotropics: new microfossil evidence from San Andrés, Tabasco, Mexico', *Proceedings of the National Academy of Sciences* (Washington, DC) 104: 6870–75

Pohl, M D, Pope, K O, Jones, J, Jacob, J S, Piperno, D R, deFrance, S D, Lentz, D L, Gifford, J A, Danforth, M E and Josserand, K (1996) 'Early agriculture in the Maya lowlands', *Latin American Antiquity* 7(4): 355–72

Pope, K O, Pohl, M D E, Jones, J G, Lentz, D L, Von Nagy, C, Vega, F J and Quitmyer, I R (2001) 'Origin and environmental setting of ancient agriculture in the lowlands of Mesoamerica', *Science* 292: 1370–73

Price, D P (ed) (2000) *Europe's First Farmers*, Cambridge: Cambridge University Press

Prous, A (1999) 'Agricultores de Minas Gerais', in Tenorio, M C (ed), *Pré-Historia da Terra Brasilis*, pp 345–58, Rio de Janeiro: Editora da UFRJ

Raymond, S (1981) 'The maritime foundations of Andean civilization: a reconsideration of the evidence', *American Antiquity* 46: 806–21

Resende, E T and Prous, A (1991) 'Os vestigios vegetais do Grande Abrigo de Santana do Riacho', *Arquivos do Museu de Historia Natural* (Belo Horizonte, Brazil) XII: 87–111

Roosevelt, A C, Lima da Costa, M, Lopes Machado, C, Michab, M, Mercier, N, Valladas, H, Feathers, J, Barnett, W, Imazio da Silveira, M, Henderson, A, Sliva, J, Chernoff, B, Reese, D S, Holman, J A, Toth, N and Schick, K (1996) 'Palaeoindian cave dwellers in the Amazon: the peopling of the Americas', *Science* 272: 373–84

Rostain S (1994) 'The French Guiana coast: a key-area in prehistory between the Orinoco and Amazon rivers', in *Between St. Eustatius and the Guianas*, pp 53–97. Publication of the St. Eustatius Historical Foundation 3, St. Eustatius: St. Eustatius Historical Foundation

Sandweiss, D and Moseley M (2001) 'Amplifying importance of new research in Peru', *Science* 294: 1651–53

Sanjur, O, Piperno, D R, Andres, T C and Wessel-Beaver, L (2002) 'Phylogenetic relationships among domesticated and wild species of *Cucurbita* (Cucurbitaceae) inferred from a mitochondrial gene: implications for crop plant evolution and areas of origin', *Proceedings of the National Academy of Science* (Washington, DC) 99: 535–40

Sauer, C O (1952) *Agricultural Origins and Dispersal*, New York: American Geographical Society

Scheel-Ybert, R (2001) 'Man and vegetation in southeastern Brazil during the late Holocene', *Journal of Archaeological Science* 28: 471–80

Schmitz, P I (1991) 'Migrantes da Amazonia: a tradicao Tupiguarani', in Kern, A (ed), *Arqueologia Prehistorica do Rio Grande do Sul*, pp 295–330, Porto Alegre, Brazil: Mercado Aberto

Shady, R, Haas, J and Creamer, W (2001) 'Dating Caral, a Preceramic site in the Supe Valley on the central coast of Peru', *Science* 292: 723–26

Smith, B D (1992) *Rivers of Change: Essays on Early Agriculture in Eastern North America*, Washington, DC: Smithsonian Institution Press

Smith, B D (1997a) 'The initial domestication of *Cucurbita pepo* in the Americas 10,000 years ago', *Science* 276: 932–34

Smith, B D (1997b) 'Reconsidering the Ocampo Caves and the era of incipient cultivation in Mesoamerica', *Latin American Antiquity* 8(4): 342–83

Smith, B D (2001a) 'Documenting plant domestication: the consilience of biological and archaeological approaches', *Proceedings of the National Academy of Science* (Washington, DC) 98: 1324–26

Smith, B D (2001b) *The Emergence of Agriculture*, New York: Scientific American Library

Smith, B D (2005) 'Reassessing Coxcatlan Cave and the early history of domesticated plants in Mesoamerica' *Proceedings of the National Academy of Science* (Washington, DC) 102: 9438–45

Sonnante, G, Stockton, T, Nodari, R O, Becerra Velazquez, V L and Gepts, P (1994) 'Evolution of genetic diversity during the domestication of common-bean (*Phaseolus vulgaris* L.)', *Theoretical and Applied Genetics* 89: 629–35

Spencer, C S, Redmond, E M and Rinaldi, M (1994) 'Drained fields at La Tigra, Venezuelan llanos: a regional perspective', *Latin American Antiquity* 5: 119–43

Spooner, D M, McLean, K, Ramsay, G, Robbie, W and Bryan, G J (2005) 'A single domestication for potato based on multilocus amplified fragment length polymorphism genotyping', *Proceedings of the National Academy of Science* (Washington, DC) 102: 14694–99

Stahl, P (2004) 'Greater expectation', *Nature* 432: 561–62

Stothert, K, Piperno, D R and Andres, T C (2003) 'Terminal Pleistocene/Early Holocene human adaptation in coastal Ecuador: the Las Vegas evidence', *Quaternary International* 109–110: 23–43

Towle, M (1961) *The Ethnobotany of Pre-Columbian Peru*, Chicago: Aldine

Turner, C and Machado, L (1983) 'A new dental wear pattern and evidence for high carbohydrate consumption in a Brazilian Archaic skeletal population', *American Journal of Physical Anthropology* 61: 125–30

Ugent, D, Pozorski, S and Pozorski, T (1981) 'Prehistoric remains of the sweet potato from the Casma Valley of Peru', *Phytologia* 49: 401–15

Ugent, D, Pozorski, S and Pozorski, T (1982) 'Archaeological potato tuber remains from the Casma Valley of Peru', *Economic Botany* 36: 182–92

Ugent, D, Pozorski, S and Pozorski, T (1984) 'New evidence for ancient cultivation of *Canna edulis* in Peru', *Economic Botany* 38: 417–32

Ugent, D, Pozorski, S and Pozorski, S (1986) 'Archaeological manioc (*Manihot*) from coastal Peru', *Economic Botany* 40: 78–102

Walker, J (2004) *Agricultural Change in the Bolivian Amazon*, Pittsburgh, PA: University of Pittsburgh, Latin American Archaeology Publications

Westengen, O T, Huaman, Z and Heun, M (2005) 'Genetic diversity and geographic pattern in early South American cotton domestication', *Theoretical and Applied Genetics* 110: 392–402

Wilson C, Simpson I A and Currie E J (2002) 'Soil management in pre-Hispanic raised field systems: micromorphological evidence from Hacienda Zuleta, Ecuador', *Geoarchaeology* 17: 261–83

Wilson, H D (1988) 'Quinoa biosystematics I: domesticated populations', *Economic Botany* 42: 461–77

CHAPTER 10

Keepers of Louisiana's Levees: Early Mound Builders and Forest Managers

Gayle J. Fritz

INTRODUCTION

North American Indians developed many different ways of procuring and producing food by practicing strategies that cover the broad zone from foraging to agricultural economies (Ford 1985; Minnis 2003; Minnis and Elisens 2000; Smith 2001). Of special interest are the sophisticated methods of enhancing resource productivity along the Pacific Coast that enabled late precontact and postcontact period societies to sustain the highest estimated population densities on the continent (50–100 persons/km^2 according to Ubelaker 1988) without reliance on domesticated plants or animals (Ames 1999). Fire was used systematically to create clearings for tubers and other underground plant foods, berries, fiber plants, and in California to maintain groves of high yielding oak trees (Lewis 1973; Turner 1999). The degree to which landscapes were modified and managed has inspired ethnoecologists to compare these regions to agricultural ones. Peacock and Turner (2000: 133), for example, quote a Lil' Wat (Lillooet) elder, Baptiste Ritchie, who used the phrase 'just like a garden' to describe the frequently burned mountainsides of British Columbia where dense patches of berries and edible tubers could once be found; the same authors suggest that 'indigenous peoples of the [interior] Plateau were essentially cultivators who managed and maintained plant resources to ensure a predictable, productive and continued supply of culturally significant plant species' (Peacock and Turner 2000: 165). Florence Shipek (1989: 159) uses 'intensive plant husbandry' to describe the way native Kumeyaay people of southern California modified their ecosystems 'by substituting species desired by humans for food, medicine, and technology for unused species.'

Given the urgent desire of anthropologists and ethnobiologists to appreciate fully the complexity of these alternative subsistence regimes and to deny no measure of achievement to indigenous peoples who implemented them for millennia, terminology becomes problematic. Archaeologists have used terms such as 'complex' or 'affluent' (Ames 1999; Zvelebil 1996) hunter-gatherers or foragers when discussing sedentary groups including those of the Pacific Coast who became territorial, sociopolitically hierarchical and monetary, while intensifying the harvesting and cultivation of native animals and plants and implementing socially controlled exchange mechanisms (Ames 1985). Others (eg Deur 2002 and authors cited above) stress the appropriateness of terms such as 'cultivators' that would bring intensive resource managers closer to equal footing with farmers who rely heavily on domesticated crops. While people who create and maintain carefully tended, highly anthropogenic landscapes without reliance on domesticates are no less 'advanced' than agriculturalists, and arguably superior if they avoid adverse ecological impacts associated with large-scale soil tillage, it seems necessary to me to maintain descriptive labels distinguishing fundamentally different modes of subsistence. Therefore, I use the term 'agriculture' in the sense of Harris (1996: 4): an economy in which crop production dominates over wild plant food use and which is 'based largely or exclusively on the cultivation of domesticated plants.'

In this paper, the focus is on another region, the Tensas Basin of northeastern Louisiana in the lower Mississippi Valley, where non-domesticated plants and animals were for millennia the primary foods for densely packed, territorial people. Although mound construction in Louisiana began no later than 3500 BC, and people not far to the north participated in crop production before 1000 BC, we have no firm evidence for adoption of domesticated food plants in the study region until shortly prior to AD 1000. Instead, plant-derived subsistence was based primarily on nuts and fruits from native tree species. When herbaceous seed-bearing plants are represented, they appear to be non-domesticated types. Maize (*Zea mays* ssp. *mays*) is not documented before AD 900, and serious use of maize seems not to have occurred until c AD 1100 or even later at some sites. Pre-maize food procurement probably resembles most closely Harris' (1996: 4) category of wild plant-food production and Smith's (2001: 30) 'intensive resource utilization and low-level food production of non-domesticates', although a few domesticated plants may have been propagated.

Smith's (2001: 33–34) discussion of low-level food production as encompassing a 'richly heterogeneous group' of 'extremely variable, successful long-term socioeconomic solutions, fine-tuned to a wide range of local cultural and environmental contexts' highlights the challenges of dealing with particular archaeological examples. A major hurdle is the ambiguity inherent in the fragmentary archaeobotanical record. In the lower Mississippi Valley, several plants, including the native gourdy squash (*Cucurbita pepo* ssp. *texana*), are represented by specimens that cannot be classified with confidence as either wild or domesticated[1]. Therefore, although ancient inhabitants of Louisiana can be categorised with confidence as low-level food producers, exactly where on Smith's broad landscape, or on which side of his 'mountain range' of domestication they fall, is difficult to determine. Furthermore, as more and more archaeologists practice flotation recovery in the lower Mississippi Valley, we are beginning to recognise intraregional variability in the timing of adoption of maize and degree of dependence upon it and native seed crops.

My central goal is to summarise archaeobotanical evidence from the lower Mississippi Valley, after placing it in environmental and cultural context, and to present this region as a case study of long-term, intensive management of undomesticated plants, at times possibly including the planting of native seeds and supplemented by domesticated specialty crops. Close scrutiny of the evidence is crucial given past assumptions that agriculture began early in this region, forming the economic base of mound-building societies. I rely heavily on assemblages from five sites in Tensas Parish, Louisiana, which were excavated under the direction of T R Kidder during the late 1980s and early 1990s (Kidder and Fritz 1993; Kidder et al 1993). The plant remains (Table 10.1) were analyzed by me and students at Washington University in St. Louis; several of us had also participated in their excavation and flotation.

WANTING EARLY MOUND BUILDERS TO BE FARMERS

Many papers in this volume are concerned with recognising early agriculture in regions of the world where evidence was formerly lacking or not accepted as valid. That is not the issue in the lower Mississippi Valley, the focus of this case study. In fact, a reverse bias has been operating in this region, primarily due to the existence of ancient mounds and other monumental earthworks, along with evidence for long-distance trade extending back in time for at least 3500 years. Until about 1980, archaeologists suspected that the lower Mississippi Valley supported some of the earliest agricultural economies in North America (see Fritz and Kidder 1993 for a historical overview). This was partly because the region is so close to Mexico, a long-accepted center of domestication. Emissaries, colonists or traders from Mesoamerica were seen as likely conveyors of seeds and knowledge along the coast of the Gulf of Mexico or across gulf waters, using the Mississippi River as a natural highway on their way to spread the benefits of civilization to less advanced societies in eastern North America. Maize (*Zea mays* ssp. *mays*) was envisioned as the primary food crop, with common beans (*Phaseolus vulgaris*) coming along at the same time for complementary protein, and Mexican squashes and pumpkins (*Cucurbita pepo* ssp. *pepo*) providing additional nutrients and flavors.

By 1953, Poverty Point in northeastern Louisiana was recognised as a major Late Archaic (c 1600–1000 BC) trading center with one very large mound, several smaller ones and six semicircular earthworks more than 1 km in diameter (Gibson 2000) (see Figure 10.1 for location of Poverty Point and other sites mentioned in the text). This strengthened opinions that some sort of early agricultural florescence occurred here. Whether or not the earthworks had Mesoamerican roots, a society this large and complex could not be envisioned as nonagricultural, and 'agriculture' in temperate parts of the New World meant corn, beans, and squash. Ford and Webb (1956: 146) wrote:

> the economy of the inhabitants of the Poverty Point site must have been based on agriculture.... Without a stable basic food so large a population could not have been concentrated in one locality, and certainly the surplus labor necessary to undertake constructions on the scale accomplished by these people would not have been available.

Table 10.1 Densities, ubiquities and percentages of key plant taxa from five sites (Blackwater, Emerson, Jolly, Osceola and Reno Brake) in Tensas Parish, Louisiana

Time period (AD)	*200–400*		*400–700*	*700–900*		*900–1200*								*1400-1500*	
	Reno Brake Issaquena	*Osceola Issaquena-Md F*	*Reno Brake Troyville*	*Osceola-Early CC Mound B Unit 0249*	*Osceola-Early CC Mound B Unit 0351*	*Osceola-Late CC Mound B Unit 0351*	*Osceola-Late CC Mound B Unit 0249*	*Osceola-Late CC Mound A*	*Osceola-Late CC Mound C*	*Osceola-Late CC Mound E*	*Osceola-Late CC Mound F*	*Jolly-Late CC*	*Blackwater-Late CC*	*Emerson-Plaquemine 1992*	*Emerson-Plaquemine 1991*
# samples	6	1	4	11	5	4	6	9	8	6	7	8	30	8	9
# litres	71	76	42	116	53	56	71	104	109	71	86	70	379	172	120
wood weight (g)>2 mm	11.24	7.43	2.96	69.58	22.84	14.18	9.31	5.14	18.02	11.62	22.95	18.25	88.33	40.05	52.18
acorn density (>1.4 mm)	8.35	0.09	8.71	20.59	2.04	3.13	2.68	0.61	6.72	8.72	3.97	1.13	18.87	1.43	6.41
% acorn ubiquity	100	100	100	100	100	100	100	89	100	100	100	88	97	100	100
pecan density (>1.4 mm)	0.18	0.03	0.55	1.73	0.43	0.07	0.10	0.09	0.19	0.23	0.47	0.26	1.10	0.02	0.03
% pecan ubiquity	83	100	100	100	100	50	33	44	100	67	57	75	67	38	11
nutshell density	8.54	0.12	9.26	22.32	2.47	3.20	2.77	0.69	7.02	9.00	4.44	1.39	19.07	1.52	6.53
maize density	0	0	0	0	0	0.48	0.42	0	0.28	0.17	0.06	1.00	0.12	2.81	9.3
% maize ubiquity	0	0	0	0	0	100	67	0.0	75	50	43	63	40	100	100
# *Cucurbita* rind	0	1	0	39	2	0	0	3	1	0	1	1	25	0	0
% *Cucurbita* ubiquity		100		64	40			25	13		14	13	10		

# seed total	87	7	39	798	175	17	27	16	124	40	34	21	178	65	68
seed density	1.23	0.09	1.08	6.88	3.30	0.30	0.38	0.15	1.14	0.56	0.4	0.30	0.47	0.38	0.57
% starchy seeds	9.2	14.3	5.1	40.4	34.3	29.4	7.4	6.3	2.4	32.5	8.8	42.9	2.8	7.7	11.8
% fruit seeds	60.9	42.9	69.2	26.3	22.9	41.4	25.9	43.8	37.1	40.0	38.2	23.8	38.8	24.6	30.9
chenopod count	7	0	2	0	5	0	0	0	2	9	0	2	2	2	0
maygrass count	0	1	0	0	52	5	2	1	1	4	1	6	3	3	0
knotweed count	0	0	0	0	3	0	0	0	0	0	2	1	1	0	0
little barley count	1	0	0	0	0	0	0	0	0	0	0	0	0	0	0
sumpweed count	2	0	0	0	17	0	0	0	4	4	4	1	45	2	10

density=number of fragments per litre of soil floated of acorn shell, pecan shell or maize (seed density is number of seeds per litre).
ubiquity=% of samples in which the plant type is represented.
% starchy seeds and % fruit seeds=% of total seed count contributed by specimens of those types (starchy seeds include chenopod, maygrass, knotweed and little barley).

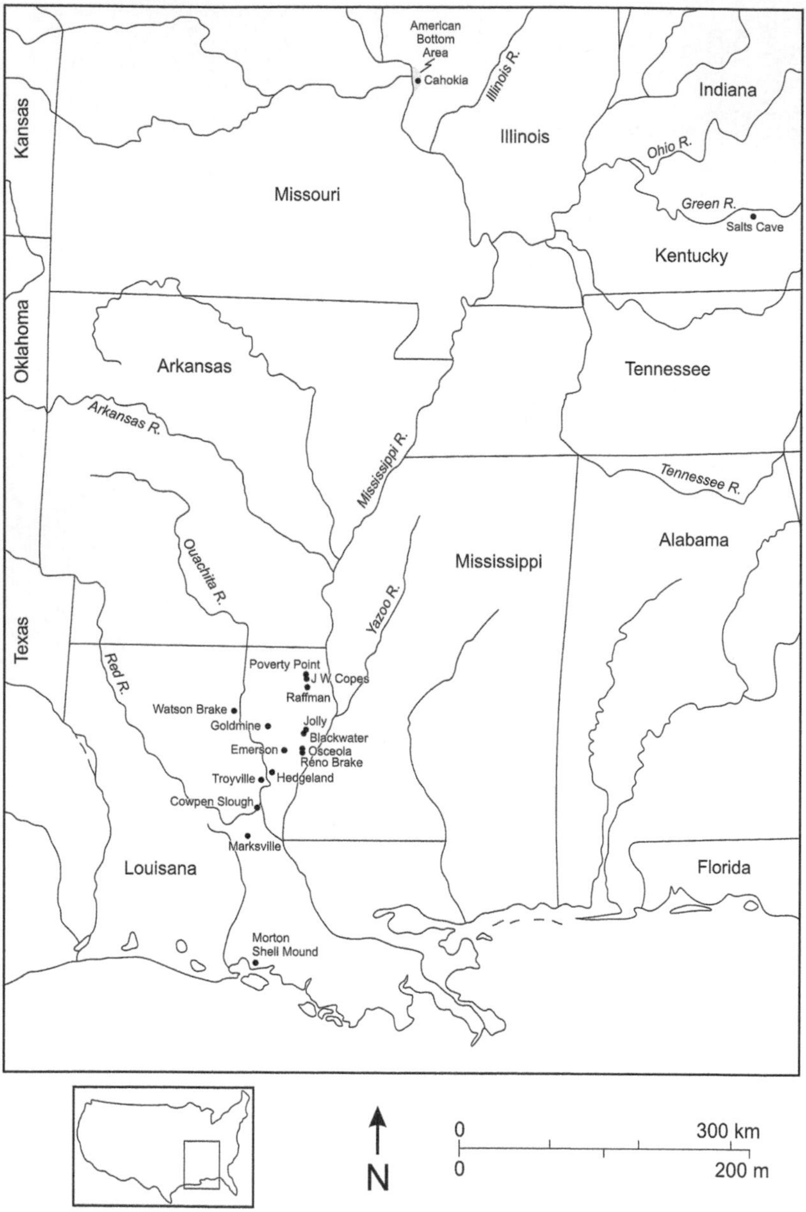

Figure 10.1 Map of the central and lower Mississippi Valley showing locations of sites mentioned in the text.

Willey and Phillips (1958: 146) viewed Poverty Point as the earliest culture in their Formative Stage, with the New World Formative defined 'by the presence of agriculture, or any other subsistence economy of comparable effectiveness, and by the successful integration of such an economy into well-established, sedentary village life'. Rather than explore the possibility that riverine northeastern Louisiana might have afforded an effective subsistence economy different from, but comparable to, agriculture, Willey and Phillips (1958: 156) concluded that despite the lack of evidence for agriculture at Poverty Point, 'it is quite impossible to imagine any other adequate economic basis in the geographic setting.'

The association of mound building with agriculture in the lower Mississippi Valley has persisted as a dominating theme that recurs over and over again as archaeologists struggle to understand cultural developments in this dynamic region. For Poverty Point and earlier time periods, however, there is no longer informed debate about production of maize, beans, or other Mexican crops, because none have turned up during efforts to recover Archaic subsistence remains. Furthermore, archaeologists today do not believe that contacts with Mexico played a significant role, if any role at all, in the activities conducted at early mounds in Louisiana. Although Poverty Point was contemporaneous with early Olmec civilization and individuals may have traveled between the two regions, the many non-perishable trade items carried from far away to Poverty Point—including galena, copper, chert, and steatite—came from the north and northeast rather than from Mexico (Gibson 2000).

A major breakthrough in understanding the roots of mound building in the lower Mississippi Valley came with verification of at least eight sites with mounds and other earthworks predating Poverty Point and extending back to at least 3500 BC (Russo 1996). The largest and most fully studied of these sites is Watson Brake, where 11 mounds, the tallest of which is seven meters high and 165 meters in diameter, were built around a circular plaza. Some of these mounds were connected by an artificial ridge one meter in height. In the absence of pottery or any local source of stones useful for stone-boiling, baked clay rods were used for cooking, probably in pits dug into the earth (Saunders et al 1997). This technological tradition continues at Poverty Point, where concentrations of fist-sized, variously shaped clay balls are found in pits indicating that they were preheated and used to cook food in earth ovens.

The realization that the tradition of building mounds in southeastern North America had deep roots added to mounting subsistence evidence to form the current consensus that Archaic peoples in the lower Mississippi Valley relied upon rich aquatic and terrestrial resources, qualifying them as affluent fisher-hunter-gatherers or low-level food producers with few if any domesticated plants (Gibson 2000; Milner 2004). Unlike their counterparts in California or the Pacific Northwest, lower Mississippi Valley societies shifted to maize farming by the time Europeans first encountered them. Still, archaeobotanical evidence indicates that reliance on domesticated plants came later here than elsewhere in the Eastern Woodlands, possibly partly due to successful maintenance of orchard-like oak groves on levees of alluvial bottomlands. This appears to have been the case in northeastern Louisiana.

THE LOWER MISSISSIPPI VALLEY: ENVIRONMENT AND PLANT FOOD RESOURCES

The major food resources in the Mississippian embayment of northeastern Louisiana and western Mississippi are concentrated in arcuate oxbow lakes and bayous and on the levees that border them. Even before they began cultivating crops, people lived on natural levees and nearby Pleistocene terrace remnants such as Macon Ridge, where Poverty Point is located, because of their elevation, proximity to fish and other aquatic animals and plants, and for the oaks (*Quercus* spp.), hickories (*Carya* spp.), pecans (*Carya illinoensis*), and other nut and fruit-bearing trees that grow on well-drained soils (see Table 10.2 for a list of economic plants discussed herein). Vines including grape (*Vitis* spp.), bramble (*Rubus* spp.) and greenbrier (*Smilax* spp.) grow profusely on the edges of levees where sunlight reaches them. Persimmon (*Diospyros virginiana*) is a member of the levee community, and palmettos (*Sabal minor*) also prefer the less clayey and better drained soil. The useful native bamboo called river cane or giant cane (*Arundinaria gigantea*) grows along with palmetto in the understorey and forms thick stands along the margins of back swamps and bayous. Animals including white-tailed deer (*Odocoileus virginianus*), turkey (*Meleagris gallopavo*) and raccoon (*Procyon lotor*) are drawn to levees because of the mast concentrations. Living in the lakes and bayous are fish, frogs, turtles, snakes, muskrats, waterfowl and several species of aquatic plants that bear edible starchy tubers. American water lotus (*Nelumbo lutea*), spatterdock (*Nuphar luteum*), and duck potato (*Saggitaria* spp.) are abundant in the area today and probably were in the past.

Levees are separated by extensive clayey backswamp zones dominated by sweetgum (*Liquidambar styraciflua*), southern hackberry (*Celtis laevigata*) and ash (*Fraxinus* spp.) trees rather than nut-bearing species. The buckshot clay soils of the backswamps are undesirable for agriculture, although expanses of them were clearcut during the late 1900s, to the regret even of some modern farmers who have struggled to produce crops on them. The patchy, crescent-shaped or serpentine patterning of desirable territories has implications for social differentiation during late precontact times (Kidder 1992, 2002). After indigenous groups permanently occupied levees and terraces flanking the best fishing bayous and lakes, a competitive tradition and perceived need for intensified resource management and exchange probably accompanied the development of increasingly hierarchical polities.

Subsistence of the Archaic Mound Builders

Pre-Poverty Point Subsistence (3500–1600 BC)

Pecans, acorns and seeds of wild-growing plants including *Chenopodium* were recovered by flotation at Watson Brake. Animal bones include those of catfish, drums, suckers, other fish, deer, turkey, small mammals, turtles, mussels, and aquatic snails (Saunders et al 1997). Other than this, little direct evidence exists for mid-Holocene resource use in the lower Mississippi Valley.

Table 10.2 Economic plants mentioned in the discussion of lower Mississippi Valley subsistence

Domesticated plants	
Maize	*Zea mays* ssp. *mays*
Eastern squash/gourd	*Cucurbita pepo* ssp. *texana* var. *ovifera*
Mexican squash/gourd	*Cucurbita pepo* ssp. *pepo*
Bottle gourd	*Lagenaria siceraria*
Bean	*Phaseolus vulgaris*
Sunflower	*Helianthus annuus* var. *macrocarpus*
Sumpweed	*Iva annua* var. *macrocarpa*
Chenopod (thin-testa)	*Chenopodium berlandieri* ssp. *jonesianum*
Trees, shrubs and vines	
Oak (acorns)	*Quercus* spp. (multiple species)
Hickory	*Carya* spp.
Pecan	*Carya illinoensis*
Persimmon	*Diospyros virginiana*
Palmetto	*Sabal minor*
Grape	*Vitis* spp.
Bramble	*Rubus* spp.
Hawthorn	*Crataegus* sp.
Blueberry	*Vaccinium* sp.
Herbaceous plants	
*Maygrass	*Phalaris caroliniana*
*Little barley	*Hordeum pusillum*
*Erect knotweed	*Polygonum erectum*
Sumpweed (wild)	*Iva annua*
Chenopod (wild)	*Chenopodium* spp.
Barnyard grass	*Echinochloa* sp.
Purslane	*Portulaca oleracea*
Greenbrier	*Smilax* spp.
Water lotus	*Nelumbo lutea*
Spatterdock	*Nuphar luteum*
Duck potato	*Saggitaria* spp.

*These taxa are included in the Eastern Agricultural Complex (along with thin-testa chenopod, sunflower, sumpweed and ovifera squash) where they are found in the midcontinent in appropriate depositional contexts. In the lower Mississippi Valley, their status as intentionally sown or otherwise cultivated seeds is unclear.

Poverty Point Subsistence (1600–900 BC)

Archaeobotanical analyses to date of flotation-recovered plant remains from Poverty Point, the nearby J W Copes site, and the Cowpen Slough site in Catahoula Parish reveal no maize or beans (Jackson 1989; Ramenofsky 1986; Shea 1978; Ward 1998). Native nuts (acorns, hickories and pecans) and native fruits (especially persimmons and grapes) dominate archaeobotanical assemblages. Rind fragments of what is probably a small, gourd-like squash are fairly common, and these are thought to represent the native eastern *Cucurbita pepo* ssp. *texana* gourd. These gourds might have been gathered wild, as

Louisiana is within their natural range (Decker-Walters et al 1993), but they could have easily been cultivated to ensure availability and enhance productivity. In this heavily fishing-oriented region, small gourds would have made good net floats (Fritz 1999; Hart et al 2004). Other implements, such as cups, spoons, rattles, and so on, were probably made from the hard gourd rind, and the edible flowers and seeds were likely to have been consumed. Seeds of herbaceous plants are not ubiquitous in Poverty Point contexts, but Ward (1998) reported 39 little barley (*Hordeum pusillum*) seeds from eight of her 40 analyzed flotation samples, which is a greater number of annual seeds than is typical of this time period. No cultigen sunflower, sumpweed, or chenopod seeds have been found at Archaic mound centers or associated habitation sites, although very small quantities of wild or weedy-type specimens of chenopod and sumpweed are reported (Jackson 1989; Shea 1978; Ward 1998).

So far, then, we have a tradition of people living in a resource-rich environment who engaged in long-distance exchange (but not with Mesoamericans) and sometimes built mounds and elaborate earthworks for communal gatherings and possible ritual enactments. By 1500 BC, settlements such as the J W Copes site may have been long-term and virtually sedentary. Pepo gourds or squashes are present and may have been cultivated, but consumption of the seeds, if practiced at all, would have been one of many purposes served by availability of this camp-following species. All other resources appear to have been collected from the wild.

DOMESTICATION OF NATIVE CROPS IN EASTERN NORTH AMERICA

The situation was quite different in more northerly stretches of the Mississippi Valley and other major stream systems including the Illinois, Ohio, Green, and Tennessee rivers. The first crops were the native eastern squash/gourd (*Cucurbita pepo* ssp. *texana* var. *ovifera*), sunflower (*Helianthus annuus* var. *macrocarpus*), sumpweed (*Iva annua* var. *macrocarpa*), and thin-coated chenopod (*Chenopodium berlandieri* ssp. *jonesianum*) (Smith 1992). Decades of archaeological, archaeobotanical, and now recent molecular research document independent domestication of these plants between 2500 and 1500 BC, with even earlier cultivation and range extension. Ovifera squash, now recognised as a different branch than its Mexican relative, *C. pepo* ssp. *pepo*[2] shows signs of domestication by 2300 BC (Decker-Walters et al 1993; King 1985), although it was probably planted far outside its natural range by 3500 BC, if not earlier (Asch 1994; Fritz 1999; Hart and Sidell 1997). Seeds of both sunflower[3] and sumpweed had been enlarged through cultivation and selection by the end of the third millennium BC (Asch and Asch 1985; Crites 1993). Selection of much thinner coated chenopod populations was accomplished by 1500 BC (Smith and Cowan 1987).

By late Poverty Point times, 3000 years ago, some people living in the Midwest riverine area to the north of the lower Mississippi Valley depended more heavily than before on domesticated chenopod, sunflower, sumpweed, and larger seeded ovifera squashes, as demonstrated by stored rockshelter caches in Arkansas and Kentucky (Fritz 1997; Gremillion 1994). The dietary significance of these native crops is clearly manifested in the Early Woodland Salts Cave palaeofecal assemblage, dating primarily to the middle of the first millennium BC, where cultigens constitute approximately two-thirds of the fecal contents (Yarnell 1969). Hopewellian and other Middle Woodland societies of the early first millennium AD left assemblages of charred plant remains that document considerable production of this same Eastern Agricultural Complex, which by the first millennium BC included maygrass (*Phalaris caroliniana*), little barley, and erect knotweed (*Polygonum erectum*) (Asch and Asch 1985; Fritz 1993; Smith 1992; Wymer 1993). None of these last three plants was indisputably domesticated, but all were processed and deposited in refuse pits and middens with known domesticates. Maygrass occurs in archaeological sites far beyond its natural range, and both maygrass and erect knotweed were stored in dry rockshelters along with other crops (Fritz 1994). At that time, 2000 years ago, maize started to appear in the Eastern Woodlands in very small amounts and at widely scattered sites (Crawford et al 1997). This maize probably came across the Plains where peripatetic Hopewellians met and traded with southwestern farmers, some of whom had already been growing it for 2000 years (Diehl 2005).

The entire first millennium AD (even after Hopewellian mound building ceased at about AD 300–500) across most of the Midwest and Midsouth saw increased commitment to the production of native crops. Maize became much more common toward the end of the millennium. Some archaeologists suspect it was widespread by AD 500 or so, and that it might be missing from both flotation and stable carbon isotope studies due to several complex factors (Hart 1999). This needs serious testing, but for now evidence indicates that earliest intensification of maize occurred at AD 800 in the American Bottom region of the central Mississippi River Valley (Johannessen 1988). After AD 1000, many groups living to the south and east of the central riverine zone of pre-maize agriculture made the transition from foraging to maize farming. This shift coincided more or less with the rise of Mississippian chiefdoms across the Southeast (Milner 2004).

THE WOODLAND PERIOD IN THE LOWER MISSISSIPPI VALLEY: TCHEFUNCTE, MARKSVILLE, ISSAQUENA AND TROYVILLE

Back in the lower Mississippi Valley, a unique pathway unfolded. The Early Woodland tradition (Tchefuncte, 800–200 BC) included construction of low burial mounds, but little of the old intraregional exchange that characterised Poverty Point, and no signs of native seed crop production. The presence of both *Cucurbita* sp. and *Lagenaria siceraria* (bottle gourd) rind at Morton Shell Mound near the Gulf Coast leads me to believe that gourds were being cultivated, as they probably had been for some time, primarily for non-food purposes (Byrd 1976). T R Kidder's ongoing research at the multicomponent

Raffman site, in the alluvial bottomlands just southeast of Poverty Point, produced new Tchefuncte subsistence data. A flotation-recovered assemblage from six sampled contexts (a total of 54.5 litres of soil floated) included 25 pieces of acorn shell, three of pecan shell, three persimmon seed fragments, and one purslane seed (*Portulaca oleracea*) (Trachtenberg 1999). More material is needed from this period, but it appears that the Archaic pattern of heavy reliance on local nuts and fruits persisted.

The lower Mississippi Valley has its own Middle Woodland burial mound tradition, known as Marksville, which flourished between 100 BC and AD 200 (Kidder 2002). Trade goods moved between Louisiana and the Havana Hopewell region of central Illinois during Marksville times, but the lower Mississippi Valley people evidently did not adopt the domesticated seed plants grown by their exchange partners to the north. Insufficient flotation and archaeobotanical analysis of Middle Woodland contexts has been conducted to rule out some level of production of Eastern Agricultural Complex crops, but work to date, including a recent report on 177 litres of soil from 18 samples from the Marksville site itself shows reliance on hickory nuts, pecans, acorns, persimmons and palmetto fruits (Roberts 1999), with no potential domesticates. A smaller assemblage (84 litres of soil floated from eight contexts) from the late Marksville component at the Raffman site was dominated by acorn shell, with low frequencies of pecan shell and persimmon seeds along with one maygrass seed (Trachtenberg 1999). Maygrass grows wild in this region today and is not necessarily indicative of agriculture, especially when recovered infrequently and in isolation from other Eastern Agricultural Complex crops.

Archaeobotanical assemblages from northeast Louisiana show higher ubiquities and densities of acorn than pecan shell by the Middle Woodland period, in spite of the fact that acorn shell is more fragile and considered one of the under-represented plant parts (Yarnell and Black 1985). Intensification of acorns might have accompanied population increase, territorial circumscription and improvements in ceramic technology. The few *Cucurbita* rind fragments from Raffman were not from zones with clear stratigraphic integrity, but it is likely that gourds were used for multiple purposes during the Middle Woodland, as they had been during the Early Woodland and Archaic.

Lower Valley participation in Hopewellian-style mortuary rituals and exchange networks fell off after AD 200 or 300, with the culture following Marksville known as Issaquena. Flotation of Issaquena deposits has been conducted at the Reno Brake and Osceola sites, although most occupation at Reno Brake occurred during the Late Woodland period (Troyville, AD 400–700) and most activity at Osceola took place during Coles Creek times (AD 700–1200) (Kidder and Fritz 1993). The Issaquena archaeobotanical assemblage from Reno Brake, analyzed by the author, came from six flotation samples (71 litres floated). At Osceola, one large Issaquena feature under Mound F yielded material from 76 litres of matrix floated. Acorn is the most abundant food plant at both sites, with pecan, grape, persimmon and palmetto well represented

(Table 10.1). Fruits (persimmon, palmetto, grape and bramble) constitute 60.9% of the Reno Brake seed assemblage, out of a total Issaquena component seed count of 87. The same samples contain only seven chenopod and one little barley seed, none of which indicates anything but wild or weedy plants, for a frequency of 9.2% 'starchy seeds'. Few seeds (a total of 7) came from the Issaquena feature under Mound F at Osceola, including three fruit seeds (two persimmon and one grape) and one maygrass.

A new and in some ways more complex platform mound-building culture called Troyville developed at about AD 400, with the largest mound at the type site measuring 80 feet high. Communal burial mounds at several sites, including Gold Mine and Reno Brake, indicate strong territorial ties (Kidder 2002). Troyville culture was contemporaneous with Late Woodland societies in the central Mississippi and lower Illinois River Valley that were depositing impressive quantities of indigenous crop seeds in middens and pit features. A comparison of seed density (number of seeds per litre of soil) and frequency of selected starchy seed taxa in the overall seed assemblage highlights the contrast between the two regions. Seed density at Reno Brake during Troyville times is 1.1 seeds per litre, compared to 6.4 seeds per litre for the Mund Phase (AD 500–600) component at the Mund site in the American Bottom area (Johannessen 1988). Maygrass, chenopod and erect knotweed contributed 94% of the total seed count at American Bottom components of that time, with chenopods having very thin testas (8–22 microns thick) characteristic of domesticated rather than wild populations, the knotweed showing signs of cultivation, and the maygrass being grown far outside its natural range. Chenopod, which at Reno Brake shows no signs of domestication, is the only one of these taxa from Troyville samples, and it constitutes only 5% of a seed assemblage dominated by persimmon, palmetto and grape.

COLES CREEK PERIOD: AD 700–1200

The Coles Creek culture began to develop between AD 700 and 800 and lasted for more than 400 years, with numerous mound-and-plaza centers built in the Tensas Valley of Louisiana and southern Yazoo Valley of Mississippi. Trade goods indicating outside contact are few, but the platform mounds are interpreted by most archaeologists as centers of competing kin groups whose leaders were gaining increasing power over their followers (Kidder 2002). Clear evidence of differential status is scanty even though numerous strategically located cemeteries have been excavated, but the distribution of earthworks across the landscape and intrasite mound layouts indicate that hierarchical relations were probably more entrenched than before.

If nascent, Mississippian-like, hereditary chiefships were developing or had already formed, we might expect a marked upswing in agricultural activities. AD 800 was the point at which maize was intensified by Emergent Mississippian farmers in the American Bottom area surrounding Cahokia (Johannessen 1988). T R Kidder and I began working on Coles

Creek subsistence remains with the expectation that either maize or Eastern Agriculture Complex crops would be plentiful (Fritz and Kidder 1993; Kidder and Fritz 1993). Maize, however, cannot be documented until late Coles Creek times, after AD 900, and even then the absolute quantities remain low for several hundred years. Ubiquities (% samples in which it occurs) of maize in the Late Coles Creek (AD 900–1200) assemblages from Osceola, Blackwater and Jolly vary between 40% of 30 samples at Blackwater to 100% of the four samples from one test unit (0351) in Mound B at Osceola (Table 10.1). In other Late Coles Creek contexts at Osceola, ubiquities of maize vary from 0 (Mound A) to 75% (Mound C), with an overall average of 55.8% for that component. Absolute counts and densities of maize at Osceola, Blackwater and Jolly are low, with densities ranging between 0.1 and 1.0 fragment per litre. At the same time, for seven early Mississippian Lohmann and Stirling phase (AD 1000–1200) components in the American Bottom, ubiquities for maize average 81.3% (Johannessen 1988).

Most Coles Creek sites have few if any native cultigens, with the exception of *Cucurbita pepo* squash/gourd, which remains thin-walled and therefore might have been gathered wild, although it may well have been cultivated. Sumpweed is relatively common (Table 10.1), but wild-sized. Chenopod is abundant in Early Coles Creek samples from Mound B at Osceola, but with a very low proportion of truncate, thin-testa fruits. Maygrass and native panicoid grasses, including barnyard grass (*Echinochloa* sp.), are also numerous in the Early Coles Creek assemblage from Osceola, but with no compelling reason to suspect cultivation, much less domestication. Early Coles Creek samples from Mound B at Osceola clearly show higher densities of seeds than do their late Coles Creek counterparts from Osceola, Jolly and Blackwater (an average of 5.1 as compared to an average of 0.5), and starchy seeds outnumber fruits in the early samples due primarily to high frequencies of maygrass, thick-testa chenopod, and cheno-am seeds. Starchy seeds make up a relatively low percentage of Late Coles Creek seed assemblages from Osceola, Jolly and Blackwater, ranging from lows of 2.4% at Osceola Mound C and 2.8% at Blackwater to highs of 32.5% at Osceola Mound E and 42.9% at Jolly. By way of contrast, the lowest frequency of chenopod, maygrass and knotweed seeds in the American Bottom during the AD 800–1200 time frame (looking at 16 Dohack through Stirling phase components) is 77%; the highest is 97% (Johannessen 1988: 150).

One early Coles Creek mound site in Louisiana stands out as an exception in the lower Mississippi Valley. At Hedgeland on the Tensas River, archaeologists found pits that yielded several hundred cultigen-looking chenopod seeds along with erect knotweed (which is uncommon at other Coles Creek sites) and domesticated sunflower (Roberts 2004). I interpret this unusual assemblage as reflecting intraregional variation and demonstrating that non-participation, low-level participation, or even somewhat higher participation in Eastern Agricultural Complex food production could be choices made by individuals or local kin groups, depending on their needs, backgrounds and preferences. At all early Coles Creek components – even Hedgeland – maize is absent and nuts are abundant, with acorn far surpassing pecan, and thick-shelled hickories showing up on and near upland zones. Native fruits are very common, led by persimmon and palmetto and followed by grape, bramble berries, hawthorn (*Crataegus* sp.), and blueberry (*Vaccinium* sp.).

PLAQUEMINE: AD 1200–1500

Maize is both abundant and ubiquitous after AD 1200, far surpassing its Coles Creek densities, demonstrating that Plaquemine people in the lower Mississippi Valley had finally become fully agricultural (Kidder et al 1993). At the single component Emerson site (AD 1400–1500), all flotation samples from units excavated in 1991 and 1992 contained maize, with maize densities of 9.3 (1991) and 2.8 (1992) far exceeding the one fragment per litre maximum of Coles Creek sites in the Tensas project area (Table 10.1). The heightened commitment to agriculture did not involve increased use of native seed crops, although small amounts of maygrass, chenopod, and wild-sized sumpweed were recovered from Emerson. Acorn densities (6.4 and 1.4 fragments per litre) remain within the range of variation of earlier sites, indicating that oak groves were not sacrificed to make room for maize fields. Pecan densities and ubiquities at Emerson are lower than those of most earlier components, but frequencies of fleshy fruits remain high. These neighbors of Mississippians display the mixed farming-foraging adaptation typical of most Eastern Woodlands nations at contact.

DISCUSSION

The lower Mississippi Valley is not the only place in North America where sedentary, ranked societies developed in the absence of classic agriculture as defined by the economic dependence on fully domesticated plants. The Calusa chiefdoms of southwest Florida had no maize agriculture up until contact, although specialty plants such as papayas (*Carica papaya*) were probably grown in home gardens (Newsom and Scarry in press). Native Californians and their counterparts in the Pacific Northwest are other examples, as discussed above, with California being particularly comparable to the lower Mississippi Valley due to shared reliance on acorns. All of these societies had access to rich aquatic resources, and all were expert fisherfolk. Coles Creek (and earlier) middens are full of fish bones (Kidder and Fritz 1993), and the combination of aquatic and terrestrial resources in such a warm but still temperate environment probably had much to do with the minor role of domesticated plants before AD 1200. Still, Louisiana mound builders could have planted more Eastern Agricultural Complex crops between 1500 BC and AD 1000, and they eventually became maize-based agriculturalists when they joined the late prehistoric Mississippian political scene.

Even so, earlier subsistence strategies were not necessarily steps along a continuum from foraging to farming, and those who practiced them were certainly not less sophisticated or 'advanced' than their more agricultural neighbors. Rather, theirs was a stable, successful, and evidently sustainable pattern of harvesting nuts, fruits and seeds that complemented a diet high in fish, turtle, deer, small mammals and wild fowl. Early mound builders of the lower Mississippi Valley probably managed orchard-like groves of oak, pecan and fruit-bearing trees, maintained clearings for grasses and other seed plants, and built long-term fishing stations. Meetings and other events held at the mound centers may have included negotiations to settle disputes over land claims and to establish priorities concerning land use. The landscape could have been just as altered or just as humanised as were those of Midwestern gardeners and farmers who depended far more on domesticated annuals.

For thousands of years, economic strategies in the lower Mississippi Valley were neither purely foraging nor agricultural. Except for domesticated gourds (whose presence is likely but not certain), foods came from resources that were wild but capable of being enhanced in productivity through clearing, thinning, planting, transplanting and nurturing in other ways. Harris (1989, 1996) calls the systematic application of practices such as these 'wild-plant production', a pattern distinguished from wild plant *procurement*. The level of resource management that I envision for northeast Louisiana by Poverty Point times and possibly sooner falls well within Smith's (2001) borders of low-level food production, providing a case study that highlights the heterogeneity of relevant examples. In this case, questions linger about the presence or absence of domesticates during much of the sequence. Although domestication is described as a mountain range in Smith's conceptual landscape, dividing low-level food producers who have no domesticates from those who do, whether or not the gourds were genetically altered, probably made little difference on the metaphorically and geomorphologically low-contour topography of the lower valley.

Assuming that domesticated gourds were grown locally, Late Archaic and many Woodland groups in the lower Mississippi Valley were comparable to counterparts in Early and Middle Holocene Oaxaca (Smith 2000) by harvesting all plants from the wild except for cucurbits. By Middle Woodland times, groups in Louisiana were like Hopewellian groups in the midcontinent in their heavy reliance on managed groves of nuts and fruits. Major differences, however, set these groups apart. Late Archaic population density was higher and mobility was lower in resource-rich Louisiana than in Early or Middle Archaic Oaxaca, and mound centers were being built in the lower Mississippi Valley for ritualised, communal trade fairs (Jackson 1991). And, unlike their lower valley contemporaries, Hopewellian mound builders in Ohio and Illinois grew enough domesticated chenopod, sunflower and sumpweed, along with other suspected native seed crops, to fill archaeologically visible storage features, even though reliance on domesticates probably fell near the low end of, or possibly even below, the 30–50% range for agriculture in Smith's (2001) scheme. Native domesticates had filtered into the lower Mississippi Valley by Coles Creek times (after AD 700), as seen at Hedgeland, and maize was grown by AD 1000. I infer from the low frequencies and densities of cultigen-type native seeds and maize, however, that most Coles Creek populations remained low-level food producers, continuing to manage oak trees, pecans and hickories (when living on or near non-bottomland soils), along with persimmons and other native fruit-producing species, for the bulk of their plant foods. Still, maize production was being intensified as Coles Creek shifted into Plaquemine culture, marking the first serious transition towards agriculture in that part of the lower Mississippi Valley.

It has been quite a struggle to decouple mound building and agriculture in this region, given the old bias that Formative societies depended on food production to build public monuments and support hierarchical political structures (cf Willey and Phillips 1958). Pepo squash/gourd may or may not have been cultivated at Poverty Point and other Archaic mound centers, but even if it was, production of domesticates was on a very low level and did not lead to scheduling conflicts or initiate positive feedback mechanisms pushing the system towards agricultural intensification. Coles Creek people had access to domesticated chenopod and sunflower, but most chose not to plant them or else to use them far less than did their neighbors to the north. When dependence on maize finally occurred, it accompanied the opening up of previously inward-looking societies, more interactions than before with increasingly powerful Mississippian polities to the east, and a radical shift in settlement patterning with larger Plaquemine mound centers surrounded by smaller farmsteads (Kidder 1998).

Along with foragers and mixed fisher-gatherer-farmers in other regions of the New World, lower Mississippi Valley societies manifest what Wade Davis (1995: 47) describes as a 'profoundly different way of living with the forest'. Attempts to understand how historic patterns evolved by studying the archaeological and palaeoenvironmental records should avoid ethnocentric tendencies that place a higher value on agriculture than on foraging or managed production of wild food plants. Certainly we should not fail to recognise any region where people participated in plant domestication or other early agricultural activities. However, it seems equally biased to latch onto and stretch any enigmatic piece of evidence or to broaden categories so as to classify societies as 'not just hunter-gatherers'. In the lower Mississippi Valley, a rich and complex cultural tradition unfolded on a background of long-term harvesting and landscape management, with relatively minor presence of domesticated plants until a few hundred years before the invasion of Europeans.

ACKNOWLEDGMENTS

T R Kidder directed the excavations at Reno Brake, Osceola, Jolly, Blackwater and Emerson, and he helped in every way possible to recover adequate flotation samples and ensure that their contexts be understood. Reca Jones of Monroe, Louisiana, deserves special thanks for the many days she spent volunteering to help with flotation at Cross Keys Plantation in St. Joseph,

Louisiana. I am sincerely grateful to Jocelyn Turner for making the map for this paper.

NOTES

1. Most Cucurbitaceae (squash/gourd family) remains recovered by flotation are thin rind fragments that fall below Smith's (2000) 2 mm cutoff for demonstrably domesticated *Cucurbita pepo*. Seeds are rarely reported from archaeological sites in the lower Mississippi Valley, and none have been described as longer than 11 mm, the lower boundary for clearly domesticated *C. pepo* seeds. Although wild specimens are not known to exceed these sizes, domesticates can and even today frequently do have rind thinner than 2 mm, making it impossible to say for sure that very thin charred rind fragments came from wild plants.
2. Numerous molecular studies support the separate evolutionary pathway of the domesticated eastern squash, *Cucurbita pepo* ssp. *texana* var. *ovifera* (formerly classified as *C. pepo* ssp. *ovifera* var. *ovifera*), but the geographic origin of that crop within a broad Gulf Coastal or interior southeastern region is still uncertain (Decker-Walters et al 2002; Paris et al 2003; Sanjur et al 2002; Wilson et al 1992).
3. Domesticated sunflower has also been reported from the San Andrés site on the coast of Tabasco, Mexico, directly dated to 2300 BC, which is approximately the same age as the oldest larger-than-wild specimen from eastern North America (Lentz et al 2001). This introduces the possibility that sunflower was independently domesticated in both eastern North America and Mexico, although it evidently did not survive as a food or ornamental crop in later Mesoamerican civilization (Heiser 2001). Recent DNA evidence points to a Midwestern US origin of domestication for modern sunflowers (including accessions from Mexico), but because sunflowers are genetically amenable to multiple domestication, further research is in order (Harter et al 2004).

REFERENCES

Ames, K M (1985) 'Hierarchies, stress, and logistical strategies among hunter-gatherers in northwestern North America', in Price, T D and Brown, J A (eds), *Prehistoric Hunter-Gatherers: The Emergence of Cultural Complexity*, pp 155–80, Orlando: Academic Press

Ames, K M (1999) 'Myth of the hunter-gatherer', *Archaeology* September/October: 45–49

Asch, D L (1994) 'Aboriginal specialty-plant cultivation in eastern North America: Illinois prehistory and a post-contact perspective', in Green, W (ed), *Agricultural Origins and Development in the Midcontinent*, pp 25–86. Office of the State Archaeologist Report 19, Iowa City, IA: University of Iowa

Asch, D L and Asch, N B (1985) 'Prehistoric plant cultivation in west-central Illinois', in Ford, R (ed), *Prehistoric Food Production in North America*, pp 149–203. Anthropological Paper 75, Ann Arbor, MI: Museum of Anthropology, University of Michigan

Byrd, K M (1976) 'Tchefuncte subsistence: information obtained from the excavation of Morton Shell Mound', *Southeastern Archaeological Conference Bulletin* 9: 70–75

Crawford, G W, Smith, D G and Bowyer, V E (1997) 'Dating the entry of corn (*Zea mays*) into the lower Great Lakes region', *American Antiquity* 62: 112–29

Crites, G D (1993) 'Domesticated sunflower in fifth millennium BP temporal contexts: new evidence from middle Tennessee', *American Antiquity* 58: 146–48

Davis, W E (1995) 'Ethnobotany: an old practice, a new discipline', in Schultes, R E and von Reis, S (eds), *Ethnobotany: Evolution of a Discipline*, pp 40–51, Portland: Dioscorides Press

Decker-Walters, D S, Walters, T W, Cowan, C W and Smith, B D (1993) 'Isozymic characterization of wild populations of *Cucurbita pepo*', *Journal of Ethnobiology* 13: 55–72

Decker-Walters, D S, Staub, J E, Chung, S M, Nakata, E and Quemada, H D (2002) 'Diversity in free-living populations of *Cucurbita pepo* (Cucurbitaceae) as assessed by random amplified polymorphic DNA', *Systematic Botany* 27: 19–28

Deur, D (2002) 'Rethinking precolonial plant cultivation on the Northwest Coast of North America', *The Professional Geographer* 54(2): 140–57

Diehl, M W (2005) 'Morphological observations on recently recovered early agricultural period maize cob fragments from southern Arizona', *American Antiquity* 70: 361–75

Ford, R I (ed) (1985) *Prehistoric Food Production in North America.* Anthropological Paper 75, Ann Arbor: Museum of Anthropology, University of Michigan

Ford, J A and Webb, C H (1956) *Poverty Point, a Late Archaic Site in Louisiana.* Anthropological Papers 46(1), New York: American Museum of Natural History

Fritz, G J (1993) 'Early and Middle Woodland Period paleoethnobotany', in Scarry, C M (ed), *Foraging and Farming in the Eastern Woodlands*, pp 39–56, Gainesville, FL: University Press of Florida

Fritz, G J (1994) 'In color and in time: prehistoric Ozark agriculture', in Green, W (ed), *Agricultural Origins and Development in the Midcontinent*, pp 105–26. Office of the State Archaeologist Report 19, Iowa City, IA: University of Iowa

Fritz, G J (1997) 'A three-thousand-year-old cache of crop seeds from Marble Bluff, Arkansas', in Gremillion, K J (ed), *People, Plants, and Landscapes: Studies in Paleoethnobotany*, pp 42–62, Tuscaloosa, AL: University of Alabama Press

Fritz, G J (1999) 'Gender and the early cultivation of gourds in eastern North America', *American Antiquity* 64: 417–29

Fritz, G J and Kidder, T R (1993) 'Recent investigations into prehistoric agriculture in the Lower Mississippi Valley', *Southeastern Archaeology* 12: 1–14

Gibson, J L (2000) *The Ancient Mounds of Poverty Point*, Gainesville, FL: University Press of Florida

Gremillion, K J (1994) 'Evidence of plant domestication from Kentucky caves and rockshelters', in Green, W (ed), *Agricultural Origins and Development in the Midcontinent*, pp 87–104. Office of the State Archaeologist Report 19, Iowa City, IA: University of Iowa

Harris, D R (1989) 'An evolutionary continuum of people-plant interaction', in Harris, D R and Hillman, G C (eds), *Foraging and Farming: The Evolution of Plant Exploitation*, pp 11–26, London; Unwin Hyman

Harris, D R (1996) 'Introduction: themes and concepts in the study of early agriculture', in Harris, D (ed), *The Origins and Spread of Agriculture and Pastoralism in Eurasia*, pp 1–9, Washington, DC: Smithsonian Institution Press

Hart, J P (1999) 'Maize agricultural evolution in the Eastern Woodlands of North America: a Darwinian perspective', *Journal of Archaeological Method and Theory* 6: 133–80

Hart, J P and Sidell, N A (1997) 'Additional evidence for early cucurbit use in the northern Eastern Woodlands of the Allegheny Front', *American Antiquity* 62: 523–37

Hart, J P, Daniels, R A and Sheviak, C J (2004) 'Do *Cucurbita pepo* gourds float fishnets?', *American Antiquity* 69: 141–48

Harter, A V, Gardner, K A, Falush, D, Lentz, D L, Bye, R A and Rieseberg, L H (2004) 'Origin of extant domesticated sunflowers in eastern North America', *Nature* 430: 201–5

Heiser, C (2001) 'About sunflowers…', *Economic Botany* 55: 470–71

Jackson, H E (1989) 'Poverty Point adaptive systems in the lower Mississippi Valley: subsistence remains from the J W Copes site', *North American Archaeologist* 10: 172–204

Jackson, H E (1991) 'The trade fair in hunter-gatherer interaction: the role of intersocietal trade in the evolution of Poverty Point culture', in Gregg, S A (ed), *Between Bands and States*, pp 265–86. Center for Archaeological Investigations Occasional Paper 9, Carbondale, IL: Southern Illinois University

Johannessen, S (1988) 'Plant remains and culture change: are paleoethnobotanical data better than we think?, in Hastorf, C A and Popper, V S (eds), *Current Paleoethnobotany*, pp 145–66, Chicago: University of Chicago Press

Kidder, T R (1992) 'Coles Creek Period social organization and evolution in northeast Louisiana', in Barker, A and Pauketat, T (eds), *Lords of the Southeast: Social Inequality and the Native Elites of Southeastern North America*, pp 145–62. Archeological Papers of the American Anthropological Association 3, Washington, DC: American Anthropological Association

Kidder, T R (1998) 'Mississippi period mound groups and communities in the lower Mississippi Valley', in Lewis, R and Stout, C (eds), *Mississippian Towns and Sacred Spaces: Searching for an Architectural Grammar*, pp 123–50, Tuscaloosa, AL: University of Alabama Press

Kidder, T R (2002) 'Woodland Period archaeology of the lower Mississippi Valley', in Anderson, D and Mainfort, R (eds), *The Woodland Southeast*, pp 66–90, Tuscaloosa, AL: University of Alabama Press

Kidder, T R and Fritz, G J (1993) 'Subsistence and social change in the lower Mississippi Valley: the Reno Brake and Osceola sites, Louisiana', *Journal of Field Archaeology* 20: 281–97

Kidder, T R, Fritz, G J and Smith, C J (1993) 'Emerson', in Kidder, T R (ed), *1992 Archaeological Excavations in Tensas Parish, Louisiana*, pp 110–37. Archaeological Report 2, New Orleans: Center for Archaeology, Tulane University

King, F S (1985) 'Early cultivated cucurbits in eastern North America', in Ford, R (ed), *Prehistoric Food Production in North America*, pp 73–98. Anthropological Paper 75, Ann Arbor, MI: Museum of Anthropology, University of Michigan

Lentz, D L, Pohl, M E D, Pope, K O and Wyatt, A R (2001) 'Prehistoric sunflower (*Helianthus annuus* L.) domestication in Mexico', *Economic Botany* 55: 370–77

Lewis, H T (1973) *Patterns of Indian Burning in California: Ecology and Ethnohistory*. Anthropological Papers no 1, Soccoro, CA: Ballena Press

Milner, G R (2004) *The Moundbuilders: Ancient Peoples of Eastern North America*, London: Thames and Hudson

Minnis, P E (ed) (2003) *People and Plants in Ancient Eastern North America*, Washington, DC: Smithsonian Books

Minnis, P E and Elisens, W J (eds) (2000) *Biodiversity and Native America*, Norman, OK: University of Oklahoma Press

Newsom, L A and Scarry, C M (in press) 'Homegardens and mangrove swamps: Pineland archaeobotanical research', in Walker, K J and Marquardt, W H (eds), *The Archaeology of Pineland: A Coastal Southwest Florida Village Complex, AD 50–1710*. Institute of Archaeological and Paleoenvironmental Studies Monograph 4, Gainesville, FL: University of Florida

Paris, H S, Yonash, N, Portnoy, V, Mozes-Daube, N, Tzuri, G and Katzir, N (2003) 'Assessment of genetic relationships in *Cucurbita pepo* (Cucurbitaceae) using DNA markers', *Theoretical and Applied Genetics* 106: 971–78

Peacock, S L and Turner, N J (2000) '"Just like a garden": traditional resource management and biodiversity conservation on the interior Plateau of British Columbia', in Minnis, P and Elisens, W (eds), *Biodiversity and Native America*, pp 133–79, Norman, OK: University of Oklahoma Press

Ramenofsky, A F (1986) 'The persistence of Late Archaic subsistence–settlement in Louisiana', in Neusius, S W (ed), *Foraging, Collecting, and Harvesting: Archaic Period Subsistence and Settlement in the Eastern Woodlands*, pp 289–312. Center for Archaeological Investigations Occasional Paper 6, Carbondale, IL: Southern Illinois University

Roberts, K R (1999) 'Plant remains', in McGimsey, C R, *Excavating the Past: Archaeology and the Marksville Site (16AV1)*, pp 72–81. Annual Report of the Regional Archaeology Program, Management Unit III, Baton Rouge, LA: Louisiana Division of Archaeology

Roberts, K R (2004) 'Plant remains', in Ryan, J (ed) *Data-Recovery Excavations at the Hedgeland Site (16CT19), Catahoula Parish, Louisiana*, pp 177–99, Baton Rouge, LA: Coastal Environments, Inc

Russo, M (1996) 'Southeastern Archaic mounds', in Sassaman, K E and Anderson, D G (eds), *Archaeology of the Mid-Holocene Southeast*, pp 177–99, Gainesville, FL: University Press of Florida

Sanjur, O I, Piperno, D R, Andres, T C and Wessel-Beaver, L (2002) 'Phylogenetic relationships among domesticated and wild species of *Cucurbita* (Cucurbitaceae) inferred from a mitochondrial gene: implications for crop plant evolution and areas of origin', *Proceedings of the National Academy of Science* (Washington, DC) 99: 535–40

Saunders, J, Mandel, R D, Saucier, R T, Allen, T, Hallmark, C T, Johnson, J K, Jackson, E H, Allen, C M, Stringer, G L, Frink, D S, Feathers, J K, Williams, S, Gremillion, K J, Vidrine, M F and Jones, R (1997) 'A mound complex in Louisiana at 5400–5000 years before the present', *Science* 277: 1796–99

Shea, A B (1978) 'Botanical remains', in Thomas Jr, P and Campbell L (eds), *The Peripheries of Poverty Point*, pp 245–60. New World Research Reports of Investigation 12, Pollack, LA: New World Research

Shipek, F C (1989) 'An example of intensive plant husbandry: the Kumeyaay of southern California', in Harris, D R and Hillman, G C (eds), *Foraging and Farming: The Evolution of Plant Exploitation*, pp 159–70, London: Unwin Hyman

Smith, B D (1992) 'Prehistoric plant husbandry in eastern North America', in Cowan, C W and Watson, P J (eds), *The Origins of Agriculture: An International Perspective*, pp 101–09, Washington, DC: Smithsonian Institution Press

Smith, B D (2000) 'Guilá Naquitz revisited: agricultural origins in Oaxaca, Mexico', in Feinman, G M and Manzanilla, L (eds), *Cultural Evolution: Contemporary Viewpoints*, pp 15–60, New York: Plenum

Smith, B D (2001) 'Low level food production', *Journal of Archaeological Research* 9: 1–43

Smith, B D and Cowan, C W (1987) 'Domesticated *Chenopodium* in prehistoric eastern North America: new accelerator dates from eastern Kentucky', *American Antiquity* 52: 355–57

Trachtenberg, S (1999) 'Macrobotanical remains from the Raffman site (16MA20)', unpublished MA thesis, Tulane University

Turner, N J (1999) '"Time to burn": traditional use of fire to enhance resource production by aboriginal peoples in British Columbia', in Boyd, R (ed), *Indians, Fire, and Land in the Pacific Northwest*, pp 185–218, Corvallis, OR: Oregon State University Press

Ubelaker, D H (1988) 'North American Indian population size, AD 1500–1985', *American Journal of Physical Anthropology* 77: 289–94

Ward, H P (1998) 'The paleoethnobotanical record of the Poverty Point culture: implications of past and current research', *Southeastern Archaeology* 17: 166–74

Willey, G R and Phillips, P H (1958) *Method and Theory in American Archaeology*, Chicago: University of Chicago Press

Wilson, H D, Doebley, J and Duvall, M (1992) 'Chloroplast DNA diversity among wild and cultivated members of *Cucurbita* (Cucurbitaceae)', *Theoretical and Applied Genetics* 84: 859–65

Wymer, D A (1993) 'Cultural change and subsistence: the Middle Woodland and Late Woodland transition in the mid-Ohio Valley', in Scarry, C M (ed), *Foraging and Farming in the Eastern Woodlands*, pp 138–56, Gainesville, FL: University Press of Florida

Yarnell, R A (1969) 'Contents of human paleofeces', in Watson, P (ed), *The Prehistory of Salts Cave, Kentucky*, pp 41–54. Reports of Investigations 16, Springfield, IL: Illinois State Museum

Yarnell, R A and Black, M J (1985) 'Temporal trends indicated by a survey of Archaic and Woodland plant food remains from southeastern North America', *Southeastern Archaeology* 4: 93–107

Zvelebil, M (1996) 'The agricultural frontier and the transition to farming in the circum-Baltic region', in Harris, D R (ed), *The Origins and Spread of Agriculture and Pastoralism in Eurasia*, pp 323–45, Washington, DC: Smithsonian Institution Press

CHAPTER 11

Modeling Prehistoric Agriculture through the Palaeoenvironmental Record: Theoretical and Methodological Issues

Deborah M. Pearsall

INTRODUCTION

The goal of this paper is to discuss theoretical and methodological issues surrounding the interpretation of assemblages of microfossils such as pollen, phytoliths and particulate charcoal in terms of the impact of prehistoric agriculture on forested environments (see Pearsall 2004 for fuller discussion). My focus will be on such applications in forested settings in the lowland neotropics and I will discuss the types of plant microfossil signatures most relevant for detecting early agriculture. Such palaeoenvironmental studies have great potential for giving us a long-range perspective on the interactions between humans and the environment, but also present challenges that must be overcome before this potential can be fully realised. After describing the approach and discussing key issues, I will present an example of how records of past vegetation may provide insight into management practices and may potentially document practices that are sustainable and non-sustainable in long-term perspective. The theoretical and methodological issues presented here are one way in which palaeoenvironmental studies can link past landscapes to present land use and provide information useful to planners as well as scholars from other disciplines.

THE APPROACH

The earliest indications of prehistoric agriculture in forested environments in the neotropics are often environmental disturbances identified in palaeoenvironmental records (eg Bush and Colinvaux 1994; Dull 2004; Goman and Byrne 1998; Islebe et al 1996; Leyden et al 1998; Piperno et al 1991a, 1991b; Piperno and Jones 2003; Veintimilla 1998, 2000). Among the indicators are high particulate charcoal concentrations from clearance activities, reduction or disappearance of pollen or phytoliths from mature forest species, increases in fossil indicators of weeds and other species indicating open habitats, and occurrence and increased abundances of domesticated plants and other useful species. Cores extracted from naturally accumulating sediments (lakes, swamps, alluvial deposits) contain pollen that falls in the coring locality (the regional pollen rain) and phytoliths transported with sediments from the watershed. Cores thus capture regional patterning in vegetation – the composite picture of human and natural processes on the landscape. Researchers have identified the onset of swiddening (7000–4500 BP) and intensive agriculture (4000–2300 BP) in the neotropics based on combinations of microfossil indicators (eg Pearsall 1995; Piperno and Pearsall 1998). The New World is home to many other biomes besides forests, from the deserts of coastal Peru to the high-elevation grasslands of the Andes. I take as my subject the forested neotropics, since it is in the drier tropical forests that agriculture first appears (Piperno and Pearsall 1998). Studying agriculture through palaeoenvironmental records is also an active area of research in the tropics and subtropics of the Old World (eg Flenley and Butler 2001; Genever et al 2003; Haberle 1994; Haberle et al 2001; Kealhofer and Piperno 1994; Zhao and Piperno 2000).

Implicit in this approach is the understanding that agriculture is a phenomenon that leaves an imprint on the environment: 'Thus I define agriculture as environmental manipulations within the context of the human coevolutionary relationship with plants' (Rindos 1984: 100). Agriculture is a set of behaviors in people – planting, weeding, harvesting, storing – that modify the environment encountered by plants. In forested environments, these behaviors include creating larger patches where sun-loving plants may grow. Over time, agriculture leads to increased dependency on domesticated plants, ie plants that have undergone genetic changes (Harlan 1992). Intensification, the process by which the yield per unit of land and/or labor of an existing resource base is increased, is an integral aspect of agricultural evolution (Morrison 1996).

Intensification of agriculture during prehistory is often interpreted as following a trajectory of agricultural evolution formulated first by Boserup (1965). From this perspective, extensive practices (long fallow swiddening) preceded and evolved into intensive practices (short fallow, annual cropping, multicropping). An alternative model exists, proposed by Denevan (2001) and articulated in part earlier by Sauer (1952) and Lathrap (1970), that agriculture began as annual cropping in floodplain environments and developed later in forested uplands as short fallow. In this model, long fallow is a late development associated with the introduction of efficient forest clearance tools and the displacement of populations from preferred habitats. Clearly these two perspectives have different implications for interpretation of environmental records as well as regional archaeology (ie Stahl 2002; see also Morrison 1996 for in-depth discussion of intensification, specialization and diversification in agriculture).

The interpretation of extensive or intensive cropping practices provides only a very coarse-grained perspective on what were likely complex interrelationships between tropical forest peoples and landscapes in the past.

Examining ethnobotanical and ethnographic literature from the Amazon, for example, reveals that a variety of traditional agricultural practices exist today (eg Anderson and Posey 1989; Balée 1994; Boom 1989; Carneiro 1983; Johnson 1983; Moran 1993; Posey 1984; Ruddle 1974; Salick 1989; Vickers and Plowman 1984). Many of these practices can be broadly characterised as swiddening in upland (non-alluvial) settings, but there is considerable variation in criteria used to select sites for clearance, in the length of time fields are cropped, the mix of crops grown, the practices adopted after field 'abandonment,' years left fallow and how fire, manuring, crop rotation, mulching and other practices are used to maintain soil fertility. Few observations can be made of traditional farming on better soils, such as river floodplain habitats, which potentially yield for extended periods, since such lands are rarely cropped using traditional methods (Moran 1993).

There is a range of beliefs and perceptions about the environment among contemporary traditional tropical forest agriculturalists which are linked, directly or indirectly, to subsistence practices, including agriculture. As Balée (1989, 1994; Balée and Gély 1989) and others (Fleck and Harder 2000; Fujisaka et al 2000; Hamlin and Salick 2003; Posey 1982; Shepard et al 2001) have demonstrated, tropical forest agriculturalists can manipulate and manage critical resources such as soil to increase the carrying capacity of land. However, practices that constitute active management for the preservation of biodiversity (ie conservation) in the eyes of an outsider may have another rationale, such as enhancement of game, for the local population. Farmers also make decisions that do not enhance biodiversity. The point is that the rationale and motivation behind traditional agricultural practices should not be assumed to be conservationist (or the opposite), but considered as topics for research.

It follows that prehistorically, in any given environment, there were likely a range of beliefs and perceptions about the environment, which led directly or indirectly to a variety of agricultural practices. These practices, in turn, modified vegetation on local and regional scales, thereby altering the diversities and abundances of microfossils preserved in sedimentary deposits. It would be very interesting to science and potentially valuable to scholars and planners if we could work backwards from the microfossils to the actual practices that permitted people to occupy the land for centuries or millennia, ie practices that were sustainable, even if we are unable to get at the underlying beliefs and perceptions that supported the system.

To continue this discussion, I first present a brief review of the nature of tropical forest agriculture that focuses on the constraints and potential of agriculture in this environment. I then consider several methodological issues that affect our ability to investigate the nature of past agricultural systems. These include distinguishing natural disturbance from human-induced disturbance and regrowth of fallows, identifying deforestation produced by tephra (volcanic ash) fall, determining fallow length from environmental records and identifying floodplain edge and gallery forest disturbance.

CONSTRAINTS AND POTENTIAL OF TROPICAL FOREST AGRICULTURE

There are two dominant characteristics of tropical ecosystems: a complexity of biological interactions and a deficiency of plant mineral nutrients in the soil (Kellman and Tackaberry 1997). Biotic complexity creates special problems when humans attempt to create simpler artificial systems such as agricultural fields. In essence agriculture is a new type of ecosystem for the forested tropics. Low diversity, herbaceous, nutrient-demanding crops, which are inefficient at competing with other plants or animal predators, are substituted for a high diversity, well-adapted woody flora. Farmers traditionally reduce risk by planting a range of varieties of crops – local varieties, or landraces – and by selecting crops that do well under local conditions (Denevan 2001). In a sense, shifting cultivation avoids the problems of maintaining low diversity cropping systems; fields are 'abandoned', ie allowed to go fallow, when weed competition, pest infestations and declining soil fertility depress productivity. Cropping on annually renewed alluvial lands provides another solution; pests and weeds are swept away in annual floods that deposit new soil.

In tropical forest ecosystems, nutrients essential for plant growth are bound up in plant biomass (Kellman and Tackaberry 1997). Unless these are released, insufficient soil nutrients may limit crop growth. Burning of vegetation releases non-volatile mineral nutrients (carbonates, phosphates and silicates of nutrient cations) stored in plant tissues to the soil. These are soluble and enter the soil when the rains begin. This raises soil pH and reduces aluminum levels, creating more favorable conditions for crops. The same effect is achieved by in situ decay of cut vegetation left as mulch (Thurston 1997). Burning has the added advantage of sterilising the soil, killing overseasoning pests and weed seeds. Many, but not all, alluvial soils are highly fertile. The abundance of weatherable minerals, and the resulting soil fertility, depends on the geology of the watershed (Moran 1993). Further, higher terraces that experience less frequent flooding generally have poorer and more weathered soils than terraces closer to the river.

Fields in a shifting cultivation system are allowed to go fallow when greater yields for the same labor input can be achieved by making a new field (Kellman and Tackaberry 1997). Potential factors affecting yields and labor input include declining soil fertility, build-up of insect pests and invasion by weeds. Available technology (steel tools, insecticides and the like) and labor are also factors. The role of insect pests in the decision to fallow a field is unclear; some studies indicate that pest outbreaks are episodic, rather than increasing with years a plot is used (Kellman and Tackaberry 1997). Some crops, such as the root crop yuca (*Manihot esculenta*), are resistant to insect damage. High crop diversity in a field can reduce pest damage, but not just diversity per se. The 'right' mix of crops is needed, or the farmer may just provide pests with tasty alternative hosts (Kellman and Tackaberry 1997).

Few data support dropping soil fertility as a cause for field abandonment (Kellman and Tackaberry 1997). This is not to say that fertility does not decline over time; rather that other factors, especially weed invasion, lead to the decision to fallow a field that still may be fertile enough to grow crops. Today adding fertilizer increases yields on permanently cropped fields, and there are ethnohistoric accounts of fish or fish heads, as well as guano, being used as maize fertilizer by traditional agriculturalists (Denevan 2001). Periodic firing of fields sterilises soil and releases some nutrients back into it, allowing for longer cropping cycles. Intercropping with nitrogen-fixing crops, such as beans and peanuts, can also help sustain field fertility (Denevan 2001).

The negative effect of weeds, especially invading grass species, in agricultural fields is basically competitive, ie undesirable species (weeds) utilise moisture, soil nutrients and space at the expense of desirable species. Traditional agriculturalists keep weeds from getting established through burning – a hot burn will kill weed seeds in the soil – or through maintaining deep mulch in the field (Denevan 2001; Thurston 1997). In established fields, two effective methods of weed control are periodic weeding and interplanting crops to include those that develop rapidly and cover soil to shade out weeds.

A useful way to view this process of forest clearance and regrowth is from the perspective of 'disturbance or interference' ecology (Goldammer 1992). From this viewpoint, human manipulation of vegetation is a natural process in the broad timescale. Certainly it has been part of the ecology of the neotropics for 12,000 years or more (Piperno and Pearsall 1998). If we can avoid subjective valuation of contemporary forest destruction, shifting cultivation can be understood as a type of small- to large-scale disturbance that is a key process in speciation and plays an important role in evolution (Goldammer 1992). As Balée's (1994) work among the Ka'apor demonstrates, the 'serial forests' created are highly valued for culturally useful plants.

In much of the neotropics, populations have reached a point where uncleared forestland is scarce (Moran 1993). Fields must be kept in production longer and/or fallow periods shortened. It is a widely held view (see, for example, Kellman and Tackaberry 1997) that continuous fallow shortening will fail to achieve a new equilibrium level of organic matter or nitrogen in tropical soils, lead to local extinctions of woody taxa and eventually reach a point where woody fallow does not reestablish. Derived savanna and collapse of the agricultural system follow. A total of 50 people/km^2 is considered the maximum carrying capacity of shifting cultivation systems, based on constraints of cropping time and fallow periods (Kellman and Tackaberry 1997). Essentially, then, intensification of shifting cultivation is not considered to be environmentally sustainable. In other words, it is not possible to practice agriculture indefinitely without ongoing deterioration of the environment – especially the soil resources – in which it exists.

There is some evidence, however, that in-field fire, as documented by Posey (1984, 2002; Anderson and Posey 1989) among the Kayapó, is an effective traditional technique for managing soil during the cultivation cycle. In-field burning allows the productive life of a field to be extended.

Other effective, if labor-intensive, weed control measures include mulching, shading out weeds with cover crops, and hand weeding using cutting tools such as the machete (today) and mattock and *macana* (prior to contact) (Denevan 2001).

There are also soils of higher fertility in tropical forests. These include anthropogenic soils – old settlements with soils enriched by manure, ash, garbage and the like – which are sought out as field sites today (Balée 1994; Denevan 2001). Abandoned settlements, or abandoned sectors of occupied villages, could have been farmed during prehistory as well. Soil fertility in small plots, such as house gardens, can also be maintained through manuring. House gardens play a minor role in food production today in the neotropics, largely because houses are frequently moved as field locations shift. If fields were not moved frequently, but grew outwards from a core kept in production for long periods, houses could be more permanent and gardens larger, more intensively cultivated and more productive. And as Lathrap (1970) and others have argued, the house garden makes a great experimental plot. Finally, some river floodplain soils are highly fertile – the 'white water' rivers of the northwest Amazon and coastal Ecuador being two examples (Moran 1993).

In his 2001 book, *Cultivated Landscapes of Native Amazonia and the Andes*, William Devevan argues that the *extensive* nature of contemporary shifting cultivation systems, ie a few years' cropping with 20+ years of fallow, is not a good model for the prehistoric situation, but is an artifact of the machete and reduced population numbers. In other words, shorter fallow *is* feasible on a sustainable basis in the forested tropics, but is just not commonly practiced today. Denevan argues that once a clearing was made at great effort using stone tools, it would be kept open and used intensively, with only short fallow. Farmers could also take advantage of natural openings and seek out secondary growth with softer wood. There is some evidence for the feasibility of this cropping scheme from the Kayapó in eastern Brazil, who use fields for 5–6 years, then fallow them for 8–11 years (Posey 1984, 2002). Management practices include in-field burning, composting and mulching and polycropping, as mentioned earlier.

Essentially, Denevan (2001) argues that *extensive* systems of shifting cultivation were too labor intensive, and not less so, to be a common strategy in the tropical lowlands during prehistory. Most prehistoric agriculture was semi-permanent (short fallow) or permanent. Polycultural plantings were more protective of soils than monoculture or zonal plantings, and therefore more sustainable.

River floodplain habitats are environments in the lowland tropics with great potential for intensive, prehistoric agriculture. We have few observations of these systems in traditional form, due to depopulation at contact and the attractiveness of these lands to European settlers. It is also difficult to generalise, since floodplains are heterogeneous and their fertility depends on the geology of the watershed (Moran 1993). Floodplains do have a predictable form, however, which can help us reconstruct their use.

In essence floodplains can be conceived of as zonal environments (Denevan 2001). Next to the active channel, sand bars and mud flats (*playas*) are exposed only at lowest water levels. These are bounded by high levees, with associated swamp forests on the backslope, which are inundated only by the highest floods. Farther back from the channel are low levees, which are exposed when normal flood levels drop. Landscapes are unstable. As the river migrates across the floodplain, land is created in one place and destroyed in another. Crops can be cultivated annually on *playas* and low levees after floodwater recedes; the risks of recurrent flooding are outweighed by the benefits of fertile sediments and destruction of unwanted vegetation and pests. On higher levees, soils are not annually renewed – although still of recent alluvial origin – and weeds and pests will invade, but floods are less of a hazard. Short-fallow cultivation is most likely in such environments. In Amazonia, villages are typically found on fringing river bluffs, rather than within the floodplain (Denevan 2001).

Other features of contemporary floodplain environments are river terraces, comprising alluvium laid down in the Pleistocene and Holocene and now being downcut by active river channels (Moran 1993). River terrace soils are thus relatively young, but are not annually renewed. There are few data on agricultural use of terraces today. In large river systems, old terrace remnants may be far from the active river channel and difficult to access by canoe. In the case of the Jama River Valley, discussed below, terraces are comprised of sediments laid down during human occupation of the valley (and earlier) and were favored settlement localities in spite of evidence for episodic inundation.

In a floodplain environment, crop mix must be matched to the pattern of river fall and rise (Figure 11.1). Permanent orchard crops, such as palms and fruit trees, would be restricted to the highest, rarely flooded ground. Fast growing, nutrient-demanding crops, such as maize (*Zea mays*), beans (*Phaseolus vulgaris*), peanuts (*Arachis hypogoea*), tobacco (*Nicotiana* spp.), ajis (*Capsicum* spp.) and squashes (*Cucurbita* spp.), would be sown on *playas* and low levees. Higher levees would be good locations for root crops with longer growing seasons such as yuca or manioc (*Manihot esculenta*), *achira* (*Canna edulis*), arrowroot (*Maranta arundinacea*), and *llerén* (*Calathea allouia*), as well as cotton (*Gossypium* spp.), intercropped with maize.

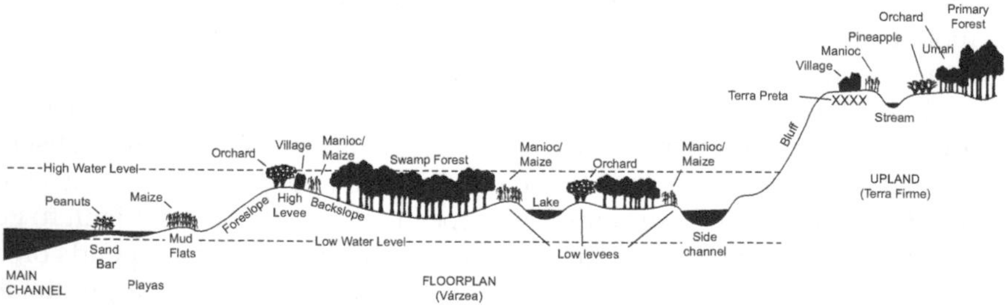

Figure 11.1 River floodplain cross-section (after Denevan 1996).

In summary, ethnographic and ethnohistorical records for the New World suggest that during prehistory humans devised a wealth of approaches not just for living within the constraints and the potential of tropical forest environments, but for shaping those habitats to their needs. Creating and maintaining agricultural fields, which by their nature are new types of ecosystems for the forested tropics, required developing ways to deal with weed competition, pest infestations and declining soil fertility. Technology was limited to relatively simple stone and wooden tools, labor to the effort of humans. Riverine habitats with rich, annually renewed soils presented opportunities for high, sustainable yields, but also the challenges of managing a dynamic landscape. Upland habitats, which can vary greatly in fertility, topography and rainfall, required other types of active management practices. The challenge for us, as scholars of agricultural origins and evolution, is to identify these agricultural practices through the archaeological and geological records.

METHODOLOGICAL CHALLENGES FOR IDENTIFYING PREHISTORIC AGRICULTURE

Establishing Linkages between Fossil Indicators and Agricultural Practices

Observations of forests, regrowing fallows and fields of various ages in lowland neotropical settings suggest that there are distinctive combinations of species and genera that correspond to the outcomes of discrete management practices. Balee's (1994) research among the Ka'apor has documented, for example, that the species composition of regrowing fallows is very different from other secondary forest. Similarly, Salick's (1989; Hamlin and Salick 2003) work among the Amuesha has shown how the species composition of fields evolves through time as different sequences of plants are grown. In theory it should be possible to model various stages of land use (field, young fallow, old fallow, secondary forest) and different strategies of cropping (successive planting, mixed planting, sector planting, monocropping, use of cover crops, mulching) by the microfossil assemblages left behind, and to distinguish among these practices.

In practice one would need to analyze plant microfossil assemblages from multiple soil samples from well-documented forest, fallow and field plots and establish key indicator species and combinations of species that produce these microfossils and that are sensitive to the ecological differences created by the different practices. Another approach would be to get at the underlying factors represented by the taxa, ie moisture, canopy openness, soil fertility, drainage and so on, which would let one apply the method in settings in which specific taxa had not been studied (ie to develop transfer functions).

This process would result in data that could then be used to simulate the impact on vegetation of different agricultural practices at time depths (runs of the simulation), population levels and landscape parameters that correspond to realistic cases (eg Brodt 1992; Rasmussen and Møller-Jensen 1999; after Denevan 2001). Predicted levels of forest clearance and their corresponding microfossil assemblages could then be produced for any point in the simulation. This would be a powerful interpretive tool to apply to palaeoenvironmental sequences in order to identify the kinds of management practices that produce the observed trends in data. We are currently exploring the utility of agent-based modeling (Kohler and Gumerman 2000) for simulating tropical forest agriculture.

Recognising Sustainable and Non-sustainable Practices

As discussed above, rapidly growing human populations, in combination with the pressures of economic development (eg Moran 1993), have led to intense pressure on tropical forests and the decline of mature forest stands as land is kept in cultivation for longer periods. Shortening of fallow period can result in local extinctions of woody taxa and establishment of permanent grasslands (Kellman and Tackaberry 1997). Such lands are no longer workable using traditional agricultural practices. Is this the inevitable result of intensification of swidden agriculture?

Some authorities believe that shifting cultivation, if properly practiced, can be sustainable (Harwood 1996). I have already mentioned some of the traditional management practices that might make this possible. From a development perspective, the key is to look at the fundamental biological processes at work in shifting cultivation and to enhance those that regenerate the soil and lead to fuller use of land, water and biotic resources for enhanced performance, ie those that result in intensification and increased carrying capacity. Smith et al (1999) discuss two such strategies: short-rotation improved fallows (ie practices aimed at keeping fallow length short and result in less secondary forest and reduced cutting of primary forest); and enriching secondary forest development (ie practices that build on the natural regeneration in secondary forests and lead to more secondary forests and use of forest products). An example of the former strategy might be the practice of using legume-cover crops in an annual rotation with maize (Gliessman 1988). The cover crop inhibits weeds, adds organics and increases nitrogen. Thereby, a system of a few years of maize cultivation followed by 5 to 8 years of fallow is converted into a system in which maize can be cropped in alternate seasons with no traditional fallow. There is some evidence that in-field fire, mulching and hand-weeding practices, such as those documented by Posey (1984, 2002) among the Kayapó, are effective traditional techniques for managing soil during the cultivation cycle and for extending the productive life of a field. Productivity may also be enhanced by selection of anthropogenic soils – old settlements with soils enriched by manure, ash and garbage – for cultivation (Balée 1994; Denevan 2001). An example of the latter strategy (enriching secondary forest development) might be traditional practices that maintain the productivity and diversity of useful arboreal resources in fallows.

I advocate an approach that uses the insights of anthropologists, ethnobotanists and agronomists to determine the kinds of management practices that enhance soil regeneration and permit sustainable agriculture. We can then identify the individual practices and model their fossil indicators. We might, for example, be able to identify the timing and extent of burning and the kinds of weeds, or levels of weediness, as separate management strategies, in combination with the lack of indicators of non-sustainable practices.

These considerations raise the question of what non-sustainable practices look like in the palaeoenvironmental record. Extinctions of woody species, establishment of savanna and deterioration of soil resources (erosion of topsoil) are among the key indicators (Kellman and Tackaberry 1997). These phenomena (at least in their advanced stages) appear to be identifiable in palaeoenvironmental records through fossil signatures of vegetation and changes in sedimentation, but only in the absence of indicators of 'natural' forces.

Recognising Climate, Volcanism and Other Non-human Impacts on Forests

One productive approach for distinguishing 'natural' and human-induced disturbance arises from the discovery that natural succession in a tropical forest results in different, and predictable, patterns of regrowth of secondary forest than the pattern produced by regrowth of old fallows. Balée (1994), for example, presents data from his fieldwork among the Ka'apor to illustrate how regrowing fallows differ from secondary or mature forest in terms of species composition, notably through the presence of many useful arboreal taxa in fallow plots. There are also cases in which anthropogenic forests appear to become the new climax vegetation; for example, establishment of nearly monotypic forests of peach palm or forests with high concentrations of other nut species (Balée 1989).

How could these differences between natural and human-initiated succession be modeled for application to interpreting palaeoenvironmental records? Comparative soil samples taken from fallows of documented age, such as those researched by Balée, and from forests in stages of 'natural' succession would allow development of microfossil signatures for natural and human-initiated succession in specific environments. One might also sample areas where clearcut logging has occurred or in hurricane-damaged areas. If comparative sampling is not possible, then it might be feasible to work from the literature on succession in tropical forests and the nature of fallows and to approach the problem from a modeling perspective.

Volcanism has been a major agent of landscape modification and vegetation change in numerous areas of the neotropics. A number of studies have been done of the effects of volcanic ash fall on vegetation, including that of Eggler (1948, 1963) and Rees (1979) on the 1943–1945 activity of Paricutín volcano in the state of Michoacán, Mexico. To summarise the key findings, ash eroded very rapidly on hill slopes, with erosion during the rainy season being the main avenue for redeposition of ash. In areas of ash fall of about 50 cm or less, buried plants regenerated through the ash in a few years, but many trees and shrubs later succumbed. New plants would not establish on primary ash deposits unless organic matter or soil was mixed in. Ash killed all aboveground portions of plants when it accumulated to a minimum depth of 0.7 to 2.0 meters, and some plants,such as small trees, were killed by much less ash. Breakage and starvation – because ash either covered leaves or caused leaf fall – had greater impact on tree mortality than depth of ash around the base.

How can die off of tropical forest due to volcanic ash fall be distinguished from clearance by farmers for agriculture? If preserved tephras – redeposited ash deposits – are present in the stratigraphy of an environmental sequence, can vegetation change be assumed to result only from ash fall if agriculture was practiced? One productive approach might be to consider the scale of tree die off in each case (ash fall versus agricultural clearance), relative to specific topographic features of the environment (Figure 11.2). For example, on flattish terrain (wide river terraces and flat expanses of uplands) ash fall would lead to widespread tree die off, with primary and secondary forest trees, and even open-area plants, affected by deep burial in ash that would remain in place (or only wash from active water channels) (Figure 11.2A). Recovery would proceed very slowly. Agriculture on flattish terrain such as river alluvium would, by contrast, lead to a reduction in primary forest trees, but secondary forest taxa, open-area taxa and useful plant species would increase (Figure 11.2C). There would also likely be differences in scale of

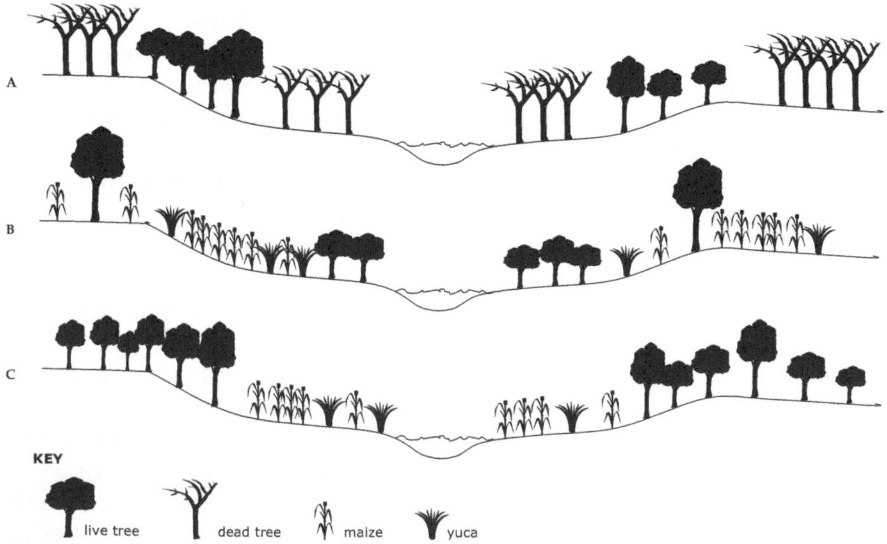

Figure 11.2 Three profiles of the same river valley showing different patterns of forest die-off and clearance. A. volcanic ash fall leaves trees alive on slopes but dead on flatter ground; B. shifting cultivation in uplands leaves trees intact along the river but cleared for crops away from the river; C. cultivation on alluvial terraces shows the opposite clearance pattern.

effects on forests. While both ash fall and agriculture would affect alluvial forests, for example, the effects of ash fall would be much more widespread.

On sloping terrain, such as uplands above river drainages, ash fall would have much less effect due to rapid erosion of ash from slopes (Figure 11.2A). Trees that were killed would resprout from bases; succession would proceed relatively rapidly and result in reestablishment of climax forest. Agriculture in uplands would result in reduction in primary forest taxa, and depending on how long land stayed in production, a slower regrowth with increasing amounts of land in various stages of fallow (Figure 11.2B). Again, modeling these processes might be a productive approach.

INVESTIGATING AGRICULTURAL PRACTICES FROM THE PALAEOENVIRONMENTAL RECORD: THE JAMA CASE

As part of research conducted by James A. Zeidler and myself on the evolution of agriculture and the development of complex societies in Manabí province, Ecuador, Veintimilla analyzed phytolith deposition in a profile of deep alluvium exposed by downcutting of the Jama River in the vicinity of San Isidro, the central place of the Jama River Valley (Pearsall 2004; Veintimilla 1998, 2000; Zeidler and Pearsall 1994) (Figure 11.3). Along the main channel of the Jama River, as well as its major tributaries, deep alluvial deposits were laid down that became prime localities for prehistoric (and modern) human occupation and agriculture. Preserved as remnant terraces today, these deposits measure some 7 to 8 m in height above the current river channel in the vicinity of San Isidro. The profile studied by Veintimilla is called the Río Grande profile, and was described and sampled by Donahue and Harbert (1994).

The Jama River flows northwest into the Pacific from headwaters in a range of coastal hills, passing through a zone of considerable ecological diversity (Zeidler and Kennedy 1994). The lower reaches of the valley fall into the dry tropical climatic zone (average rainfall less than 1000 mm/year), while the interior portion of the valley, including the Río Grande locality, is classified as semihumid tropical climate (1000–2000 mm/year of rain). Most of the inland part of the valley falls into the humid premontane forest life zone, although no intact primary forest exists today (Zeidler and Kennedy 1994).

The record of vegetation change documented in the Río Grande profile provides some support for the primacy of floodplain agriculture and short-fallow swiddening in the valley. Shifting cultivation does appear fairly early in the Jama sequence, however. After an overview of the phytolith patterns documented by Veintimilla, I will review the evidence relevant to interpreting prehistoric agricultural practices in the valley.

The Río Grande profile consists of 9 sediment horizons, for a total depth of 8.83 m, and includes three volcanic ash (tephra) layers and two soil horizons. Based on correlation of the tephras to the dated master sequence at San Isidro, the Río Grande profile can be divided into the chronological units shown in Table 11.1. The major agency of phytolith deposition in the profile is runoff of

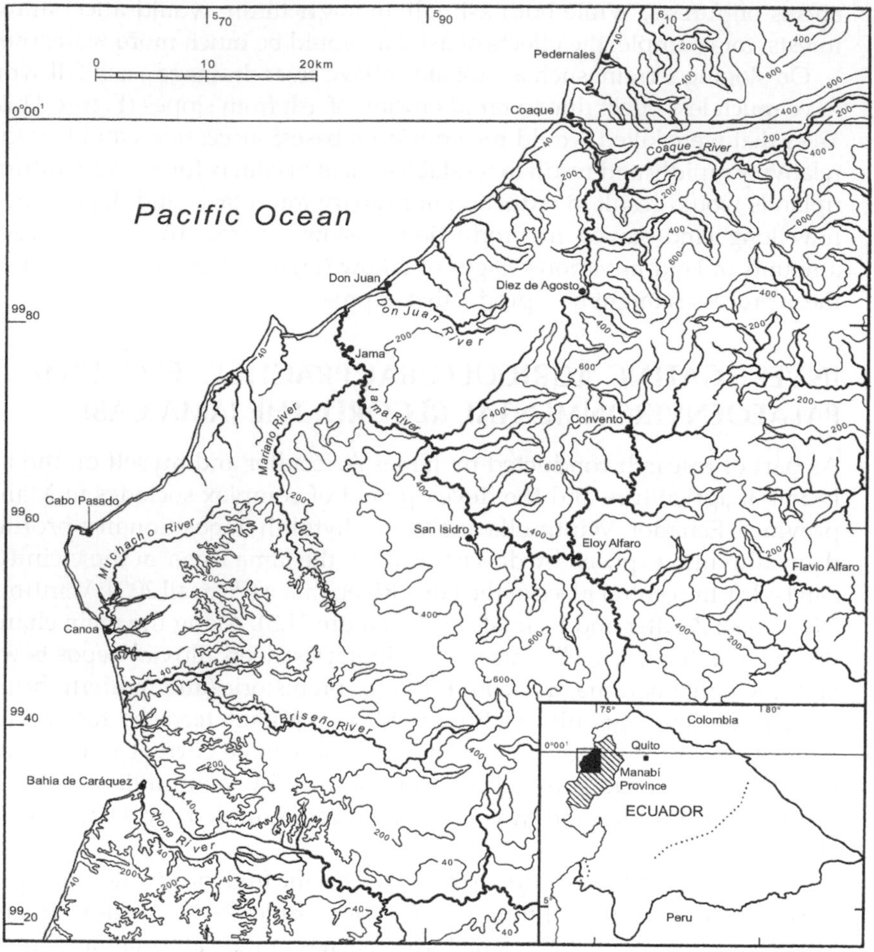

Figure 11.3 Map showing Ecuador, Manabí Province, and the Jama Valley (after Zeidler and Pearsall 1994: Figure 1.1).

Table 11.1 Chronological units for the Río Grande profile and numbers of phytolith samples analysed (after Veintimilla 1998).

Deposit 8	(1 sample)	Bioturbated deposit intrusive to Deposit 6
Deposit 7	(2 samples)	Historic and modern occupation
Deposit 6	(2 samples)	Muchique 2 and after
Tephra III		c AD 400
Deposit 5	(1 sample)	Muchique 1
Deposit 4	(1 sample)	Muchique 1
Tephra II		c 350 BC
Deposit 3	(1 sample)	Late Piquigua through Tabuchila
Tephra I		c 1955 BC
Deposit 2	(3 samples)	c 2525–1955 BC (early Piquigua)
Deposit 1	(4 samples)	c 2859–2525 BC

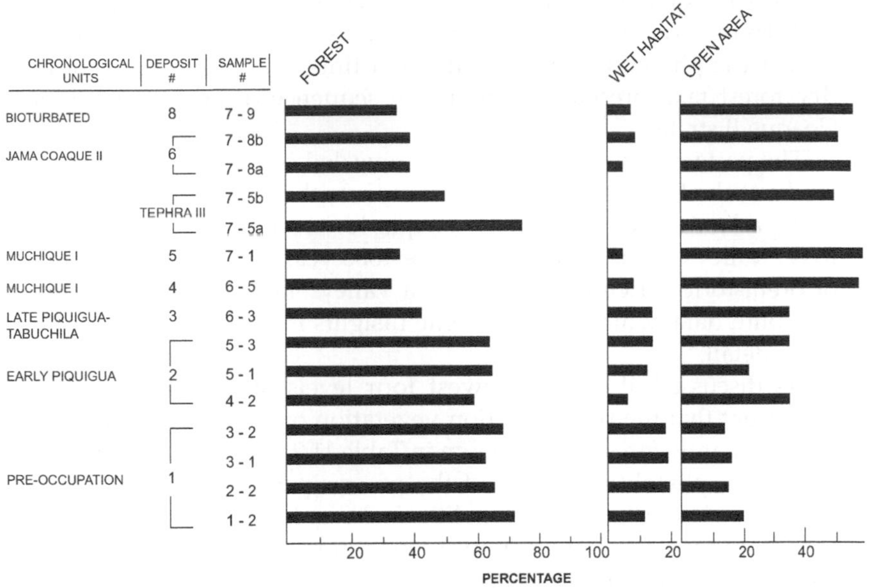

Figure 11.4 Summary phytolith data from the Río Grande profile (after Veintimilla 1998: Figure 7.15). Phytolith data are total counts of one slide-mount.

surface soil, and the phytoliths contained within it, from the watershed of the Jama River upstream of the sampling locality.

The Río Grande profile records profound changes in vegetation during the occupation of the valley (Figure 11.4). Before documented presence of people in the valley and into the early Piquigua occupation, forest dominated the valley. Among the arboreal indicators are Arecaceae (palm), Chrysobalanaceae, Dichapetalaceae, *Cleidium*, *Cordia alliodora*, and *Celtis* phytoliths, as well as cystoliths and schlerids (generalised arboreal) (Veintimilla 1998). Following Tephra I, occupation by Piquigua populations continued briefly, then the valley was abandoned for around 500 years.

The valley was reoccupied by people producing Tabuchila pottery. This period of time – Late Piquigua through Tabuchila – is encompassed in Deposit 3. Forest indicators drop to just over 40%, while open-area indicators remain at levels observed in Deposit 2. The valley was abandoned immediately after the Tephra II ash fall, for an occupation hiatus of about 250 years. Deposits 4 and 5 are contemporary with the Muchique 1 phase reoccupation of the valley. During this time, arboreal phytoliths are documented at the lowest percentage for the profile and open-area taxa, predominantly grasses, come to dominate for the first time.

Tephra III blanketed the valley around AD 400, but did not lead to its abandonment. Occupation continued with the Jama – Coaque II cultural tradition. Veintimilla analyzed two samples from the Tephra III deposits and found mostly arboreal phytoliths. It is unlikely that forest regrew; however, very few phytoliths were present due to the very rapid sedimentation rate of these deposits.

There is a slight increase in forest taxa in the two samples contemporaneous with Jama–Coaque II. This is largely due to an increase in palm phytoliths. Asteraceae phytoliths occur for the first time, with *Heliconia*. A number of dry forest taxa present earlier in the sequence disappear in the Jama – Coaque II strata.

The pattern of vegetation change revealed by Veintimilla's (1998, 2000) study of the Río Grande profile documents the emergence of an agroecology – a landscape managed by people for crop production – in the Jama River Valley over the course of some 4000 years. But what was the nature of prehistoric agriculture in the Jama Valley? Table 11.2 presents the grass phytolith data, which provide some insights into agricultural practices, in more detail.

As discussed above, the lowest four levels of the Río Grande profile document that the preoccupation vegetation of the Jama River Valley was predominantly forested. As shown in Table 11.2, grasses make up a low percentage of phytoliths in each of the preoccupation levels (7.4–14.9%). For comparison, note that grass phytoliths make up from 19.0% to 28.6% of phytoliths from mature (closed canopy) forest in the moister areas of the valley today. These comparative data do not come from primary forests (none exist in the valley today), but are the closest available analogues.

In the first level representing early Piquigua times, grasses jump to 30.2% of the total phytolith sum, and range from 17.0% to 30.2% during this period. Note which grasses increase in frequency: Bambusoideae (mostly *Guadua* and *Chusquea*), not panicoid grasses, which remain at around preoccupation levels. Bamboos are a common component of streamside vegetation in the valley today and can deposit abundant phytoliths, as seen in comparative sample #779 (44.4% bamboo phytoliths), which represents a secondary riverine forest situation. Bamboos are a much less significant component of phytolith assemblages from other secondary forests or from mature forests in the valley. The continued low levels of panicoid grasses and the increase in bamboo phytoliths suggest that initial clearance for crops by the first inhabitants of the valley took place in alluvial forests, rather than upland forest.

After the valley was reoccupied following Tephra I in the Late Piquigua–Tabuchila period, there was a significant increase in grass phytoliths in the Río Grande profile. Bamboos made up 23.4% of the assemblage and panicoid grasses jumped from preoccupation levels to 14.8%. As discussed above, the bamboo data suggest ongoing disturbance of alluvial habitats. The panicoid increases, however, suggest opening of forests back from the river for the first time. Note how the levels of panicoid grass occurrences in the Late Piquigua–Tabuchila sample compare to the levels in comparative soil samples 786 and 800, both taken from non-alluvial secondary forests (Table 11.2). If clearance took place only in alluvial settings, I would not expect to see this rise in panicoid indicators. The first evidence for settlements in non-alluvial areas of the valley dates to the Tabuchila phase, and occupation is dense along the Jama River and its major tributaries. Since vegetation had likely recovered from the Tephra I event after a hiatus of 500 years, these

Table 11.2 Grass phytolith data from comparative samples and the Río Grande profile. Sample 7-9, which is bioturbated, and the Tephra III samples, which have little deposition of phytoliths, are omitted. Percentages are calculated using the total sum (grass short cells plus non-grass diagnostics) (after Veintimilla 1998 and laboratory documents)

Río Grande Profile

Chronological Unit	*Sample #*	*% Grasses*	*% Panicoid*	*% Bambusoideae*
Jama–Coaque II	7-8b	50.2	27.0	13.8
	7-8a	55.1	32.0	11.3
Muchique 1	7-1	58.2	22.4	28.5
	6-5	54.7	24.3	27.4
Late Piquigua–Tabuchila	6-3	41.3	14.8	23.4
Early Piquigua	5-3	17.4	4.8	9.5
	5-1	20.5	5.1	10.1
	4-2	30.2	2.4	21.1
Preoccupation	3-2	8.2	5.2	1.3
	3-1	7.4	2.2	0.7
	2-2	13.5	5.6	2.2
	1-2	14.9	0	4.5

Comparative Surface Samples

Location	*Sample #*	*% Grasses*	*% Panicoid*	*% Bambusoideae*
River-edge bamboo stand	779	80.9	9.6	44.4
Secondary forest, Ramirez	786	52.4	32.8	15.2
Secondary forest, Cerro La Sanchez	800	31.2	24.5	3.4
Mature forest, El Quemado	788	28.6	8.4	13.4
Mature forest, La Chicha	795	20.8	8.5	10.0
Mature forest, La Humidad	799	19.0	4.4	12.2

shifts are attributable to human activity – namely occupation and cultivation in both alluvial and upland settings.

These trends continue during Muchique 1, when the valley was reoccupied following the Tephra II event. Further increases in panicoid grass levels and a drop in forest indicators attest to clearance away from the river and establishment of secondary forests such as those represented by comparative samples 786 and 800. Bamboo occurrence also increases, showing that alluvial lands and gallery forests continued to be disturbed.

The levels dating to Jama–Coaque II and later, after Tephra III, show continued increases in panicoid grass occurrence in the context of disappearance of some forest taxa, appearance of open-area indicators such as Asteraceae, and appreciable settlement in the uplands. Interestingly, bamboo phytoliths decline in occurrence. This is the time of maximum extent of settlement in alluvial settings, with site middens occupying virtually every patch of alluvium in the middle reaches of the valley (Zeidler and Pearsall 1994).

Can we draw any inferences about length of cropping and fallow cycles from these data? Comparative soil samples indicate that 'mature' dry tropical forest in the Jama Valley has a background level of panicoid grass phytoliths of 4.4 to 8.5%. Preoccupation levels of the Río Grande profile have no more than 5.6% panicoid phytoliths. I suggest that panicoid occurrence in the 4.5 to 8.5% range can be used as an indicator of mature, closed canopy forest (minimally disturbed), while occurrences above 8.5% (conservatively) indicate more open forest invaded by panicoid grass taxa. The comparative sample from the secondary forest at Cerro La Sanchez, for example, has 24.5% panicoid grass phytoliths in the context of an estimated canopy tree cover of 50% (Pearsall field notes). By Muchique 1 times and continuing into Jama–Coaque II, phytolith assemblages from the Río Grande profile are very similar to the Cerro La Sanchez pattern. From this I infer that the watershed of the valley was characterised by large expanses of open, secondary forest with panicoid grasses being a significant seasonal component of the understory. The continued and relatively steady presence of a forest component suggests a short fallow pattern.

Was the agroecology of the Jama Valley deteriorating into savannah in the uplands, or had it stabilised as short fallow? Without having palaeonvironmental samples from the latest phases of occupation of the valley, it is difficult to say. One indication of stability is the fact that panicoid grass levels actually fall slightly between the early and later Jama–Coaque II samples, rather than increase. On the other hand, the decline in bamboo phytolith occurrence during this period suggests severe disruption in gallery forests. One might speculate that this was due to the shear density of settlement in this zone. We do know from observations of twentieth-century settlers to the valley that forest completely regrew during the centuries the valley was largely abandoned.

CONCLUSION

In this chapter I have discussed theoretical and methodological issues surrounding the interpretation of microfossil assemblages to identify prehistoric agriculture and to weigh its impact on past environments. I suggest that the potential exists for investigating agricultural management practices and to reveal what were likely complex interrelationships between tropical forest peoples and landscapes in the past. This requires creating microfossil analogs for various kinds of practices, lengths of fallow and 'natural' disturbance indicators. These are challenging tasks that involve considerable basic research as well as formal modeling.

I have framed this discussion in the context of considering an alternative model of agricultural evolution in the lowland tropics proposed by Denevan (2001) and articulated in part earlier by Sauer (1952) and Lathrap (1970).

Agriculture began as annual cropping in floodplain environments and developed later as short fallow in forested uplands. As part of this discussion, I considered several methodological issues, including distinguishing natural disturbance from human-induced disturbance and regrowth of fallows, identifying deforestation produced by tephra fall, determining fallow length from environmental records and identifying floodplain edge and gallery forest disturbances.

I suggest that the Jama River case lends support to Denevan's model that agriculturalists moving into a new habitat and dependent on stone-tool technology will favor use of annually renewed lands requiring minimal clearing. Early occupation of the valley is oriented towards the largest pockets of alluvium along the Jama River and its major tributaries (Zeidler and Pearsall 1994), and there is no evidence for shifting cultivation at this date in the Río Grande profile. Further, there is some indication that once agricultural plots were opened in upland forest areas, they were maintained under a short fallow regime. In other words, we do not see a gradual drop in forest cover that might be expected if plots were allowed to regrow for 20+ years. Rather, arboreal indicators drop fairly abruptly to levels suggesting open-canopy forests. This drop occurs before there is evidence for extensive habitation in the uplands. Farmers were not 'forced' to shorten fallow in response to population pressure, rather they chose to keep fields in cultivation because this was labor efficient. The mix of crops documented at archaeological sites in the valley suggests polycropping was practiced (Pearsall 2004).

The Jama data do not provide us with a definitive answer to the question, was this agricultural system ultimately sustainable? To test whether the cropping practices could have been sustained indefinitely without ongoing deterioration of environment (eg soil erosion, savanna invasion, salinization), and at what population level, requires fine-grained, well-dated palaeoenvironmental sequences from throughout the valley, and these need to be closely tied to the archaeological sequence. Future research may allow us to do such sampling.

Researchers throughout the neotropics have advanced our knowledge of prehistoric agriculture through the analysis of sediment cores. The potential for getting the long view on people-plant relationships in the lowland neotropics and for documenting practices that are sustainable (or non-sustainable) is considerable. I believe this is one way in which palaeoenvironmental studies can link past landscapes to present land use and provide information that is useful both to scholars and planners interested in the conservation of tropical forest resources and to the populations dependent on those resources.

REFERENCES

Anderson, A B and Posey, D A (1989) 'Management of a tropical scrub savanna by the Gorotire Kayapó of Brazil', *Advances in Economic Botany* 7: 159–73

Balée, W (1989) 'The culture of Amazonian forests: resource management in Amazonia: indigenous and folk strategies', *Advances in Economic Botany* 7: 1–21

Balée, W (1994) *Footprints of the Forest: Ka'apor Ethnobotany – The Historical Ecology of Plant Utilization by an Amazonian People*, New York: Columbia University Press

Balée, W and Gély, A (1989) 'Managed forest succession in Amazonia: the Ka'apor case', *Advances in Economic Botany* 7: 129–58

Boom, B M (1989) 'Use of plant resources by the Chácobo', *Advances in Economic Botany* 7: 78–96

Boserup, E (1965) *The Conditions of Agricultural Growth*, Chicago: Aldine

Brodt, S B (1992) 'A simulation modeling analysis of shifting cultivation and deforestation', unpublished master's thesis, University of California–Davis

Bush, M B and Colinvaux, P A (1994) 'Tropical forest disturbance: palaeoecological records from Darien, Panama', *Ecology* 75(6): 1761–68

Carneiro, R L (1983) 'The cultivation of manioc among the Kuikuru of the upper Xingu', in Hames, R B and Vickers, W T (eds), *Adaptive Responses of Native Amazonians*, pp 65–111, New York: Academic Press

Denevan, W M (2001) *Cultivated Landscapes of Native Amazonia and the Andes*, Oxford: Oxford University Press

Donahue, J and Harbert, W (1994) 'Fluvial history of the Jama River drainage basin', in Zeidler, J A and Pearsall, D M (eds), *Regional Archaeology in Northern Manabí, Ecuador*, vol 1: *Environment, Cultural Chronology, and Prehistoric Subsistence in the Jama River Valley*, pp 43–57. University of Pittsburgh Memoirs in Latin American Archaeology, Pittsburgh, PA: University of Pittsburgh, Department of Anthropology

Dull, R A (2004) 'An 8000-year record of vegetation, climate, and human disturbance from the Sierra de Apaneca, El Salvador', *Quaternary Research* 61(2): 159–67

Eggler, W A (1948) 'Plant communities in vicinity of the volcano Paricutin, Mexico, after 2½ years of eruption', *Ecology* 29: 415–36

Eggler, W A (1963) 'Plant life of Paricutin volcano, Mexico, 8 years after activity ceased', *American Midland Naturalist* 69: 38–68

Fleck, D W and Harder, J D (2000) 'Mates Indian rainforest habitat classification and mammalian diversity in Amazonian Peru', *Journal of Ethnobiology* 20(1): 1–36

Flenley, J R and Butler, K (2001) 'Evidence for continued disturbance of upland rain forest in Sumatra for the last 7000 years of an 11,000 year record', *Palaeogeography, Palaeoclimatology, Palaeoecology* 171(3–4): 289–305

Fujisaka, S, Escobar, G and Veneklaas, E J (2000) 'Weedy fields and forests: interactions between land use and the composition of plant communities in the Peruvian Amazon', *Agriculture, Ecosystems, and Environment* 78: 175–86

Genever, M, Grindrod, J and Barker, B (2003) 'Holocene palynology of Whitehaven Swamp, Whitsunday Island, Queensland, and implications for the regional archaeological record', *Palaeogeography, Palaeoclimatology, Palaeoecology* 201(1–2): 141–56

Gliessman, S R (1988) 'Local resource use of systems in the tropics: taking pressure off the forests', in Almeda, F and Pringle, C M (eds), *Tropical Rainforests: Diversity and Conservation*, pp 53–70, San Francisco: California Academy of Sciences and Pacific Division, American Association for the Advancement of Science

Goldammer, J G (1992) 'Tropical forests in transition: ecology of natural and anthropogenic disturbance processes – an introduction', in Goldammer, J G (ed), *Tropical Forests in Transition: Ecology of Natural and Anthropogenic Disturbance Processes*, pp 1–16, Basel: Birkhauser

Goman, M and Byrne, R (1998) 'A 5000-year record of agriculture and tropical forest clearance in the Tuxtlas, Veracruz, Mexico', *Holocene* 8(1): 83–89

Haberle, S (1994) 'Anthropogenic indicators in pollen diagrams: problem and prospects for late Quaternary palynology in New Guinea', in Hather, J G (ed), *Tropical Archaeobotany: Applications and New Developments*, pp 172–201, London: Routledge

Haberle, S G, Hope, G S and van der Kaars, S (2001) 'Biomass burning in Indonesia and Papua New Guinea: natural and human induced fire events in the fossil record', *Palaeogeography, Palaeoclimatology, Palaeoecology* 171(3–4): 259–68

Hamlin, C C and Salick, J (2003) 'Yanesha agriculture in the upper Peruvian Amazon: persistence and change fifteen years down the "road"', *Economic Botany* 57(2): 163–80

Harlan, J R (1992) *Crops and Man*, 2nd ed, Madison, WI: American Society of Agronomy, Crop Science Society of America

Harwood, R R (1996) 'Developmental pathways toward sustainable systems following slash-and-burn', *Agriculture, Ecosystems, and Environments* 58: 75–86

Islebe, G A, Hooghiemstra, H, Brenner, M, Curtis, J and Hodell, D A (1996) 'A Holocene vegetation history from lowland Guatemala', *Holocene* 6(3): 265–71

Johnson, A (1983) 'Machiguenga gardens', in Hames, R B and Vickers, W T (eds), *Adaptive Responses of Native Amazonians*, pp 29–63, New York: Academic Press

Kealhofer, L and Piperno, D R (1994) 'Early agriculture in Southeast Asia: phytolith analysis evidence from the Bang Pakong Valley, Thailand', *Antiquity* 68: 564–72

Kellman, M and Tackaberry, R (1997) *Tropical Environments: The Functioning and Management of Tropical Ecosystems*, London: Routledge.

Kohler, T A and Gumerman, G J (eds) (2000) *Dynamics in Human and Primate Societies: Agent-Based Modeling of Social and Spatial Processes*, New York: Oxford University Press

Lathrap, D (1970) *The Upper Amazon*, New York: Praeger

Leyden, B W, Brenner, M and Dahlin, B (1998) 'Cultural and climatic history of Cobá, a lowland Maya city in Quintana Roo, Mexico', *Quaternary Research* 49: 111–22

Moran, E F (1993) *Through Amazonian Eyes: The Human Ecology of Amazonian Populations*, Iowa City, IA: University of Iowa Press

Morrison, K D (1996) 'Typological schemes and agricultural change', *Current Anthropology* 37: 583–608

Pearsall, D M (1995) 'Domestication and agriculture in the New World tropics', in Price, T D and Gebauer, A B (eds), *Last Hunters–First Farmers: New Perspectives on the Prehistoric Transition to Agriculture*, pp 157–92, Santa Fe, NM: School of American Research Press

Pearsall, D M (2004) *Plants and People in Ancient Ecuador: The Ethnobotany of the Jama River Valley*, Belmont, CA: Wadsworth/Thomson Learning

Piperno, D R and Jones, J G (2003) 'Palaeoecological and archaeological implications of a late Pleistocene/early Holocene record of vegetation and climate from the Pacific coastal plain of Panama', *Quaternary Research* 59(1): 79–87

Piperno, D R and Pearsall, D M (1998) *The Origins of Agriculture in the Lowland Neotropics*, San Diego: Academic Press

Piperno, D R, Bush, M B and Colinvaux, P A (1991a) 'Palaeoecological perspectives on human adaptation in Central Panama, I: the Pleistocene', *Geoarchaeology* 6(3): 201–26

Piperno, D R, Bush, M B and Colinvaux, P A (1991b) 'Palaeoecological perspectives on human adaptation in Central Panama, II: the Holocene', *Geoarchaeology* 6(3): 227–50

Posey, D A (1982) 'The keepers of the forest' *Garden* (New York) January/February: 18–24

Posey, D A (1984) 'A preliminary report on diversified management of tropical forest by the Kayapó Indians of the Brazilian Amazon', *Advances in Economic Botany* 1: 112–26

Posey, D A (2002) *Kayapó Ethnoecology and Culture*, in Plenderleith, K (ed), London: Routledge

Rasmussen, K and Møller-Jensen, G (1999) 'A generic model of shifting cultivation', *Geografisk Tidskrift* (Copenhagen) special issue 1: 157–64

Rees, J D (1979) 'Effects of the eruption of Paricutin volcano on landforms, vegetation, and human occupancy', in Sheets, P D and Grayson, D K (eds), *Volcanic Activity and Human Ecology*, pp 249–92, New York: Academic Press

Rindos, D (1984) *The Origins of Agriculture: An Evolutionary Perspective*, Orlando, FL: Academic Press

Ruddle, K (1974) *The Yukpa Cultivation System: A Study of Shifting Cultivation in Colombia and Venezuela*. Ibero-Americana 52, Berkeley: University of California Press

Salick, J (1989) 'Ecological basis of Amuesha agriculture, Peruvian upper Amazon', *Advances in Economic Botany* 7: 189–212

Sauer, C O (1952) *Agricultural Origins and Dispersals*, New York: American Geographical Society

Shepard, G H, Yu, D W, Lizarralde, M and Italiano, M (2001) 'Rain forest habitat classification among the Matsigenka of the Peruvian Amazon', *Journal of Ethnobiology* 21(1): 1–38

Smith, J, van de Kop, P, Reategui, K, Lombarda, I, Sabogal, C and Diaz, A (1999) 'Dynamics of secondary forests in slash-and-burn farming: interactions among land use types in the Peruvian Amazon', *Agriculture, Ecosystems, and Environments* 76: 85–98

Stahl, P W (2002) 'Paradigms in paradise: revising standard Amazonian prehistory', *The Review of Archaeology* 23(2): 39–50

Thurston, H D (1997) *Slash/Mulch Systems: Sustainable Methods for Tropical Agriculture*, Boulder, CO: Westview Press

Veintimilla, C I (1998) Analysis of past vegetation in the Jama River Valley, Manabí Province, Ecuador, unpublished masters thesis, University of Missouri

Veintimilla, C I (2000) 'Reconstrucción palaeo-ambiental y evolución agrícola en el Valle del Río Jama, Provincia de Manabí, Ecuador', *Revista del Museo Antropológico del Banco Central del Ecuador, Guayaquil* 9: 135–51

Vickers, W T and Plowman, T (1984) *Useful Plants of the Siona and Secoya Indians of Eastern Ecuador*, Chicago: Field Museum of Natural History

Zeidler, J A and Kennedy, R (1994) 'Environmental setting', in Zeidler, J A and Pearsall, D M (eds), *Regional Archaeology in Northern Manabí, Ecuador*, vol 1: *Environment, Cultural Chronology, and Prehistoric Subsistence in the Jama River Valley*, pp 13–41. University of Pittsburgh Memoirs in Latin American Archaeology, Pittsburgh, PA: Department of Anthropology, University of Pittsburgh

Zeidler, J A and Pearsall, D M (eds) (1994) *Regional Archaeology in Northern Manabí, Ecuador*, vol 1: *Environment, Cultural Chronology, and Prehistoric Subsistence in the Jama River Valley*. University of Pittsburgh Memoirs in Latin American Archaeology, Pittsburgh, PA: Department of Anthropology, University of Pittsburgh

Zhao, Z and Piperno, D R (2000) 'Late Pleistocene/Holocene environments in the middle Yangtze River Valley, China and rice (*Oryza sativa* L.) domestication: the phytolith evidence', *Geoarchaeology* 15: 203–22

CHAPTER 12

Chronicling Indigenous Accounts of the 'Rise of Agriculture' in the Americas

Matthew P Sayre

Theories on the 'rise of agriculture' in the Americas have often drawn upon analogies from Southwest Asia and other semi-arid areas where archaeological evidence of agriculture was first evident and where the term originated. Agricultural histories from across the globe, and discussed here with respect to the Americas, prompt a reconsideration of the application of such traditional, often Eurasian-derived concepts of agriculture and its origins to different regions. Indeed, ethnohistorical narratives associated with early agriculture in the Americas require us to rethink the ways in which we transplant the formal categories, ie words, and the concepts that they connote, ie meanings, to other parts of the world.

'Agriculture' is an Indo-European word derived from *agros* (field) that does not have a literal translation in many languages indigenous to the Americas. For example, in the Quechua language, Runa Simi, a literal translation of agriculture would be derived from the word for field – *chakra*. However, the actual practice of agriculture is more generally referred to as a nurturing activity, ie *tarpuy*. *Tarpuy* bridges dichotomies that typify traditional conceptual frameworks for interpreting agriculture, such as nature/culture and wild/domesticated. Having acknowledged some limits to vocabulary, I will discuss problems with transplanting terms and associated connotations, such as 'agriculture' and 'horticulture', to the Americas.

Post-processual representations of agriculture are often predicated on long-established Eurasian concepts and frameworks of thought. In *The Domestication of Europe* (1990: 270), Hodder describes a landscape of change that swept across Europe, with changes in the social outlook allowing for the successful importation of agriculture and the subsequent domestication of society. Hodder's work is focused on the history of Europe and he rightly warns that his perspective should not be uncritically applied to other areas of the world. The theoretical portion of the work rests on the dichotomous relationship established between 'culture' and 'nature', which serves as a template for his analysis of the domestication of plants. Hodder argues that by bringing wild plants into the *domus* people managed to domesticate the plants and themselves at the same time. Hodder states, 'the *domus* was where the wild was brought in and controlled or where culture was separated from the natural' (Hodder 1990: 53). This 'implied an increasing use of plants and animals separated from their natural state' (Hodder 1990: 97). Although Hodder's conceptual framework is only outlined here, it is used heuristically to illustrate the ways in which agriculture is conceived, and the dichotomous thinking that often lies behind such concepts.

INDIGENOUS NARRATIVES

In this paper four different indigenous accounts of agricultural practice are introduced in order to examine some of the diversity of thought regarding agricultural origins and to see whether or not these stories contain evidence of broad changes in social systems that arose through increased contact with domesticated plants. Three of the accounts focus on the origins of crops and the fourth details present and past agricultural practices in Amazonia, focusing on the Ka'apor. Three of the four accounts originate in regions considered by some scholars to be 'centers of origin' (Diamond 2002: 703; Sauer 1950).

Although these accounts are to varying degrees influenced by postcontact events, I do not wish to tack between elements that appear to be 'purely' precontact and elements that are postcontact in origin. I will follow the theoretical terrain outlined by Eduardo Kohn (2002) in his analysis of the historical 'validity' of oral traditions. He explicitly resists using these narratives as an 'imprinting of the present on the past' (Kohn 2002: 562). Rather Kohn (2002: 562) sees them as 'partially autonomous – neither fully structuring nor fully structured by current understandings of the past'.

The case studies are presented in an attempt to let local peoples speak for themselves about how the use of domesticated plants affected their societies. However, it must be acknowledged that these are hardly first-hand accounts. The printed word inevitably distorts a story that was recounted at a particular point in time, in a particular social context and with a distinct local audience in mind.

Case One: The Hopi

The Hopi people of southwestern North America have thrived in their arid homeland by skillfully manipulating the scant aquatic resources at their disposal. Harold Courtlander described in his book, *The Fourth World of the Hopi* (1971), how the Hopi people believe they received the gift of agriculture from supernatural beings. Hopi views on the origins of crops are embedded in broader cosmological beliefs.

In the Third World (ie the world before this world), people obtained their current form after losing their webbed fingers and tails. They were instructed by the Spider Grandmother to be good. So they planted their corn and waited for it to thrive, but the corn needed warmth. After receiving the gifts of fire and plants indigenous to the Americas, such as the sunflower, they were told to depart for the upper or Fourth World, ie the world we live in. As the people emerged from the *sipapuni* (a sacred opening point between the worlds), they were given the gift of various strains of maize.

After all of the other tribes had chosen their various strains of maize, the Hopi were left to choose the last lot:

> So the leader of the Hopi picked it up, saying, 'We were slow in choosing. Therefore we must take the smallest ear of all. We shall have a life of hardship, but it will be a long-lasting life. Other tribes may perish, but we, the Hopis, will survive all adversities. Thus, the Hopis became the people of the short blue corn' (Courtlander 1971: 30).

After the various tribes departed, the Hopi were given special instructions by the Spider Grandmother: 'Remember the *sipapuni*, for you will not see it again. You will go on long migrations. You will build villages and abandon them for new long migrations' (Courtlander 1971: 32). Later they were told:

> But the short blue corn that you chose at the *sipapuni* will be your guide. If you reach a certain place and your corn does not grow, or if it grows and does not mature, you will know that you have gone too far. Return the way you came, build another village and begin again (Courtlander 1971: 33).

The immediate lesson discernible from this account is that the introduction of a staple did not lead to a completely sedentary lifestyle. In fact, the people were instructed to listen to the demands of their crop, which they received in an entirely domesticated form, and to follow its instructions. Eventually they reached the 'center' and there they found the place where they were meant to reside for perpetuity.

In this narrative, it is difficult to construct the nature/culture dichotomy, as the people residing in their villages were instructed to listen to the demands and lessons of the outside world. They were instructed neither to be stewards of the land, nor to tame it. Here is a clear example of a profound distinction between western and Native American views of the world. There is a long western tradition concerned with understanding the place of human beings in the universe. For example, the concept of stewardship in the Judaeo-Christian tradition promotes the idea that humans are unique in the universe and that they are responsible for maintaining/controlling the planet. This thread runs through philosophy and anthropology, such as Berkeley's (1998) immaterialist concept of Being and Kant's formulation of anthropology (Rabinow 1988). The biblical version of stewardship is not prevalent in Native American societies, as illustrated by the Hopi story in which the Hopi are not meant to control nature. Rather, the concept of 'taking care of things' (Pimbert 1994: 3 cited in Gonzales 2000) is more relevant.

Case Two: The Inca

The Inca people of the Andes ruled a significant portion of western South America at the time of Spanish contact. Although there is still debate about the timing and trajectory of their conquests, there is no debate over whether they were one of many state societies that existed in the region (D'Altroy 2002; Lumbreras 1981; Morris and Thompson 1985; Rowe 1946; Stanish 2001).

Sabine MacCormack (1991) in *Religion in the Andes* chronicled the Inca account of the gift of agriculture to the people, *Runakuna*, as the Incas called themselves. The account she used is from the chronicle of Antonio de la Calancha, a member of the Augustinian order. Calancha was a historian of his order who depicted Augustinians' contacts with a pagan 'other'. Calancha was substantially influenced by the works of 'El Inca' Garcilaso de la Vega, a chronicler of mixed indigenous and Spanish descent who fought for the native cause and who was 'both the first, and for some centuries the only, person to grasp that Andean concepts of the holy differed radically from European and Christian ones' (MacCormack 1991: 337). Calancha made personal visits to Pachacamac and in his job as a converter/inquisitor read several Jesuit descriptions of the 'idols' that had been worshipped there (MacCormack 1991: 375). Additionally, he was familiar with islands of the sun and the moon and other sites sacred to the Inca.

About origins, MacCormack states:

> In the beginning, Pachacamac made a man and a woman, but there was no food. The woman prayed to the Sun, who did not send food but impregnated her with a boy-child. Pachacamac became jealous and dismembered the child, making maize out of the child's teeth, yucas out of the ribs and bones, and fruits and vegetables out of the flesh. But the mother grieved for her child and prayed to the Sun until he made another boy-child out of the dead one's umbilical cord. The second child was Vichama. Pachacamac, in revenge, killed the mother and created human beings, along with the lords who were to rule over them (1991: 60–61).

In anger, Vichama turned the first people into stone. Later he and the Sun repented and left the lords as stone *huacas*. With the help of the Sun, Vichama created new people of three different classes from the metals gold, silver and bronze. This portion of the account appears to have broad similarities with Plato's (2000) description of how an ideal society should consist of three natural classes, an argument that Calancha, aware of the intellectual debates of the early seventeenth century, may have incorporated into his text.

The story states that the crop plants survived the initial anger of Vichama and Pachacamac. These crops were staples and were seen as being of the same essence as the people themselves. Crops sprang from people; they were not brought in from the outside and domesticated. The crops were seen as essential for human life and without them the people would not thrive. The crops also needed people to survive. Maize would come to occupy a position of primary importance; it had its own section in the Qorikancha (sacred golden court) where the Inca himself would oversee its planting for the production of the sacred beverage *aqha/chicha*.

The account of Pachacamac reflects the coastal nature of the site. Crops of highland importance such as potatoes and quinoa are not depicted in the account of the origins of humans and crops. There is no division between those crops that have undergone genotypic and phenotypic change as a result of human action, and those that have not. Crops, such as fruit trees, that may not have been completely dependent on humans for their continued survival are given the same treatment as full domesticates because they provide the same nutritional support. People and crops are seen to create civilization together. The people would not be fully civilised without the plants, yet it is also evident that the crops sprang from the people who created hierarchy at the same point in time. Plants and people are intertwined in a relationship of mutual interdependence.

Case Three: The Tolupan

The Tolupan people of modern Honduras have lived in the neotropical region of Central America for centuries. There is only one community that maintains its original language and customs. Although the following account, drawn from Anne Chapman's ethnography (1992), is partially postconquest in origin, it does contain significant information bearing on the past. In this account, Chapman records how people are given the important crop plants at various points in time:

> During the time of the First Nation, Nowpwinapu'u [Master of Maize] brought maize to the world. Until then, the only food was the little roots of the First Nation. They seed by themselves. Now they belong to the white-collared peccaries. The Indians no longer have the right to eat those little roots (1992: 120).

The story continues:

> Chir Tsutsus brought the first *frijoles* [beans] to this world. She brought them from the caves of another nation. She harvested white beans from her head, from her hair, from above. These were for the men, food for the men. From below, from her skirt, she harvested black beans for the women, so that the women could eat (Chapman 1992: 121).

Later a child told his father:

> Hey, papa, if you kill me, you will have camotes, *chayotes* [a vine fruit], to eat. They'll ripen here. If you will only kill me, I will be born again and tobacco will grow out of my body and from my blood will sprout plants to eat – yuca, *malanga* [an edible root], bananas, *chayotes*... (Chapman 1992: 121).

The child returned to the other world with Tomam, a supreme deity, the 'Father-lord of the universe.' When they returned Tomam, gave an order to his [own] family:

> Take care of this kid for me. Don't harm him in the least. Let him go naked, as he pleases. With him, I am going to send seeds to the world of the Mortals. I will plant them in him. This is why I have brought him here (Chapman 1992: 121).
>
> In late April, the kid returned to the Earth bringing seeds of bananas, of yuca, malanga, chayote, onions, tobacco. He took them all to his father, planted them and [disappeared again (Chapman 1992: 121–122).

The agents in this account had knowledge of important economic crops of other regions. Although they were late in receiving them, crop introductions are still credited to divine origin. In the First World, before people had maize, they survived on wild plants, as did the wild animals. This did not mean that people were completely wild, rather, they could learn to grow maize and other crops and eventually abandon earlier foods. However, people were not credited with prior knowledge of how to grow the crops. They had to be taught when to plant and when to harvest. Notable is the fact that in this tale people were living in villages, in a sedentary fashion, when they received the crop plants. Undoubtedly the arrival of maize and other domesticates was embedded in social practices and everyday lives, with resultant transformations. However, people did not consider themselves wild before they began practicing agriculture, rather the social meaning of certain foods changed as a result of its adoption, but their sedentary lifestyle and social patterns may not have dramatically altered.

Case Four: The Ka'apor

William Balée's (1989, 1993, 1994) work among the Ka'apor people of the Brazilian Amazon has revealed many botanical practices of people living in this region of the tropical forest. He has documented their extensive botanical knowledge, as well as the linguistic divisions that exist in the local language between domesticated and non-domesticated plants (Balée 1989, 1994). There are clear structures in the language that parallel the Linnaean concept of genus. However, domesticates are routinely segregated from non-domesticates, even if they could presumably be classified under the same genus label. The concept of 'domestication' is significant to the Ka'apor and underlies their ethnobotanical conception and classification of plants.

Balée (1994) expands this analysis in an attempt to create a comparative ethnobotany. The division between agriculturalists and hunter-gatherers is crucial to his project, as is the relative degree of hierarchy that exists in a society. He finds the differences between hunter-gatherers and farmers to be clearly evident linguistically. Surprisingly, 'small-scale agrarian societies tend to have much more extensive inventories of plant (and animal) names than do foragers' (Balée 1994: 207). This is not an isolated observation; Brown (1985) made a convincing cross-cultural argument for this point. However, to most outside observers, foragers, who spend most of their day in contact with 'wild' plants, would seemingly be more likely to have a thorough knowledge, or naming system, for these plants. It is this fact combined with other ethnographic data that led Balée to state that many Amazonian groups have undergone 'agricultural regression'. According to Balée's model of agricultural regression, a group may initially have a crop inventory that includes maize, bananas, sweet manioc, bitter manioc and other crops. Over time, and due to ecological, cultural or other stresses, they may move into a second phase characterised by the lack of bitter manioc. The next phase could solely consist of maize production and finally the group may end up following an almost completely nomadic lifestyle.

The choice of maize as the last staple to be retained by a group of people is somewhat debatable; presumably, yuca or tubers would be less labor intensive and offer greater caloric returns. It may well be that hunter-gatherers were just as concerned with the symbolic or cultural properties of maize as they were with caloric returns. The model of regression in and of itself offers many lessons for theorists attempting to clarify the ebb and flow of agriculture among communities in the Americas. We must acknowledge the important role of semi-domesticates in the transition between the various phases of agricultural regression and, moreover, that the use of domesticated plants does not fully 'domesticate' the people.

The return to a foraging lifestyle also highlights other preconceived notions about life in the tropical forest. The forest need not be conceived of as a limited resource that is in danger of overharvesting. Rather the forest is resource abundant and with proper care will provide generously. Such human practices as depositing seeds around base camps, carefully gathering resources and continually moving locations ensure that the forest will continue to provide (Politis 1996).

The horticultural practices of various Amazonian groups illustrate how domesticated plants can be exploited in a variety of ways by different societies (Balée 1994; Descola 1996; Rival 2002). The mere encounter with domesticated plants does not exert a unilinear pressure upon a group of people. Rather the act of bringing domesticated plants into the house can eventually lead to a rejection of that practice. Indeed, the replacement of fully domesticated plants with semi-domesticates that grow near the a settlement or territory, can occur after the introduction of an agricultural lifestyle (see Minnis 2003; Waisberg and Holzkamm 1993). If the returns gained from tending a full domesticate are not worth changes in social life, then these plants may be abandoned for less demanding options, such as the return to gathering or a greater dependence upon semi-domesticates. For this reason, there needs to be greater flexibility and precision in defining what entails an 'agricultural lifestyle' if we are to fully chronicle the variety of practices encountered in the Americas.

DISCUSSION

Although there are certainly notable differences between agricultural and non-agricultural peoples, these differences are sometimes difficult to classify discretely (Kimber 1976). The most glaring difference is that agricultural peoples have the potential to produce larger caloric returns per unit of land, although farmers often continue to gather. For early farmers domesticated plants are likely to have initially been a small part of their dietary inventory which must be cared for and watched, like many wild resources. Limited use of domesticates may make them less visible in the archaeological record and make the diversity of early agricultural practices more difficult to document.

Anthropologists who work in the Americas must be wary of importing theories and classifications from other parts of the world where there appears to be solid evidence of sedentism before agriculture, then complete dependence upon domesticates as population numbers rose, and the subsequent spread of agricultural peoples. This story has been shown to be simplistic even for the center of domestication where it originated (Zeder 1996). In the case studies presented we have seen that the Hopi did not cease to migrate or disperse once they were introduced to agriculture. Similarly, Balée did not see agriculture as an irreversibly transformative process that swept across Amazonia.

Theorists should also be especially wary of importing structuralist, particularly dichotomous, concepts and terms derived from Old World data into arguments about the spread of agriculture in the Americas. The forced separation of nature and culture logically leads to the concept that the introduction of domesticated plants leads to domesticated people who subsequently lead radically different lives.

Hodder has acknowledged that the house, or *domus*, need not be based in opposition to the wild (Hodder and Cessford 2004: 20). The house was certainly an important locus of social training and indoctrination in both agricultural and non-agricultural societies. The case studies above do not support the idea that the arrival of agriculture necessarily leads to the domestication of society. At times domestic plants came before what we may conceive of as the *domus* and did not lead to predictable changes in social life, eg sedentism.

There are theoretical concepts that are more conducive to the American situation. The idea of a 'commitment' to agriculture (after Owen 2003) seems relevant. While the term 'commitment' presumes consciousness, this need not mean that agriculture necessarily began as an overtly conscious decision (contra Rindos 1984). Rather, Owen suggests that at later points in time the dependence on, and commitment to, agriculture of people in the Americas changed. This is a flexible idea that appears to fit the actual data to a greater degree than the structuralist concepts that have been applied in Europe. A 'commitment' to agriculture acknowledges that people constantly vacillated and balanced the many pragmatic and symbolic options available to them.

Scholars such as Harlan (1992), Piperno and Pearsall (1998) and Smith (1995) have all shown that agriculture may have independently arisen in many regions of the Americas. As they note, there does not appear to be a single center of agriculture in the Americas; rather, multiple groups of people were open to the inclusion of new crops into their local subsistence regimes. These scholars make the point that all peoples did not passively accept agriculture as it spread out from a core area. Rather, the Americas demonstrate that 'nature' and 'culture' were intertwined or existed along a sliding scale in a dialectical relationship, rather than in diametric opposition.

The diverse origins of many Native American staples (Harlan 1992; Hastorf 1999; Lentz 2000; Piperno and Pearsall 1998) and the diverse ecological backgrounds of Native American crops should prompt an analysis of underlying motivations. The ecological data does not support a transition to sedentism immediately before or after the beginnings of agriculture (Piperno and Flannery 2001). Thus, there may not have been a swift or dramatic change towards cultivating the seeds of annual plants; rather, small-scale communities may have relationally constructed parts of their identities around the plants that they nurtured. These plants were not as easily harvested in large swathes and there was little incentive to gather large populations in elaborate village settings. This way of life, particularly common in neotropical areas, was profoundly different to that of modern peoples and may be a limiting factor on how we conceive of early agriculture.

Most inhabitants of the 'First World' envision agriculture to be an activity that consists of caring for tidy rows of a single crop that will be systematically harvested by machines. The serious disjuncture between this vision and the reality of agriculture in the preconquest Americas is profound. Agriculture, as practiced now and in the past, may not have reduced local diversity. Fields and agroforestry practices can serve as reservoirs of variety. These accounts demonstrate the importance of listening to local accounts of agriculture and farming in order to further frame research around the broad goal of expanding our concept of domestication and advancing the historical knowledge of local practices. As our knowledge of discrete agricultural practices increases it will become clear that the use of domesticated plants did not necessarily domesticate people.

REFERENCES

Balée, W (1989) 'Nomenclature patterns in Ka'apor ethnobotany', *Journal of Ethnobiology* 9(1): 1–24

Balée, W (1993) 'Indigenous transformation of Amazonian forests: an example from Maranhao, Brazil', *L'Homme* 33(2/4): 231–54

Balée, W (1994) *Footprints of the Forest: Ka'apor Ethnobotany – The Historical Ecology of Plant Utilization by an Amazonian People*, New York: Columbia University Press

Berkeley, G (1998) *A Treatise Concerning the Principles of Human Knowledge*, Dancy, J (ed), New York: Oxford University Press

Brown, C H (1985) 'Mode of subsistence and folk biological taxonomy', *Current Anthropology* 26(1): 43–64

Chapman, A (1992) *Masters of Animals: Oral Traditions of the Tolupan Indians, Honduras*, Philadelphia: Gordon and Breach

Courtlander, H (1971) *The Fourth World of the Hopi: The Epic Story of the Hopi Indians as Preserved in their Legends and Traditions*, Albuquerque, NM: University of New Mexico Press

D'Altroy, T N (2002) *The Incas*, Malden, MA: Blackwell

Descola, P (1996) *In the Society of Nature: A Native Ecology in Amazonia*, London: Cambridge University Press

Diamond, J (2002) 'Evolution, consequences, and future implications of plant and animal domestication', *Nature* 418: 700–07

Gonzales, T A (2000) 'The culture of the seed in the Peruvian Andes', in Brush, S (ed), *Genes in the Field: On-Farm Conservation of Crop Diversity*, pp 193–216, New York: Lewis Publishers

Harlan, J R (1992) *Crops and Man*, 2nd edition. Madison, WI: American Society of Agronomy and Crop Science Society of America

Hastorf, C A (1999) 'Cultural implications of crop introduction in Andean prehistory', in Gosden, C and Hather, J (eds), *The Prehistory of Food: Appetites for Change*, pp 35–58, London: Routledge

Hodder, I (1990) *The Domestication of Europe: Structure and Contingency in Neolithic Societies*, Oxford: Basil Blackwell

Hodder, I, and Cessford, C (2004) 'Daily practice and social memory at Catalhoyuk', *American Antiquity* 69(1): 17–40

Kimber, R G (1976) 'Beginnings of farming? Some man-plant-animal relationships in central Australia', *Mankind* 10(3): 142–50

Kohn, E (2002) 'Infidels, virgins, and the black robed priest: a backwoods' history of Ecuador's montaña region', *Ethnohistory* 49(3): 545–82

Lentz, D (ed) (2000) *Imperfect Balance: Landscape Transformations in the Precolumbian Americas*, New York: Columbia University Press

Lumbreras, L G (1981) *Arqueología de la América* Andina, Lima: Editorial Milla Batres

MacCormack, S (1991) Religion in the Andes: *Vision and Imagination in Early Colonial Peru*, Princeton, NJ: Princeton University Press

Minnis, P (ed) (2003) *People and Plants in Ancient Eastern North America*, Washington, DC: Smithsonian Institution Press

Morris, C and Thompson, D E (1985) *Huánaco Pampa: An Inca City and its Hinterland*, London: Thames and Hudson

Owen, B (2003) 'The development of agriculture in the coastal Osmore Valley', paper presented at 68th annual meeting of the Society for American Archaeology, Milwaukee, WI

Pimbert, M (1994) 'The need for another research paradigm', *Seedling* 11: 20–32

Piperno, D and Flannery, K (2001) 'The earliest archaeological maize (*Zea mays* L.) from highland Mexico: new accelerator mass spectrometry dates and their implications', *Proceedings of the National Academy of Science* (Washington, DC) 98(4): 2101–03

Piperno, D and Pearsall, D (1998) *The Origins of Agriculture in the Lowland Neotropics*, New York: Academic Press

Plato (2000) *The Republic*, Ferrari, R G F (ed) and Griffith, T (trans), New York: Cambridge University Press

Politis, G (1996) *Nukak*, Santafé de Bogotá, Colombia: Instituto Amazónico de Investigaciones Científicas

Rabinow, P (1988) 'Beyond ethnography: anthropology as nominalism', *Current Anthropology* 3(4): 355–64

Rindos, D (1984) *The Origins of Agriculture: An Evolutionary Perspective*, Orlando, FL: Academic Press

Rival, L (2002) *Trekking Through History: The Huaoroni of Amazonian Ecuador*, New York: Columbia University Press

Rowe, J H (1946) 'Inca culture at the time of the conquest', in Steward, J (ed), *Handbook of South American Indians*, pp 183–330. Bureau of American Ethnology Bulletin no 143, vol 2, Washington, DC: US Government Printing Office

Sauer, C O (1950) 'Cultivated plants of South and Central America', in Steward, J (ed), *Handbook of South American Indians*, pp 487–543. Bureau of American Ethnology Bulletin no 143, vol 6, Washington, DC: US Government Printing Office

Smith, B (1995) *The Emergence of Agriculture*, Washington, DC: Smithsonian Institution Press

Stanish, C (2001) 'The origin of state societies in South America', *Annual Review of Anthropology* 30: 41–64

Waisberg, L G and Holzkamm, T E (1993) 'A tendency to discourage them from cultivating: Ojibwa agriculture and Indian Affairs administration in northwest Ontario', *Ethnohistory* 40(2): 175–211

Zeder, M (1996) 'After the revolution: post-Neolithic subsistence in northern Mesopotamia', *American Anthropologist* 96(1): 97–126

CHAPTER 13

Starch Remains, Preservation Biases and Plant Histories: An Example from Highland Peru

Linda Perry

INTRODUCTION

The arid coast of Peru is famous for its remarkably well-preserved desiccated archaeological macroremains, and reports from this region are frequently punctuated by photographs of spectacular ancient specimens of food plants such as maize and potato (eg Towle 1961; Ugent et al 1982). The seasonally wet and freeze-prone Andean highlands of southern Peru, however, are characterised by the poor preservation of organic matter, and botanical macroremains are rarely recovered from archaeological sites in this region. The lack of availability of archaeobotanical macroremains coupled with the paucity of systematic palaeoethnobotanical studies in the southern Andes have, unfortunately, left the histories of important plant food crops in large part to speculation.

A recent study of part of a small circular house at the Late Preceramic site of Waynuna in southern highland Peru combined the analyses of archaeological starches and phytoliths extracted from soil samples and lithic tools (Perry et al 2006). The resulting palaeoethnobotanical data extend the southern highland Peruvian record of maize (*Zea mays*) by more than one thousand years, demonstrate the local production and processing of maize, and provide solid evidence for the deliberate movement of plant foods by humans from the tropical forest to the highlands. The articulation of the starch and phytolith data sets effectively illustrates the potential of multiproxy microfossil analyses in delineating ancient plant use and migration in the southern Andes and attests to the great potential of this strategy for other areas with similar preservation biases.

The results from Waynuna also point to issues in the field of starch analysis that merit further exploration. Through the detailed examination of the Waynuna starch data that now include the analysis of an additional artefact, this paper will focus upon three main themes: the utility of starch analyses in understanding plant histories; the implications of the role of lithic materials in facilitating the preservation of starches in archaeological contexts; and the potential benefit of multiproxy archaeobotanical studies in the Andes.

SITE DESCRIPTION

The site of Waynuna (c 3600 to 4000 cal BP) is located in the Cotahuasi Valley of the department of Arequipa in southern Peru (Figure 13.1). During May and June of 2004, test excavations were conducted by a team led by Daniel H Sandweiss of the University of Maine, and the excavators exposed a portion of a small, Preceramic house (Perry et al 2006). The house rests upon a natural terrace located within walking distance of several other terraces, and within half a kilometre of a seasonal stream (Perry et al 2006). At 3625 m in elevation, the site lies at the intersection of what are today two distinct ecological zones (Pulgar Vidal 1987; Trawick 2002). A zone characterised by the heavily irrigated cultivation of maize extends 1300 m below Waynuna. Above the terrace, a cooler zone now used for tuber cultivation stretches up the mountainside to 4000 m. The presence of abandoned agricultural terraces in the immediate vicinity of Waynuna testifies to the utility of the land for intensive plant cultivation from at least Middle Horizon times (c 1200–950 cal BP) (Perry et al 2006).

Excavation of the Preceramic strata yielded few artefacts save quantities of obsidian debitage (Perry et al 2006). Three lithic fragments obviously derived from grinding stones (Figure 13.2) and a large flake tool (Figure 13.3) were recovered from Level 3, a thick homogeneous level that was probably fill. The clearly defined floor (Level 5) was directly underneath the use-deposit, Level 4 (Perry et al 2006). Two stratigraphically equivalent, ashy lenses (6 and 7i) separated by a protrusion of the sterile surface (7) occurred below the floor. Excavators also unearthed an arched stone house wall (Perry et al 2006: Figure 1). Circular houses constructed with local stone are typical of the Late Preceramic in the Central Andes (Malpass and Stothert 1992).

METHODS

Excavations included the collection of unscreened sediment samples from each stratigraphic level for the express purposes of microfossil analysis

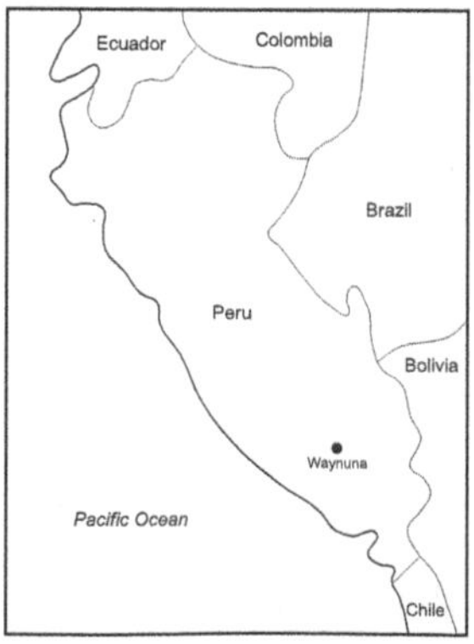

Figure 13.1 Map of Peru showing location of Waynuna.

Figure 13.2 Groundstone tool fragments, catalogue nos 29, 11 and 30 (from left).

Figure 13.3 Flake tool, catalogue no 10.

(Perry et al 2006). Lithic artefacts, including three fragments of larger groundstone tools and one flake, were bagged separately and unwashed (Figures 13.2 and 13.3). A cursory examination of the flake by Anthony J Ranere of Temple University indicated that it was probably used as a tool (pers comm 2005). Nine soil samples and the four lithic tools were chosen for analysis. Dolores Piperno of the Smithsonian Institution performed phytolith analyses on the same soil samples that were tested for starch remains (Perry et al 2006). The methods for starch extraction were modified from those published by other researchers (eg Fullagar et al 1998) and are described in detail elsewhere (Perry et al 2006: supplementary information).

RESULTS

A total of 1361 starch granules was recovered with 654 extracted from the soil samples and 707 from the lithic tools (Figure 13.4; Table 13.1). The overwhelming majority of the starches are derived from maize with 1226

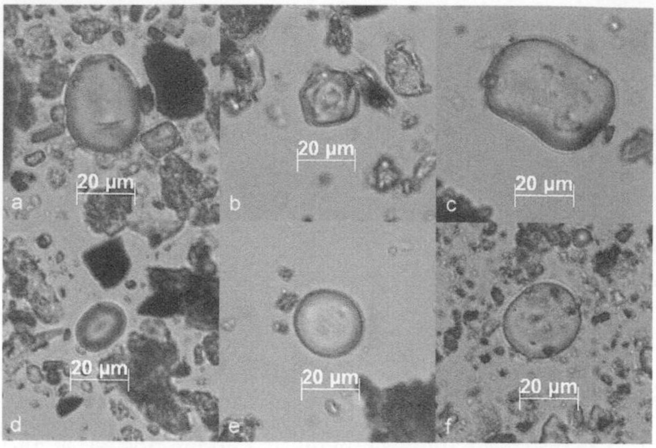

Figure 13.4 Starch granules recovered from sediments and tools: (a) arrowroot (*Maranta* sp.), catalogue 10, ventral side; (b) maize (*Zea mays*) showing damage characteristic of grinding, catalogue 29, sonication; (c) probable potato (cf *Solanum* sp.), catalogue 30, broken side; (d) unidentified legume (Fabaceae), catalogue 10, ventral side; (e) unidentified starch Type 2, catalogue 29, broken side; (f) unidentified starch Type 3 showing dark hilum and navel-like structure, catalogue 36.

definitively identified starch granules and 49 probable maize granules from all examined contexts. Further categorization of the maize granules includes 649 granules with a soft endosperm, 310 granules that are clearly derived from a hard endosperm variety, and 249 granules that exhibit damage (Figure 13.4b).

Five granules of arrowroot (*Maranta* sp.) starch were recovered from a groundstone tool fragment and the flake (Figure 13.4a); three granules that are most likely potato (cf. *Solanum* sp.) occurred on a single groundstone fragment from Level 3 (Figure 13.4c), the probable fill deposit (Table 13.1). The remaining starch remains have been categorised into diagnostic but as yet unidentified starch categories, and those that are not likely to be identified due to the absence of any distinctive morphological features.

Seven granules of starch from an unidentified legume (Fabaceae) were recovered from soil samples from the floor deposits as well as from a groundstone tool and the flake (Figure 13.4d). Thirteen granules of Unknown Type 2 were recovered throughout the column and from three of the four examined lithics (Figure 13.4e); twenty-two granules of Unknown Type 3 were recovered from the lithics with a single granule occurring in the soil sample from the culturally sterile level (Figure 13.4f). Twenty-five unidentifiable granules were also recovered.

Numbers of starch granules recovered separately from the ground and broken sides of the groundstone fragments were also recorded along with those collected using sonication (Table 13.2). Catalogue number 29 yielded more starch from the broken side, catalogue number 11 yielded fairly equivalent numbers of starch granules from each side, and catalogue number 30 yielded more starch from the ground side. Notably, the numbers of

Table 13.1 Numbers of recognisable starch granules recovered from soil and lithic samples

Catalogue Number and Level	*Maranta* sp.	cf. *Solanum* sp.	*Zea mays*	*Zea mays* (soft)	*Zea mays* (hard)	*Zea mays* (damaged)	cf *Zea mays*	Unknown Legume	Unknown Type 2	Unknown Type 3	Unidentified	**Total Maize and cf Maize**	**Total Other**	**Grand Total**
Soil Samples														
30: 3b			5	14	42	22	11				3	94	3	**97**
31: El. 2D-2				6	15	25	3		2			49	2	**51**
32: 4				2	29	16					2	47	2	**49**
33: 5 primera			6	13	24	16	2	1	1		1	61	3	**64**
33: 5 segunda			1	50+	50+	50+	6	4	1		3	157+	8	**165+**
34: 6			1	50+	9	16						76+		**76+**
36: 7i			3	19	24	14	15		1	1		75	2	**77**
35: El 2D-3			1	9	7	9	2		1		1	28	2	**30**
37: 7 (sterile)			1	6	16	16	6					45		**45**
Total (soils)	**0**	**0**	**18**	**169**	**216**	**184**	**45**	**5**	**6**	**1**	**10**	**632**	**22**	**654**
Lithic tools														
29: 3a Groundstone				11	53	29	3		3	1	2	96	6	**102**
11: 3b Groundstone	4			104	28	17	1	1		7	3	150	15	**165**
30: 3b Groundstone		3		115	8	18			1	5	6	141	15	**156**
10: 3a Flake	1			261+	5	1		1	3	8	4	267+	17	**284**
Total (tools)	**5**	**3**		**4911**	**94**	**65**	**4**	**2**	**7**	**21**	**15**	**654**	**53**	**707**
Grand total	**5**	**3**	**18**	**660**	**310**	**249**	**49**	**7**	**13**	**22**	**25**	**1275**	**75**	**1361**

The designation 'cf' indicates a tentative identification. Starch granules designated as 'cf *Zea mays*' are very likely derived from maize; however, inability to view the granules in three dimensions, absence of diagnostic features or severe damage prevented secure identification. When soil sample slides contained starch in abundance, counts for each taxonomic category were made up to fifty granules after which a designation of '50+' was given and counting stopped. Because residues on tools may reveal more about human behaviour than do soil-borne starches, tool slides were counted up to 250 with a similar designation of '250+' to indicate higher numbers.

starch granules recovered from the sonic cleaning could effectively change these results in either direction if points of origin were known. When the species of plants are considered, neither arrowroot nor the probable potato starches were recovered using sonication. These starches are considerably larger in size than all the other starches including maize and the unknowns.

Table 13.2 Provenance of starch granules on stone tools

Catalogue Number, Level and Side of Tool	*Maranta* sp.	cf. *Solanum* sp.	*Zea mays* (soft)	*Zea mays* (hard)	*Zea mays* (damaged)	cf *Zea mays*	Unknown Legume	Unknown Type 2	Unknown Type 3	Unidentified	**Grand Total**
Groundstones											
29 ground side				5	7			1			**13**
11 ground side			19	10	4				2	1	**36**
30 ground side		1	51	6	4			1		4	**67**
Ground side total	**0**	**1**	**70**	**21**	**15**	**0**	**0**	**2**	**2**	**5**	**116**
29 broken side			5	29	13	1		2	1		**51**
11 broken side	4		16	6	2				2		**30**
30 broken side		2	8	2	5				3		**20**
Broken side total	**4**	**2**	**29**	**37**	**20**	**1**	**0**	**2**	**6**	**0**	**101**
29 sonication			6	19	9	2				2	**38**
11 sonication			69	12	11	1	1		3	2	**99**
30 sonication			56		9				2	2	**69**
Sonication total	**0**	**0**	**131**	**31**	**29**	**3**	**1**	**0**	**5**	**6**	**206**
Flake											
10 dorsal side			11	3					5		**19**
10 ventral side	1		250+	2	1		1	3	3	4	**265**
Total (tools)	**5**	**3**	**491+**	**94**	**65**	**4**	**2**	**7**	**21**	**15**	**707**

DISCUSSION

Plant Histories

Prior to this work, the earliest conclusive evidence for maize from the central Andean highlands dated to c 2500 cal BP (Burger and Van der Merwe 1990). The starchy remains of maize were found in every examined context at Waynuna, and the association of these residues in multiple stratigraphic units with charcoal that dated from c 3600 to 4000 cal BP adds at least another one thousand years to the history of maize cultivation in the southern Peruvian Andes (Perry et al 2006). Maize starches extracted from stone tools have also been reported from sites in the Argentinean Andes to the south that date to c 4500 BP (Babot 2004: 206). The prevailing model of maize migration in western South America marks the entrance of this important crop plant in Ecuador about 7000 cal BP (Piperno and Pearsall 1998), and these data coordinate quite well with this model.

Given that the maize remains from Waynuna were found in great abundance and indicate the presence of multiple cultivars, it is likely that the site corresponds to a point in time after the initial arrival of maize in the region. The combination of these two studies indicates that starch analyses will be the most productive means to trace maize migration through the Andean highlands. As data from early sites accumulate, researchers should be able to flesh out the exact path of maize through this region.

The extraction of both maize cob and leaf phytoliths from the same sediments that yielded the starches provides good evidence that this crop was being cultivated very close to the excavated house (Perry et al 2006). In addition to demonstrating that maize was grown on-site, the microfossil data also provide solid evidence for the processing of maize into flour at Waynuna (Perry et al 2006). An experiment in which maize was ground into flour using stone tools yielded starch granules that exhibited diagnostic damage (Babot 2003). The researcher concluded that, with careful analysis, ancient starch granules that exhibit the same morphological features that are created by grinding can be identified and used as evidence of this processing activity (Babot 2003). For this study, *maíz mote,* hulled corn that is used for the production of *masa* flour in Peru, was soaked and then ground in a porcelain mortar and pestle. When mounted and examined with a microscope, many of the starch granules in this freshly milled flour displayed damage identical to that of significant numbers of granules recovered from sediments at Waynuna (Table 13.1).

The percentage of archaeological maize starch granules that exhibited damage consistent with that produced by grinding was as high as 51%. The probability that the damage is due to grinding is also corroborated by the biological nature of maize starch. The surface of maize starch granules is pockmarked by small pores that allow the infiltration of digestive enzymes (Fannon et al 1992). Upon penetrating the granule, these enzymes effectively digest the starch from the inside out (Fannon et al 1992; Leach and Schoch 1961). This unique property of maize starch insures that surface damage would not result from the digestive enzymes of soil-borne microorganisms but from an external mechanical force such as grinding. Thus, maize was being ground into flour at Waynuna, most likely with groundstone tools such as those that yielded the fragments sampled in this study.

The maize starch assemblage contains morphological variants typical of both hard and soft endosperm types of corns. Flour maize is characterised by soft endosperm starch, and flint and popcorns typically yield hard types. These different categories of maize starch have been successfully identified in archaeological sites in Ecuador (Pearsall et al 2004), Panama (Dickau 2005; Piperno et al 2000), Venezuela (Perry 2004, 2005), and now Peru (Perry et al 2006). Many varieties of Peruvian maize, however, are characterised by a combination of both hard and soft starch types in the same kernel. Thus, to distinguish between ancient varieties that are present only as residue, the relative proportions of each type of starch should be considered.

In the soil samples, the starch assemblages extracted from catalogue numbers 30, 31, 32 and 37 are dominated by hard endosperm maize. Catalogue number 34 is dominated by soft endosperm maize, and numbers 33, 35 and 36 have relatively equivalent quantities of each type. Among the tools, hard maize dominates the assemblage from catalogue number 29, and the other three lithics yielded a preponderance of soft endosperm maize. Thus, the starch data indicate the presence of at least two and possibly three varieties of maize. Long-term temporal variation in the use of these varieties is difficult to assess given that two clearly different assemblages occur on the tools, and the lithics were all excavated from the same stratigraphic level. The stratigraphic equivalence of the tools may indicate that the cultivation of different varieties was contemporaneous.

In addition to the maize starch, residues of arrowroot were extracted from two of the lithic tools. Again, buttressing the starch evidence are phytoliths diagnostic of the plant family Marantaceae (Perry et al 2006). These conical silica bodies are typical of those that occur in the bracts and flesh of arrowroot rhizomes, and they can also be found in other species of the Marantaceae (Piperno 2006). Species of the lowland family Marantaceae do not grow at the high elevation of Waynuna, and because starch remains from only a single member of this family, arrowroot, were recovered, there is a high probability that the phytoliths are from this cultivar. Arrowroot itself cannot be grown above 1000 m (Piperno and Pearsall 1998), so the plant, in the form of rhizomes or flour, could only have arrived at the site via human transport.

Arrowroot starch has been recovered from lowland tropical archaeological sites in Panama (Dickau 2005; Piperno and Holst 1998; Piperno et al 2000) and Venezuela (Perry 2004, 2005). Notably, despite the presence of a diverse assemblage of roots and tubers in the rich macrofossil record typical of sites on the coast of Peru, no remains of this lowland domesticate have been excavated from any coastal site (see Piperno and Pearsall 1998; Towle 1961). Further, a search through museum catalogues of pottery from the coastal Peruvian Moche culture will quickly yield photographs of ceramic pots in the form of important root crops including potatoes, canna and manioc, but no representations of arrowroot are seen. It is highly probable that the arrowroot arrived at Waynuna from the Amazonian rainforest to the east (Perry et al 2006). This hypothesis is further supported by the presence of a nearby interzonal track from the east that passes by sites contemporaneous with Waynuna. Excavations at these sites have yielded obsidian artefacts from the Alca source near Waynuna, thus indicating movement of goods between the sites (Burger et al 2000).

Tropical forest products have been documented at low-elevation western Peruvian sites prior to this find, although their migration histories are less clear than the arrowroot from Waynuna. A femur identified as monkey was excavated at La Paloma, a Middle Preceramic site south of Lima on the coast (Reitz 1988), and bird feathers possibly derived from a tropical species were reported from the Late Preceramic site of El Paraíso on the coast near Lima (Quilter 1985). Artefacts that are believed to have been constructed from chonta wood, a tropical forest palm product, were found at La Galgada, a Late Preceramic to Initial Period site situated at about 1100 m in a canyon bottom on the western slopes of the Andes in north-central Peru (Grieder et al 1988). Most recently, several tropical forest products have been reported from the Late Preceramic site of Caral, 200 km north of Lima in the coastal Supe Valley (Shady 2005). In all of these cases, the forest species could have been brought from the Amazonian forest across the Andean cordillera, or, as was suggested by one author, they could have been transported down the coast from northern Peru or southern Ecuador (Reitz 1988). The arrowroot starches and phytoliths from Waynuna at 3625 m in southern Peru are the earliest empirical evidence for movement of goods and/or people from eastern lowland zones into the Andean highlands (Perry et al 2006).

Perhaps the least surprising find in the starch analyses are the starchy remains that are likely derived from potato. Potato is one of the most important members of the Andean root crop complex (Ugent 1970), and it is commonly cultivated at this elevation (Pulgar Vidal 1987). The starch granules of potato have been recovered from desiccated tubers from the coastal Peruvian Casma Valley at sites dating to between c 4000 cal BP and 3000 cal BP (Ugent et al 1982). Thus, it appears that there was widespread cultivation of potatoes in Peru at least by Late Preceramic times. The starches from this assemblage bear similarity to those that have been described and photographed from wild species of potato (Ugent and Verdun 1983); however, the lack of comparative materials from wild specimens of *Solanum* has resulted in a tentative identification.

Finally there are the unknown starches that can be identified with further comparative work. Starch from an unidentified legume occurred in the same stratigraphic context and in larger numbers than the arrowroot and probable potato. This starch is not derived from common bean (*Phaseolus*), *tarwi* (*Lupinus*), *basul* (*Erythrina*) or *pacay* (*Inga*).

Unknowns 2 and 3 are discussed together due to their similar morphological structures. This morphological relationship leads the author to believe that they are likely confamilial or congeneric species. Unknowns 2 and 3 occur in fairly large numbers, particularly on the lithic tools. Unknown 2 has been extracted from the surfaces of lithic tools from Panama (Dickau 2005) and Venezuela (Perry 2001). Further, when colleagues have sent photographs of unknowns to the author for an opinion, this starch was almost always represented. There are only three granules of potato starch in this entire assemblage, and the significance of potato in ancient Andean subsistence is undisputed. These unidentified starches occurred in much larger numbers. Thus, it appears that there was a very important and widespread ancient starch-bearing plant that is currently unidentified.

Preservation Biases

Preservation biases frequently work against archaeologists who study organic remains; however, the sheer numbers of starch granules recovered from these samples testify to the extraordinary preservation of starch remains at Waynuna. Of particular interest are the quantities of starches extracted from the lithic tools. Each tool yielded soil in negligible quantities compared to the 20 ml of sediment sampled from each stratigraphic level. Nonetheless, the numbers of starch granules exceeded those recovered from the sediment samples (Table 13.1). This result supports previous suggestions that lithic materials play an important role in the preservation of ancient starch grains (Perry 2004; Piperno et al 2000).

The starch data also raise issues that merit further exploration. The first issue is the differential recovery of starch types from the lithics and the sediment samples; the second concerns the recovery methods used on the lithics and the interpretation of the results.

The starches of potato and arrowroot were recovered from the lithic tools but not from the sediment samples, while starches of maize, the unidentified legume, and two other unknowns were successfully extracted from both the soils and the tools. While there is no other class of remains from the probable potato, phytoliths of the Marantaceae were recovered from the sediment sample derived from the same context that contained the tool bearing the arrowroot starch. The presence of another class of microfossils that are very likely derived from the arrowroot plant can be interpreted as an indication that arrowroot macroremains or starches were probably deposited in the soil. If there were ancient starches of these plants in the soils, it is important to examine why the sampling methods failed to recover them.

One possibility is that these starches did not survive in the soils, and that they require a microenvironment such as that provided by a lithic tool to survive. Another possibility is that they did survive, and an insufficient quantity of sediment was sampled. Given that the numbers of maize starches recovered from the tools far outnumbered the probable potato and arrowroot, more soil may be required to recover the latter starches as well as others with similar characteristics. However, the unknowns were recovered in relatively small quantities from both contexts. Clearly, further experimentation is in order both to address the possible differential effects of burial on various starches and to delineate the quantities of sediment necessary for effective recovery of all soil-borne starches.

The methods used to extract the starches from the groundstone fragments loosely followed protocols used by researchers in Panama where stepwise extractions were performed to assess whether starch remains were associated with tool use (Piperno et al 2000). First, washes of each side of the tool were made. These residues were examined, and then the entire tool was subjected to a sonic extraction to shake loose all remaining starches. Maize starches were recovered via both methods, and quantities of damaged maize granules were also extracted using both washes and sonication. Despite the presence of maize on the broken sides of the tools, the extraction of the starches using sonication is an excellent indicator of the close association of these starches with the lithic material. Thus, a solid argument can be made that the maize remains are residues from plant processing with these specific groundstone tools. The same can be said for Unknown 3, which was extracted using a combination of methods, and the unidentified legume, which was extracted using sonication only.

The relationship between the groundstone tool fragments and the remains of the probable potato and arrowroot is more difficult to interpret. These starches were extracted from the washes only. The starches of both arrowroot and potato are significantly larger in size and differ in basic morphology from those of maize, and there were no crevices visible to the eye in the groundstone tools. Thus, it is unclear whether these plants were processed with these tools in a manner that did not imbed starches deeply in the stone, or if there simply were no pockets in the tool large enough to accommodate the starches. The second factor that bears consideration is the fact that the starchy remains of arrowroot were recovered only from the broken side of the groundstone fragment, and the probable potato was recovered from both the ground and the broken side of the same tool.

These tool fragments may have been repurposed and used to peel the rhizomes or tubers of the plants in question. While arrowroot must be pounded to extract the starch (Purseglove 1972), the author is unable to find any indication in the literature that potatoes are processed with stone tools prior to use as food. However, desiccated specimens of ancient peeled potatoes have been recovered from the arid coast of Peru (Ugent et al 1982).

Like the probable potato, Unknown 2 was extracted in the washing step from both the ground and broken side of a groundstone tool. Because this starch is of similar size and morphology to Unknown 3, and Unknown 3 was recovered using both washes and sonication, the data may indicate a closer relationship between the tools and Unknown 3. It is not known exactly how much contact stone and starch must have for the starch to adhere. It may be that the washes are extracting starches that were deposited through handling by 'dirty' hands or through contact with flour when the tool broke during use, and sonication may be extracting those residues embedded during actual processing. Again, experimentation can clarify these issues.

The final possibility is that the starches extracted with the washes only are derived from plant material that came into contact with the tool fragments during deposition in the midden. An archaeologist working in the Andes of Argentina extracted faecal spherulites from groundstone tools that also bore the remains of maize and other food plants (Babot 2004). Because it is highly unlikely that people would have been grinding animal waste, the source of these spherulites, on food-related implements, Babot concluded that the manure had come into contact with the tools either peri or postdeposition (Babot 2004). Camelid manure was present both on the surface and throughout the stratigraphy of the sites. The low numbers of spherulites and occurrence in only 29% of the sampled artefacts led to the conclusion that contamination from the surrounding midden material was low, but was present in these rich volcanic soils (Babot 2004).

The results of the Argentine study coupled with the benefit of hindsight effectively illustrate the importance of incorporating stepwise extraction procedures in microfossil analyses, particularly in areas with excellent preservation. The Panamanian archaeologists performed a dry brush extraction followed by a wet brushing, then a more intensive extraction of microfossils from the crevices of the tools (Piperno et al 2000). If a dry brushing step had been performed on the lithic tools in this study, a better assessment of the association of the starches with the tools might have been possible. At this point, the association of the arrowroot, the probable potato and all other starches washed from the broken sides of the tools can be explained both by tool use and by other contact with plant materials pre-, peri-, or postdeposition.

Irrespective of the original source of the starches recovered from the lithics, the results of these studies add to mounting evidence that stone materials buried within a soil matrix provide a protective environment that increases the likelihood that ancient starches will preserve. It remains to be determined whether or not this effect applies only to microfossils lodged in the cracks and crevices of the tools, or, if in cases of unusually good preservation such as at Waynuna, there also exists a shadow effect that helps protect and preserve starches in the immediately adjacent sediments. Further studies of archaeological soil-borne starches and experimentation with the burial of starches in soils and on lithic tools may assist us in the understanding of these processes.

The Utility of Starch Analysis

The data from Waynuna demonstrate that the analysis of soil-borne starches can be a key strategy for documenting the presence of important crop plants at a given site. The archaeological macroremains of maize did not preserve in the soils at Waynuna; however, diagnostic phytoliths and starches survived in abundance. Phytolith analyses allow for the identification of maize cobs and leaves and can provide evidence for the cultivation of maize on-site. Starch granules of maize can be used to discern between hard and soft endosperm types and may allow us to trace differences in use pattern through time, while the diagnostic damage of maize starch documents ancient processing techniques. The benefit of combining these two lines of evidence is clear.

The macroremains of arrowroot were also absent at Waynuna, and the processing of this plant by pounding or grating decreases the likelihood that they will be recovered except in cases of unusual preservation. The phytoliths from the bracts of this starchy rhizome are diagnostic only to family, so the understanding of the ancient use and migration of this species will be largely dependent upon the incorporation of starch analyses into archaeobotanical studies. The presence of the phytoliths of the Marantaceae in the level that contained the tools bearing arrowroot starch and not in any other context testifies to the stratigraphic stability of these microfossils in dense organic deposits.

Macroremains of potato have been excavated only from sites that are characterised by exceptional preservation. Desiccated potatoes have been recovered from sites on the coast of Peru where conditions are extremely arid (Ugent et al 1982). Wet specimens were collected from Monte Verde in Chile where waterlogging created an anaerobic environment (Ugent et al 1987).

Potato is currently the fourth most important food plant on the planet, and it likely held a significant role in the ancient diet of western South America. Starch analyses will allow archaeologists to trace the ancient use of this plant.

In addition to adding information to the databases of these crop plants, the starch analyses hint at things to come. The appearance of widespread unknowns provides a good indication that starch analyses will allow us to identify plants that are unavailable to the researcher through the analysis of macroremains, phytoliths or pollen.

In addition to providing evidence of the presence of food plants at a site, microfossils that are extracted from tools allow for a more accurate assessment of tool function than would the morphological features of the lithic material alone. The fragments of groundstone tools recovered from Waynuna were very likely derived from larger artefacts used to grind maize, and the broken pieces may have been repurposed for the peeling of arrowroot or potato. The larger flake appears to have been used for the peeling or cutting of arrowroot as well as the processing of maize. Until the time that the unknowns have been identified, further interpretation will not be possible. Nonetheless, it appears that every tool was multifunctional.

The results of the microfossil analyses at Waynuna indicate that starch analyses will be put to best use when used in multiproxy analyses that include the study of all other classes of plant remains that have survived at a given site. Different lines of evidence both support one another and fill in the gaps left by preservation biases.

CONCLUSION

The excellent preservation of microfossils at Waynuna permitted the extraction of starches and phytoliths from both sediments and lithic tools. The combination of starch and phytolith analyses provided corroborative evidence of maize and pinpointed arrowroot as the member of the Marantaceae present on site. The extraction of two classes of microremains of arrowroot from a single context also testifies to the stratigraphic stability of the microfossils at this site. Starch analysis alone is responsible for the discovery of the potato remains and the evidence of maize grinding, and these results are crucial to the interpretation of tool function. Future investigations by this author will include stepwise extraction procedures that will allow for more detailed interpretation of tool histories.

Investigations at Waynuna uncovered evidence of on-site production and processing of maize, confirmed the presence of a mixed seed and root crop subsistence economy in the southern Andes, and documented the movement of plant products from the eastern lowland forest to the highlands (Perry et al 2006). Because Waynuna is situated in an area that receives sun all day, both maize and potatoes could be grown on site (Dan Sandweiss, pers comm 2005). The arrowroot remains, however, provide solid empirical evidence that humans were moving from the eastern lowlands to the highlands along the track that passes by the nearby Alca obsidian source.

Other resources undoubtedly accompanied these travellers, and future research in this region should flesh out the understanding of interzonal interaction in southern Peru.

The association of maize microfossils with radiocarbon dates of 3600–4000 cal BP extend the record of maize in the southern Peruvian Andes by at least one thousand years (Perry et al 2006). When these data are combined with those from Ecuador and Argentina, the resulting pattern supports the thesis that maize entered South America approximately 7000 cal BP and migrated down the western side of the continent.

The results from Waynuna testify to the promise of starch analyses for the recovery and identification of botanical remains from plants that would otherwise be invisible in the archaeological record. The incorporation of starch work within archaeological investigations will ensure the understanding of the histories of important root crops like arrowroot and potato. The results from Waynuna demonstrate that different lines of evidence both support one another and fill in the gaps left by the preservation biases that are inherent in each class of botanical remains. Starch studies will be best integrated with other analytical techniques – for example, the study of phytoliths, pollen or macroremains. This holistic approach to archaeobotanical analysis will provide a more complete assemblage and, therefore, a more accurate understanding of ancient behaviours.

ACKNOWLEDGMENTS

The excavation was funded by a grant from the H John Heinz III Charitable Trust. Laboratory support was provided by the Smithsonian Institution Archaeobiology Laboratory. Peru's National Institute of Culture and the village of Huillac gave their support and cooperation to the project. Many thanks are extended to my collaborators, Dan Sandweiss, Dolores Piperno, Michael A Malpass, Kurt Rademaker, Adán Umire and Pablo de la Vera. The excavation crew included Kurt Rademaker, Adán Umire, Louis Fortin, Benjamin Morris and Oswaldo Chozo.

REFERENCES

Babot, M del P (2003) 'Starch grain damage as an indicator of food processing', in Hart, D M and Wallis, L A (eds), *Phytolith and Starch Research in the Australian-Pacific-Asian Regions: The State of the Art*, pp 69–81, Canberra: Pandanus Books

Babot, M del P (2004) 'Tecnología y utilización de artefactos de moliendaen el noroeste prehispánico', unpublished PhD thesis, Universidad Nacional de Tucumán

Burger, R L and Van der Merwe, N J (1990) 'Maize and the origin of highland Chavin civilization: an isotopic perspective', *American Anthropologist* 92: 85–95

Burger, R L, Mohr Chávez, K L and Chávez, S J (2000) 'Through the glass darkly: Prehispanic obsidian procurement and exchange in southern Peru and northern Bolivia', *Journal of World Prehistory* 14: 267–362

Dickau, R (2005) 'Resource use, crop dispersals, and the transition to agriculture in prehistoric Panama: evidence from starch grains and macroremains', unpublished PhD dissertation, Temple University

Fannon, J E, Hauber, R J and BeMiller, J N (1992) 'Surface pores of starch granules', *Cereal Chemistry* 69: 284–88

Fullagar, R, Loy, T and Cox, S (1998) 'Starch grains, sediments and stone tool function: evidence from Bitokara, Papua New Guinea', in Fullagar, R (ed) *A Closer Look: Recent Australian Studies of Stone Tools*, pp 49–60, Sydney: Archaeological Computing Laboratory, University of Sydney

Grieder, T, Bueno M A, Smith Jr, C E and Malina, R M (1988) *La Galgada, Peru: A Preceramic Culture in Transition*, Austin: University of Texas Press

Leach, H W and Schoch, T J (1961) 'Structure of the starch granule II: action of various amylases on granular starches', *Cereal Chemistry* 38: 34–46

Malpass, M A and Stothert, K E (1992) 'Evidence for Preceramic houses and household organization in western South America', *Andean Past* 3: 137–63

Pearsall, D M, Chandler-Ezell, K and Zeidler, J A (2004) 'Maize in ancient Ecuador: results of residue analysis of stone tools from the Real Alto site', *Journal of Archaeological Science* 31: 423–42

Perry, L (2001) 'Prehispanic subsistence in the middle Orinoco Basin, Venezuela: starch analyses yield new evidence', unpublished PhD dissertation, Southern Illinois University, Carbondale

Perry, L (2004) 'Starch analyses reveal the relationship between tool type and function: an example from the Orinoco Valley of Venezuela', *Journal of Archaeological Science* 31: 1069–81

Perry, L (2005) 'Reassessing the traditional interpretation of "manioc" artifacts in the Orinoco Valley of Venezuela', *Latin American Antiquity* 16(4): 409–426

Perry, L, Sandweiss, D H, Piperno, D R, Rademaker, K, Malpass, M A, Umire, A and de la Vera, P (2006) 'Early maize agriculture and interzonal interaction in southern Peru', *Nature* 440: 76–79

Piperno, D R (2006) *Phytoliths: A Comprehensive Guide for Archaeologists and Paleoecologists*, Lanham, MD: Altamira Press

Piperno, D R and Holst, I (1998) 'The presence of starch grains on prehistoric stone tools from the humid neotropics: indications of early tuber use and agriculture in Panama', *Journal of Archaeological Science* 25: 765–76

Piperno, D R and Pearsall, D M (1998) *The Origins of Agriculture in the Lowland Neotropics*, San Diego: Academic Press

Piperno, D R, Ranere, A J, Holst, I and Hansell, P (2000) 'Starch grains reveal early root crop horticulture in the Panamanian tropical forest', *Nature* 407: 894–97

Pulgar Vidal, J (1987) *Geografía del Perú*, 9th ed, Lima: PEISA

Purseglove, J W (1972) *Tropical Crops: Monocotyledons*, vol 1, London: Longman

Quilter, J (1985) 'Architecture and chronology at El Paraiso, Peru', *Journal of Field Archaeology* 12: 279–97

Reitz, E J (1988) 'Faunal remains from Paloma, an Archaic site in Peru', *American Anthropologist* 90: 310–22

Shady, R (2005) *La Civilización de Caral – Supe: 5000 Años de Identidad Cultural en el Perú*, Lima: Instituto Nacional de Cultura/Proyecto Especial Arqueológico Caral-Supe

Towle, M A (1961) *The Ethnobotany of Precolumbian Peru*, Chicago: Aldine

Trawick, P B (2002) *The Struggle for Water in Peru: Comedy and Tragedy in the Andean Commons*, Stanford, CA: Stanford University Press

Ugent, D (1970) 'The potato', *Science* 170: 1161–66

Ugent, D and Verdun, M (1983) 'Starch grains of the wild and cultivated Mexican species of *Solanum*, Subsection Potatoe', *Phytologia* 53: 351–63

Ugent, D, Pozorski, S and Pozorski, T (1982) 'Archaeological potato tuber remains from the Casma Valley of Peru', *Economic Botany* 36: 182–92

Ugent, D, Dillehay, T and Ramírez, C (1987) 'Potato remains from a late Pleistocene settlement in south-central Chile', *Economic Botany* 41: 17–27

CHAPTER 14

Emerging Food-Producing Systems in the La Plata Basin: The Los Ajos Site

José Iriarte

The La Plata Basin is a large (13,000,000 km^2) and little explored river system that, together with the adjacent littoral Atlantic region, is beginning to reveal an early and long sequence of unique and elaborate cultural trajectories (De Blasis et al 1998; Iriarte et al 2004; López 2001; Peixoto et al 2001; Schmitz et al 1998). Bounded to the north by Amazonia, to the east by the Southern Brazilian Highlands, to the west by the eastern Andean slopes, and to the south by pampa grasslands, this river basin was a corridor that traversed and connected major zones of ecological diversity and cultural complexity (De Blasis et al 1998; Erickson 1995, Heckenberger et al 1999; Ottonello and Lorandi 1987; Tarrago 2002; Walker 2004; Wust and Barreto 1999). It accounts for 40% of the continental wetlands of South America (Neiff 2001) and encompasses vast expanses of seasonal tropical/subtropical forests, *Araucaria* forest and subtropical/temperate grasslands. It constituted a geographical enclave where major cultural traditions from the tropical forests such as the Arawak (Acosta y Lara 1956; Métraux 1934; Nordensköld 1930; Torres 1934), the Tupi-Guarani (Brochado 1984; Noelli 1998; Schmitz 1991; Sušnik 1975), the eastern Andean slopes (Ottonello and Lorandi 1987; Tarrago 2002), the Southern Brazilian Highlands (Beber 2005; Noelli 2000; Schmitz 2002; Schmitz and Becker 1991), and the southern Pampas (Rodríguez 1992; Serrano 1972) converged and interacted during pre-Hispanic times, providing us with a unique opportunity to study cultural interactions at a broad geographical scale.

Despite its importance, the La Plata Basin has only received sporadic archaeological attention since the seminal works of Torres (1911), Lothrop (1932), and the later studies of Brochado (1984), Caggiano (1984) and Rodríguez (1992, 2002, 2005), among others. The study of the plant component of pre-Hispanic subsistence and economy in the prehistory of the La Plata Basin is at a very early stage. Few projects have systematically applied archaeobotanical recovery techniques and, thus, there is a paucity of primary data to provide direct evidence of prehistoric plant use and economy. As a result, human-plant-landscape interactions have remained largely inferential, and many archaeologists working in the region (eg Schmitz et al 1991) have extended the historic view of profoundly transformed, highly mobile groups of hunter-gatherers-fishers to much of the pre-Hispanic period. This perception has also been reinforced by characterizations of the region as marginal (Meggers and Evans 1978; Steward 1946; Steward and Faron 1959), which underscores the notion that these peripheral areas were deficient versions of cultures closer to civilization centers in the Andes or Mesoamerica. Implicit in these assumptions is the perception that these simpler cultures were held back or had somehow failed to achieve the higher levels of complexity accomplished by Andean and Mesoamerican chiefdoms and states, as well as the notion that these cultures were isolated, static, unchanging or developing at a slow pace.

The southern grasslands of South America also have been long perceived as inimical to exploitation by early farmers lacking steel plow technology. This may be true for a large portion of the Pampas; however, a more nuanced view of the different environments encompassing the La Plata Basin indicates that the practices of flood-recessional farming in the vast expanses of wetlands and slash-and-burn along the major river forest corridors must have played major roles in early food-producing economies in the region.

In addition, similar to other regions of lowland Central and South America with a perennially or seasonally humid climate, palaeoethnobotanical research focusing on the recovery and interpretation of macrobotanical remains has long been hampered by poor preservation of these types of plant remains (Piperno and Pearsall 1998). The same appears to be true in the study of early Formative societies that emerged in southeastern Uruguay around 4000 BP. Despite the systematic application of fine-mesh sieving in all previous excavations in the region (Bracco et al 2000a) and the implementation of an intensive flotation program (Iriarte 2003a), no macrobotanical remains other than carbonized wood, palm nut endocarps (*Butia capitata, Syagrus romanzoffiana*) and an isolated find of a squash seed (José López, pers comm 2002) have been recovered from archaeological sites (López 2001). These results indicated that different approaches are needed in order to rigorously study the prehistoric plant economy.

Multidisciplinary investigations at the Los Ajos mound complex, wetland of India Muerta, southeastern Uruguay, revealed that during an increasingly dry mid-Holocene, at around 4190 BP, the site became a permanent circular plaza village, and its inhabitants subsisted on a mixed economy, adopting major crop plants such as maize (*Zea mays* L.) and squash (*Cucurbita* spp.) (Figure 14.1a). The community-focused excavations show that the architectural plan of Los Ajos during the following Ceramic Mound Period (CMP) (3000–2500 BP) experienced the formalization and spatial differentiation of its inner precinct, establishing a new and independent architectural tradition for South America. The unprecedented cultural sequence at Los Ajos challenges the traditional view that the La Plata Basin was inhabited by groups of hunters and gatherers for much of the pre-Hispanic era (Meggers and Evans 1978; Schimtz et al 1991) and endorses previous research indicating that the southern sector of the Laguna Merín Basin was a locus of early emergent complexity (Bracco et al 2000a; López 2001). This chapter summarizes the results of botanical and faunal analyses from Los Ajos and discusses the

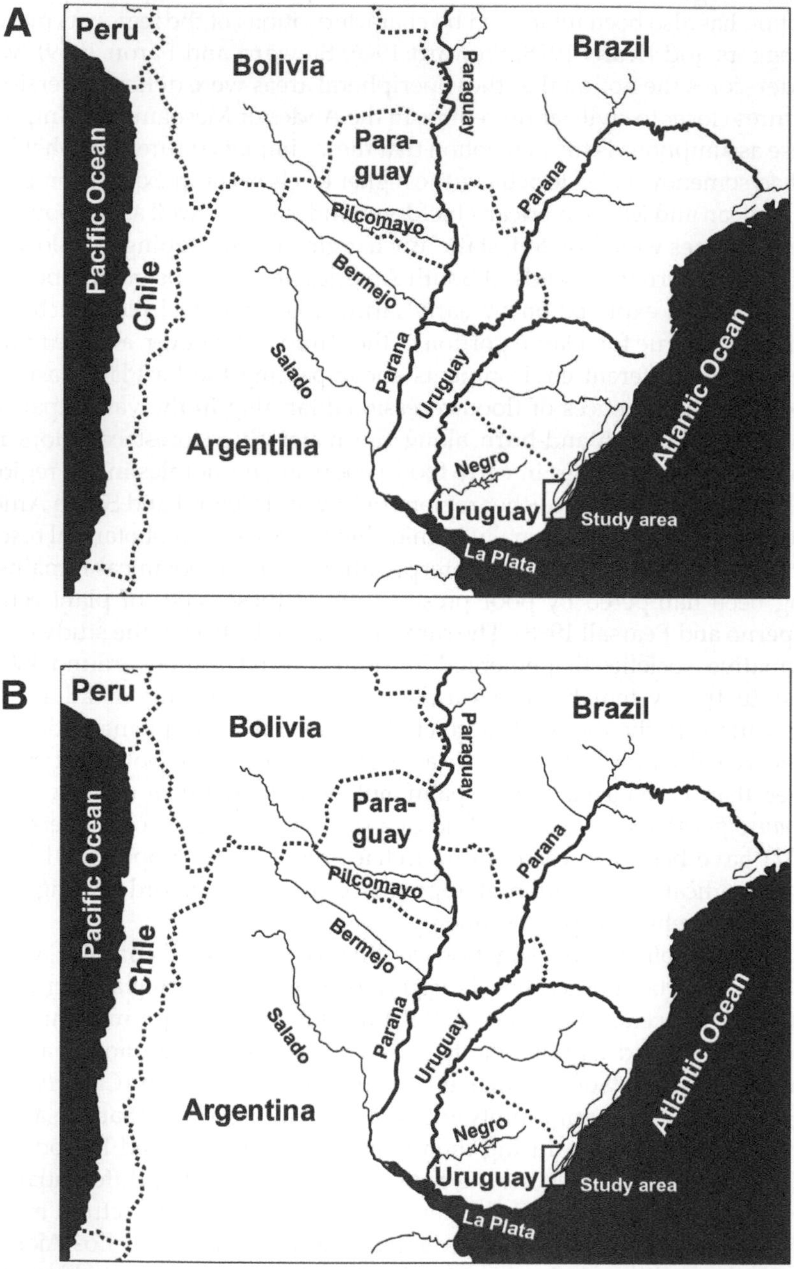

Figure 14.1 A. Map of the southeastern sector of La Plata Basin and the adjacent littoral region showing major regions of archaeological investigation. B. Map of the southern sector of the Lake Merín Basin showing archaeological sites mentioned in the text. Key: 1, Los Ajos; 2, Estancia Mal Abrigo; 3, Puntas de San Luis; 4, Isla Larga; 5, Los Indios; 6, Potrerillo.

implications of these findings for the investigation of the early dispersal of crops and the emergence of native food-producing systems in the region.

OVERVIEW OF CHRONOLOGY AND SETTLEMENT

At a time when coastal Peruvian Late Preceramic societies were building massive monuments and the Amazonian Formative was starting, major regional developments were taking place in the La Plata Basin. In southeastern Uruguay, around 4000 BP, groups of Preceramic foragers coping with a changing mid-Holocene environment began to live in well-planned villages along wetland margins, adopted a mixed economy incorporating major cultigens such as maize and squash, and later on (3000–2500 BP) adopted ceramics and began to build ceremonial architecture.

The mound-building pre-Hispanic cultures dating back to c 4000 BP – locally referred to as 'Constructores de Cerritos' in Uruguay and the Umbu (Archaic Preceramic) and Vieira (ceramic) traditions in southern Brazil – extend across large areas of coastal and inland wetlands and grasslands along the Atlantic coast between around 28º and 36º S latitude (Figure 14.1A). The study region in southeastern Uruguay is characterized by a patchwork of closely packed environments including wetlands, wet prairies, grasslands, riparian forests, large stands of palms and the Atlantic coast. It has a subtropical humid climate with high average temperatures of 21.5º C (70.7º F) during the summer and low average temperatures of 14.8º C (58.6º F) during the winter. Total annual rainfall averages 1123 mm (PROBIDES 2000) (Figure 14.1B).

The 'Constructores de Cerritos' are divided into two main periods: a Preceramic Mound Period (hereafter PMP), which begins around 4190 BP and ends with the appearance of ceramics in the region around 3000 BP; and a Ceramic Mound Period (hereafter CMP), which extends from around 3000 BP to the contact period in the seventeenth century (Bracco et al 2000a; López 2001; Iriarte 2003a).

In the southern sector of the Merín Lake, the coastal plain is dissected by at least three Late Quaternary marine terraces running parallel to the lake margin, which create different types of wetlands (Bracco et al 2000b; Montaña and Bossi 1995). Due to its recent marine origin, the area below the lower and most recent marine terrace (5 masl) is characterized by saline wetlands bearing a low density of archaeological sites. In contrast, the higher inland (15 masl) freshwater wetlands that occur around 30 km from the lake shore exhibit the largest and spatially most complex mound sites. This is particularly so in the more stable locations of the landscape, such as flattened hill spurs adjacent to the wetland floodplains, which are secure from seasonal flooding and have immediate access to the resource-rich and fertile wetlands. There, mound complexes are large, numerous, and spatially complex, exhibiting varied mounded architecture, geometrically arranged in circular, elliptical, and horseshoe formats enclosing a central communal space. These sites also display, in general, vast outer sectors exhibiting more disperse and less formally integrated mounded architecture. One of these sites, Los Ajos, which has been subject to intense archaeological investigation since the early 1990s, contains one of the earliest, most developed and best dated PMP components in the region.

The Los Ajos site

The first excavations at the Los Ajos mound complex were carried out by Bracco (1993) in the early 1990s and consisted of a block excavation in Mound Alfa, a test unit in Mound Beta and a few opportunistic test units in off-mound areas. This work established the mid-Holocene age of the earthen mounds in the area. The PMP component at Los Ajos yielded five dates between c 3950 and 3350 BP (Bracco 1993; Bracco and Ures 1999). Our new excavation program consisted of the placement of a block excavation in Mound Gamma, a test unit in Mound Delta, two trench transects articulating mound and off-mound areas, and a 50 m interval, systematic transect sampling strategy of test units to target off-mound areas (Figure 14.2A) totaling 305 m^2 of excavated areas. The renewed community-focused excavations revealed that Los Ajos is one of the largest and most formally laid-out sites in the study area and covers about 12 ha (Iriarte 2003a, 2006a; Iriarte et al 2004). The inner precinct includes six flat-topped, quadrangular platform mounds (called 6, Alfa, Delta, Gamma, 4 and 7) closely arranged in a horseshoe formation and with a height above ground level of 1.75 to 2.5 m (Figure 14.2B). Two dome-shaped mounds (called Beta and 8) frame the central oval plaza, which measures 75 by 50 m. The formal and compact inner precinct contrasts with more dispersed and informally arranged peripheral sectors, which include two crescent-shaped rises (named TBN and TBS), five circular and three elongated lower dome-shaped mounds, borrow pits and a vast off-mound area bearing subsurface occupational refuse.

The broad contemporaneity of radiocarbon dates, comparable artefact assemblages and similarities in PMP stratigraphy at mounds Alfa, Delta and Gamma indicate that a series of major social and economic changes took place at Los Ajos during the PMP around 4000 BP. At this time, the inhabitants of Los Ajos began to live in a planned circular household-based community and partitioned the site into a number of discrete functional areas, characterized by the placement of residential units around a central plaza area. Stratigraphic and artefactual data indicate that during the PMP, mounds grew as a result of multiple overlapping domestic occupations, where a wide range of activities associated with food preparation, consumption, stone tool production and maintenance took place (Iriarte 2003a, 2006a).

LOS AJOS BOTANICAL AND FAUNAL DATA

Initially, preliminary phytolith and starch grain analyses were carried out in three multimound sites of southeastern Uruguay: Isla Larga, Los Indios and Estancia Mal Abrigo (Iriarte et al 2001) (Figure 14.1b). Starch grains

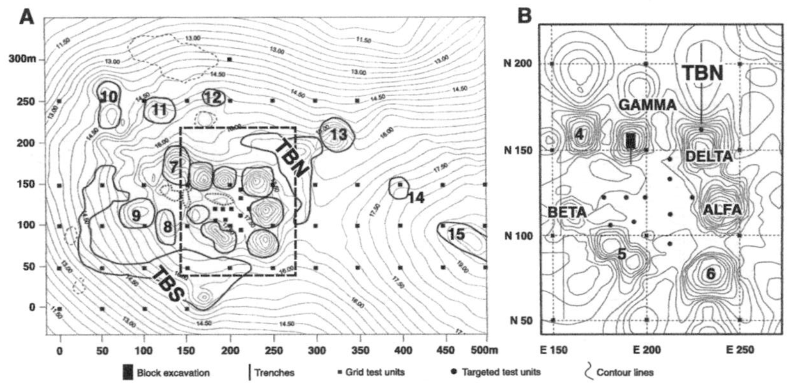

Figure 14.2 Topographic and planimetric maps of Los Ajos. A. Topographical and planimetric map of Los Ajos. Contour intervals 0.33 m. Thick lines show perimeter of mounds. B. Topographical map of the inner precinct of Los Ajos. Contour intervals 0.2 m.

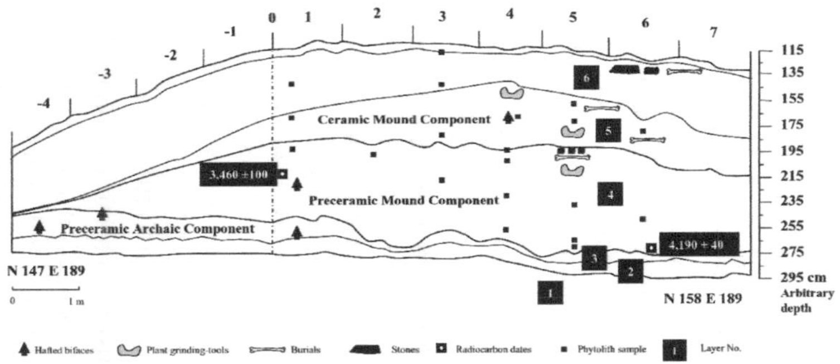

Figure 14.3 West wall stratigraphic sketch of Mound Gamma block excavation and part of trench. Vertical exaggeration x2.

from maize kernels were documented in contexts dating to around 3600 BP at Isla Larga and at c 2800 BP at Los Indios. At Isla Larga, starch grains from *Phaseolus* were derived from contexts dated to c 3050 BP (Figure 14.5A), and starch grains from achira (*Canna* sp.) rhizomes were evidenced from contexts dated to c 3660 BP (Figure 14.5B) (see below). In light of these results, a more ambitious research design was tailored to investigate the identification, chronology and contextual associations of the cultigens at the site of Los Ajos in greater detail.

Because phytolith and starch grain analyses are recent additions to palaeoethnobotanical studies in southeastern South America, particular emphasis was placed on establishing a modern plant reference collection for the region and testing the feasibility of identifying maize leaf decay through a technique developed by Piperno and Pearsall (Holst et al 2004; Pearsall 2000; Piperno 2006), which is based on the size and three-dimensional morphology of crossshaped phytoliths. These analyses demonstrated that an application of multivariate (linear discriminant function) analysis together with qualitative and other assessments of cross-shaped phytolith assemblages, as originally described by Piperno and Pearsall (Piperno 2006), can be successfully used to distinguish the presence of maize leaves in the grasslands of southeastern Uruguay (Iriarte 2003b).

Phytolith samples at Los Ajos were taken from profiled walls and features from Mound Gamma and the central part of the trench at TBN (Figure 14.3). During excavation, all potential plant-processing tools, ie those with ground and/or polished surfaces, were immediately wrapped in aluminum foil and set apart for phytolith and starch grain residue analysis. Phytolith extractions from modern plants, modern soils and archaeological sediments, together with the analysis of residues on archaeological stone tools, followed standard procedures used at the Smithsonian Tropical Research Institute (Piperno 2006; Piperno et al 2000a).

Maize cob phytoliths and starch grains were recovered from residues on plant grinding tools and selected archaeological sediments corresponding to the undisturbed occupational midden of the PMP (Figures 14.4–5). The earliest appearance of maize was marked by ruffle- and wavy top rondels (Figure 14.5C) diagnostic of maize cob phytoliths, which were recovered in the lowermost sector of the Preceramic Mound Component in Gamma Mound at 255–260 cm deep, 15 cm above the earliest date for this component of c 4190 BP. Their identification followed criteria that have allowed the discrimination of phytoliths from maize cobs in North America (Bozarth 1993; Mulholland 1993) and the neotropics (Holst et al 2004; Pearsall et al 2003; Piperno and Pearsall 1993). In the PMP, maize starch grains (Figure 14.5D) were also documented from a mano recovered at 210–215 cm deep; a radiocarbon date of c 3460 BP was obtained 5 cm below the tool (Figure 14.4A). These data indicate that Mound Gamma was an area in which maize was prepared and consumed. As mentioned earlier, maize starch was also recovered from unmixed PMP contexts dating to c 3600 BP at Isla Larga (Iriarte et al 2001), another site in the region (Figure 14.1).

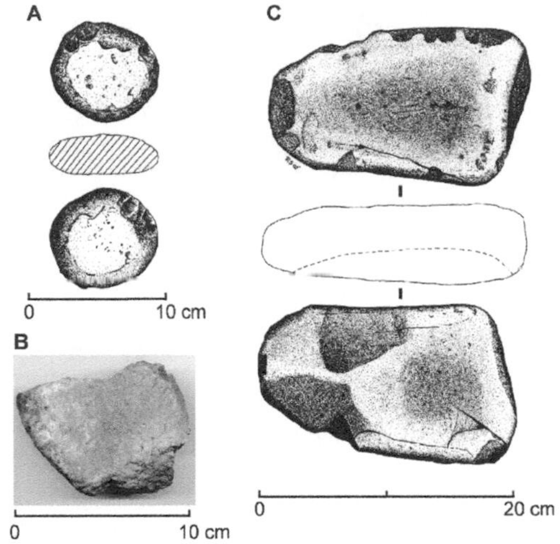

Figure 14.4 Plant-grinding tools from Mound Gamma, Los Ajos.

Few large, spherical, scalloped phytoliths diagnostic of domesticated squash (*Cucurbita* spp.) (Figure 14.5E) were recovered from the lower part of the PMP at Mound Gamma, at 255 cm deep, 15 cm above the earliest dated context of c 4190 BP. Size and three-dimensional morphology of *Cucurbita* phytoliths recovered throughout Mound Gamma are consistent with the presence of domesticated species of *Cucurbita* at the site (Piperno et al 2000b; Piperno and Stothert 2003). The absence of cultigens in the lowermost PMP levels of Mound Gamma indicate that cultigens were adopted shortly after evidence for early village life appears at Los Ajos.

During the succeeding CMP, maize and squash continued to be grown. Maize cob phytoliths were present in the selected sediments analyzed and maize starch grains and cob phytoliths were recovered from the residue of two milling stone bases recovered from Mound Gamma (Figure 14.4B and C). Domesticated *Cucurbita* phytoliths were also present throughout the CMP component. In addition, a maize leaf phytolith signature was recorded in the basal levels of the midden refuse in the TBN crescent-shaped rise dated to around 1660 BP, indicating that this was an area where maize may have been planted and/or husked (Figure 14.5F). The rich, fertile, organic soil of the TBN domestic refuse provided an ideal compost soil to grow corn and squash. The combination of the silty loam texture with the high content of midden nutrients of this anthropic soil made these disturbed areas around settlements especially attractive for planting and experimenting with these new cultigens.

As mentioned above, previous preliminary studies documented the presence of beans (*Phaseolus* spp.) (Figure 14.5A) at the Isla Larga site in contexts dating to c 3050 BP and at the Los Indios site around 1170 BP (Iriarte et al 2001). Additional starch grain and phytolith data from Isla Larga and Los Indios documented the presence of achira (*Canna* spp.) (Figure 14.5B) at c 3660 BP and leren (*Calathea* spp.) at c 1190 BP (Iriarte et al 2001). This evidence

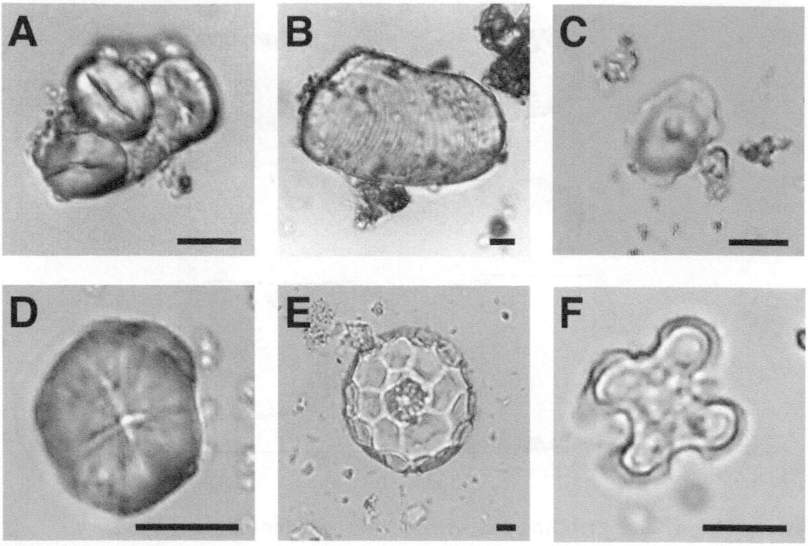

Figure 14.5 Selected phytoliths and starch grains. A. *Phaseolus* sp. starch grains (Los Indios, Mound III, 120–130 cm). B. *Canna* sp. starch grain (Isla Larga, Mound I, 275 cm deep). C. *Zea* sp. specific ruffle-top rondel phytolith (Los Ajos, Mound Gamma, 155–160 cm deep, sector 3/E). D. *Zea mays* starch grain (Los Ajos, Mound Gamma, 155–160 cm deep, sector 3/E). E. *Cucurbita* sp. spherical scalloped phytolith (Los Ajos, Mound Gamma 180–185 cm deep, sector 3/B). F. *Zea mays* large Variant 1 cross-shaped phytolith (Los Ajos, TBN, 170–175 cm deep, sector 7). Scale bar =10 mm.

indicates that wild tubers and possibly root crops played important roles in the economy of the early Formative societies in southeastern Uruguay. Root crops not only constitute an important source of carbohydrates, proteins and minerals, but are easy to procure, particularly, at wetlands, which provide an abundant and year-round supply of underground tubers and rhizomes (Gragson 1997; Lathrap 1970; Sauer 1952). A great diversity of tubers from the Cannaceae (eg *Canna glauca*), Marantaceae (eg *Calathea* spp.), Araceae (*Spathicarpa hastifolia Taccurum, weddellianum*), Typhaceae (*Typha* spp.) and Solanaceae (*Solanum hieronymi*), among others, were historically gathered by native groups in the extensive wetlands and seasonal tropical/subtropical forests of the La Plata Basin (Métraux 1946a; Schmeda-Hirschmann 1998). Thanks to new recovery techniques, notably, starch grain analysis, we are beginning to document the history of these important crops (see Iriarte chapter 9 for an overview).

The exploitation of palms is evidenced by the recovery of palm nut endocarps from butia (*Butia capitata*) and pindo (*Syagrus romanzoffiana*) in addition to the presence of abundant palm phytoliths in the shape of spinulose spheres in the basal PMP at Los Ajos, Isla Larga, and Estancia Mal Abrigo (Iriarte et al 2001). Dense stands of oligarchic *Butia* palm groves, whether wild, encouraged or cultivated, constitute an extremely rich seasonal resource for prehistoric populations living in the area (see also Noelli 2000). Palm fruits are available between December and April and can be stored for long periods of time. López and Bracco (1992) calculated that assuming a mean density of 200 specimens/ha, palm forests can produce up to 2000 kg of fruits per hectare, from which 1730 kg correspond to the pulp and the remaining 270 kg to the fruit nut (endosperm). The pulp is rich in glucose while the nut is rich in lipids, which are essential to the diet. The consumption of palm nuts is indirectly evidenced by the ubiquitous presence of nut-cracking stones – locally called 'rompe-cocos' – in both PMP and CMP contexts across the region (Bracco et al 2000a; López 2001). These artefacts have small holes pecked in the center of one or more of their surfaces into which a palm fruit kernel or nut could be placed and cracked with a light blow with another stone. There is ample ethnographic (eg Oliveira 1995) and experimental replicative data (Ranere 1980) indicating that these artefacts are suited to opening palm kernels. The manufacture of palm flour has also been extensively recorded in ethnographies of groups of the La Plata Basin (Meggers 2001; Schemeda-Hirshmann 1998). Extensive palm groves growing around marshes were primary food sources, from which the heart, shoots, kernels, and trunk starch of palms were eaten, their sap brewed into fermented beverages and their ash used as a substitute for salt (Métraux 1946a; Schemeda-Hirshmann 1994a, 1994b). Palm starch from several species (*Acrocomia aculeata, Arecastrum romanzoffianum, Cocos paraguayensis, Copernica cerifera*) was obtained from the trunk by many indigenous groups of the Chaco and is also reported for the Kaingang of the Southern Brazilian Highlands (Ambrosetti 1846; Meggers 2001; Métraux 1946b; Schemeda-Hirshmann 1998). In the Lagoa dos Patos region, the pre-Hispanic production of palm starch flour is suggested in an early chronicle (1605–1607) by Jesuit Jerónimo Rodríguez (Cesar 1981: 25). He observed that the native groups of the region: 'divide the year into four parts, three months they eat corn, the other three months they eat beans, the other three manioc, and the remaining three months, *they eat flour that they make from a certain palm*' (translation and emphasis mine). To what extent are the large tracts of monospecific stands of different palm species that grow in the region natural or managed? This is a question that remains to be explored in the future and should be aided by palaeoenvironmental reconstruction.

The analysis of faunal remains at Los Ajos from both the PMP and CMP demonstrate that these groups obtained large portions of their diet from animal resources. The faunal assemblage is dominated by large and medium-sized mammals; the most abundant is deer (*Ozotoceros bezoarticus, Mazama gouazoubira*), and it also includes otter (*Myocastor coypus*), capybara (*Hydrochoerus hydrochaeris*), opposum (*Didelphis albiventris, Lutreolina crassiculata*) and armadillos (*Dasypus* sp., *Euphractus sexintus*). Small mammals such as rodents (*Cavia* sp., *Holochilus brasiliensis*, Cricetidae), fishes, and birds are minimally represented in the faunal assemblage, but their contribution to the diet of Los Ajos people is difficult to assess due to preservation, sampling, recovery bias that led to an underestimation of their role, and in turn, to an over-representation of medium to large and large mammals (see Iriarte 2003a).

The adoption of a mixed economy is paralleled by the appearance at the Los Ajos site of plant grinding tools consisting of milling stone bases and manos in conjunction with a gradual change in the chipped stone technology towards a more generalized and expedient technology. In comparison with the Preceramic Archaic component (Early Holocene to c 4190 BP), the lithic assemblage of the PMP shows: (a) a gradual impoverishment of the technology, such as the abandonment of finely manufactured bifacial projectile points; (b) a greater diversity of tool types; and (c) the appearance of plant-processing tools at Los Ajos (Iriarte and Marozzi in press). The PMP tool assemblage recovered at Los Ajos is characterized by a generalized, non-specific assemblage that includes a broad range of different tool types displaying a wide variety of edge angles including flake knives, endscrapers, wedges, notches, point/borers and hafted bifaces. Preliminary microscopic use wear analysis carried out on a small sample of selected tools indicates that they were used for a range of domestic activities, possibly including the extraction of starch from soft plants (Iriarte 2003a). As in other parts of the Americas, these changes toward a more diverse and expedient lithic technology appear go hand in hand with a shift to a more diverse plant-oriented economy (eg Ranere 1980; Rossen 1998). The gradual change in lithic technology throughout the sequence may be the result, in part, of the continued importance of wild faunal resources in the early Formative diet.

The combined archaeological and palaeoecological data show a major reorientation of settlements amidst a changing mid-Holocene environment. Our associated palaeoecological data (Iriarte et al 2004; Iriarte 2006b) indicates the mid-Holocene between c 6620 and 4020 BP was a period of significant climate change marked by increasing aridity. This drying trend appears to have acted as an important catalyst for the reorientation of settlements towards the topographically higher freshwater wetlands. The archaeological investigations at Los Ajos indicate that during this period, local pre-Hispanic populations did not disperse (eg disaggregate into smaller groups with increased mobility) or move to other regions, but oriented their settlement towards the upper freshwater wetlands, where they established more permanent communities in strategic locations (Iriarte 2003a; Iriarte et al 2004). Increased sedentism appears to have been a response to local resource abundance in wetland areas in the face of regional resource scarcity produced by the drying trends of the mid-Holocene.

The remarkable correspondence between the larger mound complexes and the most fertile agricultural lands in the region (Iriarte 2003a; Iriarte et al 2004) suggests that this population practiced flood-recessional farming. During the spring and summer months, organic soils are exposed on the wetland margins. The superficial peat horizons are highly fertile, hold moisture and are easy to till; the floodwater of the Cebollatí River periodically inundates the area and replenishes the soils with nutrients (Juan Montaña, pers comm 2000), making the India Muerta wetlands ideal for wetland margin seasonal farming.

Water-recessional systems involving the cultivation of seasonally inundated surfaces are profuse in ethnographic and historical accounts of the

La Plata Basin (eg Métraux 1946a; Ottonello and Lorandi 1987). This practice must have played a crucial role in local pre-Hispanic economies in the La Plata Basin, in particular, for the cultivation of fast-growing crops such as maize, gourds, beans and sweet potatoes. Flood-recessional farming would have presented opportunities and challenges to Los Ajos people. While it requires minimal labor input and provides great yields in fertile soils, it also presents high risks since the water table may fluctuate in response to seasonal, annual or interannual precipitation and loss by evaporation (Doolittle 2001; Turner and Denevan 1985).

Overall, the combined microbotanical and faunal analyses along with technology and settlement pattern data indicate that the early Formative societies emerging in southeastern Uruguay shortly after c 4190 BP gradually adopted a mixed economy. In doing so, they combined a variety of subsistence activities including hunting, fishing and gathering of wild resources with the growing of corn, squash and, possibly, domesticated beans and tubers (Iriarte et al 2004).

IMPLICATIONS

The multidisciplinary results from Los Ajos have several implications for our understanding of the dispersal of cultigens and food-producing practices in the La Plata Basin and contribute significantly to the ongoing debate about the timing and nature of the incorporation of domesticated crops in American indigenous economies (Piperno and Pearsall 1998).

First, Los Ajos phytolith and starch grain data mark the earliest occurrence in southeastern Uruguay of at least two domesticated crops, maize (*Zea mays* L.) and squash (*Cucurbita* spp.), and document for the first time the presence of beans (*Phaseolus* spp.) and the tubers achira (*Canna* sp.) and leren (*Calathea* sp.). These findings indicate that the early Formative societies of southeastern Uruguay practiced a diversified economy including farming, hunting, gathering and fishing shortly after they started to build well-planned villages around 4000 BP.

Second, this novel data call into question the generally assumed view that cultigens and agriculture were brought by the Amazonian Tupi-Guarani immigrants around 2000 BP into a region inhabited by groups of foragers who were adapted to a diversity of environments (Ottonello and Lorandi 1987; Rodríguez 1992; Schmitz 1991; Schmitz et al 1991). Los Ajos data show that cultigens were introduced and became integrated into local food economies much earlier than previously thought.

This new evidence should come as no surprise. The cultigens documented in southeastern Uruguay during the mid-Holocene were domesticated elsewhere in the Americas several millennia earlier (see Iriarte chapter 9 for an overview). For example, the early dates of maize in Central America (Iriarte chapter 9; Piperno et al 2000a; Piperno and Flannery 2001) and northwestern South America (Pearsall et al 2004) show that this cultivar was potentially available to people in other parts of South America by around the sixth millennium BP. Similarly, domesticated varieties of squash were available since the tenth millennium BP as documented by studies in Mexico (Smith 1997) and Ecuador (Piperno and Stothert 2003). Several studies also show that domesticated plants spread quickly between their centers of origin and other areas, such as in the case of maize and manioc (eg Piperno et al 2000a; Pohl et al 1996; Pope et al 2001). Last but not least, there are no major geographical barriers to the spread of these cultigens into the La Plata Basin. Unlike other regions of the world, where the tropics are separated from the subtropical/temperate zones by major geographical barriers such as deserts, ie the Sahara in Africa, the major river corridors draining into the La Plata Basin may have facilitated the exchange of crops among different populations at a very early date.

Third, Los Ajos data indicate that the study of the early dispersal of cultigens and native food-producing systems should not be restricted to the practice of shifting agriculture along tropical/subtropical seasonal forest. It shows that the use and manipulation of wetlands was an earlier, more important and more frequent activity than previously thought (Pohl et al 1996; Siemens 1999). It also demonstrates the potential of grasslands and wetlands for pre-Hispanic cultural development – areas historically viewed as marginal (Stahl 2004).

Southeastern Uruguay adds another case to the diverse array of early food-producing economies in the Americas. It contributes to the growing realization that with more advanced archaeobotanical analysis, many groups that were initially considered hunter-gatherers may shift into an early food-producing category. The data from Los Ajos forces us to review initial characterizations of the Umbu-Vieira Tradition as simple hunter-gatherers (Schmitz et al 1991) and later conceptualizations of the 'Constructores de Cerritos' as complex hunter-gatherers (Bracco et al 2000a; López and Bracco 1992).

The systematic application of microfossil botanical analyses is already revolutionizing our knowledge of early food production in the Americas as the case from Los Ajos illustrates. In the future, the integration of palaeoecological and archaeological data aided by advanced archaeobotanical techniques will be a prerequisite to understanding early food-producing practices in the past in the La Plata Basin and elsewhere. As more sites are discovered and analysed, more surprises will be awaiting us.

ACKNOWLEDGMENTS

Fieldwork at Los Ajos between 1999–2001 was supported by grants from the National Science Foundation, Wenner-Gren Foundation for Anthropological Research, Smithsonian Tropical Research Institute and the University of Kentucky Graduate School. I also received support from the Comisión Nacional de Arqueología, Ministerio de Educación y Cultura, Uruguay, and the Rotary Club of Lascano, Rocha, Uruguay. I am grateful to Tim Denham, Tom Dillehay, Dolores Piperno and Luc Vrydaghs for valued suggestions, comments and editorial advice on earlier drafts of this manuscript. Figure 14.1

was prepared by Seán Goddard. Ricardo Chong prepared the other four figures. Drawings in Figure 14.4 were drafted by Rafael Suarez.

NOTE

1. Throughout the chapter, I follow Piperno and Pearsall's (1998: 6–7) general definitions for food production, horticulture and agriculture.

REFERENCES

Acosta y Lara, E (1956) *Los Chana–Timbues en la Banda Oriental*, Montevideo, Uruguay: Anales del Museo de Historia Natural

Ambrosetti, J A (1846) *Los Indios 'Caingua' del Alto Paraná (Misiones)*. Boletín del Instituto Geográfico, vol 15, Buenos Aires: Instituto Geográfico

Beber, M V (2005) O sistema do assentamento dos grupos ceramistas do planalto sul-brasilero: o caso da Tradição Taquara/Itararé, in *Documentos 10: Arqueologia no Rio Grande do Sul, Brasil*, pp 5–125, São Leopoldo, Brazil: Instituto Anchietano de Pesquisas

Bozarth, S R (1993) 'Maize (*Zea mays*) cob phytoliths from a central Kansas Great Bend aspect archaeological site', *Plains Anthropologist* 38: 279–86

Bracco, R (1993) 'Proyecto Arqueología de la Cuenca de la Laguna Merín', unpublished report presented to PROBIDES (Programa de Conservación de la Biodiversidad y Desarrollo Sustentable en los Humedales del Este), Rocha, Uruguay

Bracco, R and Ures, C (1999) 'Ritmos y dinámica constructiva de las estructuras monticulares. Sector sur de la cuenca de la Laguna Merín–Uruguay', in López, J M and Sans, M (eds), *Arqueología y Bioantropología de las Tierras Bajas*, pp 13–33, Montevideo, Uruguay: Universidad de la República, Facultad de Humanidades y Ciencias

Bracco, R L, Cabrera, L and López, J (2000a) 'La prehistoria de las Tierras Bajas de la Cuenca de la Laguna Merín', in Duran, A and Bracco, R (eds), *Arqueología de las Tierras Bajas*, pp 13–38, Montevideo, Uruguay: Ministerio de Educación y Cultura, Comisión Nacional de Arqueología

Bracco, R L, Montana, J, Panarello, H and Ures, C (2000b) 'Evolución del humedal y ocupaciones humanas en el sector sur de la cuenca de la Laguna Merín', in Duran, A and Bracco, R (eds), *Arqueología de las Tierras Bajas*, pp 99–116. Montevideo, Uruguay: Ministerio de Educación y Cultura, Comisión Nacional de Arqueología

Brochado, J P (1984) 'An ecological model for the spread of pottery and agriculture into eastern South America', unpublished PhD dissertation, University of Illinois–Urbana

Caggiano, A M (1984) *Prehistoria del NE Argentino, Sus Vinculaciones con la R. O. del Uruguay y sur de Brasil*. Antropología no 38, São Leopoldo, Brazil: Instituto Anchietano de Pesquisas

Cesar, G (1981) *Primeros Cronistas do Rio Grande do Sul*, Porto Alegre, Brazil: Universidade Federal de Rio Grande do Sul

De Blasis, P, Fish, S K, Gaspar, M D and Fish, P R (1998) 'Some references for the discussion of complexity among the Sambaqui moundbuilders from the southern shores of Brazil', *Revista de Arqueología Americana* 15: 75–105

Doolittle, F (2001) *Cultivated Landscapes of Native North America*, Oxford: Oxford University Press

Erickson, C L (1995) 'Archaeological perspectives on ancient landscapes of the Llanos de Mojos in the Bolivian Amazon', in Stahl, P (ed), *Archaeology in the American Tropics: Current Analytical Methods and Applications*, pp 66–95, Cambridge: Cambridge University Press

Gragson, T L (1997) 'Use of underground plant organs and its relation to habitat selection among the Pume Indians of Venezuela', *Economic Botany* 51: 377–84

Heckenberger, M J, Petersen, J B and Neves, E G (1999) 'Village size and permanence in Amazonia: two archaeological examples from Brazil', *Latin American Antiquity* 10: 353–76

Holst, I, Piperno, D, Pearsall, D, Benfer, R, Iriarte, I, Chandler-Ezell, K, Jones, J, Zhao, J, Kealhofer, L, Cooke, R, Ranere, A, Zeidler, J and Scott Cummings, L (2004) '25 Years of maize phytolith research in North, Central, and South America', paper presented at the 69th annual meeting of the Society for American Archaeology, Montréal, Canada

Iriarte, J (2003a) 'Mid-Holocene emergent complexity and landscape transformation: the social construction of early Formative communities in Uruguay, La Plata Basin', unpublished PhD dissertation, University of Kentucky

Iriarte, J (2003b) 'Assessing the feasibility of identifying maize through the analysis of cross-shaped size and three-dimensional morphology of phytoliths in the grasslands of south-eastern South America', *Journal of Archaeological Science* 30: 1085–94

Iriarte, J (2006a) 'Landscape transformation, mounded villages, and adopted cultigens: the rise of early Formative communities in south-eastern Uruguay', *World Archaeology* 38: 644–663

Iriarte, J (2006b) 'Vegetation and climate change since 14,810 ^{14}C yr BP in southeastern Uruguay and implications for the rise of early Formative societies', *Quaternary Research* 65(1): 20–32

Iriarte, J and Marozzi, O (in press) 'Análisis del material lítico del sitio Los Ajos', Proceedings of the XI Congreso Nacional de Arqueología, Salto, Uruguay

Iriarte, J, Holst, I, López, J M and Cabrera, L (2001) 'Subtropical wetland adaptations in Uruguay during the mid-Holocene: an archaeobotanical perspective', in Purdy, B (ed), *Enduring Records: The Environmental and Cultural Heritage of Wetlands*, pp 61–70, Oxford: Oxbow

Iriarte, J, Holst, I, Marozzi, O, Listopad, C, Alonso, E, Rinderknecht, A and Montaña, J (2004) 'Evidence for cultivar adoption and emerging complexity during the mid-Holocene in the La Plata Basin', *Nature* 432: 614–17

Lathrap, D W (1970) *The Upper Amazon*, Praeger: New York

López, J M (2001) 'Las estructuras tumulares (cerritos) del litoral Atlántico Uruguayo', *Latin American Antiquity* 12: 231–55

López, J M and Bracco, R (1992) 'Relaciones hombre-medio ambiente en las poblaciones pre-históricas del este del Uruguay', in Ortiz-Troncoso, L and Van der Hammen, T (eds), *Archaeology and Environment in Latin America*, pp 259–82, Amsterdam: University of Amsterdam

Lothrop, S K (1932) 'Indians of the Parana Delta, Argentina' *Annals of the New York Academy of Sciences* 33: 77–232

Meggers, B J (2001) 'The mystery of the Marajoara', *Amazoniana* (Kiel, Germany) 16: 421–40

Meggers, B J and Evans, C (1978) 'Lowland South America and the Antilles', in Jennings, J (ed), *Ancient Native Americans*, pp 543–91, San Francisco: Freeman

Métraux, A (1934) 'El estado actual de nuestros conocimientos sobre la extensión primitiva de la influencia Guarani y Arawak en el continente sudamericano', *Actas y Trabajos Científicos del XXV Congreso Internacional de Americanistas* 1: 181–90, Buenos Aires, Argentina: Universidad de la Plata

Métraux, A (1946a) 'Ethnography of the Chaco', in Steward, J (ed) *Handbook of South American Indians*, vol 1: *The Marginal Tribe*, pp 197–370. Washington DC: Smithsonian Institution

Métraux, A (1946b) 'The Caingang', in Steward, J (ed), *Handbook of South American Indians*, vol 1: *The Marginal Tribes*, pp 445–76. Washington DC: Smithsonian Institution

Montaña, J and Bossi, J (1995) *Geomorfología de los Humedales de Humedales de la Cuenca de la Laguna Merín en el Depto. de Rocha*, Montevideo, Uruguay: PROBIDES (Programa de Conservación de la Biodiversidad y Desarrollo Sustentable en los Humedales del Este)

Mulholland, S C (1993) 'A test of phytolith analysis at Big Hidatsa, North Dakota', in Pearsall, D and Piperno D (eds), *Current Research in Phytolith Analysis: Applications in Archaeology and Palaeoecology*, pp 131–45. Philadelphia, PA: MASCA

Neiff, J J (2001) 'Diversity in some tropical wetland systems of South America', in Gopal, B, Junk, W J and Davis, J A (eds), *Biodiversity in Wetlands: Assessment, Function, and Conservation*, vol 2, pp 157–86, Leiden: Backhuys Publishers

Noelli, F (1998) 'The Tupi: explaining origin and expansions in terms of archaeology and of historical linguistics', *Antiquity* 277: 648–63

Noelli, F S (2000) 'A ocupação humana na região sul do Brasil: arqueologia, debates e perspectivas', *Revista do Museu de Arqueologia e Etnologia* 44: 218–269

Nordensköld, E (1930) *Ars America: L'Archaeologie du Basin de l'Amazonas*, Paris: G Van Oest

Oliveira, J E (1995) 'Os argonautas Guato: aportes para o conhecimento dos assentamentos e da subsistência dos grupos que se estabeleceram nas áreas inundáveis do Pantanal Matogrossense', unpublished MS thesis, Pontificia Universidade Catolica do Rio Grande do Sul, Brazil

Ottonello, M and Lorandi, A M (1987) *Introducción a la Arqueología y Etnología*, Buenos Aires: EUDEBA

Pearsall, D (2000) *Palaeoethnobotany: A Handbook of Procedures*, New York: Academic Press

Pearsall, D M, Chandler-Ezell, K and Chandler-Ezell, A (2003) 'Identifying maize in neotropical sediments and soils using cob phytoliths', *Journal of Archaeological Science* 30: 611–27

Pearsall, D M, Chandler-Ezell, K and Zeidler, J A (2004) 'Maize in ancient Ecuador: results of residue analysis of stone tools from the Real Alto site', *Journal of Archaeological Science* 31: 423–42

Peixoto, J L, Bezerra, M A, Mozeto, A A and Hilbert, K P (2001) 'Evolução das grandes lagoas e a ocupação humana no Pantanal Mato-Grossense durante o Holoceno', *Boletim de Resumos, Mudanças Globais e o Quaternário, VIII Congreso da Abequa*, pp 438–39, Mariluz, Brazil: ABEQUA

Piperno, D R (2006) *Phytoliths: A Comprehensive Guide for Archaeologists and Paleoecologists*, Lanham, MD: Altamira Press

Piperno, D R and Flannery, K V (2001) 'The earliest archaeological maize (*Zea mays* L.) from highland Mexico: new accelerator mass spectrometry dates and their implications', *Proceedings of the National Academy of Sciences* (Washington, DC) 98: 2101–03

Piperno, D R and Pearsall, D (1993) 'Phytoliths in the reproductive structures of maize and teosinte: implications for the study of maize evolution', *Journal of Archaeological Science* 20: 337–62

Piperno, D R and Pearsall, D (1998) *The Origins of Agriculture in the Lowland Neotropics*, Academic Press: New York

Piperno, D R and Stothert, K E (2003) 'Phytolith evidence for early Holocene *Cucurbita* domestication in southwest Ecuador', *Science* 299: 1054–57

Piperno, D R, Ranere, J A, Holst, I and Hansell, P (2000a) 'Starch grains reveal early root crop horticulture in the Panamanian tropical forest', *Nature* 407: 894–97

Piperno, D R, Andres, T C and Stothert, K E (2000b) 'Phytoliths in *Cucurbita* and other neotropical Cucurbitaceae and their occurrence in early archaeological sites from the lowland American tropics', *Journal of Archaeological Science* 27: 193–208

Pohl, M D, Pope, K O, Jones, J, Jacob, J S, Piperno, D R, deFrance, S D, Lentz, D L, Gifford, J A, Danforth, M E and Josserand, K (1996) 'Early agriculture in the Maya lowlands', *Latin American Antiquity* 7: 355–72

Pope, K O, Pohl, M E D, Jones, J, Lentz, D L, Von Nagy, L, Vega, F J and Quitmyer, I R (2001) 'Origin and environmental setting of ancient agriculture in the lowlands of Mesoamerica', *Science* 292: 1370–73

PROBIDES (2000) *Plan Director de la Reserva de Biosfera Bañados del Este*, Rocha, Uruguay: PROBIDES (Programa de Conservación de la Biodiversidad y Desarrollo Sustentable en los Humedales del Este)

Ranere, A (1980) 'Stone tools and their interpretation', in Linares, O and Ranere, A (eds), *Adaptive Radiations in Prehistoric Panama*, pp 118–45. Peabody Museum Monographs no 5. Cambridge, MA: Peabody Museum, Harvard University

Rodríguez, J A (1992) 'Arqueología del sudeste de Sudamérica', in Meggers, B J (ed), *Prehistoria Sudamericana: Nuevas Perspectivas*, pp 177–209, Washington DC: Taraxacum

Rodríguez, J A (2002) 'Evolución de la tecnología prehistórica en el sudeste de América del Sur', in Ledergerber, P (ed), *Formativo Sudamericano, Una Revaluación*, pp 314–27, Quito: Ediciones Abyayala

Rodríguez, J A (2005) 'Human occupation of the eastern La Plata Basin and the adjacent littoral region during the mid-Holocene', *Quaternary International* 132: 23–36

Rossen, J (1998) 'Unifaces in early Andean culture history: the Nanchoc lithic tradition of northern Peru', *Andean Past* 5: 241–99

Sauer, C (1952) *Agricultural Origins and Dispersals*, New York: American Geographical Society

Schmeda-Hirschmann, G (1994a) 'Plant resources used by the Ayoreo of the Paraguayan Chaco', *Economic Botany* 48: 252–58

Schmeda-Hirschmann, G (1994b) 'Plant salt as an Ayoreo salt source in the Paraguayan Chaco', *Economic Botany* 48: 159–62

Schmeda-Hirschmann, G (1998) 'Etnobotánica Ayoreo: contribución al estudio de la flora y vegetación del Chaco. XI', *Candollea* (Geneva) 53: 1–50

Schmitz, P I (1991) 'Migrantes da Amazonia: a tradição Tupiguarani', in Kern, A (ed), *Arqueologia Pré-histórica do Rio Grande do Sul*, pp 295–330, Porto Alegre, Brazil: Mercado Aberto

Schmitz, P I (2002) *Casas subterrâneas nas terras altas do sul do Brasil*. Pesquisas Antropologia no 58, São Leopoldo, Brazil: Instituto Anchietano de Pesquisas

Schmitz, P I and Becker, I (1991) 'Os primitivos engenheiros do planalto e suas estructuras subterrâneas: a tradição Taquara', in Kern, A (ed), *Arqueologia Pré-histórica do Rio Grande do Sul*, pp 251–93, Porto Alegre, Brazil: Mercado Aberto

Schmitz, P I, Naue, G and Becker, I (1991) 'Os aterros dos campos do sul: a tradição Vieiria', in Kern, A (ed), *Arqueologia Pré-histórica do Rio Grande do Sul*, pp 221–51, Porto Alegre, Brazil: Mercado Aberto

Schmitz, P I, Rogge, J H, Rosa, A O and Beber, M V (1998) *Aterros Indígenas no Pantanal do Mato Grosso do Sul*. Antropologia no 54, São Leopoldo, Brazil: Instituto Anchietano de Pesquisas

Serrano, A (1972) *Lineas Fundamentales de la Arqueología del Litoral*, Córdoba, Argentina: Instituto de Antropología

Siemens, A H (1999) 'Wetlands as resource concentrations in southeastern Ecuador', in Blake M (ed), *Pacific Latin America in Prehistory: The Evolution of Archaic and Formative Cultures*, pp 137–47, Pullman, WA: Washington State University Press

Smith, B (1997) 'The initial domestication of *Cucurbita pepo* in the Americas 10,000 years ago', *Science* 5314: 932–34

Stahl, P (2004) 'Greater expectations', *Nature* 432: 561–62

Steward, J (1946) (ed) *Handbook of South American Indians, vol 1 The Marginal Tribes*, Washington, DC: Smithsonian Institution

Steward, J and Faron, C (1959) *Native Peoples of South America*, New York: McGraw-Hill

Sušnik, B (1975) *Dispersión Tupí-Guaraní Prehistórica: Ensayo Analítico*, Asunción, Paraguay: Museo Etnográfico Andrés Barbero

Tarrago, M N (2002) 'El Formativo y el surgimiento de la complejidad social en el noroeste argentino', in Ledergerber, P (ed), *Formativo Sudamericano, Una Revaluación*, pp 302–27, Quito: Ediciones Abyayala

Torres, L M (1911) *Los Primitivos Habitantes del Delta del Parana*, Buenos Aires: Imprenta de Coni Hermanos

Torres, L M (1934) 'Relaciones arqueológicas de los pueblos del Amazonas con los del Río de la Plata', *Actas y Trabajos Científicos del XXV Congreso Internacional de Americanistas* 2: 191–193, Buenos Aires: Coni

Turner II, B L and Denevan, W (1985) 'Prehistoric manipulation of wetlands in the Americas: a raised-field perspective', in Farrington, I S (ed), *Prehistoric Intensive Agriculture in the Tropics*, pp 11–30. International Series 232, part I, Oxford: British Archaeological Reports

Walker, J H (2004) *Agricultural Change in the Bolivian Amazon*. Latin American Archaeology Reports, Pittsburgh, PA: University of Pittsburgh

Wust, I and Barreto, C (1999) 'The ring villages of central Brazil: a challenge for Amazonian archaeology', *Latin American Antiquity* 10: 3–23

CHAPTER 15

A Tale of Two Tuber Crops: How Attributes of Enset and Yams may have Shaped Prehistoric Human-Plant Interactions in Southwest Ethiopia

Elisabeth Anne Hildebrand

INTRODUCTION

Anthropological literature on early food production emphasises the domestication of cereals and other annual seed crops (Cowen and Watson 1992; Gebauer and Price 1992; Price and Gebauer 1995). Because some cases of early seed-crop farming, such as Southwest Asia, are especially well studied and publicised (eg Bar-Yosef and Belfer-Cohen 1989; McCorriston and Hole 1991; Moore and Hillman 1992), they may come to be regarded as typical of all instances of prehistoric plant domestication. Scholars have long recognised, however, that early human manipulation of tropical tuber crops may have followed trajectories quite distinct from those for seed crops (Harris 1969, 1972, 1973; Hawkes 1969). As data and theories accumulate regarding the gathering, early cultivation and spread of tubers in modern and prehistoric times (eg Bayliss-Smith 1996; Chikwendu and Ozekie 1989; H J Deacon 1984: 250; J Deacon 1984: 251; Denham et al 2003; Hather 1994, 1996; Mbida et al 2000, 2001; Mignouna and Dansi 2003; Neumann 2003; Peters 1996; Piperno and Pearsall 1998; Schoeninger et al 2001; Ugent et al 1982), it is important to show how they enrich global perspectives on the ways prehistoric hunter-gatherers began to manipulate, and eventually farm, useful plants.

Fundamental differences in the biology of seed versus tuber crops and their wild progenitors certainly influence how humans use them and how the plants respond. Do these differences justify a dichotomy between domestication processes for annual seed staples farmed in fields versus perennial root and tuber crops cultivated in gardens? Or does the domain of human-plant relations include a wealth of different trajectories of manipulation *within* the arena of vegeculture? Either condition – a straight seed/tuber dichotomy or a universe of possibilities among tubers – bears on how archaeologists build and test hypotheses regarding domestication events in particular areas and how global patterns of intensification are construed.

Africa is particularly well poised to contribute to our characterisation of seed versus tuber crop domestication. African peoples domesticated cereals (finger millet, pearl millet, sorghum and others), non-cereal seed crops (noog, watermelon and other cucurbits), and tuber crops (enset, multiple species of yams, anchote and others). Domestication took place in upland, lowland, wet and dry locales and supported the development of both egalitarian and hierarchical societies (Marshall and Hildebrand 2002). The immense environmental and social variation among African food producers past and present can contribute novel insights about domestication processes and their contexts and consequences.

In this chapter, I assess the degree of homogeneity among perennial, vegetatively propagated tuber plants, their cultivation systems and domestication processes by exploring interactions between humans and two such plants. Yams (*Dioscorea cayenensis* and related species) and enset (*Ensete ventricosum*) are used by Sheko and other peoples of southwest Ethiopia who have had subsistence economies oriented toward production of tuber crops. The ways people interact with both plants are diverse, including wild use, low-intensity cultivation in mixed gardens, high-intensity field monoculture and, for yams, ongoing domestication. Field observation of a broad set of human-plant interactions can help us to evaluate possible trajectories of intensification in prehistoric times and evaluate the degree to which these trajectories may have varied from plant to plant. Detailed study and comparison of specific biological attributes of yam and enset plants can reveal how these traits might have shaped strategies for use by prehistoric hunter-gatherers prior to domestication and later favored the development of certain practices of manipulation.

A Terminological Note

I employ the term 'tuber crop' in the same comprehensive sense used by Harris (1977: 208–209) for 'root and tuber crop': to refer to plants cultivated for edible underground storage organs, which include tubers, roots, bulbs, corms and rhizomes. For plants of similar use that are uncultivated, I use the term 'tuber plant.'

SEED VERSUS TUBER CROPS AND THE CULTIVATION SYSTEMS BUILT UPON THEM

Contrasts between seed and tuber-based food production systems, and their probable processes of development, have appeared since the earliest theories of agricultural origins. Shortly after Childe (1951: 59–77) postulated a Neolithic revolution of herd management and cereal domestication, Carl Sauer (1952: 24–28) proposed an alternative scenario for the domestication and spread of tropical tuber crops. One of the first global forums on early food production (Ucko and Dimbleby 1969) compared characteristics of plants used for seeds versus tubers (Table 15.1) and the cultivation systems built upon them (Table 15.2) (Harris 1969; Hawkes 1969; see also Harris 1972, 1973).

Table 15.1 Traits of plants used for seed versus tuber nutrients (with a focus on tropical taxa) that pertain to contexts and processes of domestication (sources: Hawkes 1969: 21–24; Harris 1973: 397)

Trait	*Tuber Crops*	*Seed Crops*
Climates favored by wild progenitors	Well-demarcated wet and dry seasons; humid areas with short dry season	Well-demarcated wet and dry seasons; semiarid areas with long dry season
Environment favored by wild progenitors	Ecotones at edges of low-altitude dry forest or montane areas; lightly or heavily forested lowlands and uplands	Open-canopy wood, shrub, and grassland lowlands and uplands
Habitats favored by wild progenitors	Disturbed	Disturbed; poor, thin soil on rocks
Part of the plant that is used	Underground starch organ (tuber, corm); leaves	Seeds (cereals, oilseeds); fruits (eg Cucurbitaceae)
Part of the plant that is sown/manipulated	Underground starch organ; stem, sucker	Seeds
Growth pattern	Perennials	Most annuals; a few perennials
Protein content	Low	High

Conventional Distinctions between Seed and Tuber Crops and Implications for Human Social Organisation

According to these portrayals, seed and tuber crops share a weedy ecological orientation and thrive in environments with well-demarcated wet and dry seasons. They are said to differ, however, in their natural source environments, the parts of the plant used and manipulated and the demands their growth places upon the soil (Table 15.1). These basic differences are then thought to shape the divergent configuration of seed-crop versus tuber-crop systems depicted in Table 15.2. Seed plants' small package of carbohydrates per plant, their adaptation to open-canopy areas and their growth from seed all require thorough land clearance to produce a reasonable crop. Seed plots therefore have sparse ground cover (other than the crop itself), high erosion potential and a low diversity of plant species in an unstratified configuration with an open canopy. Plants with edible tubers, in contrast, provide a larger package of carbohydrates per plant, are adapted to more closed settings and can grow from a perennial organ year after year in the same place. In tuber plots, therefore, plants can be placed less densely, require less thorough land clearance with less potential for soil erosion and can grow in environments with high plant diversity, a fair amount of stratification and a somewhat more closed canopy.

The plots of seed crops are ecologically much less complex than those of tuber crops. Seed-based farming systems are therefore said to be much less ecologically stable than tuber-based farming systems because they lack the

Table 15.2 General characterisations of similarities and differences between seed and tuber-based cultivation systems (sources: Harris 1969: 13, 1972: 188, 1973: 405). Harris (1969: 13) contrasts *conuco* (earthen mounds of buried stem cuttings for root crops) and *milpa* (seed cultivation of maize, beans and squash) farming in the American tropics. Note that the rate of evolution of manipulated taxa (last row of data) is mentioned but not endorsed by Hawkes (1969: 24), while favored by Harris (1973: 401)

Traits of Cultivation Systems	*Tuber Crops*	*Seed Crops*
Land clearance	Incomplete	Thorough
Ground cover in plots	Continuous	Sparse
Erosion potential	Low	High
Timing/purpose of tillage	For harvest; transplantation	For planting only
Protein yield of crops	Low	Higher
Depletion of soil nutrients	Slow	Fast
Protein supplements (game, fish)	Needed	Less crucial
Complexity of ecosystem of cultivated plots	High	Low
Canopy of cultivated crop	More closed	Open
Diversity of plants in cultivated plots	High	Low
Degree of stratification of plants in plots	High	Low
Ecological stability of cultivated plots	High	Low
Tendency to spread to other areas	Weak	Strong
Rate of evolution of manipulated taxa	Slow	Fast

intricate system of checks and balances that can buffer against perturbations in rainfall and temperature, pests, predators and other factors (Harris 1972: 188, 1973: 401). Being more protein-rich, seed crops require humans to procure less supplemental dietary animal protein than tuber crops. The resulting picture is of seed cultivation systems tending toward a less diverse subsistence for humans with high degrees of environmental modification and rapid rates of change and spread, contrasted with tuber cultivation systems, which are more generalised, ecologically and spatially conservative and durable.

What are the possible consequences of these alleged differences between seed and tuber-based cultivation systems for the development of human societies? The relative ecological instability of seed-crop complexes makes them prone to spread, as localised perturbations and crop failures spur farmers to migrate (Rindos 1984: 276–278). In this way, seed-crop complexes may expand across continents more quickly than comparatively stable systems of tuber-crop cultivation. Ultimately, seeds may replace tubers as staples, as suggested by Harris for prehistoric times (1972: 188–193) and observed through both recent demic movements and technological adoptions in southwest Ethiopia (Hildebrand et al 2002). Compared to tuber crops, seed crops have a higher protein content, and are easier to store and transport. These traits could make economic systems centered around seed-crop production more amenable to the specialisation of labor, the appropriation of products by a powerful minority and concentration of human populations into larger settlements or cities. Tuber crops' lower protein content and the difficulty of storage and transport might, in contrast, foster a more generalised subsistence base and labor orientation as well as decentralised human populations and power structures. Seed and tuber crops therefore appear to set forth completely different economic foundations upon which to build complex societies.

A Dichotomy?

These characterisations and their implications both depend on a neat division of crop systems into two groups: annual field crops used for and propagated by seed versus perennial garden tubers propagated vegetatively. Such a division appeared untenable even in early formulations, however:

> The tending of small domestic 'gardens' close to dwellings probably represents man's earliest system of proto-cultivation in the tropics. It would have been best suited to the cultivation and domestication of vegetatively reproduced plants and seed-reproduced perennial climbers, shrubs, and trees, whereas cereals and other herbaceous crops would have been more effectively cultivated in larger plots cleared specifically for the purpose (Harris 1973: 405).

In stressing the potential role of seed-propagated crops in early gardens before field agriculture, Harris indicates that the crucial distinction should not be based on mode of propagation (seed versus vegetative), but on context (garden versus field) and life cycle (perennial versus annual). These distinctions undermine the clear dichotomy between seed/field/annual and tuber/garden/perennial.

Even if a dichotomy between the two kinds of crops is spurious, to what extent might similar kinds of crops have followed parallel trajectories of intensification during prehistoric times? Did relations between humans and vegetatively propagated tuber plants change in consistent ways? The study of two such plants – yams and enset – used in a common area of southwest Ethiopia reveals profound differences in their biology and utility. The extent to which physical form, environment and use by humans differ between these species today can reveal potentially divergent prehistoric pathways to food production for these taxa in particular and call attention to multiple possible trajectories of intensification among tuber crops in general. These different trajectories may, in turn, have structured the development of different forms of settlement, economy and social organisation among prehistoric and present-day tuber farmers around the world.

METHODS

Data in the following sections are drawn from 27 months of fieldwork in southwest Ethiopia: July 1996, January–December 1998, January–August 1999, July–October 2000, and February 2003 (Figure 15.1). I worked primarily with the Sheko, a little-studied ethnic group whose members speak an Omotic

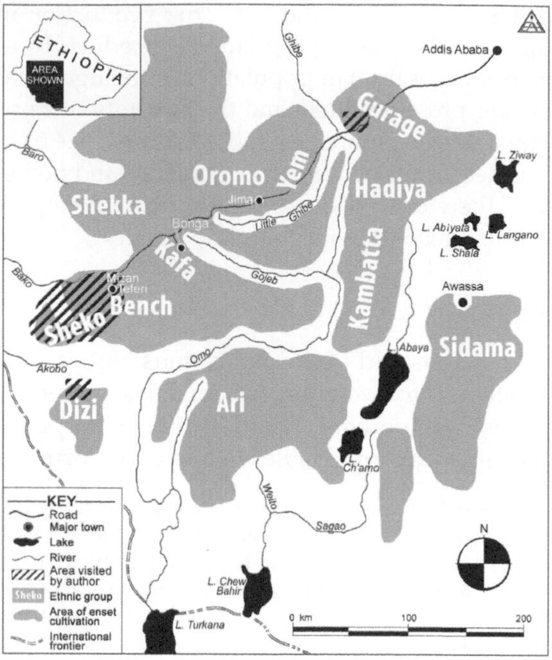

Figure 15.1 Southwest Ethiopia: ethnic groups, enset cultivation areas and locales visited by the author.

language and live in low population densities on the southwest edge of the Ethiopian Highlands (Hildebrand 2003a). Sheko lands range in elevation from 600–1800 m, receive more than 2000 mm annual rainfall and encompass wooded grassland, woodland, montane forest and swamp. Historically, the cultivation of tubers (yams, enset and taro) supported hereditary chiefdoms that were incorporated into the Ethiopian empire a century ago; the power of these chiefdoms was undercut by the socialist policies of the Dergue regime during the 1970s. I also interviewed the ethnic neighbors of the Sheko: the Dizi, Bench, Me'enit and Majangir.

Research involved learning the Sheko and Amharic languages, a year of residence with a host family in a northeast Sheko village and several surveys through Bench–Maji Zone conducted in collaboration with the zone cultural officers, during which we collected botanical specimens and folk knowledge about enset, yams, honey, useful wild plants and Sheko history and culture (Hildebrand 2003a).

In this chapter, I focus on the different biological, environmental and utilitarian attributes of enset and yams and how they may have favoured the development of certain forms of intensified human-plant interactions. Because the full range of human-plant interactions for enset and yams is not found in the study area, I refer also to my own observations among the Gurage and published practices of the Ari (Gebre Yntiso 1995; Shigeta 1990, 1991, 1996), Gurage (Shack 1966) and Sidama (Smeds 1955).

YAMS

Yams (*Dioscorea*) have a pantropical distribution. Different species were domesticated in different parts of the world; African yams were domesticated more than once by separate prehistoric human groups (Coursey 1967: 11). Among several species of wild, domestic, native and non-native yams eaten in the study area (Hildebrand 2003a; Hildebrand et al 2002), I discuss here the *D. cayenensis* species complex, a group of species that are closely related and morphologically similar to one another (Wilkin 2001: 392). *D. cayenensis* yams are the most frequently used by the Sheko and are involved in various degrees of interaction with humans, including ongoing domestication via deliberate adoptive transplantation of wild yams to domestic garden contexts (Hildebrand 2003b).

Sheko farmers have observed morphological differences among the folk taxa that belong to the *D. cayenensis* species complex (Hildebrand et al 2002) and changes in yam plants following adoptive transplantation (Hildebrand 2003b). Sheko classification of yams as wild versus domestic center more on the location of the plant (forest versus garden) than on specific morphological attributes. Future studies by *Dioscorea* specialists may evaluate morphological differences between wild and cultivated yams in southwest Ethiopia and the role of adoptive transplantation in fostering morphological change.

Morphology, Phenology and Life Cycle

Although a yam plant lives year after year in the same place, its only perennial part is a hard, rhizomatous or cormous structure at the head of the tuber (Wilkin 2001: 361, pers comm; see also Burkill 1960: 347, Figure 15). Deeper, annual portions of the tuber store the plant's energy reserves through the dry season. Tuber morphology varies according to surrounding soil conditions (Figure 15.2A–E) (Ayensu 1970: 128). Climbing foliage is annually renewed with the onset of the rains. The life cycle of a yam plant therefore consists of cyclical, seasonal transfers of energy and tissue from tuber to foliage and back again (Coursey 1967: 42–43).

D. cayenensis plants are dioecious (male and female flowers appear on separate plants) (Burkill 1960: 327). Flowers develop during the mid to late rainy season and are pollinated by flying nocturnal insects (Burkill 1960: 390; Coursey 1967: 35–36). A fruit consists of a three-lobed capsule, with one seed (winged for wind dispersal) in each lobe. If a seed-propagated *Dioscorea* plant volunteers in a garden, Sheko gladly cultivate and eat it, but farmers do not deliberately plant seeds of *Dioscorea*. Rather, they propagate yams vegetatively, by cloning. Distal, annually renewed portions of the tuber may be eaten, while the perennial crown of the tuber may be replanted or transplanted. Many oft-used wild yam plants are left to regrow in a permanent, known location. Some cultivated yams are planted in a permanent position next to trees in home gardens. In the most intensive form of cultivation, Sheko intervene in the yam life cycle via seasonal relocation

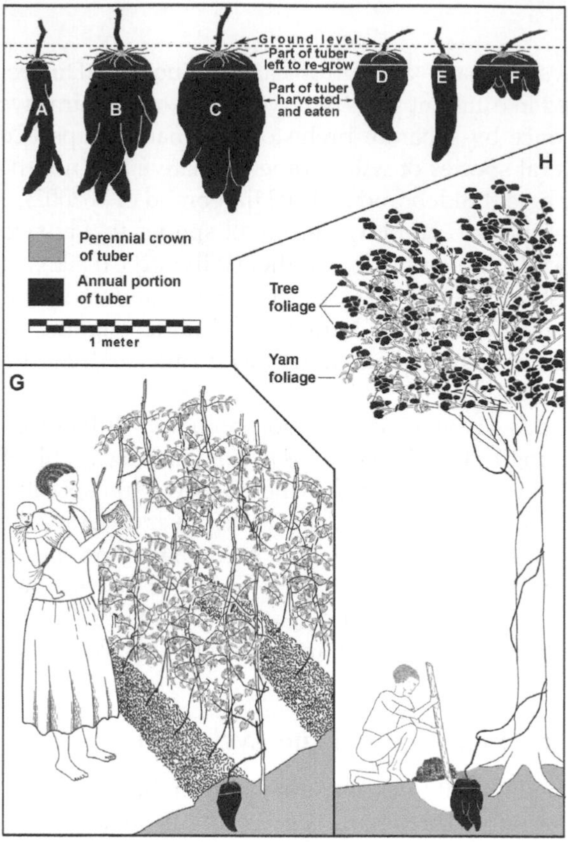

Figure 15.2 Yams: tuber morphology and contexts of growth. A. Tuber of infrequently harvested wild yam; B. tuber of frequently harvested wild yam; C. tuber of yam planted by a tree in a Sheko garden; D and E. varying forms of tubers grown by stakes in fields in Sheko, before first harvest of the season; F. small extensions of the tuber that grow after first harvest; G. rows of yams planted by stakes in field; H. man harvesting yam growing next to tree.

of the plant between shaded garden (dry season) and hoed field (wet season).

The physical form and life cycle of yams fit several generalisations about useful tuber plants: yams grow in the same place year after year, yield a large package of carbohydrates per plant and have a perennial organ. On the other hand, significant portions of the yam plant are effectively annual, including the organ used by humans (tuber).

Environment

In Ethiopia, yams of the *D. cayenensis* complex grow 550–1800 m above sea level in wooded grassland, woodland, open montane forest, secondary forest and gardens (Miège and Sebsebe 1997: 60–62). Yams are light-demanding and do poorly in primary forest but thrive in disturbed areas such as abandoned cereal fields, secondary vegetation, paths and ravines. The natural distribution of wild yam plants in a landscape is fairly scattered in any environment, but they tend to be most numerous and visible in low-altitude wooded grassland. There, they compete by twining onto trees and growing foliage high off the ground. In home gardens, cultivated yams twine onto trees, finding partial or full shade in lower reaches and full sun higher up. In nearby fields, yams twine onto stakes in full sun or partial shade.

Farmers of the study area recognise environmental constraints on the distribution of wild yams and know that disturbance of forests increases the density of yams. To seek previously unknown wild yams, they head to wooded grassland or disturbed forest.

The environmental preference of yams in the study area – for humid climates with a short dry season – are consistent with those of useful tuber plants in general. Among the source environments ascribed to tuber crops in Table 15.1 (lightly or heavily forested lowlands and uplands), yams favor a subset: lightly forested areas, which in Ethiopia occur at low altitudes. In these areas, yams thrive without human intervention. In upland or forested areas, however, yams occur in significant numbers only in the context of heavy disturbance. Weedy habitat preferences of *Dioscorea* echo those of most crop plants (seed or tuber).

Use by Humans

Sheko farmers use yams in several ways: occasional or regular harvest of wild yam plants, low-intensity cultivation next to trees in the home garden and high-intensity field monoculture of yam plants transplanted twice a year.

Men gather wild yams September–February in a variety of social contexts (individual or group, opportunistic or planned) and environments (highland forest to low wooded grassland) (Hildebrand 2003a, 2003b). After removing all of the accessible edible sections, they leave the crown of the tuber intact to regrow (Figure 15.2). Farmers have a mental map of the location, variety and palatability of wild yams in areas that they visit frequently, even hours or days distant from home. This knowledge is important, for the scattered distribution of yam plants can make the search process difficult in unfamiliar territory and the yield is 'all-or-nothing': either one finds a yam plant with a large, intact tuber, or one gains no food from yams that day.

A farmer may move a palatable yam from its natural context to his home garden, a practice I term 'adoptive transplantation' (Hildebrand 2003b). After transplantation, the adopted yam occupies a permanent position in the home garden and is treated like other yams cultivated next to trees (see below).

Families cultivate certain varieties of yams next to trees in the home garden. The ample sunlight and support for foliage mimic the forest margin and wooded grassland environments ideal for wild yams; garden contexts have the added benefit of regularly disturbed soil enriched by household waste. Each yam has a permanent place just downhill from a garden tree, onto which its stems twine. The tuber is harvested once a year, between September and February, and yields one or more meals for a large family. Harvest and replacement of the tuber are the only labor required. Provided the family does not relocate their house and garden, the perennial crown of the tuber is left in the same place from year to year.

The most intensive form of yam cultivation involves seasonal transplantation to monocrop fields of ditches with nearby stakes (Figure 15.2G). In November, after the rains cease, Sheko harvest all the edible portions of the tubers from the field and bury the perennial tuber crowns in a shady plot near the house to protect them from excessive sun. When light rains resume in March, a family clears a field, excavates ditches, breaks up the soil and plants the tuber crowns with their emergent stems. Fine soil heaped back into the ditch covers the yams and captures rain or animal waste. Biannual transplanting requires the substantial labor of digging long ditches, preparing the shady plot and moving the yams twice a year. Field yams are prone to drought or stress during transplanting and in the exposed field; a farmer may lose an entire field of yams if rain is insufficient. On the other hand, field yams grow rapidly and produce edible tubers as early as July. Tuber growth continues until the end of the rains, allowing each tuber to be harvested two or three times before its transplantation back to the shady plot (Figure 15.2F).

Use and propagation of vegetative parts of yams by the Sheko are consistent with human practices of tuber management in general. This is illustrated by the low-intensity cultivation of some yams next to trees in garden settings that have diverse taxa in a stratified configuration that is both complex and stable. The Sheko practice of monocropping yams in fields completely cleared of other vegetation, however, is more characteristic of unstable seed cropping systems: soil is fully exposed, diversity is extremely low and stratification or canopy vegetation is absent.

ENSET

The banana family (Musaceae) has two genera that provide food: *Musa* (edible fruits) and *Ensete* (edible corm). Both genera also supply important non-food materials. Farming of *E. ventricosum* in Ethiopia is the sole case of food production for the genus *Ensete*, which has six species distributed throughout tropical Africa and Asia (Lock 1993; Simmonds 1960, 1962). Peoples of the study area differentiate between wild and domestic forms of *E. ventricosum* that grow locally and are morphologically distinct (Hildebrand 2001, 2003c).

In the study area, wild enset plants grow from seed and have a fat pseudostem base and edible but unpalatable corm. Domestic enset plants have diminished seed fertility, grow from human-initiated clones and have a smaller pseudostem base, a more palatable corm and a prominent epicuticular wax bloom on abaxial leaf surfaces (Hildebrand 2001).

Morphology, Phenology and Life Cycle

All vegetative parts of enset are perennial. Aside from early and final stages of development (Figure 15.3G–K), the form of the enset plant changes little throughout its life (Figure 15.3L–N). Atop an edible subterranean corm with adventitious roots lies a zone of meristematic tissue that generates leaves; spirally wrapped leaf sheaths make up a pseudostem that supports the leaf blades. Growth to a mature height of more than 5 m takes four years in lowland areas and up to 10 years in cool highlands.

An enset plant flowers only once, at the end of its life. Female flowers develop first and are fertilised by bats and insects with pollen from another enset plant. Male flowers develop later. Enset fruits resemble bananas but are squat and bright orange and contain hard black seeds c 1 cm in diameter (>1000 per infructescence). Monkeys eat the fruits and pass the seeds in their feces; seeds also roll downhill and germinate in dense clusters in disturbed contexts.

Farmers reproduce domestic enset by cloning. Unlike banana plants, enset plants generally do not produce suckers spontaneously. To initiate sucker production, a farmer fells the pseudostem and removes the pith, which causes the plant to generate more than 100 suckers (Figure 15.3J, K). When the

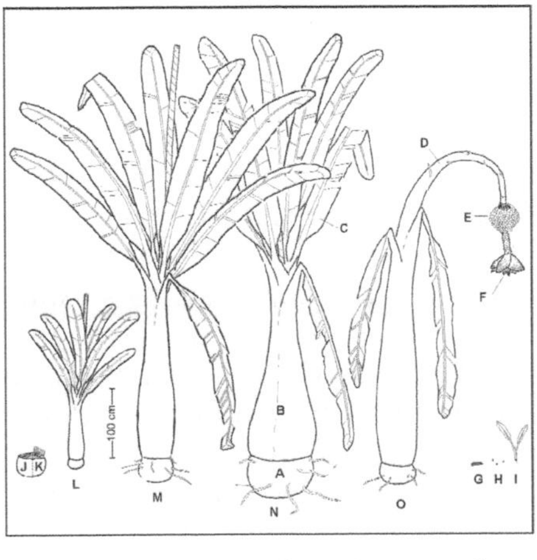

Figure 15.3 Enset morphology and growth. A. Corm, with zone of meristem at top. Main source of edible food. B. Pseudostem, made up of spirally wrapped leaf sheaths that contain edible starch. Pseudostem fibers can be made into rope. Youngest leaf grows up through center. C. Leaf blade. Can serve as wrapping, construction or roofing material. D. Peduncle, which emerges, bearing reproductive parts, at the end of enset life cycle. Rich in sugars, which may be extracted for beer. E. Female flowers and fruit. Pollination occurs before the development of male flowers (F), so enset plants are obligate outcrossers. G. Fallen fruits resemble bright orange bananas and are eaten by monkeys. H. Fallen seeds. I. Germinated seedling. J. Enset corm incised for cloning (vegetative reproduction). K. Corm with emergent suckers four months later. L. Young enset plant (wild or domestic). M. Mature domestic enset, before flowering. N. Mature wild enset, before flowering. O. Enset plant well into flowering/fruiting, the final phase of its life cycle.

suckers begin to crowd each other, the farmer splits the corm into four or more pieces and plants these near the house. Later, individual suckers are separated; these are sold, traded or given to neighbors, or transplanted into their permanent positions in the garden. The clones then take three to four years to grow to full size. Cloned enset plants may flower and bear fruit, but few of the resulting seeds appear viable (Hildebrand 2001, 2003c).

Usable parts of the enset plant include leaf blades (wrappers, roofing), pseudostem (fibers, edible starch) and corm (edible starch). Humans normally do not eat enset fruit, but may use seeds for beads or gaming pieces. Use of the corm precludes the reproduction of the plant, as it makes flowering and cloning impossible.

The morphology and life cycle of enset are highly consistent with characterisations of useful tuber plants: the entire plant is perennial and yields a large package of carbohydrates per plant. Enset differs from many tuber crops in several ways, however. It takes several years to mature and has multiple useful parts for food and other purposes. Tuber crops are normally seasonal, but the non-reproductive parts of enset are available all year, for several years. Enset plants used for food cannot, unlike many other tuber plants, contribute any genetic material to subsequent generations. The production of numerous heavy fruits and seeds just once in the life of a wild enset plant differs from many tuber crops that flower or fruit every year, and creates dense clusters of enset seedlings.

Environment

E. ventricosum thrives in tropical environments 900–2250 m above sea level; in Ethiopia it is adapted to cool highland settings, but in my experience tolerates drought and fire reasonably well (see also Simmonds 1960: 206; 1962: 31). I have observed wild enset growing in sandy streambeds in ravines, on the slopes of clayey loam and in swampy areas (see also Lock 1993: 3; Shigeta 1991: 23; Simmonds 1962: 32). Wild enset habitats appear to depend less on soil type than on the need for frequent disturbances from tree falls, stream floods, erosion or anthropogenic causes.

Simmonds considers competition with other plants 'the most conspicuous biotic factor' in the ecology of the banana family (1962: 32). Musaceae plants in shade have thinner, taller pseudostems, reduced leaf production and smaller bunches of fruit. Full shade can cause poor growth or death before flowering, although *Ensete* is more resilient than *Musa*. Wet tropical environments are rarely free of competition, as they tend to be wooded or, if frequently burned, grassy. Germination of enset therefore occurs almost exclusively in contexts of severe disturbance (Simmonds 1962: 32–33).

Once established in slopes and ravines in southwest Ethiopia, however, clusters of wild enset are fairly stable, even while competing with other weedy plants. Where systematic anthropogenic disturbance allows wild-growing *E. ventricosum* to gain a foothold, it can dominate fairly flat areas and achieve densities that shade out competitors. Stands of wild enset can be seen from several kilometers away.

Above c 2250 m elevation, environments are so cold that enset rarely reproduces by seed. Suckers from vegetatively propagated plants can grow at these elevations, however, and enset is farmed up to 3200 m (Smeds 1955: 21). The lower range of enset growth is not as flexible; I have observed that enset plants wither and die if humans attempt to move them to areas lower and drier than their natural distribution.

Like other useful tuber plants, enset needs ample rainfall and a short dry season. In the study area, its climatological requirement of moisture and its habitat requirement of an open canopy conflict, as areas with sufficient rainfall tend to be heavily forested. Like yams, wild enset in upland areas thrives only where the forest is disturbed. Unlike yams, enset cannot tolerate the dry conditions in lowland areas. Cloning technology has allowed the spread of enset into upland areas beyond its natural reproductive range, but not into hot, dry lowlands.

Use by Humans

In contrast to yams, the range of Sheko interactions with enset is fairly limited. To understand these interactions, it is necessary to consider those of other enset farmers of southwest Ethiopia. I include personal and published observations of the Gurage (200 km northeast of Sheko) and published accounts of the Ari (150 km southeast of Sheko) and Sidama (300 km southeast of Sheko). Together with Sheko practices, these include use of wild enset for non-alimentary purposes, low-intensity cultivation of enset and high-intensity farming of enset in fertilised monocrop plots.

Sheko routinely use wild enset leaves for wrapping or construction material. Most Sheko and Ari deny eating wild enset (Shigeta 1990: 102). Farmers of both ethnic groups describe cooked wild enset corms as hard, dark and unpalatable or bitter. An initial cooking experiment bore out these observations, but eating the corm produced no digestive problems. Correct characterisation of cooked enset by farmers does, paradoxically, suggest recent attempts to eat it. Southwest Sheko ate wild enset during a conflict that forced them to live in the forests for a year and found that smaller corms are more palatable than large ones.

Low-intensity cultivation of enset in gardens is documented among the Sheko (Hildebrand 2001, 2003a, 2003c) and Ari (Gebre Yntiso 1995; Shigeta 1990, 1991, 1996). About thirty large enset have a permanent place in the home garden, interspersed with other crops (eg *Cucurbita* [pumpkin], *Lagenaria* [gourd], *Coffea* [coffee]). Garden soil is inadvertently enriched by waste, night soil and dung, but no fertiliser is deliberately applied. Enset is not a staple, but supplements other grain and tuber crops, contributes to food security and sustains unexpected gatherings such as funerals. Although cooking methods are diverse (steaming or pit-baking the corm, making bread from pseudostem starch and pulverised corm and using sugary sap from the peduncle for beer), food from the plant is usually consumed within one week of harvest.

A more intensive form of enset cultivation is practiced by the Gurage (Shack 1966) and Sidama (Smeds 1955), who use it as a staple. More than 100 large enset plants occupy a dense plot behind the house, some parts of which are virtually monocropped. High population density compels Gurage farmers to maximise space by transplanting enset plants several times during their life cycles (Shack 1966: 61–62) and Sidama farmers to thin enset plants as they grow (Smeds 1955: 20). In both regions, soil fertility is maintained by frequent applications of animal dung, which necessitates the allocation of large tracts of land for pasture (Smeds 1955: 19–20). People occasionally consume enset food products immediately after harvest, but fermentation of enset pulp for months or years is the norm. Behind the house, numerous pits lined with leaf sheaths hold enset pulp while it ferments.

Patterns of enset use diverge from classic conceptions of tuber crops in several ways. Although the large, starchy corm of enset provides a resource similar to that of edible tubers, enset has many other useful parts and food is gained from aboveground organs as well. Domestic enset is propagated vegetatively, but cloning enset is more complicated than propagation of many other kinds of tuber crops and may well have begun quite late in the domestication process. In other respects, low-intensity cultivation practices for enset mirror those for other tuber crops: plants are grown in gardens that have a complex, stratified, stable ecology. In this case, however, enset forms the actual canopy for the garden. High-intensity cultivation of enset resembles systems of seed-crop production in two primary ways: soil requires the addition of fertiliser and plots are low in diversity.

REEXAMINING TRAITS ASCRIBED TO USEFUL TUBER PLANTS AND THEIR CULTIVATION SYSTEMS

Yams and enset are consistent with each other, and with traits generally associated with tuber crops, in their general climatological requirements, their preference for disturbed habitats and in the part of the plant that is sown or manipulated. Enset requires more rainfall than yams, however, and unlike yams, its uses go beyond the harvest of the tuber. Finally, yams have annually renewed tubers and foliage, including the useful part of the plant, whereas the entire enset plant is perennial. In considering the morphology, environmental preference and utility of yams and enset, it becomes clear that both taxa have traits that deviate substantially from generalisations about tuber crops – and from each other – in important ways that pertain to contexts and processes of domestication (Table 15.3).

How do these differences play out in terms of the structure, complexity and stability of yam and enset agricultural systems? In low-intensity cultivation systems, enset and yams are often planted in the same garden, so

Table 15.3 Traits of yams and enset that pertain to contexts and processes of domestication. Data are based on my own observations in parts of southwest Ethiopia where *Ensete* and *Dioscorea* appear in wild settings

Trait	*Yams*	*Enset*
Climates favored by wild plants	Humid areas with short dry season (yams require less moisture than enset)	Humid areas with short dry season (enset requires more moisture than yams)
Environment favored by wild plants	Lowland wooded savanna; forest/savanna ecotones; disturbed upland forests	Forest/savanna ecotones; disturbed upland forests
Habitats favored by wild plants	Disturbed	Disturbed
Useful part of plant	Tuber only	Corm; leaf sheaths; leaf blades
Availability	September–February	All year
Part of the plant sown/manipulated	Tuber and crown	Top of corm; meristem zone
Growth pattern	Has perennial organ, but growth of tubers and foliage is annual	All parts of plant are perennial

that their contexts of cultivation are nearly identical. Among high-intensity cultivation systems, enset and yam farming practices differ in the frequency of transplantation and the degree of vegetative cover, but may be similarly monocultural (Table 15.4). Regardless of the particular tuber crop involved, the structure, complexity and stability of agricultural systems are fairly consistent at similar levels of intensity of cultivation, but vary tremendously between high versus low-intensity cultivation systems. In fact, high-intensity cultivation of yams or enset duplicates many of the traits that make seed-based agricultural systems so ecologically unstable (Table 15.4).

Viewed through this comparative lens, the distinctions between seed and tuber crops, annual versus perennial growth and seed versus vegetative reproduction prove to be less crucial to the configuration of agricultural systems than one might think. If this is so, should we use these distinctions to categorise prehistoric domestication processes from the top down? It may be more productive to turn to the attributes of particular taxa, examine their roles in shaping human-plant interaction and build models of domestication from the ground up.

A TALE OF TWO TUBER CROPS

Yams and enset are distinct from seed crops, but also differ from one another in important ways Their physical attributes and requirements almost certainly influenced hunter-gatherer strategies for plant use prior to domestication. These traits also would have made certain forms of intensified human-plant interaction more mutually beneficial than others. In both these

Table 15.4 Similarities and differences between low and high-intensity cultivation of yams and enset, based on my own observations among the Sheko, Dizi (low-intensity enset cultivation and both low and high-intensity yam farming) and Gurage (high-intensity enset cultivation) and published descriptions of cultivation by the Sidama (high-intensity enset) and Ari (low-intensity enset)

Traits of Cultivation Systems	*Yams (low-intensity)*	*Yams (high-intensity)*	*Enset (low-intensity)*	*Enset (high-intensity)*
Land clearance	Incomplete	Thorough	Incomplete	Thorough
Ground cover in plots	Continuous	Sparse	Continuous	Medium
Erosion potential	Low	Medium	Low	Medium
Frequency, timing of tillage or transplantation	Once a year: harvest	Twice a year: planting, harvest	Once in life of plant	Several times in life of plant[1]
Protein yield of crops	Low	Low	Low	Low
Depletion of soil nutrients	Slow	Fast	Slow	Fast
Protein dietary supplements	Necessary	Necessary	Necessary	Necessary
Complexity of ecosystem of cultivated plots	High	Low	High	Low-medium[2]
Canopy of cultivated crop	More closed	Open	More closed	Closed
Diversity of plants in cultivated plots	High	Low	High	Low[2]
Degree of stratification of plants in plots	High	Low	High	Low to medium[2]
Ecological stability of plots	High	Low	High	Low
Tendency of systems to spread	Weak	Strong	Weak	Strong

[1]Described by Shack (1966: 61–62) for the Gurage and reported by locals during my 1996 visits there; Smeds (1955: 20) describes only one postcloning transplantation for enset cultivated among the Sidama, followed by progressive thinning in the garden as enset plants grew larger.
[2]Shack (1966: 59) and Smeds (1955: 19) both report the cultivation of other crops (eg coffee, barley) positioned between enset plants, but Smeds (1955: 3, 15) also describes enset cultivation as monocultural. Among the Gurage, I observed secondary crops mainly near the edges of the enset plot; the plot interior was fully shaded by enset leaves and harbored few crops.

ways, the physical traits of yams and enset shaped prehistoric processes of domestication.

How Properties of Yams and Enset may have Shaped Prehistoric Hunting and Gathering Strategies

Yams and enset differ substantially in their size and timing of availability, their environmental requirements and in their abilities to survive and reproduce amidst intensive use by humans. These differences surely would have affected prehistoric strategies for hunting and gathering, setting the foundation for modes of human-plant interaction before food production.

Morphology, Phenology and Life Cycle: Size and Timing of Availability

A wild yam plant constitutes a substantial, stationary package of carbohydrates available for a single harvest over a six-month period. The constant but scattered distribution, relatively low visibility, long duration of availability and sizeable yield makes a wild yam plant a predictable resource for a group of humans exploiting a territory that they know well, without competition. Predictability would be low, however, if multiple social groups exploited the landscape without cooperation or communication, if people were unsure of access to areas they knew well, or if they were forced to use areas previously unknown.

A wild enset plant contains a similar quantity of carbohydrates, but is available the entire year. It would, therefore, be more likely to be eaten by hunter-gatherers in seasons when more palatable carbohydrates (yams) were unavailable: March–August, or in case of failure of any major subsistence component. The uses of enset are not confined to food: the plant would have been an important year-round source of fiber and shelter materials to hunter-gatherers. Enset also differs from yams in its visibility, which would make it an easy target for use by people new to an area.

Physical differences between yam and enset plants thus laid the foundation for two completely different strategies of wild food procurement by humans. The scattered distribution of seasonally available, low-visibility yam plants would require mental maps for efficient use and might foster forms of social organisation that control or coordinate access to particular territories. Highly visible stands of perennially available, multiuse enset plants are likely to have been focal points for human residence or monitoring among prehistoric hunting and gathering societies.

Environmental Preferences of Yams and Enset

In prehistoric landscapes with minimal disturbance, yams would have occurred in many different environments in low densities, but would have been concentrated in lowland wooded grasslands. The moisture requirements of enset would have confined it to areas above 1000 m or restricted it to even higher elevations during dry periods.

The variable distribution of these two major carbohydrate sources could have led hunter-gatherers to adopt a seasonal round, occupying lowlands during the dry season, when yams were available, and shifting to the ecotones and uplands during the wet season when other carbohydrate sources were scarce. This strategy would be compatible with the seasonal and spatial distribution of other food resources of southwest Ethiopia (Hildebrand 2003a). During the wet season, fruits are most abundant in ecotones and uplands, where the best tasting honey is also found. During the dry season, fish are available in lowland streams and game is visible if grasslands are burned (naturally or purposefully). Such a seasonal system could be sustained indefinitely, assuming a balance in the need for yams and enset, a lack of competition with other groups for either lowland or upland resources, and minimal disturbance of upland habitats so that populations of enset and yams there would remain fairly limited.

Differences in the Response of Yam and Enset to Harvest by Humans

Harvesting a wild yam plant disturbs the soil around it. This may make the plant slightly more vulnerable to forest predators, but thorns usually protect the perennial crown of the tuber. Harvesting-related activities also may disturb nearby vegetation and give a yam plant greater access to light. Harvesting the corm of a wild enset plant, in contrast, kills the plant. Cutting leaves or leaf sheaths for non-food uses may slow the growth of an enset plant, but does not kill it.

A wild yam plant has a chance to reproduce every year. Harvesting its tuber does not impede its reproduction: harvest occurs after flowers and fruits have developed and subsequent years' reproductive efforts are supported by another cycle of annual growth. Wild enset, in contrast, has only one chance to reproduce at the end of its life. Cutting leaves or leaf sheaths of enset may delay reproduction, but does not prevent it. Harvesting a wild enset corm for food prevents the formation of flowers, fruit and pollen.

A wild yam left untouched grows narrow and deep, following small fissures into packed clay or silt (Figure 15.2A). It is difficult to extract and low in yield. A yam that is harvested repeatedly has its tuber covered with loose soil, facilitating growth of a larger tuber in subsequent years (Figure 15.2B). This constitutes a phenotypic response to localised conditions. Because using an enset plant for food kills the plant, a phenotypic response to corm harvest is impossible.

Intensive use of yams requires no human intervention to maintain yam populations, beyond simply replacing the tuber in the soil. Indeed, harvesting may actually increase future yields from a plant. Regular harvest of enset for food, however, carries the potential consequences of destroying an entire stand or selecting against traits held by the plants harvested. Intensive use of enset patches therefore demands a strategy to maintain enset populations and ensure reproduction of plants with desired traits by deliberately leaving certain plants to reproduce. This would require either monitoring or a cooperative approach in using the stand.

How Properties of Yams and Enset may have Structured Prehistoric Processes of Intensification

Inasmuch as the morphology, life cycle and environmental requirements of yams and enset shape their responses to human activities, these attributes would have structured the earliest modes of intensified interaction with humans. In this section, I consider variations in the extent to which disturbance can affect the distribution of enset and yams, and the more direct, deliberate ways in which humans might initiate intervention in plant reproduction and survival.

Disturbance

Humans can affect the densities and distribution of plants by disturbing natural vegetation. Disturbance varies in scale (large or small) and in source (anthropogenic versus natural). Its effects depend on its duration and regularity (consider a single tree fall or grass fire versus repeated episodes of erosion, human occupation or burning). Consequent changes in the distribution of useful plants and in related human exploitation strategies depend on the specific environmental requirements of the plants in question.

Small-scale disturbances caused by incidental human activities (clearing camps and other work areas) or nature (erosion, tree falls) create small areas where weedy species thrive. A resulting localised concentration of useful plants (yams or enset) could attract more intensive human exploitation of small, targeted areas and initiate a localised positive feedback cycle of human-plant interaction that includes regular harvesting and continual disturbance. Large-scale disturbance, whether anthropogenic (burning and clearance) or natural (brush fires, drought, or climate change) can open up vast areas to colonisation by useful weedy species and in turn lead to a generalised shift in human use of the landscape.

Sheko farmers allude to a jump in the frequency of wild *Dioscorea* if areas are disturbed. The intensity of this effect is muted in ecotonal or wooded grassland environments where *Dioscorea* already competes fairly successfully. In upland forests where growth of *Dioscorea* is impeded mainly by shade and competition, the effects of small and large-scale disturbance on wild yams are dramatic. In prehistoric times, these could have significantly expanded the spatial extent of yam-rich areas.

Because enset is confined to areas above 1000 m elevation, it occupies a more limited set of environments. Disturbance below 1000 m would not affect the distribution of enset at all. In upland areas, small-scale disturbance downhill from an enset cluster could stimulate the germination and growth of enset seeds, thereby expanding the cluster or starting a new one. This would increase the density of enset plants but not extend their distribution. Large-scale upland disturbances, especially burning, exert striking effects on the growth of enset, which is fairly tolerant of fire. Burning kills competitors; if enset plants get established in sufficient density (which easily occurs on flat areas below slopes with fruiting enset plants), they can subsequently shade out other competitors and form an essentially monocrop plot of volunteers. Such an effect was observed in two separate locations in southwest Sheko.

Yams and enset are similar in that disturbance abets their growth in upland areas, but has little effect in lowland areas. This suggests that inasmuch as disturbance played a role in prehistoric domestication of yams and enset, intensification would have taken place in upland areas that are naturally forested, rather than in wooded grassland or ecotones. The exact effects of disturbance vary between yams and enset, however. For enset, prehistoric disturbance by humans would have simply increased its density at the elevations where it already occurred naturally. For yams, the same disturbance processes would have significantly altered yam distribution by allowing the plants to colonise elevations where they were otherwise scant.

Opportunities for Human Intervention in Yam and Enset Propagation

Humans can affect plant densities through intervention in either vegetative (cuttings) or sexual (seeds) modes of propagation. Typically, seed propagation has been associated with crops with an economic value vested in seeds or fruits, whereas vegetative propagation is associated with plants with edible vegetative parts (Table 15.1). These generalisations have obvious exceptions (eg cloned bananas with edible fruit and seed-propagated leafy greens) and are based on observations of the endpoints of the domestication process. To what extent has human propagation of plants with edible underground starchy organs been limited to vegetative intervention throughout their history of interaction?

The relative efficacy of vegetative versus seed propagation by humans depends on properties of the targeted plant. Explicit consideration of modes of reproduction for yam and enset plants allows one to gauge the likelihood and possible effects of seed versus tuber-oriented intervention by prehistoric humans.

Yams and enset differ in patterns of seed dispersal. The small seeds of yams are held in capsules several meters off the ground and are scattered by wind over a significant area, so young yam plants do not compete with each other. The large, hard, heavy enset seeds drop right below the infructescence and are only dispersed farther by rolling or animal ingestion; hundreds of seedlings may germinate in a square meter and competition between seedlings is obvious.

Vegetative propagation of *Dioscorea* is a straightforward process of planting cuttings; the plant may yield the next season. Cloning of enset, in contrast, requires specialised knowledge and attention to a multistage process. Enset yields food after three years – a much longer period than that required for yields from yam clones. On the other hand, cloning enset considerably speeds the time between generations of enset plants, as one can stimulate clone production when a plant is only two years old, rather than having to wait for the plant to set seeds, which takes at least four years.

Yams and enset have dissimilar potential for vegetative versus seed-based intervention in propagation. This suggests different sequences of manipulation for yams and enset. Human intervention into yam seed propagation is unlikely because the seeds are tiny, inaccessible and disperse broadly without intervention. On the other hand, human spacing of enset seedlings is a highly plausible form of intervention, because the seeds are large, visible and fall in tight clusters. Human intervention in the simple vegetative propagation of yam plants would have occurred early in the domestication process; cloning of enset is more complex than either the cloning of yams or the spacing of naturally occurring enset seedlings, and therefore was probably a relatively late innovation.

Opportunities for Human Intervention during Non-Reproductive Phases of Yam and Enset Life Cycles

Humans can also affect the distribution of plants by moving the plant. This is not easily done for annual crops with fibrous roots. For yams, however, the storage organ (tuber plus crown) and dormant period make transplantation highly practicable. Enset plants, though not dormant, are also amenable to transplantation. The ways in which transplantation may occur differ according to the life cycles and environmental requirements of each plant taxon.

Adoptive transplantation – the movement of wild plants to domestic contexts – is one type of intervention that may have been practiced in prehistoric times. Some southwest Ethiopian farmers today move yams from lowland wooded savanna to highland settings; others move yams from forest to a nearby garden clearing (Hildebrand 2003b). Their goal is to move the plants from a distant or inconvenient location to one that is easy to access and monitor. The broad range of habitats for yams allows many relocation scenarios, eg upland to ecotone, lowland to upland, or within the same elevation zone. It also allows humans to tailor the distribution of yams to coincide more neatly with other aspects of their subsistence regime. By putting a yam plant in a location that assures regular harvest, adoptive transplantation virtually guarantees the phenotypic response of increased tuber size described above. Because the yam plant lives and regenerates its tuber year after year, the availability of its carbohydrates is effectively permanent once the plant is established.

A present-day example of adoptive transplantation of enset is not available in the field area, as Sheko draw a rigid distinction between wild (from seed) and domestic (from clone) enset (Hildebrand 2003c) and see only the latter as palatable enough to justify cultivation. Transplantation of enset within and between domestic contexts is well known across southwest Ethiopia, however. Transplantation can expand the range of cultivated enset to higher elevations where it cannot propagate by seed, but not to lower areas with insufficient moisture. Because the enset plant lives for a limited number of years and dies after being harvested for food, the availability of its carbohydrates following transplantation is impermanent and requires ongoing human intervention (disturbance to ensure self-propagation or repeated transplantation).

Despite differences in environmental requirements, it appears that for both yams and enset, transplantation involving an increase in elevation is more likely (given that yams are already dense in lowlands) and more feasible (given that enset cannot survive low arid environments) than transplantation to lower elevations. Fundamental differences emerge when one considers necessary human activities subsequent to hypothetical transplantation by prehistoric hunter-gatherers. In the case of enset, especially in extreme uplands where its growth cycle is too attenuated for sexual reproduction, additional tending or repeat transplantations might well have been necessary to maintain clusters of plants for regular use as food. Yams, once established, would require little labor input beyond the digging and consequent disturbance entailed by regular harvests.

CONCLUSION

Detailed field studies of yams and enset justify some of the distinctions conventionally drawn between seed and tuber crops. Many of the differences ascribed to seed versus tuber-crop cultivation systems, however, pertain to the intensity of cultivation methods rather than to traits of the crop plants themselves. In addition, recognition of a broad range of cultivation methods for seed and tuber crops undermines the notion of a dichotomy between seed/annual/field and tuber/perennial/garden crops. This calls into question the usefulness of generalised distinctions (between seed and tuber plants, for example) to guide discussions of intensification processes during prehistoric times.

Although enset and yams both provide large, starchy underground food sources, grow in the same area and are used by the same people, they are distinct in their morphology, life cycle, environmental requirements and utility. These differences structure human-plant interactions in fundamental ways today and would have done so prehistorically. Differences in growth and in time and place of availability would have given enset and yams highly distinct roles in a prehistoric hunting and gathering seasonal round. Because enset and yams differ significantly in their visibility and in their responses to human harvest, sustainable use by hunter-gatherers would have required separate strategies for harvesting and maintenance of wild plant populations.

The two plants also differ in their responses to deliberate and incidental human activities, such as harvesting and disturbance of vegetation, and in their susceptibility to overt manipulation of vegetative and seed-based propagation. Given that such forms of intervention often represent the initial stages of the domestication process (Harris 1989), it seems likely that enset and yam plants followed distinct pathways of intensified interaction with prehistoric humans.

The exact nature of the domestication process for yams and enset is not yet clear. This is in part because southwest Ethiopia has seen little Holocene archaeology, and the kinds of plant remains that might show prehistoric patterns of use, processing and manipulation have not yet been recovered. Building on a general model (Brandt 1984, 1996), I have suggested that enset domestication was a multistep process (Hildebrand 2003c) and proposed a tentative sequence of intensification for enset and yams in tandem in far southwest Ethiopia (Hildebrand 2003a). This model is based primarily on ethnoarchaeological data and has generated hypotheses to be tested. Alternative hypotheses should also be developed for the domestication of enset independently of yams and for domestication of yams alone. The distribution and use of these and other economically important indigenous plants in other parts of southwest Ethiopia require future consideration.

Even though southwest Ethiopia has not yet yielded all of its secrets, it has given us a fresh perspective on the fundamental ways in which biological traits of tuber crops may shape human use strategies, manipulation practices and, ultimately, pathways to food production. 'Tuber crops' and their wild progenitors do not, in fact, represent a homogeneous category of subsistence orientation in present-day or prehistoric times. Tuber cultivation systems do not have inherent ecological stability and geographical inertia, but can assume forms of intensive monocropping that, while highly productive, are also specialised, risky and prone to spread. The relations between these varied forms of subsistence economy, and the forms of social organisation they support, remain largely unexplored. Literature on early food production should recognise and emphasise the tremendous diversity in the origins and development of tuber cultivation systems and the role they played in shaping subsequent trajectories of social change throughout the world.

ACKNOWLEDGMENTS

I thank the Sheko and other farmers of southwest Ethiopia for sharing their lives and their knowledge with me. My Sheko host family was a constant source of support. Tedla Bekele, Kundisa Ferki, Birega Subsa and Yilma Miressa were wonderful colleagues in the field. I have gained much from discussions with Sebsebe Demissew and Paul Wilkin (*Dioscorea*), Fiona Marshall and Gayle Fritz (early food production), and Steve Brandt and Agazi Negash (Ethiopian prehistory). Special thanks also to Jara Hailemariam and Yonas Beyene of the Ethiopian Authority for Research and Conservation of Cultural Heritage and to Hannelore and Detlef Reuter of GTZ. Research was funded by the Wenner-Gren Foundation, the National Science Foundation, Sigma Xi, and the American Association of University Women. Crucial institutional support was provided by the Bench Zone Cultural Department, the Gurage Zone Cultural Department, the Ethiopian Ministry of Culture and Information, the Ethiopian National Museum, the Ethiopian National Herbarium and the Washington University Department of Anthropology.

REFERENCES

Ayensu, E (1970) 'Comparative anatomy of *Dioscorea rotundata* and *Dioscorea cayenensis*', in Robson, N K B, Cutler, D F and Gregory, M (eds), *New Research in Plant Anatomy*, pp 127–36. Botanical Journal of the Linnaean Society supp 1, London: Academic Press

Bar-Yosef, O and Belfer-Cohen, A (1989) 'The origins of sedentism and farming communities in the Levant', *Journal of World Prehistory* 3: 447–98

Bayliss-Smith, T (1996) 'Identifying energy thresholds in New Guinea people-plant interactions', in Harris, D R (ed), *The Origins and Spread of Agriculture and Pastoralism in Eurasia*, pp 499–537, Washington DC: Smithsonian Institution Press

Brandt, S A (1984) 'New perspectives on the origins of food production in Ethiopia', in Clark, J D and Brandt, S A (eds), *From Hunters to Farmers: The Causes and Consequences of Food Production in Africa*, pp 173–90, Berkeley, CA: University of California Press

Brandt, S A (1996) 'A model for the origins and evolution of enset food production', in Tsedeke, A, Hiebsch, C, Brandt, S A and Gebremariam, S (eds), *Enset-Based Sustainable Agriculture in Ethiopia*, pp 36–46, Addis Ababa: Institute of Agricultural Research

Burkill, I H (1960) 'The organography and the evolution of Dioscoreaceae, the family of the yams', *Journal of the Linnaean Society (Botany)* 56: 319–412

Chikwendu, V E and Ozekie, C E A (1989) 'Factors responsible for the ennoblement of African yams: inferences from experiments in yam domestication', in Harris, D R and Hillman, G C (eds), *Foraging and Farming: The Evolution of Plant Exploitation*, pp 344–57, London: Unwin Hyman

Childe, V G (1951) *Man Makes Himself*, London: CA Watts and Co

Coursey, D G (1967) *Yams: An Account of the Nature, Origins, Cultivation and Utilisation of the Useful Members of the Dioscoreaceae*, London: Longmans

Cowen, C W and Watson, P J (eds) (1992) *The Origins of Agriculture: An International Perspective*, Washington DC: Smithsonian Institution

Deacon, H J (1984) 'Excavations at Boomplaas Cave – a sequence through the Upper Pleistocene and Holocene in South Africa', *World Archaeology* 10(3): 241–57

Deacon, J (1984) 'Later Stone Age people and their descendents in southern Africa', in Klein, R G (ed), *Southern African Prehistory and Paleoenvironments*, pp 221–328, Boston: AA Balkema

Denham, T P, Haberle, S G, Lentfer, C, Fullagar, R, Field, J, Therin, M, Porch, N and Winsborough, B (2003) 'Origins of agriculture at Kuk Swamp in the Highlands of New Guinea', *Science* 301: 189–93

Gebauer, A B and Price, T D (eds) (1992) *Transitions to Agriculture in Prehistory*, Madison, WI: Prehistory Press

Gebre Yntiso (1995) 'The Ari of southwestern Ethiopia: an exploratory study of production practices', unpublished MA thesis, Addis Ababa University

Harris, D R (1969) 'Agricultural systems, ecosystems and the origins of agriculture', in Ucko, P J and Dimbleby, G W (eds), *The Domestication and Exploitation of Plants and Animals*, pp 3–15, Chicago: Aldine

Harris, D R (1972) 'The origins of agriculture in the tropics', *American Scientist* 60: 180–93

Harris, D R (1973) 'The prehistory of tropical agriculture: an ethnoecological model', in Renfrew, C (ed), *The Explanation of Culture Change: Models in Prehistory*, pp 391–417, London: Duckworth

Harris, D R (1977) 'Alternative pathways toward agriculture', in Reed, C A (ed), *The Origins of Agriculture*, pp 179–243, The Hague: Mouton

Harris, D R (1989) 'An evolutionary continuum of people-plant interaction', in Harris, D R and Hillman, G C (eds), *Foraging and Farming: The Evolution of Plant Exploitation*, pp 11–26, London: Unwin Hyman

Hather, J G (1994) 'The identification of charred root and tuber crops from archaeological sites in the Pacific', in Hather, J G (ed), *Tropical Archaeobotany: Applications and New Developments*, pp 51–63, London: Routledge

Hather, J G (1996) 'The origins of tropical vegeculture: Zingiberaceae, Araceae, and Dioscoreaceae in southeast Asia', in Harris, D R (ed), *The Origins and Spread of Agriculture and Pastoralism in Eurasia*, pp 538–50, Washington DC: Smithsonian Institution Press

Hawkes, J G (1969) 'The ecological background of plant domestication', in Ucko, P J and Dimbleby, G W (eds), *The Domestication and Exploitation of Plants and Animals*, pp 17–29, Chicago: Aldine

Hildebrand, E (2001) 'Morphological characterization of garden vs. forest-growing *Ensete ventricosum* (Welw.) Cheesman, Musaceae in Bench–Maji Zone, southwest Ethiopia', in Friis, I and Ryding, O (eds), *Biodiversity Research in the Horn of Africa Region*, pp 287–309, Biologiske Skrifter 54, Copenhagen: Danish National Academy of Sciences

Hildebrand, E A (2003a) 'Enset, yams, and honey: ethnoarchaeological approaches to the origins of horticulture in southwest Ethiopia', unpublished PhD thesis, Washington University in St. Louis

Hildebrand, E A (2003b) 'Motives and opportunities for domestication: an ethnoarchaeological study in southwest Ethiopia', *Journal of Anthropological Archaeology* 22(4): 358–75

Hildebrand, E A (2003c) 'Comparison of domestic vs. forest-growing *Ensete ventricosum* (Welw.) Cheesman, Musaceae in Ethiopia: implications for detecting enset archaeologically, and modeling its domestication', in Neumann, K, Butler, A and Kahlheber, S (eds), *Food, Fuel, and Fields: Progress in African Archaeobotany*, pp 49–70. Africa Praehistorica 15, Cologne: Heinrich Barth Institut

Hildebrand, E A, Sebsebe Demissew and Wilkin, P (2002) 'Local and regional landrace disappearance in species of *Dioscorea* L. (yams) in southwest Ethiopia: causes of agrobiodiversity loss and strategies for conservation', in Stepp, J R, Wyndham, F S and Zarger, R K (eds), *Ethnobiology and Biocultural Diversity: Proceedings of the 7th International Congress of Ethnobiology*, pp 678–95, Athens, GA: University of Georgia Press

Lock, J M (1993) 'Musaceae', in Polhill, R M (ed), *Flora of Tropical East Africa*, Rotterdam: Balkema

Marshall, F and Hildebrand, E A (2002) 'Cattle before crops: the beginnings of food production in Africa', *Journal of World Prehistory* 16(2): 99–143

Mbida, C, Van Neer, W, Doutrelepont, H and Vrydaghs, L (2000) 'Evidence for banana cultivation and animal husbandry during the first millennium BC in the forest of south Cameroon', *Journal of Archaeological Science* 27: 151–62

Mbida, C, Doutrelepont, H, Vrydaghs, L, Swennen, R, Swennen, R, Beeckman, H, De Langhe, E and Maret, P de (2001) 'First archaeological evidence of banana cultivation in central Africa during the third millennium before present', *Vegetation History and Archaeobotany* 10: 1–6

McCorriston, J and Hole, F (1991) 'The ecology of seasonal stress and the origins of agriculture in the Near East', *American Anthropologist* 93: 46–69

Miège, J and Sebsebe Demissew (1997) 'Dioscoreaceae', in Edwards, S, Sebsebe Demissew and Hedberg, I (eds), *Flora of Ethiopia and Eritrea*, vol 6: *Hydrocharitaceae to Araceae*, pp 55–62. Addis Ababa: Ethiopian National Herbarium, Addis Ababa University and Uppsala: Department of Systemic Botany, Uppsala University

Mignouna, H D and Dansi, A (2003) 'Yam (*Dioscorea* ssp.) domestication by the Nago and Fon ethnic groups in Benin', *Genetic Resources and Crop Evolution* 50(5): 519–28

Moore, A and Hillman, G (1992) 'The Pleistocene to Holocene transition and human economy in southwest Asia: the impact of the Younger Dryas', *American Antiquity* 57: 482–94

Neumann, K (2003) 'New Guinea: a cradle of agriculture', *Science* 301: 180–81

Peters, C R (1996) 'African wild plants with rootstocks reported to be eaten raw: the monocots, part III', in van der Maesen, L J G, van der Burgt, X M and van Medenbach de Rooy, J M (eds), *The Biodiversity of African Plants*, pp 665–77, London: Kluwer Academic

Piperno, D R and Pearsall, D M (1998) *The Origins of Agriculture in the Lowland Neotropics*, London: Academic Press

Price, T D and Gebauer, A B (eds) (1995) *Last Hunters–First Farmers: New Perspectives on the Prehistoric Transition to Agriculture*, Santa Fe, NM: SAR Press

Rindos, D (1984) *The Origins of Agriculture: An Evolutionary Perspective*, Orlando, FL: Academic Press

Sauer, C O (1952) *Agricultural Origins and Dispersals*, New York: American Geographical Society

Schoeninger, M J, Bunn, H T, Murray, S S and Martlett, J A (2001) 'Composition of tubers used by Hadza foragers of Tanzania', *Journal of Food Composition and Analysis* 14(1): 15–25

Shack, W (1966) *The Gurage: A People of the Ensete Culture*, Oxford: Oxford University Press

Shigeta, M (1990) 'Folk *in-situ* conservation of ensete (*Ensete ventricosum* [Welw] Cheesman): towards the interpretation of indigenous agricultural science of the Ari, southwestern Ethiopia', *Kyoto University African Study Monographs* 10(3): 93–107

Shigeta, M (1991) 'The ethnobotanical study of ensete (*Ensete ventricosum*) in southwestern Ethiopia', unpublished PhD thesis, Kyoto University

Shigeta, M (1996) 'Creating landrace diversity: the case of the Ari people and ensete (*Ensete ventricosum*) in Ethiopia', in Ellen, R and Fukui, K (eds), *Redefining Nature: Ecology, Culture and Domestication*, pp 233–68, Oxford: Berg

Simmonds, N W (1960) 'Notes on banana taxonomy', *Kew Bulletin* (London) 14: 198–212

Simmonds, N W (1962) *The Evolution of the Bananas*, London: Longmans

Smeds, H (1955) 'The ensete planting culture of eastern Sidamo, Ethiopia', *Acta Geographica* 13(4): 1–39

Ucko, P J and Dimbleby, G W (eds) (1969) *The Domestication and Exploitation of Plants and Animals*, Chicago: Aldine

Ugent, D, Pozorski, S and Pozorski, T (1982) 'Archaeological potato tuber remains from the Casma Valley of Peru', *Economic Botany* 36: 182–92

Wilkin, P (2001) 'Dioscoreaceae of South-Central Africa' *Kew Bulletin* (London) 56: 361–404

CHAPTER 16

Multidisciplinary Evidence of Mixed Farming During the Early Iron Age in Rwanda and Burundi

Marie-Claude Van Grunderbeek and Emile Roche

INTRODUCTION

People of the Urewe Early Iron Age culture first settled the Great Lakes region (East-Central Africa), including the Rwanda–Burundi Central Plateau, during the first millennium BC. The Urewe culture marks the earliest occurrence of several practices on the plateau, including iron smelting and pottery making. Urewe culture is characterized by well-made decorated earthenware (Van Grunderbeek 1988) and distinctive iron-smelting technology (Van Grunderbeek et al 2001). Archaeologists have largely assumed that the Urewe culture was based, in part, on a mixed farming economy because its material culture was suggestive of relatively high levels of economic, political and social organisation (Van Grunderbeek 1992: 74–76).

In this paper, multidisciplinary lines of evidence for early agricultural practices associated with the Urewe culture on the Rwanda–Burundi Central Plateau are presented for the first time. The research draws on complementary approaches, including archaeology, botany, charcoal analysis and palynology, which together suggest the presence of agricultural activities on Kabuye Hill, east of Butare, Hills Area (Central Plateau), Rwanda from 100–300 cal AD to 400–500 cal AD. The presence of domesticated cereal (grain or pollen) is normally taken to be direct evidence for agriculture, but in practice many archaeologists doubt the reliability of such evidence based on taphonomic considerations, ie percolation in sandy soils, which are not applicable here. In this chapter, direct evidence of agriculture, ie cereal pollen grains from secure archaeological contexts, is presented and put into a broader interpretative context by multiple lines of indirect evidence. Together these lines of evidence are indicative of a mixed farming economy. This study illustrates the value of pursuing multidisciplinary investigations of agriculture, particularly in areas of highly acidic soils that destroy most macrobotanical remains.

THE UREWE CULTURE

Urewe culture in the Great Lakes region has a strategic place both chronologically and geographically. The culture takes its name from a site in Kenya at Kavirondo Bay, northeast of Lake Victoria. The culture marks the oldest evidence of the Early Iron Age (EIA) south of the equator, dates from the first millennium BC, and is the 'cradle' for the subsequent spread of EIA-cultures, and of agriculture, eastwards in Southeast Africa up to the Limpopo River by about AD 800 cal. The social processes that produced the Urewe culture in the Great Lakes region, whether diffusion or cultural interaction, are unclear. It is noteworthy that many of the sites dated to the first half of the first millennium AD are situated near or within presently densely populated areas. These sites represent the peak of the Urewe culture, when areas with heavy soils could be exploited. This site distribution is, in part, a product of sampling; the Bantu Studies Project conducted rapid survey programs from present-day settlements. More extensive surveys away from densely settled regions are likely to extend the EIA chronology of the Great Lakes region and generate data comparable to that of the Rwanda/Burundi and Buhaya regions.

In the Great Lakes area, Urewe culture was preceded by cultures based on stone technology. The absence of stone artefacts at Urewe sites suggests there was limited interaction with the preexisting peoples. However, some innovations, sometimes coupled to changes in pottery type, may indicate subsequent interactions with, and the diffusion of innovations from, non-Urewe EIA cultures. From approximately 2000 years ago in the Buhaya area, Schmidt and Childs (1985) noted the burial of a piece of slag in a basal hole of iron-smelting crucibles. Similarly in Rwanda from 400–600 cal AD onwards, a sealed pot was buried in the base of a crucible (see Kabuye II below).

From 600–900 cal AD, roulette-decorated pottery and shaftless iron-smelting furnaces mark the advent of the Late Iron Age (LIA). Localised elements of Urewe culture persisted until 1200–1600 cal AD. Today Batwa still manufacture roulette-decorated pottery and live all over the country in small outcast communities. They are still the main pottery providers for the whole of Rwanda and Burundi society, being present at local markets to sell their production of the last three days. Iron-smelting practices were largely abandoned after World War I, when cheap iron was introduced by Europeans, although there was a short resurgence during the World War II.

HISTORY OF INVESTIGATIONS

The publication by M D Leakey (Leakey et al 1948) of ceramic material from Urewe sites in Kenya interested many researchers who subsequently investigated the Great Lakes region of eastern Africa. Among them was M Posnansky (1961, 1966, 1967, 1968), who undertook pioneering archaeological investigations of the region and whose work was continued during the 1960s by members of the Bantu Studies Project established by the British Institute of Eastern Africa in Nairobi. These studies sketched the geographic extent and an almost 2000-year antiquity for the Urewe culture in the Great Lakes region (Van Grunderbeek 1992).

About the same time, Jean Hiernaux collected comparable archaeological material in Kivu Province (Democratic Republic of Congo), Burundi and Rwanda. In major publications (Hiernaux and Maquet 1957, 1960), the ceramics are described and classified into class A, which was distinguished from later traditions, class B (roulette-decorated) and class C (still undetermined). Hiernaux identified the first EIA iron-smelting furnace crucibles, obtained radiocarbon dates (Fagan 1969: 155) and established the relationship between Urewe ceramics and iron production (Hiernaux 1960). His research has been reviewed and expanded through subsequent excavations and dating (Nenquin 1967; Van Noten 1979, 1983).

Along the western coast of the Victoria Nyanza, in the Buhaya area, Tanzania, P R Schmidt (1997) developed a research program on oral tradition during the 1970s, which led him to the Urewe culture. The dates he obtained extended the Urewe chronology to parts of the first millennia BC and AD.

We began our research on the Urewe culture in 1978 (Van Grunderbeek et al 1983, 1984). Components of our research focussed upon the chronology of Urewe culture in Rwanda, Burundi and the general Great Lakes region (Van Grunderbeek 1992), the description of Urewe ceramics of Rwanda and Burundi (Van Grunderbeek 1988) and the conceptual reconstruction of an EIA Urewe iron-smelting furnace (Van Grunderbeek et al 2001). As part of our research, we also reconstructed the pre-EIA environments of the region at c 2000 years ago (Roche and Van Grunderbeek 1987) and the subsequent effects of human degradation, initially as a result of iron smelting (Van Grunderbeek and Doutrelepont 1989) and then agriculture (Roche 1991).

RWANDA AND BURUNDI: ENVIRONMENTAL OVERVIEW

Rwanda and Burundi are situated in East-Central Africa, just below the equator. They straddle the eastern horst of the western branch of the Eastern African Rift. This mountainous region descends steeply to Lake Kivu and Lake Tanganyika to the west and gently to the Victoria Nyanza Basin to the east. Both countries enjoy equatorial mountainous climates, characterized by mild and steady temperatures, at elevations between 1200 to 2000 m above sea level. Peaks up to 3000 m have colder, alpine climates. Moderate precipitation is evenly distributed throughout the year, although there is marked seasonality with a long dry season from June to mid-September and a heavy rainy season from February to June. There are also regional variations in climate across both countries (Lebrun 1956; Sirven et al 1974; Van der Velpen 1970).

The huge differences in elevation across this relatively small area create a remarkable diversity of natural regions. Today the landscapes in Rwanda and Burundi, both densely populated countries sustained essentially by agriculture, are mainly anthropic. The effects of human impacts and climate on the environment have been differentiated using palynological reconstructions for the last 30,000 years (Roche 1998: Figure 6). In this part of

Africa, high-elevation peat bogs offer the longest uninterrupted sequences to reconstruct palaeoclimate. Only this type of site is preserved with limited outside interference, other than climatic, and most are located far away from inhabited zones. Palynologists seek sites that have not been subject to excessive aridity, which causes hiatuses in the deposits and oxidation of pollen grains and other organic matter. Emile Roche examined and linked the palynological records from archaeological sites and nearby peat bogs to study human-environment interactions.

Human interference in the environment was discernible during the first millennium BC, and became devastating during the first millennium AD. During this latter period, the effects of human activities were exacerbated by a dry climatic phase. From that time on, human impacts continued despite the return of moister climates. The cumulative effects of human activities over the last 2000 years have yielded an approximately 95% anthropic environment across both countries (Roche 1996) (Figures 16.1 and 16.3). Only small areas (4.5% of total surface area) of the original natural plant cover remain today that are of the kind that existed 3000 years ago before the coming of the Urewe culture to the Rwanda and Burundi highlands (Central Plateau) (Figure 16.2).

Prior to the advent of the Urewe culture, the distribution of natural vegetation zones largely depended on three factors: elevation, temperature and rainfall. A palaeoecological study of former vegetation communities conducted by Emile Roche combined information on remnant communities of natural plant ecology (Lewalle 1972; Liben 1962; Troupin 1966), elevation and climate to infer eight original vegetation regions (Roche and Van Grunderbeek 1987; Figure 16.3):

(1) Afro-alpine belt above 2600 m elevation, dry, although the tops are blanketed in fog;
(2) Dense mountainous rain forest, mantled slopes from 2600 to 2000 m in elevation with high rainfall;
(3) Submontane forests mark the transition from 2000 to 1700 m with less humid climates and more heterogeneous woodland assemblages;

Figure 16.1 Photograph of landscape showing degraded vegetation at Kabuye.

(4) Deciduous zambezian forest across the Malagarasi Depression (southeast Burundi), which is subject to seasonal climates;
(5) Open woodlands of the Central Highlands (Hills Area of the Central Plateau), suited to farming due to adequate moisture, relatively fertile soils, ie where bananas grow today (Allan 1965; Van Grunderbeek et al 2001), and an open and gently sloping landscape;
(6) Eastern wooded and grassy savanna of the Akagera-Bugesera, below 1500 m and with drier climates, although infested by tsetse fly noxious to cattle;
(7) Woodland covering pyroclastic deposits on the lower slopes of volcanoes to the north of Rwanda;
(8) Rusizis wooded savannas, with dry climates and infested with tsetse fly.

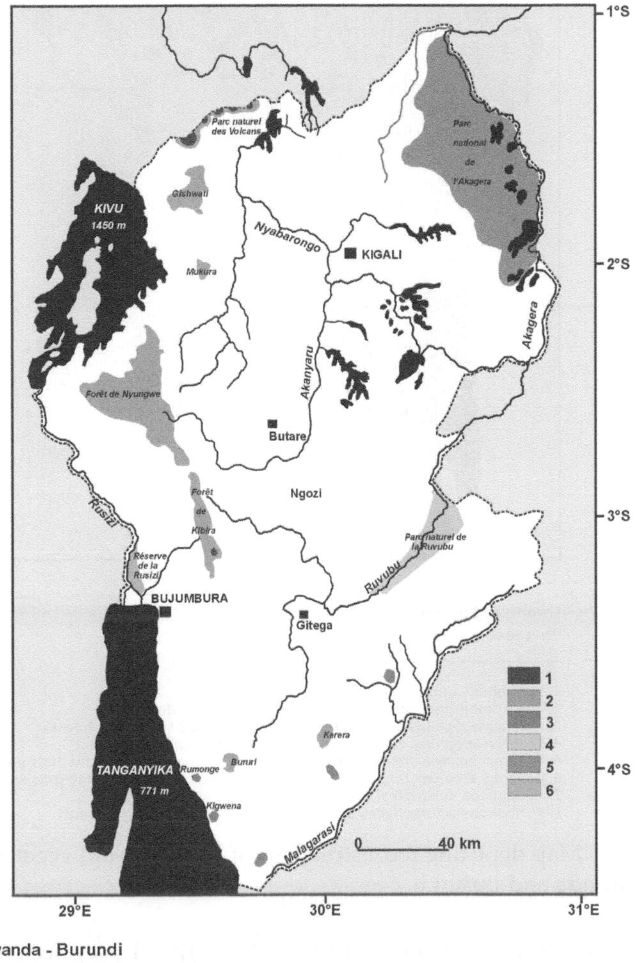

Rwanda - Burundi
Milieux naturels / semi-naturels:

1. Formations afro-alpines - Afro-alpine belt
2. Forêt afro-montagnarde, généralement secondarisée - Afro-montane forest, mostly secondarized
3. Forêt tropophile zambézienne - Deciduous zambezian forest
4. Savanes boisées des collines - Open woodlands of the Hills area
5. Savanes arborées ou herbeuses orientales - Eastern wooded and grassy savannas
6. Formations sclérophylles de la Rusizi - Rusizi's wooded savannas

Figure 16.2 Map depicting present-day vegetation communities of Rwanda and Burundi.

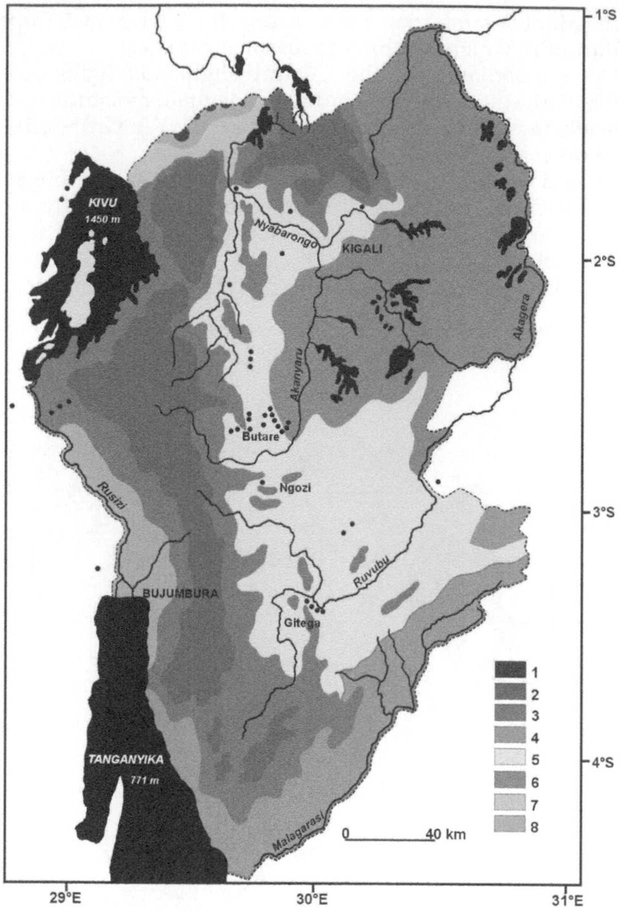

Rwanda - Burundi
Phytogéographie - Natural vegetation areas

1. Formations afro-alpines - Afro-alpine belt
2. Forêt ombrophile de montagne - Mountainous rainforest
3. Forêts mésophiles ou secondaires - Mesophilous secondary forests
4. Forêts tropophiles zambéziennes - Deciduous zambezian forest
5. Savanes boisées des plateaux centraux - Open woodlands of the Hills area
6. Savanes arborées ou herbeuses orientales - Eastern wooded and grassy savannas
7. Formations sclérophylles des laves - Lava's wooded area
8. Formations sclérophylles de la Rusizi - Rusizi's wooded savannas

Figure 16.3 Map depicting reconstructions of 2000-year-old vegetation communities for Rwanda and Burundi.

From the first millennium BC until the seventh century AD, people settled the woodland savanna of the Hills Area, ie elevations of 1500–1700 m of the Central Plateau region (Van Grunderbeek et al 1983, 1984) along the fringes of the forest (Van Grunderbeek and Doutrelepont 1989; Figure 16.4). Parts of this settlement pattern persisted until the fifteenth century AD (Van Grunderbeek 1992).

The palaeoecological reconstruction of 2000-year-old environments (Figure 16.3) shows the Hills Area (Central Plateau, vegetation region 5) was a favoured location for Urewe people, as were similar areas in the Great Lakes region, ie to the north and west of the Victoria Nyanza in Kenya, Uganda and Tanzania (Phillipson 1977; Schmidt 1974), and west of Lake Kivu in Congo (Hiernaux and Maquet 1957, 1960). The Hills Area lies between the Congo-Nile crest to the west, the Akagera-Bugesera lowlands to the northeast and the Kumoso lowlands to the southeast. The term 'plateau' is inadequate because the landscape is hilly with rounded and variable topography, including peaks of 1500 to 2000 m, and separated by a myriad of valleys, which have been subject to alluvial deposition and today have flat, marshy bottoms. The dry season lasts for three months, the average temperature being about 20° C (68° F); rainfall ranges from 1100 to 1300 mm/year.

The Hills Area has diverse soil types, which can be classified in the large latosols group, and include fine clayey kaolisols and gravelly lithosols, which are mainly red coloured due to iron oxidation. Kaolisols cover the largest surface of Rwanda and Burundi. Humic kaolisols are well drained, characterized by a marked humic horizon and derived from quartz, gneiss, slate and basalt bedrocks. They extend to the middle elevations, mantling most areas above 1500 m in the centre and west. The quartz crests develop very coarse gravel lithosols of quartz and lateritic rubble (INEAC 1963).

The ancient soils in this part of Africa are mostly ferruginous and acidic. Today they are subject to severe erosion as a result of the dense populations practicing agriculture on any available soil. Consequently few remains of the EIA, primarily pottery, iron slag, baked clay fragments, charcoal and an occasional cobblestone, are preserved. Very few structures and living floors have been found; the exceptions are deep iron-smelting crucibles and pits.

EVIDENCE OF HUNTING, FISHING AND GATHERING

Protohistoric studies suggest past subsistence practices included hunting and fishing, gathering and cattle breeding, as well as agriculture (Schoenbrun 1998: 66–68). Here, several key elements of Urewe subsistence are considered.

Faunal remains of hunting and fishing are absent at EIA sites in Burundi and Rwanda (including the Urewe culture). However, gathering activities are implied by the location of Urewe settlements in the savannas and at the edge of the forest (Van Grunderbeek and Doutrelepont 1989) (Figure 16.4). Urewe people had access to a wide range of edible fruit, berries and roots, although there is still no archaeobotanical evidence of their use. Faunal remains have been collected in shelters dating to the ninth century AD at Akameru and Cyinkomane in northern Rwanda. In these LIA sites, bones of wild game were mixed with those of domestic animals and roulette-decorated pots (Gautier in Van Noten 1983: 104–20; Van Neer 2000: 169).

For present-day Ntu languages in the area, linguists have identified terms for bow and arrow used for hunting (as distinct from a bow used to pierce

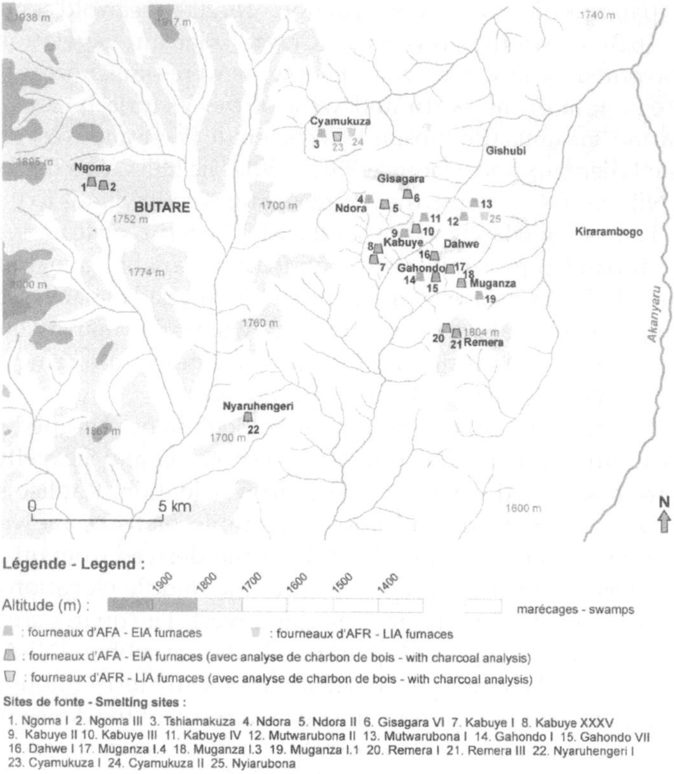

Figure 16.4 Map of Early Iron Age sites, east of Butare, Hills Area.

a vein to bleed cattle, which is practiced today by East African herdsmen) and game meat. Glottochronological estimates[1] suggest these terms date to the first millennium BC, ie to the EIA (Ehret 1998: 123–24). During the 19th and 20th centuries, however, the Banyarwanda (except the Batwa) only hunted game for fur and leather – not for meat (Maquet 1957: 77), and thus suppositions about the practice and antiquity of hunting in the past require critical evaluation and evidential verification. Similar caution is required for fishing techniques, which Ehret uses glottochronological estimates of Ntu terms to date to the first millennium BC (Ehret 1998: 123).

DIRECT EVIDENCE FOR EARLY IRON AGE AGRICULTURE

At three archaeological sites on Kabuye Hill, Rwanda (Figure 16.4), samples were collected for pollen analysis: Kabuye IV, a furnace crucible dated to 240–400 cal AD[2] (Oxcal, 95.4%), represents the earliest iron smelting on the hill; Kabuye III, a crucible dated to 420–600 cal AD[3] (Oxcal, 95.4%), occurs at the end of about two centuries of iron smelting; and Kabuye II, a crucible dated to 560–690 cal AD[4] (Oxcal, 95.4%), was an isolated structure used after a period of abandonment. Kabuye IV and III represent the

Urewe culture, whereas Kabuye II is assigned to an intrusive, but as yet undetermined EIA culture. The sample from Kabuye II was collected from a sealed pot at the base of the crucible. Pollen spectra at all three sites include cereal pollen (Figure 16.5).

After close comparison with present-day species, part of the Gramineae pollen at Kabuye IV (3rd–4th centuries AD) was identified as *Eleusine coracana* (finger millet), although it represented only a small percentage (2%) of the total pollen sum. A similar percentage of finger millet pollen was present in the Kabuye III (5th–6th centuries AD) and Kabuye II (6th–7th centuries AD) samples. Sorghum (*Sorghum bicolor*) represents 3% of total pollen spectra in the Kabuye IV and Kabuye III samples. Although highly significant, these percentages of cereal pollen do not indicate an intensive commitment to cereal cultivation. Rather, small plots were probably devoted to cereal cultivation, initially of *Eleusine*, a very fire-resistant African cereal that benefits from slash-and-burn preparation of the plot[5], and subsequently of sorghum.

Cereal pollen in an archaeological site, even from apparently secure contexts, is often considered to be intrusive by archaeologists. Cereal pollen and grains are relatively heavier than those of other plants and, consequently, are often considered to have percolated through the stratigraphy into older contexts. At the Kabuye Hill sites, percolation and other taphonomic problems are unlikely to be factors. Each site consists of a crucible for an iron-smelting furnace. These crucibles consist of a bowl dug into the soil; the crucible walls were either burned or baked to a thickness of up to 5 cm. On abandonment, the crucibles were filled with brick debris from the shaft, which baked during the smelting, together with other residues such as iron slag. Filling of the crucible imprisoned the pollen and prevented the deposition of younger, intrusive pollen by mechanical percolation or bioturbation. The Kabuye II sample was collected from an extremely secure context – a pot sealed with a potsherd from the base of the crucible.

The presence of pollen from two domesticated cereals is significant. Although both crops are likely to have an African origin (Harvey and Fuller 2005: 743), they were not domesticated in Rwanda or Burundi. Both crops were brought into the region by people, most likely by members of the Urewe culture engaged in mixed farming. Wild *Eleusine indica*, an ancestor of cultivated *Eleusine carocana*, exists in Rwanda, but the domesticated species is thought to have originated in the highlands of eastern Africa. *Sorghum bicolor* also grows in present-day Rwanda, but only as the cultivated species in gardens, as a garden escapee or crossbreed.

INDIRECT EVIDENCE FOR EARLY IRON AGE AGRICULTURE

In addition to the direct evidence of cereal pollen in secure contexts, several proxy records suggest some farming occurred during the EIA. These records fill out our interpretation of EIA subsistence and provide a more complete and vivid image of society.

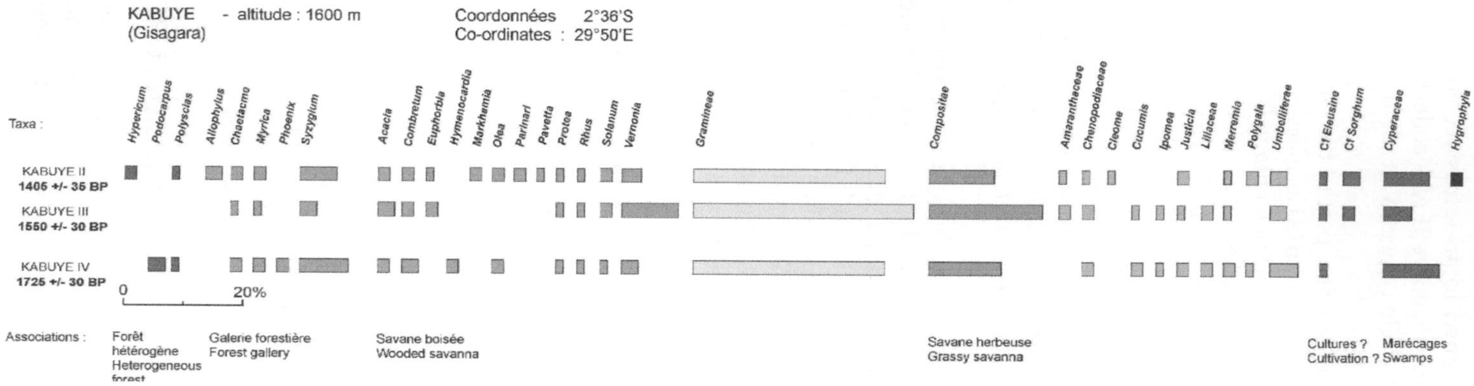

Figure 16.5 Pollen diagram for samples collected from Early Iron Age archaeological sites on Kabuye Hill, east of Butare.

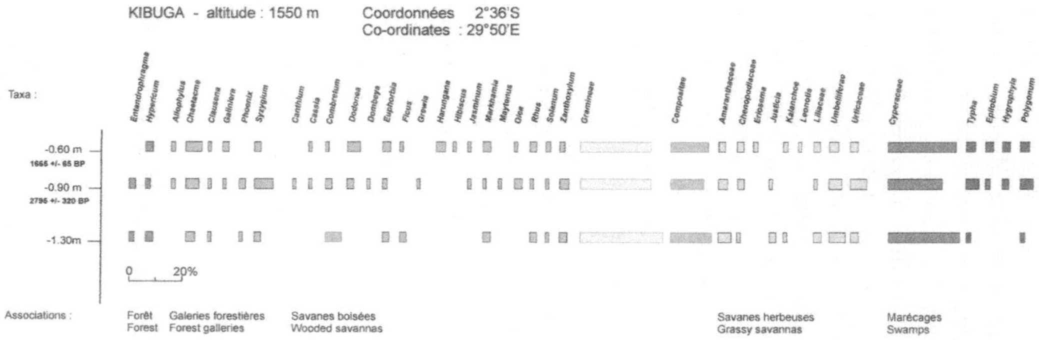

Figure 16.6 Pollen diagram for samples collected from stratigraphy exposed in the Kibuga section, east of Butare.

Environmental Advantages of the Hills Area (Central Plateau)

Today in the Hills Area (Figure 16.3), natural vegetation has been overexploited and reduced to only a few refugia on Kirarambogo Hill, east of Butare, Rwanda, and in the Karuzi region, Burundi. The principal tree species of this formerly more extensive woodland savanna are: *Acacia* spp., *Albizia adianthifolia, Parinari mobola, Entada abyssinica, Maytenus senegalensis, Protea madiensis, Erythrina abyssinica, Rhus vulgaris, Harungana madagascariensis, Ficus* sp. and *Combretum* spp.

In the centre of Rwanda, the gallery forest of *Syzygium cordatum, Macaranga schweinfurtii* and *Alchornea cordata*, just like the groves of wooded ravines with *Albizia grandibracteata, Syzygium guineense* and *Sterculia tragacantha*, no longer exist (Figure 16.1). At Karuzi, on the other hand, a few gallery forests with *Newtonia buchananii* and *Millettia dura* still enclose marshy depressions of *Cyperus papyrus* and thickets of *Syzygium cordatum, Myrica kandtiana* and *Ficus verruculosa*. In these continuously wet areas, small patches of *Pycreus mundtii* (Umurago), a salty grass, are used for cattle fodder. In this area, there are natural prairies of *Loudetia kagerensis* and *Hyparrhenia diplandra* (Pahaut and Van der Ben 1962; Habiyaremye and Roche 2003).

Of all natural regions of Rwanda and Burundi, the Hills Area was the most attractive to settlement by EIA populations. An open landscape along the savanna/forest borders offered advantages of both ecosystems: plenty of wood and iron ore close at hand for iron smelting, clay soils for ceramics, as well as diverse floral and faunal resources (Van Grunderbeek and Doutrelepont 1989). Of all the various natural regions, this land is pedologically the most favourable to agricultural activity. Climates exhibit alternating dry and wet seasons and temperate climates, which today allow three harvests a year.

The first incursions into the Hills Area probably occurred at Gasiza in Rwanda and Rwiyange in Burundi during the first part of the first millennium BC, where people could take advantage of lighter soils. The vegetation there was more susceptible to disturbance during the cold and dry period, which just preceded this period of occupation, and consequently the soils were easier to dig and trees easier to cut. The Urewe culture expanded during the first millennium AD onto the heavier and deeper soils of the Hills Area to the east of Butare (Rwanda) and of Gitega (Burundi).

Environmental Degradation on Kabuye Hill, East of Butare: The Woodland Cover

Cultivation of cereal often leads to severe environmental degradation because it entails the clearance of existing vegetation to create a plot for cultivation (Ehret 1998: 127–30) and thus, should be easier to notice. The woodland savannas that extended approximately 2000 years ago into the Butare area had a more closed canopy than the relict assemblages that survive today in central Rwanda. Human activities, which have intensified through time and the effects of which were exacerbated during dry climatic phases, have denuded these woodland savannas.

Identification of wood species out of the charcoal collected in the crucibles of iron-smelting furnaces (Van Grunderbeek et al 1984) and the study of pollen in sediments collected at nearby human occupation sites (Roche 1996) confirm that the natural environment of the Butare area during the EIA was a woodland savanna dominated by *Acacia, Canthium, Combretum, Drypetes, Grewia, Hymenocardia, Maytenus, Ochna, Olea, Parinari, Protea, Rhus, Zanthoxylum,* Gramineae and Compositae. The analyses reveal the presence of other habitats as well, notably marshy depressions with Cyperaceae (*Papyrus*), forest galleries and riverine woods with *Syzygium, Albizia, Garcinia, Landolphia* and *Myrica*. The influence of the mountainous forest to the west is indicated by the presence and eastwards spread on higher ground of forest species including *Entandrophragma, Macaranga, Podocarpus* and *Polyscias*.

By the Kabuye IV period (3rd–4th centuries AD), these vegetation assemblages were already altered. In their place, secondary assemblages dominated and comprised Acanthaceae (*Justicia*), Amaranthaceae, Chenopodiaceae, Convolvulaceae, Compositae including *Vernonia*, Cucurbitaceae, Polygalaceae (*Polygala*) and Solanaceae (*Solanum*). Increases in these taxa are likely to have spread over, and indicate the former presence of, paths, fallows and even fields. Some secondary plants, such as *Amaranthus augustifolius*, the young sprouts and leaves of which can be consumed as greens, *Cucumis* and *Cucurbita* may have been cultivated.

In Kabuye III (5th–6th centuries AD), there was an increase in *Vernonia* (shrubby plant of anthropised ecology in the composite family) and Chenopodiaceae, which could be the result of severe environmental degradation following extensive deforestation and land reclamation, but which could also be concomitant to a dry climatic period that would have encouraged the development of savanna.

Environmental Degradation on Kabuye Hill: Compositae Fallow after Cultivation

Palynological studies enabled the reconstruction of environments on Kabuye Hill during the EIA (Roche and Van Grunderbeek 1987; Van Grunderbeek et al 1983, 1984; Figures 16.4–5). A field parcel, abandoned after a period of cultivation, has a succession of characteristic fallow weed flora, which can be used as indicators of agricultural activity (L Liben, pers comm). The presence of ruderal plants, nitrophiles of degraded savanna, and secondary fallow species (such as *Chenopodium, Cleome, Cucumis, Justicia, Polygala, Solanum, Vernonia,* Amaranthaceae, Convolvulaceae and Verbenaceae) are suggest-ive of mixed farming activities in the area. Some of the plants, such as *Chenopodium, Cucumis, Justicia, Vernonia* and members of the Amaranthaceae, are edible and may have been eaten.

For Kabuye IV (3rd–4th centuries AD), pollen taxa indicate a relatively closed-canopy ecology with well-developed shrub and tree strata; tree taxa comprise 28% of the total pollen spectrum and savanna grass species 52%, while fallow taxa represent only 8% and cereals (cf *Eleusine*) represent 2%. In this landscape, human impact was minor and still sustainable for the environment.

For Kabuye III (5th–6th centuries AD), pollen reveals a significant development of shrubs, grasses and fallow habitats. A more open landscape is indicated by higher Gramineae (40%), Compositae (26%, of which 10% is *Vernonia*) and fallow (>10%) taxa, while trees species account for only 12% of total pollen. Cereals (cf *Eleusine* and *Sorghum*) account for 5% of total pollen. Taken together, palynological data indicate intense environmental degradation caused by human activities. At this time, during the 4th and 5th centuries AD, human impacts may have been exacerbated by drier climates, which would have favoured the extension of open savanna landscapes. Climatic influences on environmental change are suggested by studies of sedimentary sequences recorded in several regions of Rwanda (Roche 1996: 85).

For Kabuye II (6th–7th centuries AD), palynology indicates that the savanna had returned to a more woodland habitat with a resurgence in diversity and abundance (31%) of tree species. Grasses account for 50% of total pollen, but fallow species decline markedly to account for less than 5%. Cereal pollen (cf *Eleusine* and *Sorghum*) represents 5%, as noted for the Kabuye III sample. A moister phase during the seventh century AD was favourable to woodland resurgence.

The Kibuga Section: A Regional View

Environmental degradation is recorded in sediments that accumulated downhill from Kabuye Hill in the Kibuga Valley. The incision of a small river exposed a sedimentary section in which a sandy-peaty level, dated to AD 285 (uncalibrated), is covered by gravel enclosed in coarse sand (Figure 16.6–7). Pollen analysis of the peaty zone gave a spectrum similar to that of Kabuye IV (240–400 cal AD), that is to say, indicative of a woodland savanna environment with forest gallery, subject to some disturbance, but without having reached the level of degradation of Kabuye III (420–600 cal AD) (Figure 16.6).

The coarse sediments on top of this level did not contain any pollen, due to oxidation. By comparison with data in other borings, they are contemporaneous with and postdate Kabuye III (420–600 cal AD) and appear to result from an erosional phase following aggressive deforestation by people on the hills surrounding the valley. For Kabuye III, there is a marked divergence in vegetation types from those anticipated based on climatic influences alone; climates were warm and relatively humid at this time. Environmental degradation resulted from forest clearance for either the production of charcoal for smelting iron (rather than for household fireplaces that were usually fed with brushwood) (Van Grunderbeek et al 1983, 1984) or cereal cultivation (Ehret 1998: 127–30). Climatic cooling and reduced human activity are exhibited in the Kabuye II (560–690 cal AD) pollen spectrum, which favoured the regeneration of the environment from the sixth century AD.

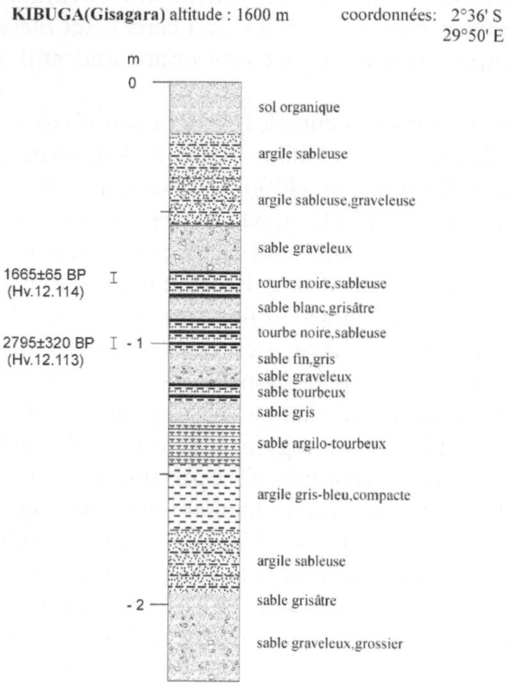

Figure 16.7 Stratigraphy of the Kibuga section.

Gramineae pollen and those revealing postagricultural fallow are present in archaeological sites on Kabuye Hill from the third century AD. However, similar human impacts appear to postdate the upper sample from Kibuga, less than one kilometre from the Kabuye sites. In cores from other valleys in the Hills area (Central Plateau), human influence is only noticeable from the seventh century AD, during an aridity peak that stressed and sensitised vege- tation to human impacts.

Therefore, to recognise the beginning of agricultural activities in an area, pollen analysis should focus on well-preserved and well-provenanced samples from the fills of archaeological features. Off-site records, even a short distance away, may not record localised agricultural activities. Localised agricultural practices may only register in regional records during periods of environmental stress, such as during an arid period. During these times, the impacts of human activities on the environment are magnified and, hence, more discernible in the palynological record.

Linguistic Evidence for Cereals, Cultivation and Cooking

Ehret (1998: 127–30) listed terms in Ntu relevant to cereal culture, together with a term meaning 'millstone'. These seem to be derivative loan words from predecessors in the Great Lakes region. Because he was unable to identify terms for granaries, Ehret presumed that grain initially constituted only a minor component of the diet. Cereals became important later, which is discernible linguistically from a distinct pair of terms adopted by Ntu-speaking peoples to refer to *Sorghum* and *Eleusine*. Schoenbrun (1998: 9) deduced that *Eleusine* cultivation, as well as the Ntu term, were derived from the 'Central Sudanic' linguistic group. Schoenbrun (1998: 72) also recognised in Ntu languages ancient terms meaning 'hoe' or 'blade', which could be interpreted to support arguments for agriculture during the EIA. Additionally, a linguistic study (Bostoen 2003–2004: 289–90, 493) noted the appearance in Ntu languages of a 'roasting bowl' (see below).

A Possible Vegecultural Component: Bananas

Considerations of agriculture should include vegeculture, as well as granoculture. Vegeculture is a kind of agriculture involving the asexual propagation of plants, which may have been practiced before the advent of Ntu-speaking peoples (Ehret 1998: 127–30). Since vegeculture can be practiced amidst existing vegetation without major clearance, its impacts are often less dramatic and less readily identifiable. Consequently, vegecultural practices usually have low environmental impacts and are more difficult to discern from the archaeological and palaeoecological records. Here we consider the potential for cultivation of two genera of banana – *Musa* and *Ensete*.

The timing of the earliest plantain cultivation (*Musa* spp. bananas) is uncertain. Plantain cultivation is considered to be of Asian origin, and we consider it unlikely to have reached Burundi and Rwanda by the EIA. Although *Musa* phytoliths dating to the beginning of the first millennium BC have been found in Cameroon, West Africa (Mbida et al 2001), these findings require confirmation and are not considered here. Linguistic data suggest plantains were introduced to the Great Lakes region within the last 1000 years (Schoenbrun 1997: 74, 80; see De Langhe, chapter 19, for an alternative view).

In contrast, some species of *Ensete*, the so-called 'false banana', are indigenous to Africa. Schoenbrun (1998: 81, note 54) gives an overview of the ritual function *Ensete* had in the ancient kingdoms of the Great Lakes region, traditions that indicate the past importance of this food and a likely long antiquity of use. It is still consumed by some tribes in the Horn of Africa (see Hildebrand, chapter 15). The fruit is inedible, but the trunk is crushed or ground into flour for consumption.

Cattle Teeth: Possible Agro-Pastoral Lifeways

Archaeological evidence indicates Urewe populations of the EIA owned cattle and were familiar with domestication. Of note are cattle teeth, one of which was collected beneath a tuyere of an iron-smelting furnace on Remera Hill (Remera I: Figure 16.4, no. 20), near Butare. The furnace crucible is dated to the 3rd–4th centuries AD (Van Noten 1983: 20, 77, Plates 31–33)[6]. Remera I is located less than 5 km from Kabuye Hill and is contemporary with Kabuye

IV. The molar fragments derive from a young adult, with estimated shoulder height of 110 cm (Van Neer 2000: 169). Animal husbandry, complementing agriculture, is thus a possibility during this period. Schoenbrun (1998: 74) identifies an Ntu term, which may go back to that epoch, meaning 'fresh milk' and suggests cows may have been milked at this time. Ntu terms, potentially dating to the first millennium BC, refer to honey and bee keeping (Ehret 1998: 125–26). Bones of domestic animals, including sheep, goat and chicken, as well as cattle, only appear in LIA sites at Akameru, Cyinkomane and Gisagara (Gautier in Van Noten 1983: 104–20; Van Neer 2000: 169). In the area east of Butare (Gisagara II, III, V and VII), cattle and an ovicaprine were identified in a disturbed LIA deposit with roulette-decorated pottery (Van Neer 2000: 169). Gisagara II was dated to the 11th–12th centuries AD[7].

Ceramics

Due to their fragility, ceramics are often considered indicators of a sedentary lifeway. Pottery of the Urewe culture is small, with a maximum height of about 30 cm; an exception is illustrated by J Hiernaux that measures about 36 cm high (Hiernaux and Maquet 1960: 45, Figure 21). The small size of Urewe pottery could reflect poor quality clays used in their manufacture, usually locally available colluvial clays[8], but could also be a function of use, about which little is known. Presumably, Urewe pots were used to store dry and/or freshly collected starch-rich plant parts, such as peas; hold liquids such as water, beer, milk and honey; be used for cooking; or serve as tableware and kitchenware. The investigation of animal and plant residues within ceramics is urgently needed to address questions of pottery function.

Urewe ceramics take a variety of forms (Van Grunderbeek 1988), but can be categorised as either pots (with a closed mouth and no handle) or bowls (an open form). Sixty percent of the total ceramic assemblage is pots, of which 10% are small or miniature versions. Bowls are 40% of the total ceramic assemblage, most of which are of wide, open type; about 1% is hemispheric and another 1% is subspherical.

The diversity of forms suggests that specialised activities and roles occurred within Urewe cultural life. This kind of differentiation may point to sedentary husbandry activities, perhaps supplemented by agriculture. It is not characteristic of nomadic or transhumant lifeways of the LIA, when roulette-decorated pottery is found in a single, multipurpose form.

Urewe pottery decoration underlines different components of the form: the rim is bevelled, the neck is covered with hatches and the shoulder bears a ribbon-like decoration of parallel incisions (channel) forming linear geometric motifs. Decoration is similar on bowls and pots. Van Grunderbeek (1988: 52) has suggested that the pot predated the bowl, because decorative motifs appear to have been transferred from the former to the latter without adaptation to the formal differences (Van Grunderbeek 1988).

A linguistic study (Bostoen 2003–2004: 289–90, 493) highlights the appearance in Ntu language of a term *-kádangò*, or 'roasting bowl'. It first occurs in the Great Lakes region, after which the topic appears in the east and the southeast along the coast. Archaeological evidence of the (chronologically later) Urewe bowl form and linguistic evidence for a 'roasting bowl', may both indicate the same phenomenon, ie a new 'cooking' habit. The development of a new culinary practice may result from a new lifestyle, or new foods, perhaps the result of agriculture.

Industrial Activity

In his study on EIA in Buhaya, Tanzania, Schmidt (1974) established a link between iron smelting and the cultivation of food. His argument rests on the economic assumption that an industrial activity requires a society to produce a food surplus (Van Grunderbeek et al 1983, 1984). Surpluses support specialised non-food producing activities and are redistributed to craftsmen and/or coworkers. Unlike Egypt and Southwest Asia, where the dawn of the Iron Age is noted through written documents and iron objects, iron metallurgy in Rwanda and Burundi is only evidenced by structures of production, and not by the manufactured objects.

The Exploitation of Inorganic Salt

Kjekshus (1977) mentions the nutritional need for inorganic salt by sedentary farmers reliant on a diet based on cereals. The consumption of cereals causes an imbalance in mineral salts other than potassium. Urewe ceramics have been collected during archaeological surveys at salt deposits in Pwaga and Nyamsunga, in the Uvinza region, Western Tanzania (Phillipson 1970: 5; Sutton and Roberts 1968: 63; Van Grunderbeek et al 2001). A radiocarbon date situates this Urewe presence at between the 3rd and 9th centuries AD. The presence of Urewe pottery at a salt deposit may indicate a nutritional need for inorganic salt, which in turn may circumstantially indicate a diet reliant on cereals.

CONCLUSION

Multidisciplinary research has provided direct and indirect evidence that the Urewe culture was based on mixed farming. The pollen research conducted in Rwanda and Burundi is unusual because it has enabled the results from high-elevation peat deposits to be compared with records from: (1) three Early Iron Age archaeological sites (crucibles of Urewe and non-Urewe iron-smelting furnaces) on Kabuye Hill; (2) an eroded bank along the Kibuga, which is downhill from Kabuye Hill; and, (3) peat deposits of the Hills area (Central Plateau) where, during the Early Iron Age, the Urewe culture developed.

The presence in archaeological sites of Gramineae pollen and post-agricultural fallow are suggestive of agricultural activity (as opposed to other activities) in the Butare area from the third century AD. At other sites in the region, however, a similar agricultural signature is visible during an aridity peak in the seventh century AD. Signatures of human impacts are not present at high-elevation bogs, which provide a record for the last 30,000 years. Consequently, to recognise early agricultural activities in an area, pollen (and other microfossil samples) should be collected from features at archaeological sites, although preservation and taphonomic conditions are rarely propitious. A palynological site even a short distance from an archaeological site may not register similar or synchronous human-induced changes.

Several lines of evidence raise the possibility of agriculture during the Early Iron Age in Rwanda and Burundi. These include:

(1) The presence of cereal pollen (finger millet and sorghum) in samples collected from secure contexts at a series of sequential iron-smelting sites on Kabuye Hill;
(2) The environmental suitability of the Hills area (Central Plateau) for agriculture;
(3) The degradation of the environment due to cultivation and soil exhaustion, illustrated by the presence of fallow Compositae in the pollen assemblages from archaeological sites;
(4) The impact of human activity on the environment by deforestation for iron smelting and cultivation;
(5) A severe erosional phase at the end of the Early Iron Age settlement in this area, interpreted to result from human activities;
(6) Linguistic studies indicating the presence of cereal cultivation and vegeculture;
(7) The presence of cattle teeth indicating a mixed farming (agro-pastoral) way of life;
(8) The presence of ceramics of varying forms, including a new roasting bowl type;
(9) An industrial activity, iron smelting, which ordinarily requires the secure food production afforded by agriculture; and,
(10) The exploitation of inorganic salt to potentially compensate for a dietary insufficiency caused by a dependence on cereals.

The presence of EIA agriculture in Rwanda and Burundi, potentially including mixed farming with cereal cultivation, is important because it contributes to our understanding of a fast and overwhelming colonization, namely the remarkable progression of a new lifeway south of the equator from the first millennium BC to the first millennium AD.

ACKNOWLEDGMENTS

The authors are grateful to Luc Vrydaghs and Tim Denham for their assistance translating and editing this chapter from the French original.

NOTES

1. Glottochronology is a system of estimating the relative age of languages based on lexical variation between parent languages, although it is problematic (Nurse 1997).
2. Kabuye IV (GrN-7905: 1725 ± 30 BP) AD 250–390 (Oxcal 68.2%) (Bronk Ramsey 1995, 2001).
3. Kabuye III (GrN-8219: 1550 ± 30 BP) AD 430–550 (Oxcal 68.2%).

4. Kabuye II (GrN-7904: 1405 ± 35 BP) AD 618–662 (Oxcal 68.2%).
5. Since the 1980s, Eleusine (finger millet) culture is disappearing in Rwanda because savanna fires are now prohibited by law to protect the environment, especially trees. Savanna fires are used by cattle herders to promote new growth at the end of the dry season, when the grass dries up. Two days after burning, green shoots regenerate the grazing lands, supplying fresh forage to the cattle. These savanna fires also burn tree shoots, destroying any tree cover.
6. 78 Remera I, GrN-9663: 1730 ± 30 BP, Oxcal AD 250–390 (68.2%) / AD 240–400 (95.4%).
7. 79 Gisagara II (GrN-9661: 925 ± 30 BP) AD 1030–1160 (Oxcal 68.2%) / AD 1020–1190 (Oxcal 95.4%).
8. The clay soils used today for the manufacture of earthenware were only deposited on the valley floors after an erosional phase initiated by forest clearance and environmental degradation during the Early Iron Age.

REFERENCES

Allan, W (1965) *The African Husbandman*, Edinburgh: Oliver and Boyd

Bostoen, K (2003–2004) 'Étude comparative et historique du vocabulaire relatif à la poterie en Bantou', unpublished PhD thesis, University of Brussels

Bronk Ramsey, C (1995) 'Radiocarbon calibration and analysis of stratigraphy: the OxCal program', *Radiocarbon* 37(2): 425–30

Bronk Ramsey, C (2001) 'Development of the radiocarbon program OxCal', *Radiocarbon* 43(2a): 355–63

Ehret, C (1998) *An African Classical Age: Eastern and Southern Africa in World History 1000 BC to AD 400*, Charlottesville: University Press of Virginia and Oxford: James Currey

Fagan, B (1969) 'Radiocarbon dates for sub-Saharan Africa-VI', *Journal of African History* 10: 149–69

Habiyaremye, F X and Roche, E (2003) 'Incidence anthropique sur le milieu montagnard du graben centrafricain: complément phytodynamique aux interprétations palynologiques', *Geo-Eco-Trop* 27(1–2): 53–62

Harvey, E L and Fuller, D Q (2005) 'Investigating crop-processing using phytolith analysis: the example of rice and millets', *Journal of Archaeological Science* 32: 739–52

Hiernaux, J (1960) Unpublished field notes for excavations undertaken in 1960 at Ndora and Tshiamakuza [Cyamuykuza]; in possession of M-C Van Grunderbeek

Hiernaux, J and Maquet, E (1957) 'Cultures préhistoriques de l'âge des métaux au Ruanda-Urundi et au Kivu (Congo belge), première partie', *Bruxelles: Bulletin de l'Académie Royale des Sciences Coloniales* (new series) 2: 1126–49

Hiernaux, J and Maquet, E (1960) 'Cultures préhistoriques de l'âge des métaux au Ruanda-Urundi et au Kivu (Congo belge), deuxième partie', *Bruxelles: Mémoire de l'Académie Royale des Sciences d'Outre-Mer, Classe des Sciences Naturelles et Médicales* (new series) 10(2): 5–102

INEAC (1963) *Carte des Sols et de la Végétation du Congo, du Rwanda et du Burundi. A. Sols*, Brussels: Institut National pour l'Étude Agronomique du Congo

Kjekshus, H (1977) *Ecology Control and Economic Development in East African History: The Case of Tanganyika 1850–1950*, London: Heinemann

Leakey, M D, Owen, W E and Leakey, L S B (1948) *Dimple-Based Pottery from Central Kavirondo.* Occasional Papers 2, Nairobi: Coryndon Memorial Museum

Lebrun, J (1956) 'La végétation et les territoires botaniques du Ruanda-Burundi', *Bruxelles: Les Naturalistes Belges* special no: 22–48

Lewalle, J (1972) 'Les étages de végétation du Burundi occidental', *Bulletin du Jardin Botanique National Belge* 42: 1–247

Liben, L (1962) *Nature et Origine du Peuplement Végétal (Spermatophytes) des Contrées Montagneuses du Congo Oriental.* Series 2, 15(3), Bruxelles: Académie Royale Belge, Classe des Sciences

Maquet, J J (1957) *Ruanda: Essai Photographique sur une Société Africaine en Transition*, Bruxelles: Elsevier

Mbida, C, Doutrelepont, H, Vrydaghs, L, Swennen, R, Swennen, R, Beeckman, H, De Langhe, E and Maret, P de (2001) 'First archaeological evidence of banana cultivation in central Africa during the third millennium before present', *Vegetation History and Archaeobotany* 10: 1–6

Nenquin, J (1967) 'Contributions to the study of the prehistoric cultures of Rwanda and Burundi', *Musée Royal de l'Afrique Centrale, Tervuren, Belgique: Annales en Sciences Humaines* 59: 257–71

Nurse, D (1997) 'The contributions of linguistics to the study of history in Africa', *Journal of African History* 38: 359–91

Pahaut, P and Van der Ben, D (1962) 'Bassin de la Karuzi: notice explicative de la carte des sols et de la végétation', *Bruxelles, Publications INEAC*: 18–47

Phillipson, D W (1970) 'Notes on the later prehistoric radiocarbon chronology of eastern and southern Africa', *Journal of African History* XI(1): 1–15

Phillipson, D W (1977) *The Later Prehistory of Eastern and Southern Africa*, London: Heinemann

Posnansky, M (1961). 'Pottery types from archaeological sites in East Africa', *Journal of African History* 2(2): 177–98

Posnansky, M (1966) *Prelude to East-African History*, Oxford: Oxford University Press

Posnansky, M (1967) 'The Iron Age in East Africa', in Bishop W and Clark, J D (eds), *Background to Evolution in Africa*, pp 629–48, Chicago: University of Chicago Press

Posnansky, M (1968) 'Bantu genesis – archaeological reflections', *Journal of African History* 9(1): 1–11

Roche, E (1991) 'Evolution des paléoenvironnements en Afrique centrale et orientale au Pléistocène supérieur et à l'Holocène: influences climatiques et anthropiques', *Bulletin de la Société Géographique de Liège* 27: 187–208

Roche, E (1996) 'L'influence anthropique sur l'environnement à l'Âge du Fer dans le Rwanda ancien', *Geo-Eco-Trop* 20(1–4): 73–89

Roche, E (1998) 'Évolution du paléoenvironnement holocène au Rwanda: implications climatiques déduites de l'analyse palynologique de séquences sédimentaires', in Demarée, G, Alexandre, J and De Dapper, M (eds), *Proceedings of the International Conference on Tropical Climatology, Meteorology and Hydrology, Brussels 22–24 May 1996*, pp 108–27, Brussels: Royal Meteorological Institute of Belgium and Royal Academy of Overseas Sciences

Roche, E and Van Grunderbeek, M-C (1987) 'Apports de la palynologie à l'étude du Quaternaire supérieur au Rwanda', *Mémoires et Travaux EPHE* (Montpellier) 17: 111–27

Schmidt, P R (1974) 'An investigation of Early and Late Iron Age cultures through oral tradition and archaeology: an interdisciplinary case study in Buhaya, Tanzania', unpublished PhD thesis, Northwestern University

Schmidt, P R (1997) *Iron Technology in East Africa: Symbolism, Science, and Archaeology*, Bloomington, IN: Indiana University Press and Oxford: James Currey

Schmidt, P R and Childs, S T (1985) 'Innovation and industry during the Early Iron Age in East Africa: the KM2 and KM3 sites of northwest Tanzania', *The African Archaeological Review* 3: 53–94

Schoenbrun, D (1998) *A Green Place, A Good Place: Agrarian Change, Gender, and Social Identity in the Great Lakes Region to the 15th Century*, Portsmouth: Heinemann and Oxford: James Currey

Sirven, P, Gotanegre, J F andPrioul, C (1974) *Géographie du Rwanda*, Brussels: De Boeck

Sutton, J E G and Roberts, A D (1968) 'Uvinza and its salt industry', *Azania* (Nairobi) III: 45–63

Troupin, G (1966) *Étude Phytocénologique du Parc National de l'Akagera et du Rwanda Oriental*, Liége: Institut National de Recherche Scientifique

Van der Velpen, C (1970) *Géographie du Burundi*, Brussels: De Boeck

Van Grunderbeek, M C (1988) 'Essai d'étude typologique de céramique urewe de la région des collines au Burundi et Rwanda', *Azania* (Nairobi) XXIII: 11–55

Van Grunderbeek, M C (1992) 'Essai de délimitation chronologique de l'Âge du Fer Ancien au Burundi, au Rwanda et dans la région des Grands Lacs', *Azania* (Nairobi) XXVII: 53–80

Van Grunderbeek, M C, Roche, E and Doutrelepont, H (1983) *Le Premier Âge du Fer au Rwanda et au Burundi: Archéologie et Environnement*, Butare, Rwanda: Institut National de Recherche Scientifique

Van Grunderbeek, M C, Doutrelepont, H and Roche, E (1984) 'Influence humaine sur le milieu au Rwanda et au Burundi à l'Âge du Fer ancien: apports de la palynologie et de l'étude des charbons de bois', *Revue de Paléobiologie* special issue: 221–29

Van Grunderbeek, M C and Doutrelepont, H (1989) 'Etude de charbons de bois provenant des sites métallurgiques de l'Âge du Fer ancien au Rwanda et au Burundi', in Hackens, T, Munaut, A V and Till, C (eds), PACT: *Révue du groupe européen d'études pour les techniques physiques, chimiques, biologiques et mathématiques appliquées à l'archéologie / Journal of the European Study Group on Physical, Chemical, Mathematical and Biological Techniques Applied to Archaeology* (Strasbourg) 22: 281–95

Van Grunderbeek, M C, Roche, E and Doutrelepont, H (2001) 'Type de fourneau de fonte de fer, associé à la culture urewe (Âge du Fer ancien), au Rwanda et au Burundi', in Descoeudres, J P, Huysecom, E, Serneels, V and Zimmerman, J L (eds), *Mediterranean Archaeology* 14: 271–98

Van Neer, W (2000) 'Domestic animals from archaeological sites in Central and West-Central Africa', in Blench, R M and MacDonald, K C (eds), *The Origins and Development of African Livestock: Archaeology, Genetics, Linguistics and Ethnography*, pp 163–90, London: University College London Press

Van Noten, F (1979) 'The Early Iron Age in the interlacustrine region: the diffusion of iron technology' *Azania* (Nairobi) XIV: 61–80

Van Noten, F (1983) *Histoire Archéologique du Rwanda*, Tervuren, Belgium: Musée Royal de l'Afrique Centrale

CHAPTER 17

The Development of Plant Cultivation in Semi-Arid West Africa

Stefanie Kahlheber and
Katharina Neumann

INTRODUCTION

Semi-arid West Africa[1] is a key area for the understanding of Africa's pathway to agriculture. Classified as a 'non-centre' of plant domestication (Harlan 1971), the vast area between Senegal and Lake Chad is the place of origin of several African crops. In the last decade, a number of systematic archaeobotanical investigations on charred fruits and grains, impressions in ceramics and charcoal from archaeological sites have become available. We present here the major results of our archaeobotanical work in Burkina Faso and Nigeria and additional evidence from Mali, Ghana, Mauritania, Senegal and Cameroon. These data, though still patchy, draw a new outline on the development of sub-Saharan plant food production[2], which is unique and distinctly different from other regions of the world.

While in the pristine agricultural centres of the Near East, China and Mesoamerica plant domestication preceded animal domestication (Smith 1998), in Africa herding of domesticated animals developed first and flourished for several millennia before crop cultivation was adopted (Hassan 2002; MacDonald 2000; Marshall and Hildebrand 2002). In addition, foraging economies had been very successful in the vast grasslands and savannas of the Sahara and sub-Saharan Africa for 2.5 million years, including the Early and Middle Holocene humid period (Neumann 2003, 2005). In contrast to farming, foraging and pastoralism have one important feature in common: the need for mobility. As the development of agriculture is usually closely linked with sedentism, the high mobility of hunter-gatherers and pastoralists in the Sahara and in the adjacent savannas postponed the evolution of farming longer than on other continents.

The palaeoenvironmental background for increasing sedentism and incipient plant cultivation is the abrupt termination of the Holocene African humid period around 3500 cal BC[3] (DeMenocal et al 2000). In the course of a few centuries or even decades, the hitherto complete plant cover of the Sahara was probably replaced by patchy vegetation, restricted to favourable areas, and floristically impoverished (Schulz 1991). The increasing desiccation of the

Sahara culminated in a severe dry spell around 2500 cal BC (Guo et al 2000) when the southern Saharan lakes dried out and human populations were forced to move southwards. During the second half of the third and the second millennium BC, population density increased in the Sahara–Sahel region between 19° N and 14° N (Vernet 2002), and a northern influence in the material culture is traceable farther south, such as in the Méma, Mali (MacDonald et al 2003), in Gajiganna, Nigeria (Breunig and Neumann 2002a), and even as far as Central Ghana (Davies 1980; Flight 1976; Watson 2003). It is in this context that domesticated plants appear for the first time in sub-Saharan Africa.

THE BEGINNING OF PLANT FOOD PRODUCTION

Archaeobotanical Evidence

Unequivocal evidence of plant food production in West Africa is traceable from the second millennium BC[4] onwards. The domesticated form of pearl millet (*Pennisetum glaucum*) appears in sites from the modern southern Sahara to the Sudanian zone (Figure 17.1) and seems to have been the preferred crop of the first West African farmers, although its roles in the economy and archaeological contexts vary considerably from site to site.

In the Sahel of Burkina Faso, sites of the second millennium BC are of small spatial extent and have a rather low amount of cultural remains (Breunig and Neumann 2002a). At Oursi West and Tin Akof, mobile groups cultivated pearl millet on a small scale (Kahlheber 2004; Kahlheber et al 2001; Neumann and Vogelsang 1996; Neumann et al 2001; Vogelsang et al 1999). It is assumed that these communities largely depended on hunting and gathering, and pearl millet was just an additional carbohydrate source in a diversified subsistence. Archaeobotanical evidence of pearl millet consists of charred caryopses (cereal grains) and of imprints in organically tempered potsherds (Figure 17.2). One caryopsis from Tin Akof has been AMS dated to 1035–915 cal BC, but the crop is already present in the first of the three occupational phases beginning around 1800 cal BC (Kühltrunk 2000). Cultivation does not seem to have altered the natural vegetation considerably, as in the charcoal samples from Tin Akof indicators of fallows or typical park savannas are absent (Neumann et al 2001).

For the Mauritanian sites of the Tichitt-Oualata tradition a more intensive cultivation of pearl millet is suggested. From the numerous imprints of domesticated pearl millet in potsherds, one has been AMS dated to 1935–1685 cal BC (Amblard 1996). Indirect evidence comprises architectural features, which have been interpreted as remnants of granaries or walls of field and garden enclosures (Amblard 1996; Amblard-Pison 1999; Munson 1971, 1976). The existence of gardens is corroborated by micropalaeontological analyses of sediments (Person et al 2001). The architectural evidence for cultivation corresponds with the development of larger settlements occupied by a sedentary population. During the height of Tichitt-Oualata around 1600 cal BC its characteristic elements, including domesticated pearl millet, also

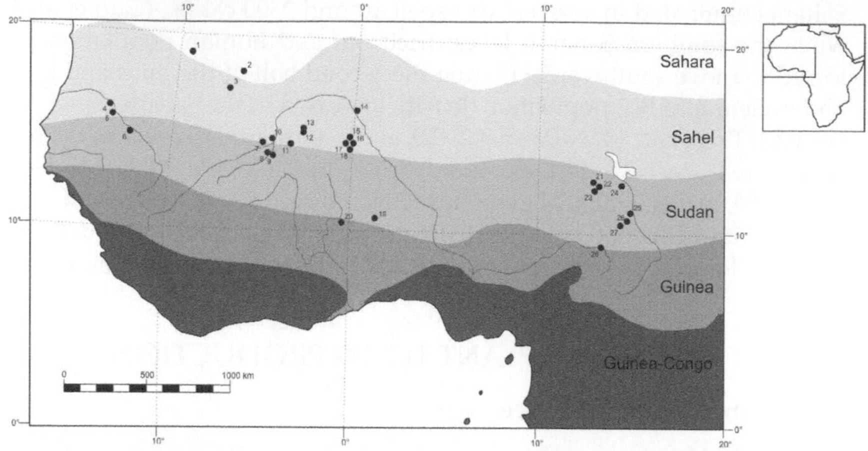

Figure 17.1 Archaeological sites in West Africa with remains of domesticated crops.

	Site/site complex	*Country*
1	Dhar Tichitt	Mauritania
2	Dhar Oulata: Oued Chebbi	Mauritania
3	Dhar Nema: Djiganyai, Bou Khzama	Mauritania
4	Cubalel/Siwré (MSV)	Senegal
5	Sincu Bara (MSV)	Senegal
6	Arondo (USV)	Senegal
7	Dia (NID)	Mali
8	Jenné-Jeno (NID)	Mali
9	Toguéré Doupwil (NID)	Mali
10	Toguéré Galia (NID)	Mali
11	Tellem (Falaise de Bandiagara)	Mali
12	Tongo Maaré Diabel (Gourma)	Mali
13	Windé Koroji Ouest I (Gourma)	Mali
14	Gao Gadei	Mali
15	Tin Akof (BF94/133)	Burkina Faso
16	Kissi (BF96/22, BF97/31, Ki 40)	Burkina Faso
17	Mare d'Oursi area: Corcoba (BF97/5), Oursi West (BF94/45), Oursi Nord (BF97/13), Oursi Ost (BF97/25), Oursi 1 (BF97/26), Oursi 2 (BF97/27), Oursi 3 (BF97/28), Oursi 4 (BF97/29), Oursi hu-beero (BF97/30), Kolèl Nord (BF97/23)	Burkina Faso
18	Saouga area: Saouga A (BF 94/120), Saouga B (BF95/7), Sirkangou (BF96/17)	Burkina Faso
19	Yohongou	Benin
20	Birimi	Ghana
21	Gajiganna (BDC)	Nigeria
22	Zilum (BDC)	Nigeria
23	Dorota, Elkido Nord (BDC)	Nigeria
24	*firki* plain: Daima, Kursakata, Mege,	Nigeria
25	Yaéré: Kayam, Mongossi	Cameroon
26	Diamaré Plain: Balda Tagamré, Goray, Jiddere Saoudjo, Salak, Tchere	Cameroon
27	Piémonts Mandara: Louggéréo, Mowo	Cameroon
28	UBV: Bé, Douloumi, Sumpa	Cameroon

abbreviations:
BDC: Bama Deltaic Complex
MSV: Middle Senegal Valley
NID: Niger Inland Delta
UBV: Upper Benue Valley
USV: Upper Senegal Valley

appear in the more southerly located Dhar Nema (MacDonald et al 2003). Whether, however, the appearance and increasing percentages of pearl millet imprints in the second millennium Tichitt ceramics are an expression of an in situ development (MacDonald et al 2003; Munson 1976) or the tradition arrived with fully developed agro-pastoral practices from elsewhere (Amblard 1996) remains a matter of discussion.

At Birimi, a site of the Kintampo complex[5] in northern Ghana, two samples of domesticated *Pennisetum glaucum* have been directly dated to 1620–795 cal BC and to 1980–1520 cal BC (D'Andrea et al 2001). As pearl millet represents by far the most common species in the archaeobotanical samples, it seems that the subsistence of northern Kintampo groups highly relied on this crop. Cultivation at Birimi might have taken place in a swidden horticulture system, in small plots of relatively uniform stands (D'Andrea and Casey 2002). Pearl millet, a storable resource for sustaining sedentary people through the long dry season, was integrated into existing subsistence systems as an effective adaptation to the seasonal dry climatic regime of the Sudanian woodland savanna.

For the Gajiganna culture in the Chad Basin of Northeast Nigeria, plant impressions in ceramics illustrate the development from a pastoral to an agro-pastoral economy where plant cultivation was added to a diversified subsistence with cattle keeping, fishing and hunting. The first Gajiganna settlers immigrated as environmental refugees from the southern Sahara into the Chad Basin around 1800 cal BC (Breunig 2004; Breunig and Neumann 2002a, 2002b; Breunig et al 1996). They were pastoralists in an environment with permanent lakes and rich wild animal and plant resources (Klee and Zach 1999; Klee et al 2004; Van Neer 2002). Around 1500 cal BC, the settlement pattern changes, and larger sites as well as storage facilities indicate permanent occupation (Breunig and Neumann 2002a,

Figure 17.2 Domesticated pearl millet (*Pennisetum glaucum*): modern cultigen (left) and imprints in potsherds from the site Tin Akof/Burkina Faso (right).

2002b; Gronenborn 1998). Domesticated pearl millet is present in plant impressions from 1300–1100 cal BC onwards, and its proportions increase until it gains absolute dominance around 1000 cal BC. At that time, farther east in the *firki* clay plains, the site Kursakata was occupied by mobile cattle herders and fishers. Some caryopses of domesticated pearl millet have been found in the Late Stone Age layers of the site, but cereal cultivation played a minor role; the use of Paniceae grasses (*Brachiaria* sp., *Digitaria* sp., *Echinochloa* sp.) and wild rice (*Oryza* sp.) was more important (Klee et al 2000; Zach and Klee 2003).

At Windé Koroji Ouest in southern Gourma, Mali, domesticated *Pennisetum glaucum* is also present in the second millennium BC. Settlement mounds point to a sedentary occupation, and pearl millet cultivation is interpreted as part of a diversified subsistence with hunting, fishing, gathering and herding. The charred finds of Windé Koroji Ouest are dated by context as early as 2180–1780 cal BC (Capezza 1997; MacDonald 1996). Ceramic sherds with imprints of pearl millet have also been reported for the site of Karkarichinkat Sud, in the lower Tilemsi Valley, Mali. The original publication of the finds consists of a short, vague comment (Smith 1984) and remains somewhat doubtful. However, as a site with some of the earliest evidence (dated to c 1400 cal BC at the latest) (Smith 1974, 1975) Karkarichinkat Sud has been repeatedly cited in the secondary literature (eg Marshall 1998; Neumann 2003).

Origin and Domestication of Pearl Millet

Until now, our knowledge of the when, where and how of pearl millet domestication has been patchy and to a large degree hypothetical. All existing archaeobotanical assemblages with pearl millet point to an introduction of the domesticated form, either at the beginning of the occupation or in later phases of the sequences. There are single finds of wild pearl millet (*Pennisetum glaucum* ssp. *violaceum*), but no archaeological site provides convincing evidence for its conscious use.

Based on the modern distribution of wild pearl millet Brunken et al (1977), Harlan (1971) and others have placed the domestication area at the southern fringe of the Sahara, in a belt across the continent. More recently, Tostain (1998) suggested from enzyme studies an area in southern Mauritania, Senegal and eastern Mali as a centre of early domestication from where the crop could have spread across the continent. This is in accordance with a cluster of early dates west of Lake Chad and later dates for other parts of Africa (Kahlheber 2004). The nearly contemporaneous appearance of domesticated pearl millet shortly after 2000 cal BC in Mauritania, northern Ghana, Mali and Burkina Faso, ie in different biogeographical zones and in varying subsistence systems, implies that to that point in time it had already spread from its centre of origin. As all these archaeobotanical finds show distinct morphological domestication features (Table 17.2, Figure 17.3), the available chronological data might thus serve as a *terminus ante quem*[6]. The small size of caryopses reported from Birimi, Tin Akof and Kursakata, however, do not necessarily

Table 17.1 Domesticated food crops represented in archaeobotanical assemblages of semi-arid West Africa (cf identification in brackets)

Site	Country	Period* (calibrated)	Reference	*Pennisetum glaucum*	*Sorghum bicolor*	*Oryza glaberrima* /sp.[+]	*Digitaria exilis*	*Eleusine coracana*	*Triticum aestivum*/ *T. turgidum* conv. *durum*	*Vigna subterranea*	*Vigna unguiculata*	*Abelmoschus esculentus*	*Hibiscus sabdariffa*	*Citrullus lanatus*
Yohongou	Benin	AD 650–1200	Petit 2002; Petit et al 2001	x	x	–	–	–	–	–	–	–	–	–
Kissi 22 (BF96/22)	Burkina Faso	AD 80–200	Kahlheber 2004	x	–	–	–	–	–	–	x	–	–	–
Kissi 40		AD 900–1050	Kahlheber 2004	x	–	–	–	–	–	–	x	–	(x)	–
Kissi 40 (BF97/31)		AD 1000–1200	Kahlheber 2004	x	–	–	–	–	–	–	(x)	–	–	–
Kolèl Nord (BF97/23)		AD 900–1200	Kahlheber 2004	x	(x)	–	–	–	–	–	–	–	–	–
Oursi Ost (BF97/25)		AD 1000–1200	Kahlheber 2004	x	–	–	–	–	–	–	–	–	–	–
Oursi Nord (BF97/13)		AD 400–1250	Kahlheber 2004	x	AD 700–800	–	–	–	–	x	x	–	x	AD 1000–1100
Oursi West (BF94/45), Final Stone Age		1200–1100 BC	Kahlheber 2004	x	–	–	–	–	–	–	–	–	–	–
Oursi West (BF94/45), Iron Age		AD 50–250	Kahlheber 2004	x	–	–	–	–	–	x	x	–	x	–
Oursi 1 (BF97/26)		AD 0–250	Kahlheber 2004	x	–	–	–	–	–	–	x	–	–	–
Oursi 2 (BF97/27)		AD 850–1000	Kahlheber 2004	x	–	–	–	–	–	–	x	–	–	–
Oursi 3 (BF97/28)		AD 850–1000	Kahlheber 2004	x	–	–	–	–	–	–	–	–	–	–
Oursi 4 (BF97/29)		AD 850–1000	Kahlheber 2004	x	–	–	–	–	–	–	–	–	–	–
Oursi hu-beero (BF97/30)		AD 1000–1200	Kahlheber 2004	x	x	–	–	–	–	x	x	–	x	–

(Continued)

Table 17.1 (*Continued*)

Site	Country	Period* (calibrated)	Reference	*Pennisetum glaucum*	*Sorghum bicolor*	*Oryza glaberrima* /sp. †	*Digitaria exilis*	*Eleusine coracana*	*Triticum aestivum*/ *T. turgidum* conv. *durum*	*Vigna subterranea*	*Vigna unguiculata*	*Abelmoschus esculentus*	*Hibiscus sabdariffa*	*Citrullus lanatus*
Saouga A (BF 94/120)		AD 750–1150	Kahlheber 2004	x	–	–	–	–	–	x	x	–	(x)	–
Saouga B (BF95/7)		AD 750–1350	Kahlheber 2004	x	–	–	–	–	–	x	x	–	–	–
Sirkangou (BF96/17)		AD 950–1050	Kahlheber 2004	x	x	–	–	–	–	–	–	–	–	–
Tin Akof		1900–900 BC	Kahlheber 2004	x	–	–	–	–	–	–	–	–	–	–
Kayam (Yaéré)	Cameroon	AD 500–1900	Otto 1996	–	x	–	–	–	–	–	–	–	–	–
Mongossi (Yaéré)		AD 400–1650	Marliac 1991; Otto and Delneuf 1998	–	–	–	–	–	–	–	x	–	–	–
Balda Tagamré (Diamaré)		AD 700–1900	Delneuf et al 1998; Otto 1996; Otto and Delneuf 1998	–	x	–	–	–	–	–	x	–	–	–
Goray (Diamaré)		AD 900–1750	Marliac 1991[1]; Otto and Delneuf 1998[2]	x[1]	x[2]	–	–	–	–	–	–	–	–	–
Jiddere Saoudjo (Diamaré)		AD 500–1900	Otto 1996	–	x	–	–	–	–	–	–	–	–	–
Salak (Diamaré)		AD 500–1950	Otto 1998; Otto and Delneuf 1998	–	x	–	–	–	–	–	x	x	x	–
Tchere (Diamaré)		AD 500–1900	Otto 1996	–	x	–	–	–	–	–	–	–	–	–
Louggéréo (Piémonts Mandara)		AD 1300–1600	Delneuf and Otto 1995	–	x	–	–	–	–	–	–	x	x	–
Mowo (Piémonts Mandara)		AD 1600	Delneuf and Otto 1995; Otto and Delneuf 1998	–	x	–	–	–	–	–	x	x	x	–

Bé (Benue)		AD 800–1700	David 1976, 1981	–	x	–	–	x	–	–	–	–	–	–
Douloumi (Benue)		AD 500–1500	David 1976, 1981	–	x	–	–	–	–	–	–	–	–	–
Sumpa (Benue)		AD 1200–1900	David 1976, 1981	–	–	–	–	x	–	–	–	–	–	–
Birimi	Ghana	1750–1100 BC	D'Andrea and Casey 2002; D'Andrea et al 2001	x	–	–	–	–	–	–	–	–	–	–
Jenné-Jeno (NID)	Mali	250 BC– AD 1400	McIntosh 1995	x	x	x	AD 400	–	–	–	–	–	–	AD 1200–1400
Toguéré Doupwil (NID)		AD 1100–1500	Bedaux et al 1978; Lange 1978	x	–	x	x	–	–	–	–	–	–	–
Toguéré Galia (NID)		AD 1150–1600	Bedaux et al 1978; Lange 1978	x	–	x	x	–	–	–	–	–	–	–
Dia (NID)		800 BC– AD 1700	Bedaux et al 2001; Murray 2004, 2005	x	AD 1000–1600	x	–	–	AD 800–1200	–	–	–	–	–
Tellem (Bandiagara)		AD 1000–1600	Bedaux 1972	x	–	–	–	–	–	–	–	–	–	–
Windé Koroji Ouest I (Gourma)		2100–1100 BC	Capezza 1997; MacDonald 1996	x	–	–	–	–	–	–	–	–	–	–
Tongo Maaré Diabel (Gourma)		AD 200–1200	Capezza 1997; MacDonald 1998	x	–	–	–	–	–	–	–	–	–	–
Gao Gadei (Niger Bend)		AD 700–1600	Fuller 2000; Insoll 2000	x	–	x	–	–	–	–	–	–	–	x
Dhar Tichitt	Mauritania	2000–500 BC	Jacques-Félix 1971; Munson 1971, 1976	c 1000 BC	–	–	–	–	–	–	–	–	–	–
Oued Chebbi (Dhar Oulata)		2000 BC–0 BC/AD	Amblard 1996; Amblard and Pernès 1989	x	–	–	–	–	–	–	–	–	–	–
Djiganyai (Dhar Nema)		1650–1500 BC	MacDonald et al 2003	x	–	–	–	–	–	–	–	–	–	–
Bou Khzama (Dhar Nema)		800–400 BC	MacDonald et al 2003	x	–	–	–	–	–	–	–	–	–	–
Gajiganna (BDC)	Nigeria	1800–800 BC	Klee et al 2004	1200–1000 BC	–	x	–	–	–	–	–	–	–	–
Zilum (BDC)		600–400 BC	Magnavita et al 2004	x	–	–	–	–	–	–	x	–	–	–
Dorota (BDC)		AD 450–600	Magnavita 2002; Kahlheber in press	x	x	–	–	–	–	–	–	–	–	–

(Continued)

Table 17.1 *(Continued)*

Site	Country	Period* (calibrated)	Reference	*Pennisetum glaucum*	*Sorghum bicolor*	*Oryza glaberrima* /sp.[†]	*Digitaria exilis*	*Eleusine coracana*	*Triticum aestivum/ T. turgidum* conv. *durum*	*Vigna subterranea*	*Vigna unguiculata*	*Abelmoschus esculentus*	*Hibiscus sabdariffa*	*Citrullus lanatus*
Elkido Nord (BDC)		AD 350–450	Magnavita 2002; Kahlheber in press	x	x	–	–	–	–	–	–	–	–	–
Daima (*firki*)		550 BC– AD 1150	Connah 1981	c AD 1100	c AD 800	–	–	–	–	–	–	–	–	–
Kursakata (*firki*)		1200 BC– AD 150	Klee et al 2000; Zach and Klee 2003	x	–	x	–	(x) c AD 100	–	–	–	c 800 BC	–	–
Mege (*firki*)		850 BC– AD 1983	Klee and Zach 1999; Gronenborn 1998, 2001	x	AD 1400–1600	x	–	–	–	–	–	–	–	–
Cubalel/Siwré (MSV)	Senegal	AD 0–1200	Bocoum and McIntosh 2002	x	–	–	–	–	–	–	–	–	–	–
Sincu Bara (MSV)		AD 0–1200	Bocoum and McIntosh 2002; McIntosh and Bocoum 2000	x	–	–	–	–	–	–	–	–	–	–
Arondo (USV)		AD 400–1000	Gallagher 1999	x	x	–	–	–	–	–	–	–	–	–

For abbreviations see Figure 1.

*Approximate period covered by archaeobotanical remains. For crops that appear later in the sequence, the radiocarbon dates of first evidence are given separately.

[†]Differentiation between grains of domesticated rice, *Oryza glaberrima*, and wild forms is difficult and often not carried out. Therefore, evidence for all *Oryza* species was combined.

indicate an early state in cultivation history, as argued by D'Andrea et al (2001) and Zach and Klee (2003). Though increases in grain size and weight are major and primary objectives in cereal breeding, high rates of cross-pollination and the sympatric distribution of wild and domesticated forms of pearl millet may have prevented rapid selection for larger grains. Instead, improved cultivation methods and a long farming experience are required. It seems that people cultivated first with comparatively simple techniques and consciously selected for special traits only in later periods.

Once domesticated, the physiological characters of pearl millet, particularly high drought resistance and fast maturation, might have favoured its cultivation. In the Sahel, pearl millet is preferably grown on light sandy dune soils, which possess a relatively high water storage capacity, are easy to cultivate and, due to their scattered woody vegetation, unproblematic to clear. All these factors enabled basic pearl millet cultivation with simple tools and limited agricultural experience, even within a mobile way of life.

The First Agricultural Communities of the Second Millennium BC: Cultivation and Risk Minimisation

The available archaeobotanical evidence suggests that plant cultivation in the second millennium BC was integrated into existing pastoral or foraging life-ways as a means for risk minimisation and for greater predictability of resource exploitation. Its development is correlated with large-scale population

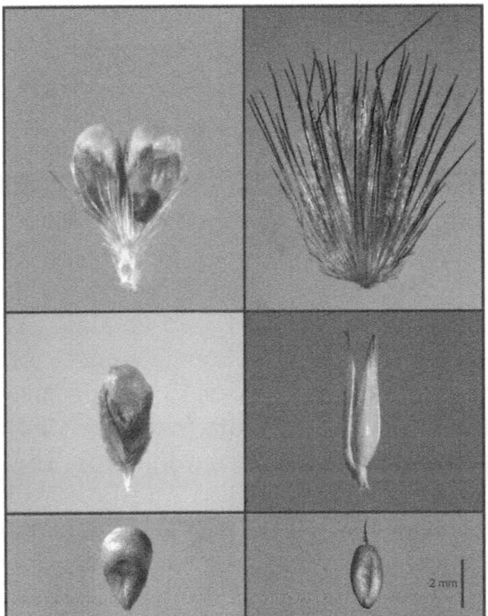

Figure 17.3 Involucres, spikelets and caryopses of modern domesticated pearl millet (*Pennisetum glaucum* ssp. *glaucum*, left) and its wild ancestor (*Pennisetum glaucum* ssp. *violaceum*, right) (scale is valid for both figures).

Table 17.2 Selected morphological characters of wild, weedy and domesticated pearl millet (after Brunken 1977)

	Wild *Pennisetum glaucum* ssp. *violaceum*	Weedy *Pennisetum glaucum* ssp. *sieberianum*	Domesticated *Pennisetum glaucum* ssp. *glaucum*
Involucrum			
callus	existent: involucrum deciduous	existent: involucrum deciduous	missing: involucrum persistent
stalk	short or missing, <0.25 mm long	short to long, 0.2–1.5 mm long	long, 1.1–25 mm long
bristles	thick, dense, longer than spikelet(s)	longer than spikelets	small in number, mostly shorter than spikelets
number of spikelets	1 (–2)	mostly 2	(1–) 2 (–9)
floral bracts	long; tightly enclosing caryopsis	almost or completely enclosing caryopsis	persistent, lush; only partly enclosing caryopsis
length of lemma of upper floret	≥5 mm (5–6.5 mm)	3.5–5.5 mm	<4 mm (1.4–4 mm)
Caryopsis			
shape	elliptic to lanceolate, truncate, dorsally compressed	elliptic to obovate, obtuse to truncate, terete or moderately dorsally compressed	obovate, obtuse to acute, terete, also angular
length	2–3 mm	2–4.5 mm	2–5.5 mm
width	1–1.5 mm	1–2.2 mm	1.6–3.2 mm
depth	0.6–1 mm	1–2 mm	1–3 mm

movements in the southern Sahara and the Sahel and increasing sedentism. Although the archaeological contexts are variable and indicate a wide range of subsistence strategies, they all display monocrop assemblages; pearl millet is the only domesticate reported for this period in semi-arid West Africa. On sites from the Tichitt-Oualata or Gajiganna traditions, where direct archaeobotanical evidence comes exclusively from ceramic impressions, this pattern may be due to the special suitability of pearl millet chaff for tempering. But the data from Tin Akof, Kursakata and Birimi suggest that pearl millet was indeed the only crop. The archaeological contexts of the Sahelian sites in Mali, Nigeria and Burkina Faso show a common pattern: full sedentism was not yet established and plant cultivation probably played a minor role in a mixed economy. Birimi and Tichitt-Oualata are both unique and have no equivalent in the settlement history of semi-arid West Africa. In Birimi, it seems that pearl millet constituted a significant resource for a sedentary population, whereas for the Tichitt-Oualata villages, there is only faint evidence for plant utilisation so far, and the presence of an agro-pastoral economy remains to be confirmed by more archaeobotanical and archaeozoological data.

ADVANCED AGRICULTURE IN THE IRON AGE

The West African Iron Age is chronologically defined from the middle of the first millennium BC to the early second millennium AD (McEachern 1997) and culturally circumscribes a large range of variations. The archaeological defining characteristic of this period – indigenous iron metallurgy – may not have played a significant role in all cultures and the initial appearance of metal technology varies substantially in timing (DeCorse and Spiers 2001). Typical features of the Iron Age are more permanent settlements and the appearance of complex social formations. Groups of large settlement mounds, furnishing long archaeological sequences, are a common Iron Age site type in several West African regions (Breunig and Neumann 2002b; McIntosh 1994).

In conjunction with sedentism and demographic growth, an intensification of food production has been postulated (DeCorse and Spiers 2001). Indeed, domesticated animals such as cattle, sheep and goat were widely raised during the Iron Age, but this had already been the case in former periods (MacDonald and MacDonald 2000). The evidence for the intensification of plant food production is quite clear. Several archaeobotanical assemblages illustrate the development of new agricultural systems distinctly different from final Late Stone Age cultivation practices. They laid the economic base for growing populations, increased division of labour and social stratification, finally resulting in the emergence of the famous West African empires.

Archaeobotanical Evidence: New Crops

The first obvious change at the beginning of the Iron Age is the diversification of crop inventories. In sites of northern Burkina Faso dating shortly after the start of the common era this is manifested in the appearance of *Hibiscus sabdariffa* (roselle) and of the pulses *Vigna subterranea* (Bambara groundnut) and *V. unguiculata* (cowpea; Figure 17.4) (Kahlheber 2004). With their high protein content, pulses are an excellent nutritional complement to a cereal-based carbohydrate diet and, to a certain extent, a replacement for animal proteins. The value of *Hibiscus sabdariffa*, a species of the family Malvaceae, consists in a multiplicity of uses: leaves serve as a vegetable, flower parts are used for drinks rich in vitamins, and the seeds are a source of vegetal fat. New crops were also found in Iron Age sites of northern Cameroon and northeastern Nigeria (Table 17.1; Delneuf and Otto 1995; Klee et al 2000; Otto 1996, 1998; Otto and Delneuf 1998; Zach and Klee 2003). From these, *Abelmoschus esculentus* (okra), whose fleshy fruits are a relished vegetable, is restricted to the broader Lake Chad region. It might well have been domesticated in this area, as *A. ficulneus*, one of its possible ancestors, is distributed there (Burkill 1985–2000).

Noteworthy, pearl millet remained the basic cereal crop and major carbohydrate source throughout the West African Iron Age. In northern Burkina Faso persistent and intensive cultivation led to an alteration of its gene pool, as grain size increased in the course of the period (Kahlheber 2004). The role of the other important African cereal, *Sorghum bicolor*, is not always clear. Sorghum has been reported for a number of West African Iron Age sites, for example in Daima, Nigeria (Connah 1981), northern Benin (Petit et al 2001), northern Cameroon (David 1976; Otto 1996) and Mali (Lange 1978; McIntosh 1995[7] (Table 17.1). Its significance, however, is in many cases difficult to assess either because selective archaeobotanical sampling methods resulted in an under-representation of smaller-grained cereals like pearl millet or because quantification of the finds has not been undertaken. However, some archaeobotanical assemblages clearly demonstrate that *Sorghum bicolor* was of minor importance. In northern Burkina Faso, the appearance of sorghum around AD 700–800 has been interpreted in terms of a local engagement in trade, and the crop might have been an imported luxury food or a remnant of passing caravans (Kahlheber 2004). In other locations the crop also occurs in minor quantities as, for example, at Arondo in the Senegal Valley (Gallagher 1999). Only in the samples from Elkido Nord and Dorota in northeast Nigeria is sorghum clearly more important than pearl millet (Kahlheber in press). With two direct AMS dates of AD 384–426 and AD 449–592, the sorghum finds are among the oldest for West Africa (Magnavita 2002). Farther east in the *firki* clay plains, today the region of dry season sorghum cropping *par excellence*, sorghum actually appeared late, such as in Mege during the 15th–16th centuries (Gronenborn 1998, 2001; Klee and Zach 1999). This late emergence of *Sorghum bicolor* in West Africa and its, for the most part, minor position in Iron Age sites are astonishing, given the fact that sorghum today is widely cultivated in the southern Sahel and Sudanian zones.

Besides *Sorghum bicolor*, a number of other plants enter the archaeobotanical record at the end of the Iron Age, in the West African Middle Age[8]. In most cases they presumably were trade items, though it is not always clear if these species were exclusively traded, or if traders introduced them and they were subsequently adopted for cultivation. Wheat (*Triticum aestivum/T. turgidum* conv. *durum*) found in Dia, Mali, (Murray 2005) and date palm (*Phoenix dactylifera*) reported for Gao Gadei, Mali (Fuller 2000), are examples of such luxurious trade goods. Cotton (*Gossypium* sp.) another such good, was confirmed by seed finds from the 14th–15th centuries from Dia and Gao Gadei as well as by textiles, for instance, from Tellem, Mali, dating between the 11th and 16th centuries AD (Bedaux et al 1991). The data for watermelon (*Citrullus lanatus*) are more ambiguous. Archaeobotanical remains of this species appear in small numbers at the beginning of the second millennium AD in trading centres such as Jenné-Jeno and Gao Gadei (Fuller 2000; McIntosh 1995) as well as in Oursi Nord, Burkina Faso. The status of *Citrullus lanatus* as an imported crop is a matter of debate as the origin and natural distribution of the species is still unknown (Wasylikowa and van der Veen 2004).

From an archaeobotanical perspective, medieval times are not only characterised by agricultural newcomers, but also by the intensified exploitation of wild plants for technological and other uses that can be correlated with crafts and trade. One example is *Acacia nilotica* from Oursi Nord whose pods and seeds serve as a tanning agent in leather fabrication (Kahlheber 2004).

Skins and leather are known to be important items in trans-Saharan trade with North Africa; therefore, the presence of *Acacia nilotica* indicates an intensification of commercial activities.

New Production Systems

The appearance of new crops goes hand-in-hand with the development of advanced cropping systems. For Burkina Faso, mixed systems have been reconstructed (Figure 17.4), with pearl millet as the main crop and *Vigna unguiculata, V. subterranea* and *Hibiscus sabdariffa* as intercrops (Kahlheber 2004). Such systems are commonly seen as a strategy to improve productivity as they allow better use of farming space and a more efficient cultivation schedule during the growing season. A further advantage of mixed cropping is a lower propensity to pests, plant diseases, climatic risks and soil erosion (Franke 1995). The capability of pulses to fix nitrogen permits the exploitation of additional marginal grounds. Soil fertility is maintained for longer and arable land can be cultivated more or less permanently, which complies ideally with the needs of a sedentary population.

More intensive land use was also permitted by the implementation of agroforestry systems, which combine the cultivation of crops and/or livestock breeding with the systematic exploitation of wild woody plants on the same plot (see Boffa 1999). The resulting park savannas are a constituent element of the modern West African landscape (Figure 17.5), and their variable species composition is correlated with different land-use systems (Pélissier 1980;

Figure 17.4 Mixed cropping with pearl millet and cowpea (top) and archaeobotanical finds of cowpea (*Vigna unguiculata*) from Iron Age Oursi hu-beero/Burkina Faso.

Pullan 1974; Seignobos 1982). The archaeobotanical assemblages from Iron Age sites in Burkina Faso illustrate the high diversity of utilised park savanna species, with *Vitellaria paradoxa* (shea butter tree; Figure 17.5), *Sclerocarya birrea* and *Adansonia digitata* (baobab) as the most important species (Table 17.3). Fruits, seeds and leaves were gathered and furnished vegetal fats, vitamins and microelements. Although these useful trees were never domesticated, they were protected during field clearance and their growth was thus greatly encouraged.

More efficient farming practices not only supported the sedentary life of a growing population but increasing yields also allowed the production of a surplus, giving security against the unpredictability of the Sahelian climate. The efficiency of farming systems favoured prosperity and hierarchical structures and laid the economic base for the medieval West African empires.

Environmental Change and the Limits of Intensified Production

The intensification of agricultural production is visible in the archaeobotanical assemblages of settlement mounds in northern Burkina Faso and northern Cameroon. In both areas, shifting cultivation systems and park savannas with shea butter trees (*Vitellaria paradoxa*) became established during the early Iron Age and the natural plant cover gradually changed into anthropic vegetation. The shifting cultivation cycle led to the dominance of woody species (mainly Combretaceae and *Piliostigma* spp.), which can produce suckers after being cut, and are therefore typical for fallows

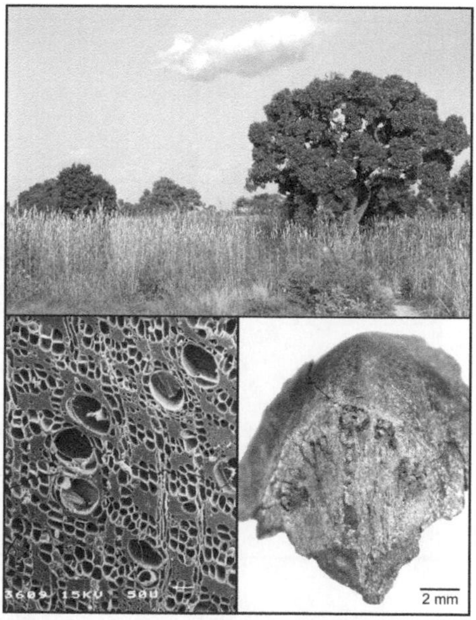

Figure 17.5 Modern park savanna with shea butter trees (*Vitellaria paradoxa*) (top), archaeological charcoal in cross-section (base, left) and a seed fragment of *Vitellaria paradoxa* from Iron Age Saouga/Burkina Faso (base, right).

Table 17.3 Woody plants represented in fruit/seed assemblages of sites in northern Burkina Faso (cf identification in brackets)

phase	site	*Acacia nilotica*	*Acacia* sp.	*Adansonia digitata*	*Balanites aegyptiaca*	*Celtis integrifolia*	*Detarium* cf *microcarpum*	*Diospyros mespiliformis*	*Faidherbia albida*	cf *Ficus* sp.	*Grewia* cf *bicolor*	*Grewia* sp.	*Lannea* sp.	*Parkia biglobosa*	*Sclerocarya birrea*	*Tamarindus indica*	*Vitellaria paradoxa*	*Vitex doniana/simplicifolia*	*Ziziphus mauritiana/ spina-christi*
Final Stone Age	Corcoba (BF97/5)			x														x	
	Tin-Akof (BF94/133)			x	(x)		(x)											x	(x)
	Oursi West (BF94/45)			x	x	x									x				
Iron Age	Corcoba (BF97/5)			x											x			x	
	Kissi 22 (BF96/22)			x	x		x								x		x	x	x
	Kissi 40			(x)	x		x								x			x	x
	Kissi 40 (BF97/31)				x										x				x
	Kolèl Nord (BF97/23)	(x)			x										x				
	Oursi 1 (BF97/26)			x	x	x									x		(x)	x	x
	Oursi 2 (BF97/27)			x		x	(x)								x		x		x
	Oursi 3 (BF97/28)			x			x								x		x		x
	Oursi 4 (BF97/29)	(x)		x			x				x				x		x		x
	Oursi hu-beero (BF97/30)	x			x	x	x					x			x			x	x
	Oursi Nord (BF97/13)	(x)	x	x	x	x	x		x	x	x	x			x		x	x	x
	Oursi Ost (BF97/25)	(x)		x											x				x
	Oursi West (BF94/45)	(x)	x	x	x	x							x		x		x	x	x
	Saouga A (BF94/120)	(x)		x	x	x	(x)	x					x		x		x	x	x
	Saouga B (BF95/7)	(x)	x	x	x	x	(x)	x					x	x	x	x	x	x	x
	Sirkangou (BF96/17)						(x)						(x)		x		x	x	

(Hahn-Hadjali 1998; Renard et al 1993). Between the middle of the first millennium and the fourteenth century AD, pearl millet production intensified in northern Burkina Faso, coupled with livestock keeping, as indicated by the presence of pasture plants (eg *Aristida* sp., *Digitaria* sp., *Eragrostis* sp., *Schoenefeldia gracilis*) and *Faidherbia albida* (Höhn 2002; Höhn et al 2004; Kahlheber 2004; Kahlheber et al 2001). In medieval Saouga, this comes to pass in the course of a short occupation period, which lasted no longer than 100–250 years (Kahlheber 1999; Neumann et al 1998). In Salak, Cameroon, sorghum cultivation with massive clearings on clay soils extended considerably between AD 1000 and 1250 (Otto and Delneuf 1998).

Some communities seem to have reached the limits of agricultural growth. This is illustrated by the archaeobotanical sequence of Oursi Nord, Burkina Faso, ranging chronologically from about AD 400 to 1200 (Kahlheber 2004). While crop plants, especially *Pennisetum glaucum*, constitute a steady proportion of the assemblages, the number of finds and the diversity of fruit trees continuously decrease. At the same time records of pasture plants from more remote habitats increase. This points to an intensification of livestock breeding and to an expansion of cultivated areas, which resulted in shorter fallow phases and soil degradation and could have provoked conflicts between farmers and herders.

In the fourteenth century AD, the settlement mounds of northern Burkina Faso and northern Cameroon were abandoned. Similar developments occurred in other West African regions, for instance the Méma region of Mali (McIntosh 1994; Togola 1993). Climatic deterioration, political instability and warfare are cited as possible reasons for the large-scale population movements (Devisse and Vernet 1993; Marliac 1991; Pelzer et al 2004). However, the land-use conflicts of a growing population might also have initiated and intensified political destabilisation.

THE FIRST MILLENNIUM BC: A TRANSITIONAL PHASE

From the available archaeobotanical data, it becomes clear that the plant food production systems of the second millennium BC and the Iron Age were distinctly different in semi-arid West Africa. This is equally valid for other cultural traits, such as settlement patterns, material culture and social organisation (Breunig and Neumann 2002a, 2002b). Numerous innovations of the Iron Age, especially metallurgy, large centralised settlements and highly productive agriculture, have their roots in the first millennium BC, which thus may be understood as a transitional phase from the final Late Stone Age to the Iron Age. However, the dynamics and innovative processes in this period are not well understood due to a lack of targeted research and the rarity of suitable sites. In Burkina Faso, the settlement sequence shows a gap in the first millennium BC, and occupation becomes archaeologically almost invisible (Breunig and Neumann 2002b). The Tichitt, Gajiganna and Kintampo traditions all ended some time after 1000 cal BC (Breunig and Neumann 2002b; MacDonald et al 2003; Stahl 1993). It seems that living conditions became more difficult for the early agricultural groups at the beginning of the first millennium BC, probably due to an environmental crisis caused by increasing aridity.

On the other hand, increasing aridity opened up new areas to settlement that had been inundated and inaccessible before: the *firki* clay plains of the inner Chad Basin, the Niger Inland Delta, and the Senegal River Valley (Breunig 2004; McIntosh 1999). Here we have the first signs of more diversified agricultural systems. In Dia, Mali, subsistence around 800–400 cal BC was largely based on the cultivation of domesticated rice (*Oryza glaberrima*) (Murray 2004). Though pearl millet was known and wild plants were intensively used (Murray 2005), rice farming must have been a crucial advantage for the colonisation of the wetland habitats of the Niger Inland Delta. In Zilum, Nigeria, in the middle of the first millennium BC, cowpea (*Vigna unguiculata*) was grown beside pearl millet as the staple crop. Zilum belongs to the latest phase of the Gajiganna culture and witnesses an organisational complexity in archaeological structures and material culture (Magnavita et al 2004). Mixed cropping of pulses and cereals presumably contributed to an increase in agricultural productivity that supported the flourishing of this settlement.

The innovations at the transition from the final Late Stone Age to the Iron Age might be seen as adaptations to climatic instability and unpredictable resources (Breunig and Neumann 2002b). At least in some regions, the environmental changes of the first millennium BC seem to have worked as a kind of experimental field that paved the way for more elaborate forms of agricultural and sociopolitical systems, which emerged in the last centuries BC and became well established in the course of the Iron Age.

LOW-LEVEL PLANT FOOD PRODUCTION IN WEST AFRICA

The fact that only small quantities of domesticates occur in archaeobotanical samples from West Africa has sometimes been attributed to preservation, site context or recovery problems. Meanwhile, a number of studies have confirmed that the low diversity of crop plants is an overall characteristic of prehistoric sub-Saharan agricultural systems (Capezza 1997; D'Andrea and Casey 2002; Fuller 2000; Gallagher 1999; Kahlheber 2004; Klee et al 2000, 2004; Lange 1978; Magnavita et al 2004; Murray 2005; Otto 1996; Zach and Klee 2003). In addition, archaeobotanical assemblages with abundant remains of wild plants demonstrate that they played a substantial role in African subsistence throughout the Holocene even in agropastoral societies. The harvest of wild edible grasses, mostly belonging to the Paniceae, has been reported from numerous sites, for example from northeastern Nigeria, Mali and Senegal (Gallagher 1999; Klee et al 2000; McIntosh 1995; Murray 2005). The exploitation of woody plants has also a long tradition, which is impressively illustrated by the rich archaeobotanical diversity of useful trees in final Late Stone Age and Iron Age sites in Burkina Faso (Table 17.3).

Ethnobotanical studies in West Africa have shown that the use of wild plants continues until modern times (eg Burkill 1985–2000). Woody plants are especially appreciated, and there is hardly a tree or shrub in West Africa without economic value. Besides the exploitation of natural populations (eg stands of wild grasses), a number of apparently wild plants undergo cultivation practices such as sowing, selective propagation and protection against predators. Prominent examples are herbaceous species serving as pot-herbs being found in most traditional West African dishes (eg *Amaranthus* spp., *Cleome gynandra*, *Corchorus* spp.). Woody plants in agroforestry systems are not planted, but selectively protected during clearing in the course of shifting cultivation cycles. Classification of these species is a matter of debate, ranging from crop status to intermediary or semi-domesticated to actually wild (eg Garine-Wichatitsky 1997; Harlan 1975). Their distribution might be heavily influenced by humans, as observed with the tree species *Adansonia digitata*, *Vitellaria paradoxa* and *Faidherbia albida* (Boffa 1999; Maranz and Wiesmann 2003; Pélissier 1980; Seignobos 1982). Nevertheless, neither morphological changes nor other domestication features have been observed. Despite their enormous benefits, these species have not been domesticated thus far and are referred to by the term 'Cinderella' species (Leakey and Newton 1994).

CONCLUSION

The earliest archaeobotanical evidence for plant cultivation in West Africa consists of pearl millet finds from Mauritania, Ghana, Burkina Faso, Mali and Nigeria dated shortly after 2000 cal BC. As all these remains of *Pennisetum glaucum* are fully domesticated, the domestication process must have taken place earlier, most probably somewhere in the southwestern Sahara. However, archaeobotanical records from the southern Sahara and the Sahel, which might illustrate the transition from wild to domesticated pearl millet, still remain to be found. Even less is known about the early history of other African crops such as *Vigna subterranea* and the Malvaceae species. *Sorghum bicolor*, although widely grown today, did not appear until Iron Age times in West Africa and was mostly of minor importance. There are a number of early records for wild sorghum in Northeast Nigeria as well as in other parts of the continent (eg Fahmy 2001; Klee et al 2000, 2004; Wasylikowa 1997). Nevertheless, domesticated sorghum was slowly adopted and turns up as late as other Iron Age crops, but does not predate the first millennium BC anywhere on the African continent (Fuller 2004). Connah (1985) and Sutton (2004) have postulated that the spread of sorghum is associated with the establishment of iron metallurgy. Only with adequate tools was it feasible to till heavy clays, which are still the preferred soils for sorghum cultivation and a substratum to which sorghum is better adapted than pearl millet. However, other Iron Age changes in plant food production, such as the development of mixed cropping, cannot be correlated with the technical advantages of metal use.

Two stages of agricultural development can be distinguished, which roughly parallel the archaeological phases of the final Late Stone Age (= West African Neolithic) and the Iron Age (Breunig and Neumann 2002b; Neumann 2003). From 2000 cal BC, Late Stone Age communities integrated pearl millet cultivation into diversified subsistence systems with hunting, gathering, fishing and pastoralism. These cultures, which might be called 'Neolithic' in a strict sense of the term (McIntosh 2001), flourished in the second millennium BC. The 'Neolithic' traditions ended during the transitional phase in the first millennium BC, most probably as a result of climatic change. The following centuries were a time of transformations and innovations, which remain poorly understood until now. People immigrated into the floodplains of the middle Niger, the Chad Basin and the Senegal Valley, and metallurgy and larger centralised settlements emerged. In the Inland Niger Delta, domesticated African rice (*Oryza glaberrima*) was cultivated, and the appearance of cowpea (*Vigna unguiculata*) in the Nigerian Chad Basin indicates the development of intensified farming. In the Iron Age, a fully agricultural economy became the dominant way of life in semi-arid West Africa. People cultivated pearl millet together with pulses and other crops in mixed systems and practised agroforestry to increase productivity and to satisfy the needs of a growing sedentary population.

Apart from pearl millet and other crops, the archaeobotanical record of the last 4000 years in West Africa has yielded numerous remains of wild plants, for example from fruit trees or edible grasses. The intensive use of wild plants and semi-domesticates was an important element of African subsistence throughout all periods and continues to be successful until modern times. Thus, West Africa can be regarded as a living laboratory for the understanding of agricultural origins. In diachronic as well as synchronic perspective, it presents numerous examples of the 'middle ground' between pure hunter-gatherers and agriculturalists largely dependent on domesticated crops. This transition, termed 'low-level food production', or 'wild plant food production' (Harris 1989, 1996; Smith 2001), has often been neglected in the archaeological literature, as it is difficult to reconstruct for prehistoric times. According to the current archaeobotanical data, many West African final Late Stone Age and even some Iron Age cultivators cannot be called 'farmers', but they practiced plant cultivation as one option in a diversified subsistence strategy. It is conceivable that some low-level food production developed long before domesticates appear in the archaeological record. Tracing and reconstructing these subsistence systems is one of the great challenges for African archaeobotany in the future.

ACKNOWLEDGMENTS

Thanks are due to Monika Heckner and Barbara Voss for graphical support, Peter Breunig, Barbara Eichhorn and Carlos Magnavita for useful critical remarks and Richard Byer for language editing. We benefited also from the editorial reviews of Luc Vrydaghs and Tim Denham. The research in Burkina

Faso and Nigeria has been largely funded by the German Research Foundation (DFG).

NOTES

1. Semi-arid West Africa is defined here as the area between the 100 mm and 1000 mm rainfall isohyets, coinciding in phytogeographical terms with the Sahelian and Sudanian vegetation zones.
2. Our terminology of plant food production as opposed to foraging follows Harris (1989, 1996). Cultivation is the sowing or transplanting of wild or domesticated plants. Domestication implies morphological and genetic changes of plants that occur under cultivation and usually result in their inability to reproduce in the wild. Farming/agriculture is largely based on the cultivation of domesticated plants.
3. The discussion on agricultural origins is hampered by the different timescales used in archaeological and palaeoenvironmental studies. We use here the conventional archaeological scale with ages indicated in calendar years, ie calibrated years BC/AD. Uncalibrated radiocarbon dates have been calibrated with Oxcal (Bronk Ramsey 1995, 2001).
4. The terminology for the period of early food production in Africa is not uniform and often contradictory. For the second millennium BC in West Africa the terms 'West African Neolithic' (McIntosh 2001) or 'Final Stone Age' (Breunig and Neumann 2002a) have been proposed, among others.
5. Seed finds of cowpea (*Vigna unguiculata*) from the Kintampo sites farther south in Central Ghana are dated as early as the second millennium BC and are currently under study (D'Andrea pers comm; Stahl 1985). It is still unknown if and how the development of plant food production at the more humid forest-savanna margins was related to final Late Stone Age pearl millet cultivation in semi-arid West Africa (see Neumann 2003).
6. Pearl millet finds in India, dating as early as 2500–2300 BC, provide another chronological marker (Fuller 2003; Fuller et al 2004). As the crop is not native to India, this probably indicates an earlier domestication in Africa.
7. Rare finds of Jenné-Jeno, Mali, attributed to the initial phase of the settlement around 250 BC– AD 400 (McIntosh 1995) urgently need confirmation concerning their age.
8. The West African Middle Age is the period of urban settlements, increasing Arabic influence and the rise of West African empires, starting around the eighth century (Huysecom 1987).

REFERENCES

Amblard, S (1996) 'Agricultural evidence and its interpretation on the Dhars Tichitt and Oualata, south eastern Mauritania', in Pwiti, G and Soper, R (eds), *Aspects of African Archaeology: Papers from the 10th Congress of the Pan African Association for Prehistory and Related Studies*, pp 421–27, Harare: University of Harare Publishers

Amblard-Pison, S (1999) 'Communautés villageoises néolithiques des Dhars Tichitt et Oulata (Mauretanie)', unpublished PhD thesis, Paris I–Panthéon Sorbonne

Amblard, S and Pernès, J (1989) 'The identification of cultivated millet (*Pennisetum*) amongst plant impressions on pottery from Oued Chebbi (Dhar Oualata, Mauritania)', *African Archaeological Review* 7: 117–26

Bedaux, R M A (1972) 'Tellem, reconnaissance archéologique d'une culture de l'Ouest africain au moyen âge: recherches architectoniques', *Journal de la Société des Africanistes* 42(2): 103–85

Bedaux, R M A, Constandse-Westermann, T S, Hacquebord, L, Lange, A G and van der Waals, J D (1978) 'Recherches archéologiques dans le Delta Intérieur du Niger (Mali),' *Extrait de Palaeohistoria* 20: 91–220

Bedaux, R M A, Bolland, R and Boser-Sarivaxévanis, R (1991) 'Les textiles Tellem du Mali: aperçu des résultats des recherches', *Bulletin de l'Institut Royal Tropical* 324: 1–31

Bedaux, R, MacDonald, K, Person, A, Polet, J, Sanogo, K, Schmidt, A and Sidibe, S (2001) 'The Dia archaeological project: rescuing cultural heritage in the Inland Niger Delta (Mali)', *Antiquity* 75: 837–48

Bocoum, H and McIntosh, S K (2002) *Fouilles à Sincu Bara, Moyenne Vallée du Sénégal*, Nouakchott, Dakar: CRIAA, IFAN, CAD

Boffa, J-M (1999) *Agroforestry Parklands in Sub-Saharan Africa*. FAO Conservation Guide 34, Ottawa: Renouf

Breunig, P (2004) 'Environmental instability and cultural change in the later prehistory of the Chad Basin', in Krings, M and Platte, E (eds), *Living with the Lake – Perspectives on Culture, Economy and History of Lake Chad*, pp 52–72. 121 Studien zur Kulturkunde, Cologne: Franz Steiner

Breunig, P and Neumann, K (2002a) 'From hunters and gatherers to food producers: new archaeological and archaeobotanical evidence from the West African Sahel', in Hassan, F A (ed), *Droughts, Food and Culture: Ecological Change and Food Security in Africa's Later Prehistory*, pp 123–55, New York: Kluwer/Plenum

Breunig, P and Neumann, K (2002b) 'Continuity or discontinuity? The 1st millennium BC crisis in West African prehistory', in Lenssen-Erz, T et al (eds), *Tides of the Desert: Contributions to the Archaeology and Environmental History of Africa in Honour of Rudolf Kuper*, pp 491–505, Cologne: Heinrich Barth Institute

Breunig, P, Neumann, K and van Neer, W (1996) 'New research on the Holocene settlement and environment of the Chad Basin in Nigeria', *African Archaeological Review* 13(2): 111–45

Bronk Ramsey, C (1995) 'Radiocarbon calibration and analysis of stratigraphy: the OxCal program', *Radiocarbon* 37(2): 425–30

Bronk Ramsey, C (2001) 'Development of the radiocarbon program OxCal', *Radiocarbon* 43(2a): 355–63

Brunken, J (1977) 'A systematic study of Pennisetum sect. Pennisetum (Gramineae)', *American Journal of Botany* 64(2): 161–76

Brunken, J, de Wet, J M J and Harlan, J R (1977) 'The morphology and domestication of pearl millet', *Economic Botany* 31: 163–74

Burkill, H M (1985–2000) *The Useful Plants of West Tropical Africa*, vols 1–5, Kew: Royal Botanic Gardens

Capezza, C (1997) 'The plant remains', in MacDonald, K (ed), *Final Report on the 1995/1996 Southern Gourma Field Season Institute of Archaeology*, pp 1–7, London: University College London, Institute of Archaeology

Connah, G (1981) *Three Thousand Years in Africa: Man and His Environment in the Lake Chad Region of Nigeria*, Cambridge: Cambridge University Press

Connah, G (1985) 'Agricultural intensification and sedentism in the firki of NE Nigeria', in Farrington, I S (ed), *Prehistoric Intensive Agriculture in the Tropics*, pp 765–85. International Series 232, part II, Oxford: British Archaeological Reports

D'Andrea, A C and Casey, J (2002) 'Pearl millet and Kintampo subsistence', *African Archaeological Review* 19(3): 147–73

D'Andrea, A C, Klee, M and Casey, J (2001) 'Archaeobotanical evidence for pearl millet (*Pennisetum glaucum*) in sub-Saharan West Africa', *Antiquity* 75: 341–48

David, N (1976) 'History of crops and peoples in North Cameroon to AD 1900', in Harlan, J R, de Wet, J M and Stemler, A B L (eds), *Origins of African Plant Domestication*, pp 223–67, The Hague: Mouton

David, N (1981) 'The archaeological background of Cameroonian history', in Tardits, C (ed), *Contribution de la Recherche Ethnologique à l'Histoire des Civilisations du Cameroun*, pp 79–98. Colloques Internationaux du CNRS 551, Paris: Ed CNRS

Davies, O (1980) 'The Ntereso culture in Ghana', in Swartz, B K and Dumett, R (eds), *West African Culture Dynamics: Archaeological and Historical Perspectives*, pp 205–25, The Hague: Mouton

DeCorse, C and Spiers, S (2001) 'West African Iron Age', in Peregrine, P N and Ember, M (eds), *Encyclopedia of Prehistory*, vol 1: *Africa*, pp 313–18, New York: Kluwer/Plenum

Delneuf, M and Otto, T (1995) 'L'environnement et les usages alimentaires en vigueur à l'époque protohistorique dans l'extrême-nord du Cameroun', in Marliac, A (ed), *Milieux, Sociétés et*

Archéologues, pp 213–26, Paris: Karthala, Office de la Recherche Scientifique et Technique de Outre-Mer (ORSTOM)

Delneuf, M, Essomba, J-M and Froment, A (1998) (eds) *Paléo-Anthropologie en Afrique Centrale: Un Bilan de l'Archéologie au Cameroun*, Paris: L'Harmattan

DeMenocal, P, Ortiz, J, Guilderson, T, Adkins, J, Sarnthein, M, Baker, L and Yarusinsky, M (2000) 'Abrupt onset and termination of the African humid period: rapid climate responses to gradual insolation forcing', *Quaternary Sciences Review* 19: 347–61

Devisse, J and Vernet, R (1993) 'Le bassin des vallée du Niger: chronologie et espaces', in *Vallées du Niger*, pp 11–37, Paris: Editions de la Réunion des Musées Nationaux

Fahmy, A G-E (2001) 'Palaeoethnobotanical studies of the Neolithic settlement in Hidden Valley, Farafra Oasis, Egypt', *Vegetation History and Archaeobotany* 10: 235–46

Flight, C (1976) 'The Kintampo culture and its place in the economic prehistory of West Africa', in Harlan, J R, de Wet, J M and Stemler, A B L (eds), *Origins of African Plant Domestication*, pp 211–21, The Hague: Mouton

Franke, G (1995) (ed) *Nutzpflanzen der Tropen und Subtropen*, 1: *Allgemeiner Pflanzenbau*, Stuttgart: Ulmer

Fuller, D Q (2000) 'The botanical remains', in Insoll, T (ed), *Urbanism, Archaeology and Trade: Further Observations on the Gao Region (Mali), The 1996 Field Season Results*, pp 28–35. British Archaeological Reports International Series 829, Oxford: Archaeopress

Fuller, D Q (2003) 'African crops in prehistoric South Asia: a critical review', in Neumann, K, Butler, A and Kahlheber, S (eds), *Fuel, Food and Fields: Progress in African Archaeobotany*, pp 239–71, Cologne: Heinrich Barth Institute

Fuller, D Q (2004) 'Early Kushite agriculture: archaeobotanical evidence from Kawa', *Sudan and Nubia Bulletin* 8: 70–4

Fuller, D Q, Korisettar, R, Venkatasubbaiah, P C and Jones, M K (2004) 'Early plant domestications in southern India: some preliminary archaeobotanical results', *Vegetation History and Archaeobotany* 13: 115–29

Gallagher, D (1999) 'Analysis of seeds and fruits from the sites of Arondo and Ft Senedebu, Senegal', unpublished senior thesis, Rice University (Houston)

Garine-Wichatitsky, E (1997) 'Sauvage ou domestique? Remarques sur l'inventaire des plantes à brèdes chez les Gimbe et les Duupa du Nord-Cameroun', in Barreteau, D, Dognin, R and Graffenried, C von (eds), *L'Homme et le Milieu Végétal dans le Bassin du Lac Tchad*, pp 311–26, Paris: Office de la Recherche Scientifique et Technique de Outre-Mer (ORSTOM) and Institut de Recherche pour le Développement (IRD)

Gronenborn, D (1998) 'Archaeological and ethnohistorical investigations along the southern fringes of Lake Chad, 1993–1996', *African Archaeological Review* 15(4): 225–59

Gronenborn, D (2001) 'Masakwa in the Chad Basin: an examination of the archaeological and historical sources', in Kahlheber, S and Neumann, K (eds), *Man and Environment in the West African Sahel: An Interdisciplinary Approach*, pp 73–84. Berichte des Sonderforschungsbereichs 268, Bd 17, Frankfurt: SFB 268

Guo, Z, Petit-Maire, N and Kröpelin, S (2000) 'Holocene non-orbital climatic events in present day arid areas of northern Africa and China', *Global and Planetary Change* 26: 97–103

Hahn-Hadjali, K (1998) 'Les groupements végétaux des savanes du sud-est du Burkina Faso (Afrique de l'ouest)', *Etudes sur la Flore et la Végétation du Burkina Faso et des Pays Avoisinants* 3: 3–79

Harlan, J R (1971) 'Agricultural origins: centers and non-centers', *Science* 174: 468–74

Harlan, J R (1975) *Crops and Man*, Madison, WI: American Society of Agronomy, Crop Science Society of America

Harris, D R (1989) 'An evolutionary continuum of people-plant interaction', in Harris, D R and Hillman, G C (eds), *Foraging and Farming: The Evolution of Plant Exploitation*, pp 11–26, London: Unwin Hyman

Harris, D R (1996) 'Domesticatory relationships of people, plants and animals', in Ellen, R and Fukui, K (eds), *Redefining Nature: Ecology, Culture and Domestication*, pp 437–63, Oxford: Berg

Hassan, F A (2002) 'Ecological changes and food security in the later prehistory of North Africa: looking forward', in Hassan, F A (ed), *Droughts, Food and Culture: Ecological Change and Food Security in Africa's Later Prehistory*, pp 321–33, New York: Kluwer/Plenum

Höhn, A (2002) 'Vegetation changes in the Sahel of Burkina Faso (West Africa) – analysis of charcoal from the Iron Age sites Oursi and Oursi-village', in Thiébault, S (ed), *Charcoal Analysis: Methodological Approaches, Palaeoecological Results and Wood Uses*, pp 133–39. British Archaeological Reports International Series 1063, Oxford: Archaeopress

Höhn, A, Kahlheber, S and Hallier-von Czerniewicz, M (2004) 'Den frühen Bauern auf der Spur – Siedlungs- und Vegetationsgeschichte der Region Oursi Burkina Faso', in Albert, K-D, Löhr, D and Neumann, K (eds), *Mensch und Natur in Westafrika – Ergebnisse aus dem Sonderforschungsbereich 'Kulturentwicklung und Sprachgeschichte im Naturraum Westafrikanische Savanne'*, pp 221–55, Weinheim: Wiley, VCH

Huysecom, E (1987) *Die Archäologische Forschung in Westafrika*. Materialien zur Allgemeinen und Vergleichenden Archäologie 33(1), Munich: Beck

Insoll, T (2000) *Urbanism, Archaeology and Trade: Further Observations on the Gao Region (Mali): The 1996 Fieldseason Results*. British Archaeological Reports International Series 829, Oxford: Archaeopress

Jacques-Félix, H (1971) 'Grain impressions', in Munson, P J, 'The Tichitt tradition: a late prehistoric occupation of the southwestern Sahara', pp 355–61, unpublished PhD thesis, University of Illinois

Kahlheber, S (1999) 'Indications for agroforestry: archaeobotanical remains of crops and woody plants from medieval Saouga, Burkina Faso', in van der Veen, M (ed), *The Exploitation of Plant Resources in Ancient Africa*, pp 89–100, New York: Kluwer/Plenum

Kahlheber, S (2004) 'Perlhirse und Baobab – Archäobotanische Untersuchungen im Norden Burkina Fasos', PhD thesis, J W Goethe-Universität Frankfurt (http://publikationen.ub.uni-frankfurt.de/volltexte/2005/561)

Kahlheber, S (in press) 'Archaeobotanical remains of Early Iron Age sites in the Bama deltaic complex of NE Nigeria', *Journal of African Archaeology*

Kahlheber, S, Albert, K-D and Höhn, A (2001) 'A contribution to the palaeoenvironment of the archaeological site Oursi in North Burkina Faso', in Kahlheber, S and Neumann, K (eds), *Man and Environment in the West African Sahel: An Interdisciplinary Approach*, pp 145–59. Berichte des Sonderforschungsbereichs 268 Bd 17, Frankfurt: SFB 268

Klee, M and Zach, B (1999) 'The exploitation of wild and domesticated food plants at settlement mounds in North East Nigeria (1800 BC to today)', in van der Veen, M (ed), *The Exploitation of Plant Resources in Ancient Africa*, pp 81–88, New York: Kluwer/Plenum

Klee, M, Zach, B and Neumann, K (2000) 'Four thousand years of plant exploitation in the Chad Basin of northeast Nigeria I: the archaeobotany of Kursakata', *Vegetation History and Archaeobotany* 69: 223–237

Klee, M, Zach, B and Stika, H P (2004) 'Four thousand years of plant exploitation in the Lake Chad Basin (Nigeria), part III: plant impressions in potsherds from the Final Stone Age Gajiganna culture', *Vegetation History and Archaeobotany* 13: 131–42

Kühltrunk, P (2000) 'Taphonomische und typologische Keramikanalyse des steinzeitlichen Fundplatzes Tin-Akoff im Norden Burkina Fasos', unpublished MA thesis, J W Goethe-Universität Frankfurt

Lange, A G (1978) 'Paléo-Ethnobotanique', in Bedaux, R M A, Constandse-Westermann, T S, Hacquebord, L, Lange, A G and van der Waals, J D (eds), Recherches Archéologiques dans le Delta Intérieur du Niger (Mali), *Retrait de Palaeohistoria* 20: 170–80

Leakey, R R B and Newton, A C (1994) 'Domestication of "Cinderella" species as the start of a woody-plant revolution', in Leakey, R R B and Newton, A C (eds), *Tropical Trees: The Potential for Domestication and the Rebuilding of Forest Resources*, pp 3–6, London: HMSO

MacDonald, K C (1996) 'The Windé Koroji Complex: evidence for the peopling of the eastern Inland Niger Delta (2100–500 BC)', *Préhistoire et Anthropologie Méditerranéennes* 5: 147–65

MacDonald, K C (1998) 'More forgotten tells of Mali: an archaeologist's journey from here to Timbuktu', *Archaeology International* 1: 40–42

MacDonald, K C (2000) 'The origins of African livestock: indigenous or imported?', in Blench, R M and MacDonald, K C (eds), *The Origins and Development of African Livestock: Archaeology, Genetics, Linguistics and Ethnography*, pp 2–17, London: University College London

MacDonald, K C and MacDonald, R H (2000) 'The origins and the development of domesticated animals in arid West Africa', in Blench, R M and MacDonald, K C (eds), *The Origins and*

Development of African Livestock: Archaeology, Genetics, Linguistics and Ethnography, pp 127–62, London: University College London

MacDonald, K, Vernet, R, Fuller, D Q and Woodhouse, J (2003) 'New light on the Tichitt tradition: a preliminary report on survey and excavation at Dhar Nema', in Mitchell, P, Haour, A and Hobart, J (eds), *Researching Africa's Past: New Contributions from British Archaeologists*, pp 73–80. Oxford University School of Archaeology Monograph no 57, Oxford: School of Archaeology, Oxford University

Magnavita, C (2002) 'Recent archaeological finds of domesticated *Sorghum bicolor* in the Lake Chad region', *Nyame Akuma* (Calgary) 57(1): 14–20

Magnavita, C, Kahlheber, S and Eichhorn, B (2004) 'The rise of organisational complexity in mid-first millennium BC Chad Basin', *Antiquity* 78(301) (http://antiquity.ac.uk/ProjGall/magnavita/index.html)

Maranz, S and Wiesmann, Z (2003) 'Evidence for indigenous selection and distribution of the shea tree, *Vitellaria paradoxa*, and its potential significance for prevailing parkland savanna tree patterns in sub-Saharan Africa north of the equator', *Journal of Biogeography* 30: 1505–16

Marliac, A (1991) *De la Préhistoire à l'Histoire du Cameroun Septentrional*, Paris: Office de la Recherche Scientifique et Technique de Outre-Mer (ORSTOM)

Marshall, F (1998) 'Early food production in Africa,' *Review of Archaeology* 19(2): 47–58

Marshall, F and Hildebrand, E (2002) 'Cattle before crops: the beginnings of food production in Africa', *Journal of World Prehistory* 16(2): 99–143

McEachern, S (1997) 'Western African Iron Age', in Vogel, J O (ed), *Encyclopedia of Precolonial Africa*, pp 425–29, Walnut Creek, CA: Altamira Press

McIntosh, S K (1994) 'Changing perceptions of West Africa's past: archaeological research since 1988', *Journal of Archaeological Research* 2(2): 165–97

McIntosh, S K (1995) *Excavations at Jenné-Jeno, Hambarketolo, and Kaniana (Inland Niger Delta, Mali), the 1981 Season*, Berkeley: University of California Publications in Anthropology

McIntosh, S K (1999) 'A tale of two floodplains: comparative perspectives on the emergence of complex societies and urbanism in the Middle Niger and Senegal valleys', in Sinclair, P (ed), *The Development of Urbanism from a Global Perspective*, Uppsala: Uppsala Universitet (www.arkeologi.uu.se/afr/projects/BOOK/Mcintosh/mcintosh.htm)

McIntosh, S K (2001) 'West African Neolithic', in Peregrine, P N and Ember, M (eds), *Encyclopedia of Prehistory*, vol 1: *Africa*, pp 323–38, New York: Kluwer/Plenum

McIntosh, S K and Bocoum, H (2000) 'New perspectives on Sincu Bara, a first millennium site in the Senegal Valley', *African Archaeological Review* 17(1): 124–78

Munson, P J (1971) 'The Tichitt tradition: a late prehistoric occupation of the southwestern Sahara', unpublished PhD thesis, University of Illinois

Munson, P J (1976) 'Archaeological data on the origins of cultivation in the southwestern Sahara and their implications for West Africa', in Harlan, J R, de Wet, J M and Stemler, A B L (eds), *Origins of African Plant Domestication*, pp 187–210, The Hague: Mouton

Murray, S S (2004) 'Searching for the origins of African rice domestication', *Antiquity* 78(300) (http://antiquity.ac.uk/ProjGall/murray)

Murray, S S (2005) '*Recherches archéobotaniques*', in Bedaux, R M A, Polet, J, Sanogo, K and Schmidt, A (eds), *Recherches Archéologiques à Dia dans le Delta Intérieur du Niger (Mali): Bilan des Saisons de Fouilles 1998–2003*, pp 386–400. Leiden: Mededelingen van het Rijksmuseum voor Volkenkunde

Neumann, K (2003) 'The late emergence of agriculture in sub-Saharan Africa: archaeobotanical evidence and ecological considerations', in Neumann, K, Butler, A and Kahlheber, S (eds), *Fuel, Food and Fields – Progress in African Archaeobotany*, pp 71–92, Cologne: Heinrich Barth Institute

Neumann, K (2005) 'The romance of farming – plant cultivation and domestication in Africa', in Stahl, A B (ed), *African Archaeology: A Critical Introduction*, pp 249–75, Malden, MA: Blackwell

Neumann, K and Vogelsang, R (1996) 'Paléoenvironnement et préhistoire au Sahel du Burkina Faso', *Berichte des Sonderforschungsbereichs 268* 7: 177–86

Neumann, K, Kahlheber, S and Uebel, D (1998) 'Remains of woody plants from Saouga, a medieval West African village', *Vegetation History and Archaeobotany* 7: 57–77

Neumann, K, Breunig, P and Kahlheber, S (2001) 'Early food production in the Sahel of Burkina Faso', *Berichte des Sonderforschungsbereichs 268* 14: 327–34

Otto, T (1996) *Phyto-Archéologie de Sites Archéologiques de l'Age du Fer du Diamaré, Nord du Cameroun: le Site de Salak. Etude de Bois et de Graines Carbonisées*, thèses et documents microfichées 151, Paris: Office de la Recherche Scientifique et Technique de Outre-Mer (ORSTOM)

Otto, T (1998) 'Essai sur l'histoire du paysage au Diamaré pour les deux derniers millénaires', in Delneuf, M, Essomba, J-M and Froment, A (eds), *Paléo-Anthropologie en Afrique Centrale: Un Bilan de l'Archéologie au Cameroun*, pp 157–64, Paris: L'Harmattan

Otto, T and Delneuf, M (1998) 'Evolution des ressources alimentaires et des paysages au nord du Cameroun: apport de l'archéologie', in Chastanet, M (ed), *Plantes et Paysages d'Afrique: Une Histoire à Explorer*, pp 491–514, Paris: Karthala

Pélissier, P (1980) 'L'arbre dans les paysages agraires de l'Afrique Noire', *Cahier de l'ORSTOM* 17(3–4): 131–36

Pelzer, C, Müller, J and Albert, K-D (2004) 'Die Nomadisierung des Sahel: Siedlungsgeschichte, Klima und Vegetation in der Sahelzone von Burkina Faso in historischer Zeit', in Albert, K-D, Löhr, D and Neumann, K (eds), *Mensch und Natur in Westafrika – Ergebnisse aus dem Sonderforschungsbereich 'Kulturentwicklung und Sprachgeschichte im Naturraum Westafrikanische Savanne'*, pp 256–88, Weinheim: Wiley, VCH

Person, A, Amblard-Pison, S, Ferré, B and Saoudi, N E (2001) 'Jardins perchés néolithiques sur le Dhar Oualata (Mauritanie)', *Afrique: Archéologie et Arts* 1: 101–09

Petit, L (2002) 'Approaching Benin's past: archaeological and historical research in north-western Benin', unpublished PhD thesis, J W Goethe–Univerität Frankfurt

Petit, L, Bagodo, O, Höhn, A and Wendt, K-P (2001) 'Archaeological sites of the Gourma- and Mékrou-Plains', *Berichte des Sonderforschungsbereichs 268* 14: 229–36

Pullan, R A (1974) 'Farmed parkland in West Africa', *Savanna* 3(2): 119–51

Renard, C, Boudouresque, E, Schmelzer, G and Bationo, A (1993) 'Evolution de la végétation dans un zone protegée du Sahel (Sadoré, Niger)', in Floret, C and Serpantie, G (eds), *La Jachère en Afrique de l'Ouest*, pp 297–305, Paris: Office de la Recherche Scientifique et Technique de Outre-Mer (ORSTOM)

Schulz, E (1991) 'Paléoenvironnement dans le Sahara central pendant l'Holocène', *Palaeoecology of Africa* 22: 191–201

Seignobos, C (1982) 'Matières grasses, parcs et civilisations agraires (Tchad et Nord-Cameroun)', *Cahiers d'Outre-Mer* 35(139): 229–69

Smith, A B (1974) 'Preliminary report of excavations at Karkarichinkat Nord and Karkarichinkat Sud, Tilemsi Valley, Republic of Mali, Spring, 1972', *West African Journal of Archaeology* 4: 33–55

Smith, A B (1975) 'A note on the flora and fauna from the postpalaeolithic sites of Karkarichinkat Nord and Sud', *West African Journal of Archaeology* 5: 201–4

Smith, A B (1984) 'Origins of the Neolithic in the Sahara', in Clark, J D and Brandt, S (eds), *From Hunters to Farmers: The Causes and Consequences of Food Production in Africa*, pp 84–92, Berkeley: University of California Press

Smith, B D (1998) *The Emergence of Agriculture*, New York: Scientific American Library

Smith, B D (2001) 'Low-level food production', *Journal of Archaeological Research* 9(1): 1–43

Stahl, A B (1985) 'The Kintampo culture: subsistence and settlement in Ghana during the mid second millennium BC', unpublished PhD thesis, University of California-Berkeley

Stahl, A B (1993) 'Intensification in the West African Late Stone Age: a view from Central Ghana', in Shaw, T, Sinclair, P, Andah, B and Okpoko, A (eds), *The Archaeology of Africa: Food, Metals and Towns*, pp 261–73, London: Routledge

Sutton, J E G (2004) 'Africa, agriculture and iron', in Oestigaard, T, Anfinset, N and Saetersdal, T (eds), *Combining the Past and the Present: Archaeological Perspectives on Society*, pp 107–18. British Archaeological Reports International Series 1210, Oxford: Archaeopress

Togola, T (1993) 'Archaeological investigations of Iron Age sites in the Méma region (Mali)', unpublished PhD thesis, Rice University (Houston)

Tostain, S (1998) 'Le mil, une longue histoire: hypothèses sur sa domestication et ses migrations', in Chastanet, M (ed), *Plantes et Paysages d'Afrique: Une Histoire à Explorer*, pp 461–90, Paris: Karthala

Van Neer, W (2002) 'Food security in western and central Africa during the late Holocene: the role of domestic stock keeping, hunting and fishing', in Hassan, F A (ed), *Droughts, Food and Culture: Ecological Change and Food Security in Africa's Later Prehistory*, pp 251–74, New York: Kluwer/Plenum

Vernet, R (2002) 'Climate during the late Holocene in the Sahara and the Sahel: evolution and consequences on human settlement', in Hassan, F A (ed), *Droughts, Food and Culture: Ecological Change and Food Security in Africa's Later Prehistory*, pp 47–63, New York: Kluwer/Plenum

Vogelsang, R, Albert, K-D and Kahlheber, S (1999) 'Le sable savant: les cordons dunaires sahéliens au Burkina Faso comme archive archéologique et paléoécologique du Holocène', *Sahara* 11: 51–68

Wasylikowa, K (1997) 'Flora of the 8000 year old archaeological site E-75-6 at Nabta Playa, Western Desert, southern Egypt', *Acta Palaeobotanica* 37(2): 99–205

Wasylikowa, K and van der Veen, M (2004) 'An archaeobotanical contribution to the history of watermelon, *Citrullus lanatus* (Thunb.) Matsum. and Nakai (syn. *C. vulgaris* Schrad.)', *Vegetation History and Archaeobotany* 14: 213–17

Watson, D (2003) 'Hunter-gatherers and the first farmers of prehistoric Ghana: the Kintampo archaeological research project', in Mitchell, P, Haour, A and Hobart, J (eds), *Researching Africa's Past: New Contributions from British Archaeologists*, pp 61–68. Oxford University School of Archaeology Monograph no 57, Oxford: School of Archaeology, Oxford University

Zach, B and Klee, M (2003) 'Four thousand years of plant exploitation in the Chad Basin of NE Nigeria. II: discussion on the morphology of caryopses of domesticated *Pennisetum* and complete catalogue of the fruits and seeds of Kursakata', *Vegetation History and Archaeobotany* 12: 187–204

CHAPTER 18

Human Impact and Environmental Exploitation in Gabon during the Holocene

Richard Oslisly and Lee White

INTRODUCTION

Starting in 1992, we conducted a multidisciplinary study of human impacts on central African forests in the Lopé National Park in Gabon. Recently, the same method was applied to the study of the forested landscapes of the Banyang Mbo Sanctuary in western Cameroon and the national park of Campo Ma'an in southwestern Cameroon. The study areas are distributed over 1000 km of central Africa from the equator (Gabon) to Cameroon's volcanic arc (Figure 18.1); they differ from each other in climate and vegetation assemblages. For a long time, people in these regions preferred occupation sites on hilltops and ridgelines overlooking open areas for protection. The research established that openings in the forest created by slash-and-burn agriculture and for procuring fuel for smelting were recolonised by specific species; among these were a number of traditional economic plants.

Following a brief introduction to the archaeology of the region, the second section of this chapter outlines research findings that have used distinctive tree species and vegetation assemblages as markers of past human occupation in the forest and the savanna/forest mosaic. The third section focuses specifically on the Iron Age (2500–1400 BP and 900–100 BP) in Lopé National Park, where multidisciplinary lines of evidence have been used to characterise the past exploitation of tree species. Recent developments throughout the African belt suggest the exploitation and, perhaps, management of wild resources might have been more broadly practiced, more complex and have a longer history than usually assumed (see Blench chapter 21; Kahlheber and Neumann chapter 17). At the end of this chapter, a working hypothesis is developed that considers the potential significance of our research to the study of the development of early African arboriculture (see Harris chapter 2, for a theoretical discussion of arboriculture with respect to agriculture).

ARCHAEOLOGICAL CONTEXT

Archaeological investigations along the Atlantic front of central Africa identify three periods: the Late Stone Age (60,000–4000 BP), the Neolithic Stage, also known as the 'Stone to Metal Period' (4000–2500 BP) and the Iron Age (2500–500 BP); the last is subdivided into the Early (2500–1400 BP) and Late (1000–500 BP) Iron Age (de Maret 1995; Oslisly and White 2000). The Late Stone Age is characterised by microlithic tools that have been recovered over a broad area extending from the Lopé (Gabon) to Monte Alen (Equatorial Guinea) and Nguti and Shum Laka (Cameroon). During the Neolithic Stage, occupation sites are typically located on the tops of small hills, have large refuse pits and yield polished stone tools and ceramics. This period corresponds more or less with a drier climatic phase (Kibangian B) recorded for all of central Africa (Maley 1992, 1993; Schwartz 1992). Representative for this period are the sites of Obogogo, Nkometou, Ndindan and Okolo in Cameroon (Atangana 1988; Claes 1985; Mbida 1992) and Riviere Denis, Okala and Lopé in Gabon (Clist 1990; de Maret 1992, 1995; Oslisly and White 2000). Subsequent large Iron Age settlements are also situated on hills and are typically encircled by large refuse pits. Archaeological assemblages of Iron Age settlements typically include abundant ceramic artefacts, polished stone tools, iron slag, charcoal and charred nuts. Obogogo, Okolo and Oliga in Cameroon and Lopé in Gabon have occupational components representative of this period (Elouga 2000; Oslisly et al 2000). With the notable exception of the Nkang site (Mbida et al 2000, 2001), remains of cultivated or domesticated crops and domestic animals have not been documented for settlements in these regions.

Abundant archaeological remains reflect the importance of the Lopé area for people in the past (Oslisly 1993; Oslisly and Peyrot 1992; Oslisly and White 2000). The oldest deposits date back to the Early and Middle Stone Age. Most of these deposits are highly acidic, disturbed and of little help to reconstruct past interrelationships between people and their environment. More evidence has been obtained for the Late Stone Age (LSA) (40,000–4000 BP), Neolithic Stage (NS) (3600–2500 BP) and Iron Age (IA) (2500–1400 BP and 900–100 BP). The LSA, NS and IA sites are located mainly on hilltops within patches of savanna. The LSA sites consist of large amounts of debitage of locally derived stone. Intensive occupation, metallurgy and cultural diversity typify the NS and IA sites. While the early IA sites (2500–1800 BP) are located within the savanna, sites dating between 1800 BP and 1400 BP are located deeper in the forest (Oslisly 2001). This distribution tracks migratory flux related to the Bantu expansion.

In Lopé, erosion along active river channels has exposed combustion pits yielding important pieces of charcoal from several tree species; sites include the Mingoue and Mombela sites, dating to 1800 and 1500 BP respectively. The results from the charcoal analysis suggest enclosed savanna (Oslisly and Dechamps 1994). Archaeological features related to iron smelting have also been located in nearby areas and date to 1700–1600 BP (Oslisly and Dechamps 1994). Human occupation was extensive in the area at this time and, presumably, metal-working populations practiced itinerant slash-and-burn agriculture and made charcoal for iron production (Oslisly 1999).

Although Early Iron Age occupation across the region is confirmed from 2500 to 1400 BP, there is some regional variability. For example, there is an absence of any known human occupation between 1350 BP and 850 BP in the middle Ogooué Valley (Oslisly 1993, 1995, 1998). Even though there are 120 radiocarbon dates for this portion of the valley, no date has been obtained for this period (Oslisly 2001; Oslisly and White 2000). This rather long period without human occupation must have involved profound social and environmental changes to patterns of human settlement and the forest/savanna mosaic in the middle Ogooué Valley.

Given the paucity of plant and animal remains at most archaeological sites in the region, alternative research designs were needed to infer past plant and animal exploitation strategies. Consequently, a multidisciplinary research strategy was designed and conducted at Lopé and other areas in the region. Correlations between present-day plant associations and former human occupation, as well as palynology and archaeobotany, were used to understand people's exploitation of plants, particularly arboreal resources, in the past.

TREES AS MARKERS OF ARCHAEOLOGICAL SITES

Research conducted in Gabon since 1992 on the relationship between people and forests (Oslisly and White 2000; White and Oslisly 1999) and since 1998 in Cameroon (Oslisly et al 2000) demonstrates that the distributions of some pioneer tree species, such as *okoumé* (*Aucoumea klaineana*) and *azobé* (*Lophira alata*) strongly correlate with prehistoric and historic occupations

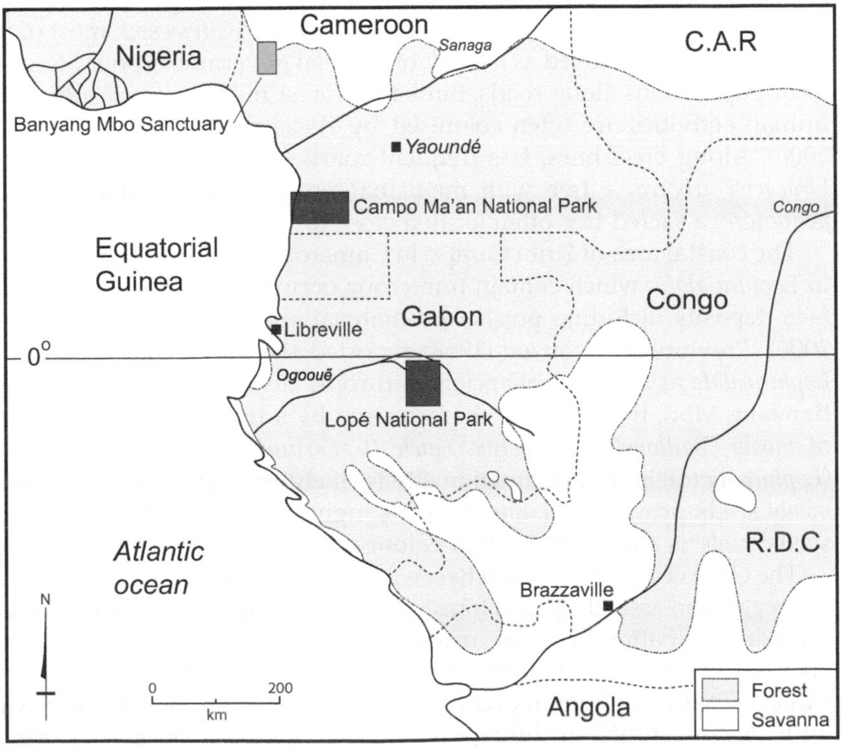

Figure 18.1 Location of study areas in Cameroon and Gabon.

Table 18.1 Distinctive tree species and assemblages marking archaeological sites in Cameroon and Gabon

Period	*Country*	*Study Area*	*Location*	*Tree Association*		*Archaeological Feature*
				Species	*Type*	
Recent	**Cameroon**			*Elaeis guineensis*	food	village
	Gabon			*Mangifera indica*	food	
Ancient	**Cameroon**	Banyang Mbo Sanctuary	Hilltop	*Lophira alata*	colonising	village
				Baillonella toxisperma	food	
				Canarium schweinfurthii	food	
		Campo Ma'an NP	Hilltop	*Lophira alata*	colonising	
	Gabon	Lopé NP	Hilltop	*Aucoumea klaineana*	colonising	village
			Altitudinal	*Dracaena arborea*	medicinal	
				Ceiba pentandra	sacred	
			Savanna Hilltops	*Millettia sanagana*	utilitarian	village
				Pentaclethra macrophylla	food	

on hilltops (Table 18.1). For example, in the north of Lopé, abandoned village sites are marked by oil palm (*Elaeis guineensis*), mango trees (*Mangifera indica*), manioc (*Manihot esculenta*) and citrus trees (*Citrus sp.*); most of these villages were abandoned as a result of colonial government policy to resettle local populations along roads. Similarly, forest hilltop sites resulting from human activities are often colonised by *Aucoumea klaineana* (White et al 2000). Along crest lines, less frequent markers of human occupation are *Dracaena arborea*, a tree with medicinal properties, and stands of *Ceiba pentandra*, a sacred tree often located close to villages.

The coastal area of Kribi Campo in Cameroon is renowned for areas rich in *Lophira alata*, which contain numerous occupation sites marked by surface deposits including potsherds, charcoal and stone tools (Oslisly et al 2006). Previously, Letouzey (1968) reported the systematic occurrence of *Lophira alata* as a marker of ancient anthropic clearings in Campo Ma'an. In Banyang Mbo, if a hilltop is characterised by a tree association composed of *moabi* (*Baillonella toxisperma*), *aiélé* (*Canarium schweinfurthii*) and *azobé* (*Lophira alata*), an old occupation site is likely to be present. Traditionally, *moabi* seeds provided oil, and people sought *aiélé* for their fruits, resin and seeds; *azobé* is a pioneer tree that colonises cleared and disturbed areas.

The observed relationship between former human occupation and distinctive vegetation assemblages in forested landscapes is also evident in savanna. Former occupation sites are readily identified from aerial photographs taken of burnt savanna subject to pluvial erosion, eg 90% of eroded areas exhibit remains of human occupation in the Lopé National Park. Savanna with woody shrubs on hilltops is associated with two species, *Millettia sanagana* and *Pentaclethra macrophylla*. *Millettia sanagana* is a traditional building material and the seeds of *Pentaclethra macrophylla* were used to prepare edible flour. Once again, hilltop plant associations include diagnostic species that reflect human activities, although they may also reflect variations in forest structure and plant composition (White et al 2000).

In this section, associations between past human activities and contemporary vegetation assemblages and successions, including economically useful plant species, have been noted at Lopé and for other areas of central Africa. In the next section, these present-day associations are tracked into the past, specifically the Iron Age of the Lopé Basin, using palynological records and multidisciplinary lines of evidence. In addition, these associations are compared against charcoal records from occupation and iron-smelting sites to infer past human selectivity in the use and potential transplantation of useful tree species during the Iron Age of the Lopé Basin.

ARBOREAL EXPLOITATION DURING THE IRON AGE OF THE LOPÉ

The Lopé Basin is a large depression of rolling hills located in the centre of the middle reaches of the Ogooué Valley (0–0°15′ S and 11–12° E). The area is characterised by a complex mosaic of savanna and rain forest. Along most watercourses, narrow gallery forests dominated by Cesalpinaceae are found. The grass *Pobeguinea arrecta* is dominant in the savanna, with scattered fire-resistant shrubs including *Crossopteryx febrifuga*, *Nauclea latifolia* and *Bridelia ferruginea*. Savanna is burnt annually by humans and repeated fires produce abrupt edges between the savanna and forest.

In the Republic of Congo, it is a well-established practice to burn during the dry season to prevent forest expansion and preserve enclosed savanna (de Foresta 1990; Schwartz et al 1990). Similarly, our research in Lopé National Park shows that unburnt savanna is quickly recolonised by the forest through a series of major plant associations that can be distinguished based on age and species composition (White et al 2000; Table 18.2). The charcoal record from cores extracted from Lake Kamalété in Lopé and the presence of occupation sites dating from the Late Stone Age to Iron Age on surrounding hilltops, including some iron-smelting ovens dated to 2300–1800 BP (Oslisly and White 2000), indicate the long-term environmental effects of, and association between, human occupation and dry season fires.

Before the more intensive human impacts of the last three millennia, vegetation assemblages resulted from a climatic anomaly marked by a yearly average rainfall of 1500 mm, with as little as 1200 mm in some years. Due to its position in the rain shadow of the Massif du Chaillu and Monts de Cristal to the west, the region is one of the driest in Gabon. Cold oceanic currents result in a cool, cloudy dry season from mid-June to mid-September with three months of drought, when rainfall is below 50 mm and mean temperature about 25° C. In the wet season, three-quarters of the rainfall occurs during violent storms resulting from easterly squall lines; rainfall is

Table 18.2 Tree species of major plant assemblages that recolonise savanna in the Lopé National Park (after White and Oslisly 1999)

Plant Species	*Savanna*	*Colonising Forest*	*Monodominant Forest*	*Marantaceae Forest*	*Mixed Forest*
Crossopteryx febrifuga	X	X			
Nauclea latifolia	X	X			
Aucoumea klaineana		X	X	X	X
Lophira alata		X	X	X	X
Barteria fistulosa		X	X	X	X
Sacoglottis gabonensis		X	X	X	
Xylopia aethiopica		X	X	X	X
Antidesma vogelianum		X	X		
Pauridiantha efferata		X	X		
Maprounea membranacea		X	X		X
Cola lizae			X	X	X
Erythroxylum manni		X	X		
Xylopia quintasii		X	X	X	X
Klainedoxa gabonensis		X	X	X	
Diospyros dendo			X	X	X
Pentaclethra eetveldeana				X	
Pentaclethra macrophylla				X	X
Hylodendron gabunense				X	
Canarium schweinfurthii				X	
Polyalthia suaveolens				X	
Dacryodes buettneri				X	X
Pterocarpus soyauxii					X
Pycnanthus angolensis					X
Celtis tessmannii					X
Testulea gabonensis					X
Scottellia coriacea				X	X

intense, accompanied by gusts of wind, and extremely erosive, particularly in well-drained areas close to the Ogooué River.

Palaeoenvironmental Reconstruction

Palaeoenvironmental reconstructions for the Lopé area employ sedimentology, $\delta^{13}C$ isotopic data of soil organic matter, palynology and charcoal analyses, and associated radiocarbon dates from archaeological and palaeoenvironmental sites (Ngomanda et al 2005; Oslisly 1999; Oslisly and White 2000; Oslisly et al 1996, 1997; Peyrot et al 2003; White et al 2000). The results of most analyses are complementary and enable coarse resolution reconstructions of palaeoenvironments during the Holocene.

The sedimentological data derive from three sedimentary sections: Lope 2, Lope 16 and Kanzamabika. The dates obtained for Lope 2 and 16 are chronostratigraphically consistent and the sections represent the entire Holocene (Table 18.3). Mineralogical analysis indicates particles originated from nearby geological formations and particle size analysis attests to a dramatic

Table 18.3 Uncalibrated radiocarbon ages for Lope 2 and 16

Depth (cm)	*Uncalibrated Radiocarbon Age*	*Lab Number*	*Reference*
Lope 2			
–40	6760±120 BP	Gif 9864	Oslisly et al 1996
60–70	9170±100 BP	Gif 9865	Oslisly et al 1996
100–110	10,320±110 BP	Gif A 95561	Oslisly et al 1996
Lope 16			
–40	5400±60 BP	Gif 10328	Peyrot et al 2003

diminution in coarse sand in each section through time. Coarse sand represents up to 30% at the base of the sequences, but less than 1% at the top; the majority of this decrease occurs near the base. Interestingly, the quartz grains at the base of the sequences are more eroded than those at the top. Changes in particle size frequencies up the sections mark a shift in the nature and intensity of erosion and deposition, the base of each section being characterised by greater hydrodynamic processes than the top (Peyrot et al 2003). The effects of hydrodynamic processes are generally considered to be greater with reduced vegetation cover.

The $\delta^{13}C$ signatures from these sites and another, Galerie de l'Aéroport (Oslisly and White 2000), provide a more complex interpretation of vegetation changes through time. Values derived from soil organic matter can be compared to those calculated for modern pioneering forests or woody savannas, savannas and mature forests in order to reconstruct vegetation assemblages in the past (Table 18.4). Changes in vegetation cover undergo changes at Lope 2 and 16 suggesting a shift from more forested to more open landscapes, while the Kanzamabika section is more suggestive of a savanna throughout the sequence (Oslisly et al 1996; Peyrot et al 2003) with forest persisting at Galerie de l'Aéroport and Mount Yindo (Peyrot et al 2003). Taken together, the $\delta^{13}C$ signatures for the Lopé suggest regionally varied changes to mosaics of vegetation through time.

In close proximity to Lope 2, two sequences have been sampled for palynological studies: the Marais 1 site (White et al 2000) and the Galerie de l'Aéroport site (Oslisly and White 2000). The bases of these sequences have been dated to 1810±70 BP (Gif 9856) and 1910±60 BP (Gif 9962) respectively (Oslisly and White 2000). The pollen spectra in both sequences show a predominance of grasses; Graminae and Cyperaceae account for 95% of total pollen sum. Pollen from pioneering species typical of the forest/savanna mosaic also occurs frequently, eg *Elaeis guineensis, Macaranga, Tetrorchidium* and *Uapaca guineensis* (Maley 1992; Oslisly and White 2001). While also dominated by a Graminae signal, the pollen data from Lake Kamalété show the persistence over the last 1300 years of a relatively stable forest-savanna mosaic (Ngomanda et al 2005).

Charcoal analysis of material originating from archaeological features (pits and combustion pits) provides evidence of pioneer (*Elaeis guineensis*), forest/savanna margin (*Sapium ellipticum* and *Erythroxylum* sp.), secondary

Table 18.4 Carbon isotope ratios for soil organic matter from palaeoenvironmental sites in the Lopé National Park and for modern vegetation assemblages. The carbon isotopic composition of plants varies according to photosynthetic pathways (C3 or C4). Graminae in savanna are C3 plants while forest trees are C4. The isotopic carbon signal of soil organic matter can indicate vegetation assemblages in the past (Mariotti 1991; Schwartz 1991)

Site	*Lope 2*	*Lope 16*	*Kanzamabika*	*Galerie de l'Aéroport*	*Mount Yindo*
Top	<−15‰	−15 to−23‰	−12 to−17‰	−25 and−28‰	−28.5 to−30‰
Base	−19.9 to−22.9‰	−23 to−24‰	−12 to−17‰	−25 and−28‰	−28.5 to−30‰
	Savanna	*Forest*	*Savanna/forest*		
Modern reference values	−10<x<−14‰	−25<x<−27‰	−14<x<−25‰		

forest (*Canarium schweinfurthii* [Burseraceae]) and mature forest (*Brachystega cf zenkeri*, *Didelotia africana* [Cesalpinaceae] and *Maranthes* sp. [Chrisobalanaceae]) tree species (Oslisly and Dechamps 1994). These findings are corroborated by others scattered throughout the Lopé of burned root systems (Mingoué and Mount Mombela) and charcoal layers exposed in road cuts on escarpments in the Lopé National Park, some of which extend unbroken for lengths exceeding 2000 m with a thickness of 15 cm (Oumoundo). Archaeological features related to iron smelting have also been located in nearby areas and date to 1700–1600 BP (Oslisly and Dechamps 1994). For Mingoué and Mombela, charcoal analysis identified *Copaïfera religiosa* (Cesalpinaceae), *Erythroxylum emarginatum* (Erythroxylaceae) and *Sapium ellipticum* (Euphorbiaceae) (Oslisly 1999). These vegetation assemblages are typical of the savanna/forest edge and date from 1760±40 BP (Gif 9968) and 1815±40 (Gif 10280) respectively. In the Oumoundo road cuts, dated from 1760±40 BP (Gif 9968), charcoal analysis identified *Allophyllus* cf *africanus* (Sapindaceae), *Psychotria* sp. (Rubiaceae), *Tessmannia* cf *lescrauwaetii* and *Gilletiodendron* cf *pierrenum* (Cesalpinaceae); all these species are typical of mature forests and pioneering fronts (Oslisly 1999).

The multidisciplinary lines of evidence for the Lopé National Park enable long-term environmental trends and human impacts to be considered. The shift in sedimentation at c 10,000 BP, ie the reduction in coarse sand deposition, corresponds to a period of increased rainfall in a landscape dominated by savanna. The sedimentation change loosely corresponds to the transition from the dry Leopoldvillian climatic phase (30,000–12,000 BP) marking the end of the Pleistocene to the wetter Kibangian A climatic phase (12,000–3500 BP). The terrestrial sedimentation record accords with ocean records along the Congolese coast (Giresse 1978). The reduction in coarser particles near the base of the sequences suggests the expansion of forests and less erosion, an interpretation supported by the $\delta^{13}C$ data.

During the drier Kibangian B climatic phase (3500–2000 BP), the Lopé landscape appears to have been a mosaic of forest and savanna. Pollen and charcoal analyses document a forest/savanna mosaic for the last 2000 years, despite the slightly wetter climates after 2000 BP. The mosaic landscape was consolidated as human occupation became more intensive, probably in response to an increased frequency in savanna fires and slash-and-burn activities. Although variations in the distribution of forest and savanna occurred during the IA (2500–500 BP), the range of species present remained comparable to the modern (Ngomanda et al 2005; Oslisly and White 2000; White et al 2000).

Archaeobotanical Findings

Taking into account fluctuating archaeological and palaeoenvironmental contexts, charcoal recovered from various features at occupation sites of different antiquity, in both savanna and forest locations, sheds the greatest light on people's use of trees in the past. Charcoal identification of arboreal species suggests that a range of traditional non-domesticated economic species were used, with significant differences according to the period and archaeological feature of origin (Table 18.4). The charcoal species recovered from refuse pits at villages are interpreted in terms of three categories: utilitarian (eg *Albizia adianthifolia*), food (eg *Canarium schweinfurthii*) and cultural (*Pterocarpus soyauxii*). Traditional economic and cultural species are usually represented by wood charcoal while the traditional food plants, including *Elaeis guineensis* (occurring at most Iron Age sites), *Coula edulis* and *Canarium schweinfurthii* are more frequently represented by endocarps. Species represented by charcoal analyses at occupation sites are typical of mature and secondary forests and pioneering fronts. However, none of the traditional food species identified by charcoal analysis (Table 18.4) appears to be a regular component of the five major plant formations prevalent in Lopé (Table 18.2). Consequently, the frequent occurrence of charcoal from economically useful species in archaeological refuse pits is not coincidental and must suggest something about the deliberate exploitation of wild arboreal resources, a regular but underestimated Old World practice (see Kahlheber and Neumann chapter 17), or the cultivation of wild woody species, ie arboriculture (see Blench chapter 21).

The evidence for oil palm (*Elaeis guineensis*) is potentially of great significance and deserves more detailed attention. Oil palm is the most important traditional source of oil and wine and is a major component of the food economy. It could have been used in either wild or cultivated form (Duke 1978, 1979). As a pioneering species, oil palm thrives in open areas in riverine forests, rivers or freshwater swamps and tolerates temporary flooding or a fluctuating water table. The cultivated forms, as well as the wild ones, require adequate light and soil moisture and are propagated from seed (Duke 1978, 1979). It has been suggested that major evidence for oil palm in the palaeoenvironmental record may indicate forest openings resulting from

Table 18.5 Charcoal recovered from features at hilltop archaeological sites (after Oslisly 1998, 1999, 2001; Oslisly and White 2000). LSA: Late Stone Age; NS: Neolithic Stage; EIA: Early Iron Age; LIA: Late Iron Age

Period	*Feature*	*Species*	*Type*	*Use*
L S A	Hearth	*Diogoa zenkeri*	Woody tissues	Utilitarian (fuel)
60,000 – 4000 BP		*Strombosiospis tetrandra*	Woody tissues	Utilitarian (fuel)
		Pterocarpus soyauxii	Woody tissues	Cultural
N S	Refuse pits	*Elaeis guineensis*	Endocarp	Food
3600 – 2500 BP		*Coula edulis*	Endocarp	Food
		Antrocaryon klaineanum	Endocarp	Food
E I A	Refuse pits	*Elaeis guineensis*	Endocarp	Food
2500 – 1400 BP		*Coula edulis*	Endocarp	Food
		Canarium schweinfurthii	Endocarp	Food
		Antrocaryon klaineanum	Endocarp	Food
	Oven	*Pterocarpus soyauxii*	Woody tissues	Utilitarian (fuel)
		Brachystega cf *zenkeri*	Woody tissues	Utilitarian (fuel)
		Didelotia africana	Woody tissues	Utilitarian (fuel)
		Albizia ferruginea	Woody tissues	Utilitarian (fuel)
		Maranthes sp.	Woody tissues	Utilitarian (fuel)
L I A	Layers	*Pterocarpus soyauxii*	Woody tissues	Cultural
900 – 500 BP		*Albizia adianthifolia*	Woody tissues	Utilitarian (fuel)
		Combretum bracteatum	Woody tissues	Utilitarian (fuel)
		Caloncoba glauca	Woody tissues	Utilitarian (fuel)
		Imenia africana	Woody tissues	Utilitarian (fuel)
	Oven	*Pterocarpus soyauxii*	Woody tissues	Utilitarian (fuel)
		Guibourtia cf *demeusii*	Woody tissues	Utilitarian (fuel)

human activity and, specifically, agricultural practices (Elenga et al 1992; Schwartz 1992; Vincens et al 1998). Nearly all IA sites in Lopé contain oil palm endocarps.

However the association of oil palm and people should not be presumed. The pollen evidence from Lake Kamalété contains a peak in oil palm pollen between 1255±75 BP (Gif 11789) and 520±110 BP (Gif 11584), a time period corresponding to the cultural hiatus of 1400 BP to 900 BP in the middle Ogooué Valley (Oslisly 1993, 1995, 1998). This oil palm peak could result from natural climatic variability, which has been documented at the regional scale for Cameroon (Maley 1999) and at several sites from western equatorial Africa (Ngomanda et al 2005).

At Lopé, although no oil palm groves were observed, the palm tree is widespread. It occurs preferentially in the gallery forests distributed along active channels, several of which are close to archaeological features related to iron smelting. The base of a section at Galerie de l'Aéroport, located in a river bank, contains a charcoal level underlying a level with abundant oil palm tree nuts. The levels are dated to 1910±60 BP (Gif 9962) and 1920±40 BP (Gif 9961) respectively (Oslisly and White 2000). This period corresponds to the end of the drier Kibangian B and the return of wetter conditions favouring forest expansion. *Elaeis guineensis*, a pioneer tree, could have benefited from this climatic change. However, the species could also have adventitiously colonised gaps in the forest created by Iron Age settlers practising slash-and-burn and producing charcoal for smelting. In other areas, the interpretation of oil palm expansion is more problematic; for example the peak in *Elaeis* pollen in the Lake Kamalété deposits occurred during the cultural hiatus of 1400 BP to 900 BP.

Generally, it has been assumed that without human intervention, elephants eat the young palm shoots, thereby limiting the regeneration of the plant. Thus the expansion of oil palm groves need not imply deliberate planting, but probably does involve another form of human intervention, whether deliberate or unintentional. It is highly likely that metallurgy and food economy were interdependent practices, even though they are often studied in isolation. Clearances for iron smelting may have been part of a general subsistence strategy and encouraged the extension of oil palm and other useful trees.

A WORKING HYPOTHESIS: EARLY IRON AGE ARBORICULTURE?

Research conducted at Lopé, Banyang Mbo and Campo Ma'an document Late Stone Age, Neolithic Phase and Iron Age sites located on hilltops. Archaeological sites are characterised by specific tree associations, including pioneering, fruit and utilitarian trees (Table 18.1). The multidisciplinary approach applied at Lopé, which can be extended to other parts of central Africa and beyond, allows the diversity of human imprints on the environment to be reconstructed. For example, if the high relative frequencies of fruit endocarps in refuse pits in Lopé during the Iron Age were taken in isolation, they could be explained by gathering activities. However, the charcoal data complemented by extensive palaeoenvironmental and botanical surveys suggest another interpretation: the unintentional and deliberate alteration of tree species composition around occupations. As well as arboreal colonisation of abandoned plots and cleared gaps, people tended and probably planted economic species. Unintentional alteration of vegetation communities and the creation of disturbed environments around settlements were probably important factors, but do not fully account for recurrent associations with economically useful trees, several of which are not pioneer species.

Taken together the archaeological and palaeoecological evidence from Lopé National Park indicate wild plant exploitation for food during the Neolithic Stage (3500–2500 BP) and are suggestive of some form of arboriculture developing as early as the Early Iron Age (2500–1400 BP), and potentially earlier. First, abandoned occupation sites across the forested landscapes of central Africa can be identified by the presence of economically useful trees. The association of trees with former occupation sites is not coincidental and suggests deliberate or accidental transplantation by people in the recent and distant past. Second, palaeoecology indicates that the frequencies of at least one traditional food plant, oil palm (*Elaeis guineensis*), varied considerably in the past, and that the distribution of this plant across the landscape in the present and the past is anomalous and may in part be a product of human management. Third, the charcoal frequencies of traditional food species at occupation sites indicate the targeting of specific arboreal resources, including some species known to mark former occupation sites, ie *Canarium schweinfurthii* and *Elaeis guineensis*. The multidisciplinary work at Lopé provides glimpses of arboreal resource exploitation and enables us to propose the hypothesis that arboriculture was practiced in central African forests since the Early Iron Age. At present, this is only a working hypothesis; the antiquity and nature of these past practices remain to be determined.

ACKNOWLEDGMENTS

The authors are grateful to Luc Vrydaghs and Tim Denham for their assistance in translating and editing this chapter from the French original.

REFERENCES

Atangana, C (1988) 'Archéologie au Cameroun méridional: étude du site d'Okolo', doctoral dissertation in archaeology, Université de Paris I Panthéon-Sorbonne

Claes, P (1985) 'Contribution à l'étude des céramiques anciennes des environs de Yaoundé', mémoire de licence, Université Libre de Bruxelles

Clist, B (1990) 'Des derniers chasseurs aux premiers métallurgistes: sédentarisation et débuts de la métallurgie du fer (Cameroun, Gabon, Guinée équatoriale)', in Lanfranchi, R and Schwartz, D (eds), *Paysages Quaternaires de l'Afrique Centrale Atlantique*, pp 458–79, Paris: ORSTOM Editions

de Foresta, H (1990) 'Origine et évolution des savanes intramayombiennes (RP du Congo): II. Apports de la botanique forestière', in Lanfranchi, R and Schwartz, D (eds), *Paysages Quaternaires de l'Afrique Centrale Atlantique*, pp 326–25, Paris: ORSTOM

de Maret, P (1992) 'Sédentarisation, agriculture et métallurgie du sud Cameroun: synthèse des recherches depuis 1978', in Essomba, J M (ed), *L'Archéologie au Cameroun: Actes du Colloque de Yaoundé*, pp 247–62, Paris: Karthala

de Maret, P (1995) 'Pits, pots and the far west stream', *Azania* (Nairobi) 29–30: 318–23

Duke, J A (1978) 'The quest for tolerant germplast', in *Crop Tolerance to Suboptimal Land Conditions*, pp 1–61. ASA Special Symposium 32, Madison, WI: American Society of Agronomy

Duke, J A (1979) 'Ecosystematic data on economic plants', *Quarterly Journal of Crude Drug Research* 17(3–4): 91–110

Elenga, H, Schwartz, D and Vincens, A (1992) 'Changements climatiques et action anthropique sur le littoral congolais au cours de l'Holocène', *Bulletin de la Société Géologique de France* 163: 239–52

Elouga, M (2000) 'Carte archéologique du nord de la Sanaga: paysage des sites et mise en évidence de la transgression forestière sur la savane (centre du Cameroun)', in Servant, M and Servant-Vildary, S (eds), *Dynamique à Long Terme des Écosystèmes Forestiers Intertropicaux*, pp 133–37, Paris: IRD-UNESCO

Giresse, P (1978) 'Le contrôle climatique de la sédimentation marine et continentale en Afrique centrale atlantique à la fin du Quaternaire: problèmes de corrélations', *Paleogeography, Palaeoclimatology, Palaeoecology* 23: 57–77

Letouzey, R (1968) *Etude Phytogéographique du Cameroun*, Paris: Edition Lechevallier

Maley, J (1992) 'Mise en évidence d'une péjoration climatique entre ca 2500 et 2000 BP en Afrique tropicale humide', *Bulletin de la Société Géologique de France* 163: 363–65

Maley, J (1993) 'The climatic and vegetational history of the equatorial regions of Africa during the Upper Quaternary', in Shaw, T, Sinclair, P, Bassey A and Okpoko, A (eds), *The Archaeology of Africa: Food, Metals and Towns*, pp 43–52, London: Routledge

Maley, J (1999) 'L'expansion du palmier à huile (*Elaeis guineensis*) en Afrique centrale au cours des trois derniers millénaires: nouvelles données et interprétations', in Bahuchet, S (ed), *L'Homme et la Forêt Tropicale*, pp 237–54, Marseille: Travaux de la Société d'Ecologie Humaine

Mariotti, A (1991) 'Le carbone 13 en abondance naturelle, traceur de la dynamique de la matière organique des sols et de l'évolution des paléoenvironnements continentaux', *Cahiers ORSTOM, Série Pédologique* 26(4): 299–313

Mbida, C (1992) 'Etude préliminaire du site de Ndindan et datation d'une première série de fosses', in Essomba, J M (ed), *L'Archéologie au Cameroun: Actes du Colloque de Yaoundé*, pp 263–84, Paris: Karthala

Mbida, C, Van Neer, W, Doutrelepont, H and Vrydaghs, L (2000) 'Evidence for banana cultivation and animal husbandry during the first millennium BC in the forest of southern Cameroon', *Journal of Archaeological Science* 27: 151–62

Mbida, C, Doutrelepont, H, Vrydaghs, L, Swennen, R, Swennen, R, Beeckman, H, De Langhe, E and de Maret, P (2001) 'First archaeological evidence of banana cultivation in central Africa during the third millennium before present', *Vegetation History and Archaeobotany* 10: 1–6

Ngomanda, A, Chepstow-Lusty, A, Makaya, M, Schevin, P, Maley, J, Fontugne, M, Oslisly, R, Rabenkogo, N and Jolly, D (2005) 'Vegetation changes during the last 1300 years in western equatorial Africa: a high-resolution pollen record from Lake Kamalété, Lopé Reserve, Central Gabon', *The Holocene* 15(7): 1–11

Oslisly, R. (1993) *Préhistoire de la Moyenne Vallée de l'Ogooué (Gabon)*. TDM Number 96, Paris: ORSTOM Editions

Oslisly, R (1995) 'The middle Ogooué Valley, Gabon: cultural changes and palaeoclimatic implications of the last four millennia', *Azania* (Nairobi) 29–30: 324–31

Oslisly, R (1998) 'Hommes et milieux à l'Holocène dans la moyenne vallée de l'Ogooué au Gabon', *Bulletin de la Société Préhistorique Française* 95(1): 93–105

Oslisly, R (1999) 'Contribution de l'anthracologie à l'étude de la relation homme/milieu au cours de l'Holocène dans la vallée de l'Ogooué au Gabon', in Maes, F and Beeckman, H (eds), *Wood to Survive*, pp 185–93. Annales Sciences Économiques du Musée Royal de l'Afrique Centrale de Tervuren 25, Tervuren: Annales du Musée Royal de l'Afrique Centrale

Oslisly, R (2001) 'The history of human settlement in the middle Ogooué Valley (Gabon): implications for the environment', in Weber, W, White, L, Vedder, A and Naugthon-Treves, L (eds), *African Rain Forest Ecology and Conservation: An Interdisciplinary Perspective*, pp 101–18, New Haven: Yale University Press

Oslisly, R and Dechamps, R (1994) 'Découverte d'une zone d'incendie dans la forêt ombrophile du Gabon ca 1500 BP: essai d'explication anthropique et implications paléoclimatiques', *Comptes Rendus de l'Académie Royale des Sciences de Paris* 318(II): 555–60

Oslisly, R and Peyrot, B (1992) 'L'arrivée des premiers métallurgistes sur l'Ogooué (Gabon)', *African Archaeological Review* 10: 129–38

Oslisly, R and White, L (2000) 'La relation homme/milieu dans la réserve de la Lopé (Gabon) au cours de l'Holocène: les implications sur l'environnement', in Servant, M and Servant-Vildary, S (eds), *Dynamique à Long Terme des Écosystèmes Forestiers Intertropicaux*, pp 241–50, Paris: IRD–UNESCO

Oslisly, R, Peyrot, B, Abdessadok, S and White, L (1996) 'Le site de Lopé 2: un indicateur de transition écosystémique ca 10,000 BP dans la moyenne vallée de l'Ogooué', *Comptes Rendus de l'Académie Royale des Sciences de Paris* 323: 933–39

Oslisly, R, Peyrot, B, Abdessadok, S and White, L (1997) 'Réponse au commentiare de D. Schwartz sur le note "Le site de Lopé 2": un indicateur de transition écosystémique ca 10,000 BP dans la moyenne vallée de l'Ogooué', *Comptes Rendus de l'Académie Royale des Sciences de Paris* 325: 393–95

Oslisly, R, Mbida, C and White, L (2000) 'Les premiers résultats de la recherche archéologique dans le sanctuaire de Banyang Mbo (sud-ouest du Cameroun)', *L'Anthropologie* 104: 341–54

Oslisly, R, Ateba, L, Betouga, R, Kinyock, P, Mbida C, Nlend, P and Vincens, A (2006) 'Premiers résultats de la recherche archéologique sur le littoral du Cameroun de Kribi à Campo', in Le Secrétariat du Congrès (ed), *Préhistoire en Afrique: Section 15: Sessions Générales et Posters, Acts of the XIVth UISPP Congress, University of Liège, Belgium*, pp 127–34. British Archaeological Reports International Series no 1522, Oxford: Archaeopress

Peyrot, B, Oslisly, R, Abdessadok, S, Fontugne, M, Hatte, Ch and White, L (2003) 'Les paléoenvironnements de la fin du Pléistocène et de l'Holocène dans la réserve de la Lopé (Gabon): approches par les indicateurs géomorphologiques, sédimentologiques et anthropogènes des milieux enregistreurs de la dépression de la Lopé', *L'Anthropologie* 107: 291–307

Schwartz, D (1991) 'Intérêt de la mesure du ^{13}C des sols en milieu naturel équatorial pour la connaissance des aspects pédologiques et écologiques des relations savane-forêt: exemples du Congo', *Cahiers ORSTOM, Série Pédologique* 26(4): 327–41

Schwartz, D (1992) 'Assèchement climatique vers 3000 BP et expansion Bantu en Afrique centrale atlantique: quelques réflexions', *Bulletin de la Société Géologique de France* 163(3): 353–61

Schwartz, D, Lanfranchi, R and Mariotti, A (1990) 'Origine et évolution des savanes intramayombiennes (RP du Congo): II. Apports de la pédologie et de la biogéochimie isotopique (^{14}C et ^{13}C)', in Lanfranchi, R and Schwartz, D (eds), *Paysages Quaternaires de l'Afrique Centrale Atlantique*, pp 314–325, Paris: ORSTOM

Vincens, A, Schwartz, D, Bertaux, J, Elenga, H and Namur, C de (1998) Late Holocene climatic changes in western equatorial Africa inferred from pollen from Lake Sinnda, Southern Congo,' *Quaternary Research* 50: 34–45

White, L and Oslisly, R (1999) 'Lopé, a window on the history of the Central African Rainforest', in Nasi, R, Amsallem, I and Drouineau, S (eds), *Gestion Durable des Forêts Denses Africaines Aujourd'hui*, pp 1–30, Libreville, Gabon: Centre de Cooperation Internationale en Recherche Agronomique pour le Développement (CIRAD)

White, L, Oslisly, R, Abernethy, K and Maley, J (2000) 'L'Okoumé (*Aucoumea klaineana*): expansion et déclin d'un arbre pionnier en Afrique centrale atlantique au cours de l'Holocène', in Servant, M and Servant-Vildary, S (eds), *Dynamique à Long Terme des Écosystèmes Forestiers Intertropicaux*, pp 399–411, Paris: IRD–UNESCO

CHAPTER 19

The Establishment of Traditional Plantain Cultivation in the African Rain Forest: A Working Hypothesis

Edmond De Langhe

INTRODUCTION

Vegetatively propagated crops may have been cultivated since remote times in the African rain forest, but their ancient history remains obscure. In the case of the bananas, the traditional agriculture in this region relies almost exclusively on a typical group of starchy bananas, the plantains. Characteristic of these plants are the long and rather slender fruits with a yellow to orange pulp. Plantains are hybrids between edible diploids deriving from the species *Musa acuminata* and the wild species *Musa balbisiana* (Simmonds 1962). They probably originated in the area in and around New Guinea (Carreel 1994; Carreel et al 2002).

Several theories have been advanced during the last century for the introduction of the banana to Africa. Since the 1960s, the theory first suggested by Simmonds (1962: 144) of an introduction around 500–600 AD[1] via Madagascar has been preferred. More recently it was suggested that the plantains were the first group to reach the continent, perhaps more than 3000 years before present, with Austronesian-speaking ancestors as the possible vector (De Langhe et al 1994–95; De Langhe and de Maret 1999). The suggestion received confirmation with the discovery in Cameroon of banana phytoliths dating to 2500 BP; it was argued that these phytoliths belonged to the plantain group (Mbida et al 2001). The phytoliths point to the possibility, first, that plantain cultivation could already have been established for some time in the Sanaga River basin area and, second, that initial plantain cultivation would have progressed across the continent from east to west over an undetermined timespan, but can be tentatively plotted in the period of 3500–2500 BP. The present contribution offers a hypothesis as to how plantains under cultivation spread across the continent.

THE GENERATION OF THE PLANTAIN CULTIVARS

In the African rain forest more than one hundred plantain cultivars are grown, the vast majority of which have not been reported elsewhere, not even in other parts of humid tropical Africa (Rossel 1998; Swennen 1990: 172). This fact points to cultivation since remote times in these forests; the numerous cultivars would subsequently have been generated through somatic mutations during the long period of cultivation (Simmonds 1966). Ethnic groups would have moved over long distances in the forest over time and carried with them the core of plantain variation, while further *ad hoc* cultivation would have caused additional and location-specific variation.

The local names for the rare non-plantain cultivars that are currently grown in traditional contexts in the rain forest generally indicate a recent and alien origin. Plantains were therefore most probably the only group of bananas in humid central Africa for a very long time until the arrival of the Portuguese along the West African coasts by the mid-fifteenth century.

FAVOURABLE ENVIRONMENTS FOR THE PLANTAINS

The plantains have no natural habitat. Since the edible diploids from which they partly derive were domesticated plants, the generation of the plantains could only have occurred at sites where farmers modified the natural vegetation. A conspicuous consequence is that plantains do not survive under a closed canopy. An abandoned plantain field will soon perish in a secondary forest.

Plantains require a humid and rather warm climate without major oscillations. The pseudostem suffers heavily from a long dry period. A dry season of more than three months is fatal. The plant will also die off after a few nights with a minimum temperature of less than 10° C. The underground parts eventually perish as well, contrary to many other bananas where the corm survives to produce a new pseudostem after serious drought or after a cold period. Plantains can survive in protected growth conditions at elevations between 1000 and 1600 msl. In these habitats, the plant develops slowly and produces smaller bunches, but suckers will grow better, due to a diminished apical dominance, so that eventual survival is ensured. This potential to survive at higher elevations is key to understanding how plantains could 'move' across the East African highland complex (Figure 19.1).

Plantains are edaphically much more demanding than most other bananas. They develop poorly ramified rooting systems with relatively few root hairs and thus require easy access to nutrients. The roots therefore manifest a high tropism towards organic material (Swennen 1984: 119). As a consequence, the rooting system will be superficial and very vulnerable on poor ferralsols where nutrients are concentrated in the shallow uppermost horizon. More nutrient-rich soils are also unpromising if they contain little organic material in the upper horizons.

The environmental conditions for reasonable productivity severely restrict suitable areas for plantains in Africa. These areas can be summarised as follows:

(a) clearings in dense forest, with limited duration (maximally three years on more fertile soils). Itinerant agriculture is the unavoidable consequence and plantains may have been the crop that forced shifts towards new land;
(b) villages in the forest with plantains grown in the backyards. This may have been the dominant cultivating system under initial, low demographic pressure;
(c) semi-deciduous tree associations at the periphery of the rain forest, where poor ferralsols are less frequent. The more open canopy would have allowed for planting without preliminary clearing in the past. Duration of the dry season is the paramount limiting factor farther north or south of the forest;
(d) in mountain zones, especially in East Africa, along soil-fertile slopes in the mountain forest stratum that are protected from cold, dry winds. In the past, clearings would not have been required as long as the plantains were sufficiently exposed to the sun along the mountain slopes.

VEGETATION AT 3500–2500 BP

The period during which initial plantain cultivation could have moved across the continent coincides with the final phase of a pronounced aridification process that started about 4500 BP (Sowunmi 2002). The natural limits of the rain forest (Mercader et al 2000) and the East African Highlands (Hamilton and Taylor 1991) would not have been very different from today (Figure 19.2). However, during the wet mid-Holocene period a semi-deciduous humid forest covered the area to the north and west of Lake Victoria. This vegetation would have survived in part until the highly destructive forest clearing that started around 2000 BP (Roche and Van Grunderbeek 1987). Farther east, rain forest galleries, nowadays interrupted in many places by drier landscapes, lined the major rivers, such as the Tana, and are found along the coastline beyond the mangrove zone.

POSSIBLE DIFFUSION ROUTES OF THE PLANTAIN ACROSS AFRICA

Taking into account the above ecological restrictions on the one hand and the phytogeographic situation around 3500–2500 BP on the other, the following hypothetical route for the anthropogenic move of the plantain across the continent can be reconstructed (Figure 19.3):

(a) From the coastal rain forest to the eastern limits of the main rain forest (ie north to the Lake Victoria). The obvious obstacles are the steppe conditions in the Eastern Rift Valley and on the plateaux of eastern Kenya and Tanzania. The latter could have been avoided in one or both of the following ways: either along the forest gallery of the Tana River, thus reaching the mountainous forest belt around Mount Kenya, or directly through the mountainous forest formations of the Usambara and Pare Hills, thus reaching the forests around Kilimanjaro. For the crop to get across the more than 100-kilometer-wide Rift

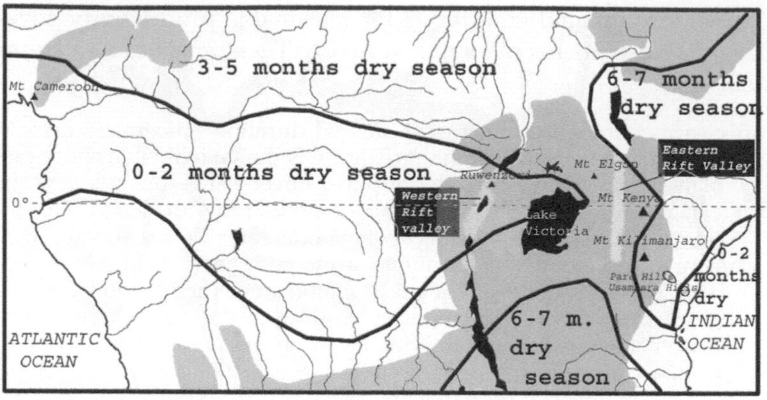

Figure 19.1 Topography and climate of central Africa. The areas shaded grey are at elevations above 1000 m.

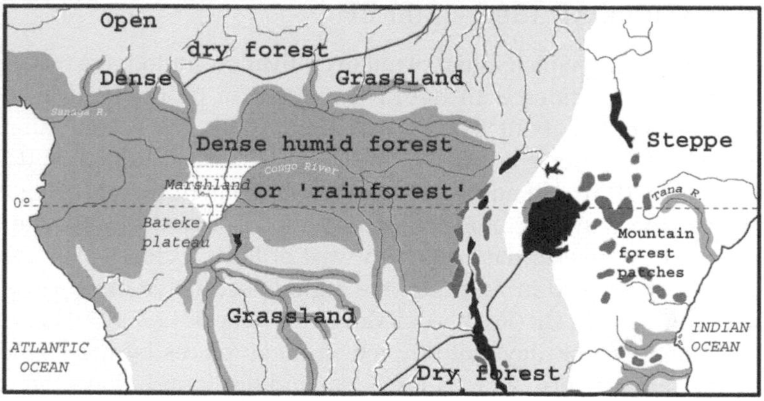

Figure 19.2 Phytogeography of central Africa.

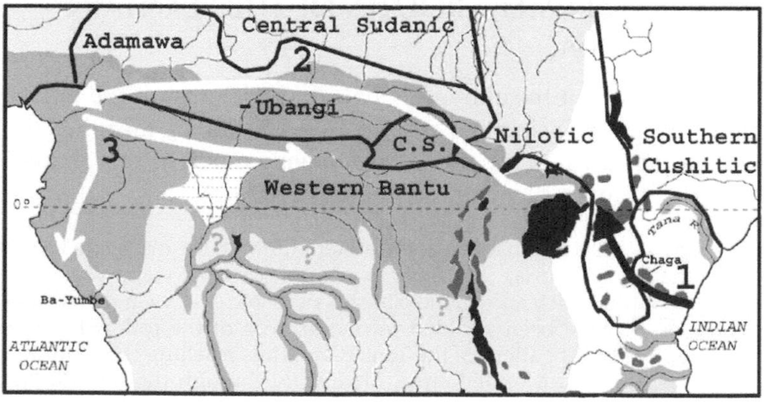

Figure 19.3 Possible route of plantain diffusion across central Africa, with major language phyla displayed.

Valley presupposes the existence of people interested in plantain as a food crop on both sides of the valley.

(b) From the eastern rain forest limits to the Sanaga River basin (present-day central Cameroon). Three possibilities for diffusion are to be evaluated:

(1) straight across the rain forest, which implies clearings in the dense forest;
(2) circumventing the rain forest on its southern side, across the transition zone from semi-deciduous to drier formations and woodland. This option has the advantage of not requiring forest clearing for the plantains and is still in the current zone of less than three months of dry weather. However, in addition to the extremely long distance, the movement would have encountered two almost intractable obstacles: the marshlands along the Zaire River and the dry and extensive Bateke Plateau;
(3) an almost straight journey through the same formations but on the northern fringes with a more open forest canopy. These areas offer the additional advantage of a near continuum of favourable habitats.

The first possibility for diffusion, option 1, clearly calls for people who fluently practised at least an initial form of vegeculture[2] after appropriate forest clearing. Options 2 and 3 do not necessarily imply forest clearing or any previous acquisition of vegecultural technology. Weening a sucker from the main pseudostem and planting it in a small, shallow hole requires no more than a variant of the digging stick, which I observed in the forest in the 1950s. These three possibilities can be narrowed down by reviewing how people lived in these geographical areas and their subsistence practices.

THE HUMAN FACTOR

Human cultures existing in the geographical areas during the period under consideration have been identified from linguistic and archaeological investigations and can be summarised as follows:

(a) *Agro-pastoralism* was established in the savannas to the north of the rain forest (Central Sudanese-speaking people), and in the northern parts of the highland Rift Valley of East Africa (Southern Cushitic and, much later, Southern Nilotic speakers). The agricultural component was based on grain crops. No linguistic traces of vegeculture, such as the African yam (*Dioscorea cayenensis-rotundata*), domesticated in western Africa, were detected in the reconstructed protoforms of these languages (Ehret 1971, 1987).

(b) *Vegeculture* dominated at the far western side of the rain forest and the drier landscapes to the north and west. Benue–Congo-speaking people, with proto(?)-Bantu speakers as a main branch, lived in and around the Sanaga River basin. The main food crop was the African yam, but some oil palm arboriculture (semi-domesticated *Elaeis guineensis*) and cultivation of grain crops, such as *voandzeou* (*Voandzeia* spp.), were probably completing the vegetal diet. The supposed expansion of (proto)western Bantu culture into the wet, dense forest may already have started by 3500 BP (Vansina 1990).

(c) *Hunting-foraging* would have been practiced over the rest of the rain forest area, probably supplemented by fishing where available (Vansina 1990). East Africa would have been predominantly populated by hunter-foragers (ie the ancestors of Khoisan speakers).

The plantain as a crop is definitely a component of vegeculture, even if the practice would be in a nascent phase. In the search for the possible diffusion paths of the plantain, people who practised either agro-pastoralism or hunting-foraging could not have played a significant role.

TENTATIVE RECONSTRUCTION OF THE INITIAL PLANTAIN ROUTE

Phase 1: Across East Africa

A major challenge to the present hypothesis lays in the supposed east-west progression of plantain cultivation in East Africa at around 3000 BP, across the dry highlands and especially the eastern Rift Valley. As was explained above, these regions were populated by either hunter-gatherers or agro-pastoralists who apparently did not take any interest in a crop that could not survive in such harsh environments. These were clearly not people who were instrumental in the transfer of the crop along suitable niches (ie the humid mountain forests), down to Mt Elgon at the eastern limits of the then rain forest zone.

However, rather systematically overlooked has been the existence in the past of people who were perhaps neither 'hunter-gatherers', nor typical agro-pastoralists. Unlike the hunter-gatherers of the steppe who are called '*wasi*' by the Bantu speakers over a large portion of East Africa[3], these people had quite different names among the Bantu in the region[4]. They left no clear traces but it has been recorded (Schanz 1913; Stahl 1964) that the first Bantu settlers met 'little men' who, according to some sources, lived in caves and 'escaped in the forest and nevermore re-appeared'. It was said that these people had cattle and banana fields (Schanz 1913). During his study of the Chaga language, Phillipson, a specialist in East African Bantu languages, became puzzled by the existence in the E-Bantu languages (Guthrie classification) of many terms linked to the banana. These terms were not innovations and Phillipson could not link them to any of the common languages in the area, ie Bantu, Southern Nilotic, or Southern Cushitic (Phillipson 1984)[5]. He advocates a search for the origin of these words, which have a regional distribution and have not yet been attributed to a source (Phillipson 1984: 219).

The probability of the coexistence at around 3000 BP of hunter-gatherers in the steppe and semi-agriculturists in the more humid areas can not be disregarded. The latter could have lived among others on the humid side of the mountain range stretching from the Usambara Hills to Kilimanjaro, as well as around mounts Kenya and Elgon. In such environments vast *Ensete* populations were thriving on the humid slopes, the remnants of which can still be observed. These people may then have been familiar with *Ensete* corm manipulation for food and would rapidly have grasped the significance of banana and taro suckers (De Langhe et al 1994–95)[6]. Since they would have been settled on the humid mountain slopes all over the region, they could have facilitated the transfer across the steppes of not only plantain but other Southeast Asian crops as well, such as taro (*Colocasia esculenta*) and the water yam (*Dioscorea alata*).

The above considerations suggest that plantains were carried across East Africa along or within the mountain forest patches, rather than along rivers such as the Tana.

Phase 2: Across Central Africa from East to West

The three possible routes of diffusion of the plantain listed above under item (b) (page 365) are reviewed here. For the movement of plantains in central Africa, the cultural configuration appears to exclude option 1 since the Bantu culture was not yet established in the dense rain forest at that time, ie 3500–2500 BP. The physically difficult option 2, circumventing the forest, would have left early traces of vegeculture in or around the southern part of the forest, which does not seem to be the case. Option 3, progression along the less dense northern side of the forest, would have been the most likely one and advocated here.

A branch of the Central Sudanese-speaking group became established in the northeastern part of the dense forest. The current descendants of these people include the Mangbetu, who are renowned for their great interest in plantains and who consequently keep a very high diversity of plantain cultivars. Their generic name for plantains, '*–bugu*', is found in a huge portion of the eastern rain forest. There is a total absence of '*–kondo*', the generic name so typical for the Western Bantu culture. The time at which their ancestors entered the forest remains obscure, but the generic name configuration could point to an establishment of Central Sudanese people in the forest before the Western Bantu. In this case, there is evidence that the plantain diffusion to the west across central Africa underwent a ramification to the south at an early stage.

Phase 3: The Subsequent Move into the Dense Rain Forest

It is suggested here that plantain, taro and water yam joined the initial vegecultural suite at about the time of the expansion of the Bantu culture into the rain forest, thereby supplying it with a significant boost. Whether cultivation was restricted to backyards or in cleared forest plots remains to be investigated, but forest clearing was no longer a problem with iron tools.

In the course of the Bantu expansion, the importance of both the African and Asian yam diminished to the advantage of the plantain (Vansina 1990). The spectacular and unique diversification in plantain cultivars was the outcome of this further expansion. Swampy areas had to be avoided and the move into the Congo Basin would consequently have been restricted to the northern side of the wet forest, with a minor branch towards the area where current Yumbe speakers are settled. The Western Bantu cultural expansion could have met with a Central Sudanese one in the eastern rain forest region; this is a point for further research.

CONCLUSION: TESTING THE HYPOTHESIS

Plantains (AAB-genome subgroup[7]) in the African rain forest, together with the east African Highland bananas (AAA-genome subgroup) and the 'Maia maoli-Popoulu-Iholena' bananas in Polynesia (another AAB-subgroup, sometimes called 'Pacific plantains'), are the only three subgroups displaying extraordinary diversity while also being cultivated far from the original wild *Musa* germplasm in south Asia and New Guinea. These groups are found nowhere else. Such diversity can thus only be explained by a long series of cumulative somatic mutations in the regions where they are currently located.

The present hypothesis places the plantain odyssey across Africa at least one millennium earlier than currently accepted. This date received its first substantial confirmation with the *Musa* phytolith findings in Cameroon dated to 2500 BP.

Even elaborate theories in support of a more recent introduction are vague as to when and how the initial plantains crossed East Africa, because there is a lack of reliable banana-type differentiation in relevant historical sources (Rossel 1998). The theories do not explain the contrast between the great diversity in plantain cultivars in the African rain forest and the meagre diversity in the regions of origin of the banana, which are equally rich in rain forest flora. Moreover, various other bananas would have accompanied plantains introduced during historical times, but no trace of such bananas in traditional rain forest agriculture is found. The investigation of linguistic data is providing indications as to how plantains may have moved in the early Bantu period (Rossel 1998), but is not of much help to a solid reconstruction of plantain history in Africa.

The hypothesis presented here circumvents the above difficulties. It assumes that the diffusion of the plantain across Africa occurred much earlier than traditionally accepted, and that it was not accompanied by other bananas. A few initial cultivars crossed East Africa in a trial-and-error way with the help of pre-Bantu people already practising vegeculture. The diffusion of these plants eventually ended in the rain forest where a more than 2000-year-long cultivation history explains the extraordinary diversity of plantains, a diversity that is found nowhere else, not even in the regions of origin on the other side of the Indian Ocean.

Musa phytoliths are the obvious tools for verifying, modifying or nullifying the present hypothesis (Figure 19.3). However, the search for such phytoliths should be restricted to the Central African part of the supposed pathway, ie phases 2 and 3. The morphological/morphometrical study of *Musa* phytoliths has not differentiated between phytoliths of plantains and of other cultivated bananas. Consequently, we cannot avoid confusion in East Africa with phytoliths of the long-popular highland bananas, which belong to the triploid *M. acuminata* genome group, since these bananas were likely already present in the Great Lakes area by the first half of the first millennium AD (Schoenbrun 1993: 52).

Musa phytoliths detected in archaeological context in the rain forest of Cameroon almost certainly belong to the plantain subgroup, since no local names point to any other banana type (as explained above).

Research into finding, differentiating and dating *Musa* phytoliths will clarify the various facets of the hypothesis proposed in this chapter. For example, a date of 3000 BP in the eastern part of the wet forest would exclude a Bantu presence there at that time, but would leave open the possibility of early Central Sudanese influence, if these language speakers had the technology to clear forest. Phytoliths found in a 3000–2500 BP stratum farther north, along the route proposed as phase 2, would strengthen the hypothesis.

This contribution uses physical, botanical and anthropological data in order to show that plantains may have diffused across the African continent more than 3000 year ago. If *Musa* phytoliths are further detected in an archaeological context comparable to that of the 2500-year-old phytoliths found at Nkang in Cameroon, the hypothesis would provide a new framework for investigating the early development of agriculture in central Africa.

ACKNOWLEDGMENTS

It is with pleasure that I express my gratitude to Tim Denham for his numerous valuable suggestions; they substantially improved the quality of the contribution.

NOTES

1. All dates reported in this chapter are uncalibrated.
2. Vegeculture: a horticultural system based on vegetative reproduction of root and tuber plants.
3. Phonology: β*asi* in Chaga; *a:*θ *i* in Gekoyo; *washi* in Shambaa. The Maasai call them *il-tórrobo*. The survivors are called Aramanik or Asax in the Chaga area.
4. β*akoniŋgo* (Chagga); β*ataɍimba* (Machame); *si*β*ira* (Gweno, Asu); *gumba* (Gekoyo). The 'ɍ' stands for the typical (semi-guttural?) 'r' common in Chaga and used in the generic name for banana, *iɍu'u*.
5. Phillipson wonders (1984: 218): 'les locuteurs du ou des parler(s) bantou ayant donné naissance aux dialectes chaga actuels ont-ils trouvé sur le Kilimanjaro un people agricole les ayant précédés et ayant déjà développé le type d'exploitation du milieu, basé sur les plantes d'origine sud-asiatique…?'
6. *Ensete ventricosum* still is a sacred plant among the traditional Bantu clans in East Africa.
7. The majority of the edible bananas are triploids. The genomes can be either 'A', derived from *Musa acuminata*, or 'B', derived from *Musa balbisiana*. Domestication of the banana started in the New Guinea area with wild *M. acuminata*, which became parthenocarpic, thus producing edible AA diploids. These diploids subsequently formed hybrids with either of the two species, thus generating the triploid combinations.

REFERENCES

Carreel, F (1994) 'Etude de la diversité génétique des bananiers genre *Musa* à l'aide des marqueurs RFLP', unpublished thesis, Institut National Agronomique, Paris-Grignon

Carreel, F, Gonzalez de Leon, D, Lagoda, P, Lanaud, C, Jenny, C, Horry, J P and Tezenas du Montcel, H (2002) 'Ascertaining maternal and paternal lineage within *Musa* by chloroplast and mitochondrial DNA RFLP analyses', *Genome* 45: 679–92

De Langhe, E and de Maret, P (1999) 'Tracking the banana: its significance in early agriculture', in Gosden, C and Hather, J (eds), *The Prehistory of Food: Appetites for Change*, pp 277–96, London: Routledge

De Langhe, E, Swennen, R and Vulsteke, D (1994–5) 'Plantain in the early Bantu world', *Azania* (Nairobi) XXIX–XXX: 147–60

Ehret, C (1971) *Southern Nilotic History: Linguistic Approaches to the Study of the Past*, Evanston, IL: Northwestern University Press

Ehret, C (1987) 'Proto-Cushitic reconstruction', *SUGIA, Sprache und Geschichte in Afrika* 8: 7–180

Hamilton, A C and Taylor, D (1991) 'History of climate and forests in tropical Africa during the last 8 million years', *Climate Change* 19: 65–78

Mbida, C, Doutrelepont, H, Vrydaghs, L, Swennen, R L, Swennen, R J, Beeckman, H, De Langhe, E and de Maret, P (2001) 'First archaeological evidence of banana cultivation in central Africa during the third millennium before present', *Vegetation History and Archaeobotany* 10: 1–6

Mercader, J, Runge, F, Vrydaghs, L, Doutrelepont, H, Ewango, C and Juan-Tresseras, J (2000) 'Phytoliths from archaeological sites in the tropical forest of Ituri, Democratic Republic of Congo', *Quaternary Research* 54: 102–12

Phillipson, G (1984) *'Gens des Bananeraies': Contribution Linguistique à l'Histoire Culturelle des Chaga du Kilimanjaro.* Cahier 16, Paris: Editions Recherche sur les Civilisations

Roche, E and Van Grunderbeek, M-C (1987) 'Apports de la palynologie à l'étude du Quaternaire supérieur au Rwanda', *Mémoires et Travaux EPHE* (Montpellier) 17: 111–27

Rossel, G (1998) 'Taxonomic-linguistic study of plantain in Africa', unpublished PhD thesis, School of Asian, African and American Studies, Leiden, The Netherlands

Schanz, J (1913) *Mitteilungen über die Besiedlung des Kilimandscharo durch die Dschagga und deren Geschichte*, Leipzig: Teubner

Schoenbrun, D (1993) 'Cattle herds and banana gardens: the historical geography of the western Great Lakes region, c AD 800–1500', *African Archaeological Review* 11: 39–72

Simmonds, N W (1962) *The Evolution of the Bananas*, London: Longmans

Simmonds, N W [1959] (1966) *Bananas*, London: Longmans

Sowunmi, M A (2002) 'Environmental and human responses to climatic events in West and west central Africa during the Holocene', in Hassan, F A (ed), *Droughts, Food and Culture: Ecological Change and Food Security in Africa's Later Prehistory*, pp 95–104, New York: Kluwer Academic/Plenum Publishers

Stahl, K M (1964) *History of the Chagga People of Kilimanjaro*, London: Mouton

Swennen, R (1984) 'A physiological study of the suckering behaviour in plantain (*Musa* cv. AAB)', unpublished PhD thesis, Katholieke Universiteit, Leuven, Belgium

Swennen, R (1990) 'Limits of morphotaxonomy: names and synonyms of plantain in Africa and elsewhere', in Jarret, R L (ed), *Identification of Genetic Diversity in the Genus* Musa, pp 172–210, Montferrier-sur-Lez, France: INIBAP

Vansina, J (1990) *Paths in the Rainforests: Towards a History of Political Tradition in Equatorial Africa*, Madison, WI: University of Wisconsin Press

CHAPTER 20

African Pastoral Perspectives on Domestication of the Donkey: A First Synthesis

Fiona Marshall

INTRODUCTION

After millions of years as hunters and gatherers, people domesticated a variety of plant and animal species during the last 12,000 years. The domestication of large mammals provided early food producers with predictable sources of animal food and raw materials and influenced the density and mobility of human settlement, as well as trade, warfare and the development of disease. Given their wide-ranging influence, surprisingly few large mammals were ever domesticated, however, and most for food. Recent zooarchaeological and genetic research has improved understanding of the region and timing of domestication of many large mammals and suggests multiple domestication events for horses, pigs, cattle, sheep and goats (Bradley and Loftus 2000; Edwards et al 2004; Ghuffra et al 2000; Hanotte et al 2002; Hienleder et al 1998; Jansen et al 2002; Larson et al 2005; MacHugh and Bradley 2001; Rosenberg and Redding 1998; Vigne et al 2000; Zeder and Hesse 2000). There are still gaps in information regarding the place and timing of domestication of major livestock species, especially from arid or tropical regions. Moreover, little is known about local contexts for domestication or the way that hunter-gatherers and early food producers combined isolated domesticates from their own and other regions in the development of new farming systems and societies. Information on the reasons for these ancient realignments of relations among animals, people and environments can provide important insights on the trajectories of ancient societies, as well as on the nature of global food systems, biodiversity and geopolitical power structures today (Clutton-Brock 1989, 1999; Diamond 1997; Harris 1996; Harris and Hillman 1989; Piperno and Pearsall 1998; Price and Gebauer 1995; Smith 1998).

Donkeys are native to the arid tropics and subtropics of northeast Africa. They were the chief source of land transport for ancient Egyptians and Sumerians and are thought to have played an important role in the development of ancient trade in western Asia and the Mediterranean. Together with mules and horses, they remained the most important form of transport in much of the world prior to the introduction of railways. In contemporary rural societies of Africa, western Asia, central and South America, donkeys are an essential pack animal (Fielding and Pearson 1991; Starkey 2000). Donkeys are, however, among the least studied of the widely distributed large domestic mammals of the world.

The African wild ass (*Equus africanus* Heuglin and Fitzinger 1866) has long been thought to be the ancestor of the domestic donkey (*Equus asinus* Linnaeus, 1758) (Brehm 1915; Clutton-Brock 1992a; Epstein 1971; Heck 1899; Murdock 1959: 105; Zeuner 1963). They hybridise readily with domestic donkeys, are highly endangered today and are found only in the Horn of Africa at 9.56–18º N (Groves 2002: 101–02; Moehlman 2002; Moehlman et al 1998; terminology follows Gentry et al 1996, 2004). Because the distribution of *Equus africanus* is primarily African and donkey bones were found in Predynastic Egyptian sites (6000–5000 BP) (Peet 1914; Petrie 1914), donkeys have historically been considered an Egyptian domesticate. Since the nineteenth century scholars have thought that they were important to the development of trade in ancient Egypt (Brehm 1915; Clutton-Brock 1992a; Epstein 1971: 397; Heck 1899; Romer 1928; Zeuner 1963). Hassan (1988: 158–59, 161, 168, 1993: 557, 559, 565) has also suggested that donkeys played a role in the distribution and integration of nomes (territorial divisions) and management of the Egyptian state. In the 1980s, however, identification of possible *E. africanus* at sites in the Levant and Arabian Peninsula and the extensive analysis of early donkeys at sites such as Tal-e Malyan in Iran (c 2800 cal BC, c 4200 BP) directed attention to western Asia as a possible context for domestication of the donkey (Clutton-Brock 1986, 1992a; Meadow and Uerpmann 1991; Uerpmann 1987, 1991; Zeder 1986).

Recently, well-preserved specimens of *Equus asinus* have been recovered and studied from Africa, chiefly from Egyptian Predynastic and Early Dynastic sites such as Abusir, Buto and Hierakonpolis (Boessneck et al 1992; Driesch 1997; van Neer pers comm), but also from the Sudanese pastoral site of Wadi Hariq (Berke 2001). Donkeys have not been a focus of zooarchaeological research in Africa, however, and the low densities of wild ass or donkey bones at archaeological sites beyond the Nile have not encouraged specialised study of, or communication about, such specimens. As a result, while Asian scholars have subjected the equid archaeological record to scrutiny, Africanist archaeologists have not yet revisited the larger issues of the context and significance of the possible domestication of the donkey in Africa.

Since the domestication of the donkey in Egypt was first proposed, there has also been a significant shift in thinking regarding the nature and origins of the earliest domesticates in Africa and the beginnings of food production on the continent. Despite the presence of early ceramics in Africa (reviewed in Close 1995) and approaches to 'non-centric' domestication of plants conceived by Harlan (1971), studies of the beginnings of food production in Africa have been much influenced by Near Eastern paradigms (Garcea 2004). It has long been expected that similar domestic grains, contemporary or slightly later domestic animals and settled early agricultural communities would be found in northern Africa (Clark 1976; Shaw 1977). Much research remains to be conducted to refine the exact timing and location of the appearance of domestic plants in Africa, but it is now clear that the sequence of domestication of plants and animals in Africa and the subsequent development of early food-producing societies differs in interesting ways from regions such as western Asia. It is accepted, largely on the basis of genetic evidence, that cattle were domesticated 9000–7500 BP in the Sahara. African plants were not, however, domesticated until significantly later, after c 4000 BP (Bradley and Loftus 2000; Gifford-Gonzalez 2005; Hanotte et al 2002; Marshall and Hildebrand 2002; Neumann 2003, 2005; Wendorf et al 2001). The earliest food producers did not cultivate domestic plants and were mobile cattle-based, wild grass-collecting herders, rather than settled agriculturalists as in Southwest Asia. Sheep and goat were introduced to Africa, possibly via the Sinai, by 7000 BP and incorporated into cattle-herding societies. Saharan pastoralists became increasingly mobile as the grasslands dried from c 7000 BP (Gautier 1987a; Marshall and Hildebrand 2002).

The early domestication of animals, late domestication of plants and continued use of wild resources influences the conceptual framework for, and trajectory of research on, domestication processes in Africa. Africanist scholars use the term 'food production' to refer to heavy reliance on domestic animals, domestic plants or both. It is understood that most early African food producers also maintained a wild resource base. This could be thought of as 'ecodiverse food production'. Because pastoralism is an early food-producing system, rather than a later development associated with cities, and domestic animals were used earlier than domestic plants in Africa north of the equator, the terms 'origins of agriculture' and 'the beginnings of food production' are not used interchangeably as they are in many other regions. Research on the beginnings of food production in Africa has focused on several major themes including indigenous versus exotic origins of African cattle, adoption of cattle and sheep and goats by Saharan hunter-gatherers, the timing of appearance of African domestic plants and increasing group mobility against a backdrop of mounting aridity and highly variable rainfall. Classical scholarship drew attention to the Nile and an ancient Egyptian context for the domestication of the donkey in Africa. Thus far, archaeologists have not seriously considered the possibility that herders of northeast African grasslands domesticated another indigenous African mammal, the donkey, 7000–6000 years ago to solve transport problems posed by climatic deterioration and their increasingly mobile way of life (Beja-Pereira et al 2004; Marshall 2003; see also Brewer et al 1994: 99; Olsen 1996: 202). Domestication of donkeys and their incorporation into a society dependant on domestic cattle, Southwest Asian sheep and goat and African wild plants, are likely to have had significant consequences for development of African pastoralism and the spread of food production in Africa. In the following section I discuss ethnoarchaeological data on the significance of donkeys for contemporary African pastoral societies and use this information to develop a pastoral hypothesis for the domestication of the donkey in Africa. I go on to review palaeontological and archaeological literature that bears on this hypothesis, comparative archaeological data from Asia and the contributions of recent genetic and linguistic studies to understanding the timing, location and process of domestication of the donkey.

INFLUENCE OF THE DONKEY ON PASTORAL SOCIETIES: ETHNOARCHAEOLOGICAL INSIGHTS

Preliminary ethnoarchaeological study and synthesis of the available ethnographic literature on eastern and northern African pastoralists such as the Maasai (Jacobs 1975), Turkana (Little and Leslie 1999), Borana (Coppock 1994), Barabaig (Klima 1970) and Tuareg (Nicolaisen 1963; Nicolaisen and Nicolaisen 1997), suggest that donkeys play a key role in the organization of contemporary mobile herding societies in Africa. I discuss here the role that biology played in making the donkey the pack animal of choice, specific tasks for which donkeys are employed and the larger implications of the use of donkeys for the organisation of contemporary and prehistoric pastoral societies. Ethnoarchaeological data are the basis of my main hypothesis that donkeys were domesticated in northeastern Africa by herders.

Cattle (*Bos taurus* Linnaeus, 1758) are much more numerous than donkeys in pastoral societies, but donkeys are the main pack animal of Africa. Camels (*Camelus bactrianus* Linnaeus, 1758) are a more recent introduction (Rowley-Conwy 1988). The main advantage of donkeys over cattle as pack animals in Africa derives from their desert heritage and the mechanical efficiency of their gait. Donkeys are faster and energetically more efficient at walking and carrying loads than cattle, especially over steep and stony terrain (Dijkman 1991; Jones 1977: 5, 11; Yousef 1991). They have a labile body temperature, tolerate high levels of desiccation and rehydrate very quickly (Maloiy 1970, 1971; Maloiy and Boarer 1971). Donkeys do not require a rest period for rumination and are able to digest while dehydrated, unlike most large mammals. It is also argued that donkeys are especially intelligent and easy to train (Jones 1977: 9, 12). Because cattle are easily dehydrated, it is difficult to keep them in good condition as pack animals in arid African rangelands. Donkeys, by contrast, are adapted to the desert and very tough.

Donkeys are used for a wide range of tasks in African pastoral societies. They are essential for moving pastoral household goods, small children and animals from one camp to another. This allows pastoral families to make frequent and at times long-distance moves with their animals and possessions. Spencer (1973: 14) notes that during the 1960s, Samburu women packed all their belongings onto donkeys for migrations approximately every five weeks. Turkana families use donkeys for dry season moves every few weeks over distances ranging from 10 to 50 km (Dyson-Hudson and Dyson-Hudson 1999: 79; Leslie and Dyson-Hudson 1999: 233; McCabe et al 1999: 108), and the Wodabe Fulani still use donkeys for seasonal migrations of more than 100 km (Pao and Palin 2002: 119). Donkeys are also important for assisting with daily tasks such as collecting household water and firewood

(Århem 1985; Coppock 1994: 153; Jacobs 1975: 408; Little et al 1999: 58; Nicolaisen and Nicolaisen 1997: 162–224). Livestock maintenance, including transport of water for young livestock (calves, kids and lambs), forage for calves and collection of salt, is also facilitated by using donkeys as pack animals (Baxter 1991: 194; Coppock 1994: 153; Jacobs 1975: 408; Klima 1970: 10; Little et al 1999: 58). Finally, even among the many African pastoral groups who do not engage in specialised trading, donkeys are used to transport goods home from trading centres (Fratkin 1991: 47; Homewood and Rodgers 1991: 145; Klima 1970: 10).

Donkeys are not generally eaten, but may provide an emergency source of meat and milk (Johnson 1999: 99; Leslie and Dyson-Hudson 1999: 233; Little et al 1999: 58; Nicolaisen and Nicolaisen 1997: 163, 164). They are, however, probably eaten more than is reported; the presence of wild ass in northeast African sites over the last 20,000 years shows that they have a long history of use for food in the region (Table 20.1). Nevertheless, the main role of the donkey in Africa today is as a pack animal. Most families do not own many donkeys, not more than six, but I know of no *mobile* pastoral groups living in semi-arid regions of Africa that do not use donkeys.

Many of the tasks for which donkeys are used are considered 'women's work', and donkeys are often considered women's animals and of relatively low status (Mohammed 1991: 187). The use of the donkey as a pack animal has far-reaching consequences, however, for organization of movement, land management and labour, as well as for human and livestock health and population dynamics. Employing donkeys to collect water from distant sources allows present-day pastoral camps to move as a unit to follow pasture far from permanent water (Jacobs 1975: 408, 417; Little et al 1999: 58, 60). Such residential mobility in turn allows conservation of dry season grazing and resources close to permanent water, a rapid response to erratic rainfall and flexible use of pastures. The role of the donkey in bringing water to small stock avoids the use of cattle for fetching water and improves the health of cattle and small stock, especially during droughts (Little et al 1999: 58). The consequent reduction in women's labour also bears on long-term reproductive health. A recent study of Oromo villagers in southern Ethiopia suggests that reducing the time and energy spent obtaining water increases women's fertility (Gibson and Mace 2002: 631, 636). Reproduction rates of both humans and livestock may be affected by using donkeys for transport of young. In summary, donkeys allow closer birth spacing and larger mobile populations of both humans and livestock, change the dynamics of women's labour and allow rapid mobile response to erratic rainfall and pastoral survival in semi-desert regions.

DONKEYS BENEATH THE RADAR: TAPHONOMIC ISSUES

Ethnoarchaeological data provide useful insights into site formation processes affecting the likelihood of archaeological preservation and identification of *E. asinus* skeletal material on African pastoral sites. Despite their

Table 20.1 Information on sites with African wild ass, *E. africanus* (Figure 20.2)

Site	*Time Period*	*Sample*	*References*
AFRICA			
Morocco			
Mugharet el Aliya	Late Pleistocene	*E. africanus*	Howe and Movius 1947; Churcher and Richardson 1978: 380, 412
Grotte de Kiefan Bel Gomari	Late Pleistocene	*E. africanus*	Romer 1928: 122; Churcher and Richardson 1978: 380, 412
Algeria			
Lac Karar	Middle Pleistocene–Holocene	*E. africanus*	Churcher and Richardson 1978: 380, 412
Djelfa	Middle–Late Pleistocene	*E. africanus*	Romer 1928: 122; Churcher and Richardson 1978: 380, 412
Oued Seguin	Below Mousterian Levels	*E. africanus*[?]	Romer 1928: 122; Churcher and Richardson 1978: 380, 412
Ternifine/Palikao	Late Pleistocene–Holocene (Ternifine: 26), zone 2	*E. africanus*[?]	Romer 1928: 122; Arabourg 1931: 174, 1952: 8; Churcher and Richardson 1978: 380, 412
Carrière des Bains	Late Pleistocene	*E. africanus*[?]	Romer 1928: 122; Arambourg 1931: 174; Churcher and Richardson 1978: 380, 412
Pointe Pescade	Late Pleistocene	*E. africanus*[?]	Arambourg 1931: 174; Churcher and Richardson 1978: 380, 412
Grotte Ali Bacha	Mousterian–Holocene	*E. africanus*	Romer 1928: 122; Churcher and Richardson 1978: 380, 412
Grotte des Troglodytes	Neolithic	*E. africanus*	Romer 1928: 122; Churcher and Richardson 1978: 380, 412
Saida	Neolithic	*E. africanus*	Romer 1928: 122
de la Tranchée	Neolithic	*E. africanus*	Romer 1928: 122
Grotte du Polygone	Holocene	*E. africanus*	Churcher and Richardson 1978: 380, 412
Abd el Kadar	Holocene	*E. africanus*	Churcher and Richardson 1978: 380, 412
Grand Rocher	Holocene, Neolithic	*E. africanus*	Romer 1928: 122; Churcher and Richardson 1978: 380, 412
Grotte de Bou Zabaouin	Neolithic	*E. africanus;* 27 dental specimens	Romer 1928: 122; Churcher and Richardson 1978: 380, 412
No. 25 Near Garet et Tarf	Upper Paleolithic	*E. africanus*	Romer 1935: 173–85; Churcher and Richardson 1978: 380, 412
Garet et Tarf	Holocene	*E. africanus*	Churcher and Richardson 1978: 380, 412

(*Continued*)

Table 20.1 (*Continued*)

Site	*Time Period*	*Sample*	*References*
Aioun Beriche	Holocene	*E. africanus*; more than 6 dental specimens	Churcher and Richardson 1978: 380, 412
El Mouhaad, Ain Beida Region	Early Holocene, Capsian Shell Midden	*E. africanus*; 30 dental specimens	Romer 1935: 173–85; Churcher and Richardson 1978: 380, 412
Site No. 12-Ain Beida region	Early Holocene, Capsian Shell Midden	*E. africanus*; lower jaw, isolated teeth	Romer 1935: 173–85; Churcher and Richardson 1978: 412
Oued Tamanrasset	Post-Pleistocene[?]	*E. africanus*	Romer 1935: 177; Churcher and Richardson 1978: 380, 412
Libya			
Ti-n-Torha East	9th millennium BP	*E. africanus*; 11 specimens	Gautier and van Neer 1977–1982: 96, 100–103, 106
Uan Muhuggiag	7–6th millennium BP	*E. africanus*; 3 specimens	Gautier 1987b: 286, 288
Hagfet et Tera	Upper Palaeolithic	*E. africanus*[?]	Churcher 1974: 371
Haua Fteah	earlier than 8000 BP, late Dabba, Lybico-Capsian	*E. africanus*[?]	Higgs 1967; Churcher 1972: 49, 1974: 374; Churcher and Richardson 1978: 412
Egypt			
Bir Sahara	Middle Stone Age	*E. africanus*	Gautier 1980, 1984
Bir Tarfawi	Middle Stone Age	*E. africanus*	Gautier 1980, 1984
Gilf Kebir, Wadi Bakht	Late Stone Age	*E. africanus* 5 teeth, 11 postcranial specimens	Gautier 1980; Gautier and van Neer 1977–1982; Uerpmann 1987: 28–29
Qâu	Holocene	*E. africanus*	Churcher 1972: 135
Dakhleh, Sheikh Muftah East	Neolithic	cf *E. africanus* or *E. asinus*; 3 tarsals	Churcher 1986: 419
Kom Ombo	c 15,000–13,000 BP	*E. africanus*	Reed and Turnbull 1969: 55–56; Churcher 1972: 49–52, 1974: 364, 367
Abu Simbel east bank	Early Palaeolithic–Late Neolithic	*E. africanus*	Uerpmann 1987
Sudan			
Wadi Shawc	5000–3000 BP	*E. africanus*	Uerpmann 1987: 28–29
Wadi Shaw	c 15,000–9000 BP Sebilian, Qadan	*E. africanus*	Gautier 1968; Churcher 1974:128, 135;
Wadi Halfa area, ASG-G-25	Abkan earlier than 5000 BP	*E. africanus*; 1 complete mandible	Perkins 1965: 57–59 Nordström 1972; Gautier 1984: 49
Singa	Middle Stone Age	*E. africanus*[?]	Bate 1951: 3; Churcher and Richardson 1978: 380, 412; Gautier 1984: 45
Abu Hugar	Middle Stone Age	*E. africanus*[?]	Bate 1951: 3; Churcher and Richardson 1978: 380, 412; Gautier 1984: 45

(*Continued*)

Table 20.1 (*Continued*)

Site	*Time Period*	*Sample*	*References*
Somalia			
Midhishi	MSA/LSA	*E. africanus*	Bunn quoted by Brandt 1986: 56
Guli Waabayo	MSA/LSA Magosian level	*E. africanus*; 1 cheek tooth	Bate 1954; Marshall pers observation
Buur Heybe	LSA levels	*E. africanus*[?]	Brandt 1986
Buur Hakeba Rifle Range site	MSA/LSA Doian level	*E. africanus*; 7 cheek teeth	Bate 1954; Brandt 1986: 66
Hargeisa H.7	MSA/LSA Somaliland Stillbay	*E. africanus*; 6 cheek teeth	Bate 1954
ASIA			
Jordan			
Ra's en Naqb	Natufian, Chalcolithic	*E. africanus*	Uerpmann 1991: 15–16
Ain Ghazal	Natufian, Chalcolithic	cf *E. africanus*; several cheek teeth, numerous long bone ends	Driesch and Wodtke 1997: 530–33
Wadi Mataha	Early Natufian	cf *E. africanus*; 1 central tarsal	Whitcher et al 2000: 43–44
Syria			
Mureybit	earlier than 10,000–9000 BP	cf *E. africanus*; 1 metatarsal	Ducos 1975, 1986
Shams ed-Din	c 5000–4500 BC, Halafian	2 first phalanges cf *E. africanus*; *E. hemionus* present	Uerpmann 1986: 256, 259
United Arab Emirates			
Hili 8	Bronze Age, c 4400 BP (c 3000 cal BC)	cf *E. africanus* or *E. asinus*; 6 cranial, 26 postcranial specimens	Uerpmann 1991: 15–16
Oman			
Ra's al-Hamra, RH-5, -6, -10	c 5500–2000 cal BC, shell middens	cf *E. africanus*; 2nd phalanx, Rh-5	Uerpmann 1991: 14
Saudi Arabia			
Ain Qannas	c 5500 BC	cf *E. africanus*; also cf *E. hemionus*, 20 specimens	Uerpmann 1991: 16; Zeder in Masry 1973: 236
Yemen			
Ash Shumah	8th millennium BP, shell date (c 6684–6475 cal BC)	*E. africanus*; 930 specimens	Cattani and Bökönyi 2002: 44–51

importance, donkeys have relatively low status among African pastoral groups, especially in comparison to cattle. For this reason, and because donkeys are better at digging for water and deterring predators than are cattle, sheep and goats, donkeys are generally managed less than other livestock.

Many pastoralists such, as Maasai, Turkana, Borana and Orma, do not herd donkeys (Coppock 1994: 111; Ensminger pers comm 2002; Little et al 1999: 56). Donkeys are the only livestock not necessarily penned at night for protection from predators. They are not kept in large numbers and seldom eaten. As a result, their bones are not likely to have accumulated in ancient pastoral settlements. Furthermore, unlike cattle, donkeys were neither ceremonially buried, nor a common subject in symbolically important rock art. Taking settlement, burial and rock art data together, donkeys are likely to be under-represented in the archaeological record. Moreover, some of the same husbandry practices that contribute to the scarcity of donkeys in archaeological sites also suggest that they will be difficult to distinguish from wild ass. African herders place a premium on the strength of donkeys, and management of the donkey is minimal. Under such conditions there is not likely to be strong selection for size decrease or morphological change. As a result, donkeys are likely to be rare and difficult to identify in early pastoral sites.

PASTORAL HYPOTHESIS FOR DOMESTICATION OF THE DONKEY IN AFRICA

Cattle were domesticated in northern Africa at c 9000–8000 BP in order to facilitate intensive mobile use of the landscape (Hanotte et al 2002; Marshall and Hildebrand 2002). As this review shows, domestication of the donkey a few thousand years later could have radically changed the scale of strategic use of mobility and the organization of early herding groups. Domestic donkeys are not present at early pastoral sites such as Nabta Playa in the eastern Sahara at c 8000 BP, Enneri Bardagué in the Tibesti, or Uan Muhaggiag in the Acacus at c 6800 BP (Gautier 1987a, 1987b). These herders, although mobile, would have been more tied to permanent water than contemporary pastoralists. At Nabta Playa, for example, the presence of visible hut floors and the structured use of space (Close and Wendorf 1992) suggest greater investment in place than is common among recent pastoralists. Without abundant rainfall, such groups would have been vulnerable to depletion of local resources such as grazing, wild plant foods and firewood. They would have also been more reliant on frequent trips by small groups, made up of young men with mobile adult cattle, for targeted grazing of green flushes and the collection of salt, wood, clay, stone and other resources. When the whole settlement moved, families would have traveled shorter distances more slowly, responded less rapidly to erratic rainfall and carried fewer possessions. Groups could not have moved far or fast as a unit without endangering the health of people and herds.

After the wet period of the early Holocene the archaeological record shows progressive climatic deterioration in northeast Africa, with especially unpredictable rainfall and some marked droughts from 7000 BP (Cremaschi 2002; Hassan 2002; Wendorf et al 2001). Sites in the Acacus, and other regions of the Sahara after this time, provide evidence of shorter stays and higher mobility (Barich 2002; Cremaschi et al 1996; di Lernia 2002; Gautier 1987b; Gautier and van Neer 1977–1982: 82). Rock art panels such as those of the Tassili in the Sahara suggest that early herders may have used cattle to carry people and other loads (Holl and Dueppen 1999; Figure 20.1). Interpretation of such images is difficult and dates are unknown, but it is possible that the use of cattle for transport was an early solution to problems of group mobility. This would not have been without its problems: climatic deterioration was rapid and rainfall increasingly unpredictable. In semi-desert conditions donkeys would have been a much more efficient transport animal than cattle. Combining ethnographic perspectives on the utility of donkeys with archaeological perspectives on climatic variability and challenges for early pastoral groups, I hypothesise that early northeast African pastoralists domesticated the donkey under conditions of increased aridity c 7000–6500 BP in order to facilitate long-distance mobility while maintaining the growth and health of human, cattle, sheep and goat populations.

AFRICAN WILD ASS AND DONKEY IN THE ZOOLOGICAL AND ARCHAEOLOGICAL RECORDS

In order to evaluate this hypothesis, I review four different lines of evidence: zoogeography, zooarchaeology, linguistics and genetics. I discuss the modern distribution of *Equus africanus* in the region, zooarchaeological data on the ancient distribution of *E. africanus* on the continent, *E. asinus* from African and Asian archaeological sites, as well as recent African linguistic and African and Eurasian genetic research that bears on this question.

The Wild Ancestor: Nomenclature, Variation and Recent Distribution

The wild ancestor of the donkey, the African wild ass (*Equus africanus*), is phenotypically variable. It is possible that four distinct subspecies existed historically in Africa (Figure 20.1) (Groves 1986, 2002; Groves and Willoughby 1981; but see Yalden et al 1986). The Somali wild ass are the largest of the African wild ass, with well-marked leg stripes and no shoulder cross (Groves 1974: 110, 1986, 2002; Groves and Willoughby 1981). The Nubian wild ass has no leg stripes, but two types of shoulder cross (Groves 1974: 109–10, 1986). Its distribution may have graded west into that of the Saharan wild ass. The latter is thought to be a smaller animal, greyer in colour, with a long, thin cross stripe (Groves 1986: 33–34, 2002: 103). Saharan wild ass probably existed in the Ahaggar, Tibesti and Fezzan, but have been little studied. An Atlas variety of African wild ass with a shoulder cross and striped legs may also have survived until c AD 300 (Antonius 1938; Clutton-Brock 1992a; Groves 1986: 33). Groves (1974: 161, 1986: 35–36, 62–65) notes that Nubian wild ass have commonly been considered the ancestor of the domestic donkey, but that there are differences in colour, marking and cranial morphology between both Somali and Nubian populations and domestic donkey. He argues that the ancestor of the donkey may be a form of wild ass now extinct (Groves 1976: 163). Museum collections of African wild ass are, however, very small and many aspects of morphological variability are unexplored.

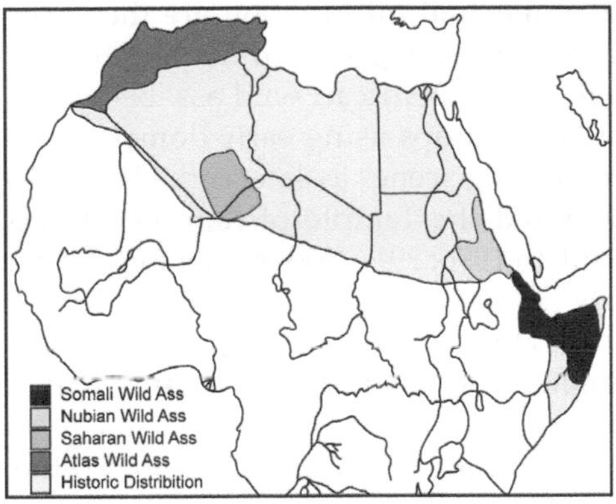

Figure 20.1 Map of historic distribution of African wild ass, *E. africanus* (after Groves 1974, 1986 and Table 20.1).

Distinguishing *E. Africanus*, Zebra and Domestic Donkeys

In order to identify African wild ass in archaeological assemblages, and to differentiate them from domestic donkey, researchers use morphological criteria developed by Groves (1974: 185, 1986: 36–37) and Eisenmann and colleagues (Divé and Eisenmann 1991; Eisenmann 1986; Eisenmann and Beckouche 1986), as well as size data from Egyptian and other sites. Distinguishing African wild ass from other wild equids is not as challenging in Africa as in western Asia, where variation in similar-sized equids is greater. In Africa, Grevy's zebra (*Equus grevyi* Oustalet, 1882) are sympatric but generally larger than African wild ass. Burchell's zebra (*Equus burchellii* Gray, 1823) are generally distributed farther south than African wild ass; although similar in size, they are significantly more robust postcranially. Eisenmann (1986; Divé and Eisenmann 1991; Eisenmann and Beckouche 1986) has done much research on the relative proportions of postcranial elements of *E. africanus*, especially metapodials, which are useful for differentiating African equids. Dental characteristics, especially those of the lower cheek teeth, are widely used to identify African wild ass (Bökönyi 1985; Churcher and Richardson 1978; Eisenmann 1986; Groves 1974; Uerpmann 1991); cranial proportions are also useful (Eisenmann 1986; see also Groves 1986). Comparative specimens of modern *E. africanus* are so rare, however, that little is known about variation in size and proportions.

Nevertheless, modern domestic donkeys are significantly smaller than *E. africanus* and most zooarchaeologists distinguish African wild ass from domestic donkeys on the basis of size. Morphological characters are also important for recognising domestication. Groves (1986: 36) argues that the occipital crest is square in wild asses and shorter and rounder in domestic asses. He also notes that the nasofrontal suture of donkeys is much straighter than that of the wild ass. Eisenmann and colleagues (Divé and Eisenmann 1991; Eisenmann and Beckouche 1986) note significant differences between the proportions of the lower limbs, especially front limbs, metacarpals and phalanges of African wild ass and donkeys.

Together, size and contextual information are the criteria most widely used by Africanist zooarchaeologists for distinguishing the bones of domestic donkey from those of African wild ass. Because African wild ass continued to be hunted by groups using early domestic donkeys – they are depicted in Egyptian hunting scenes as late as the New Kingdom as in the tomb of Tutankhamen and the Temple of Ramses III at Medinet Habou (Closse 1998: 32; Houlihan 1996: 29) – zooarchaeologists are conservative in identification of early domestic donkey. For these reasons and because ethnographic data shows that selection processes on lightly managed early domestic donkeys are likely to have been associated with limited size and morphological changes, zooarchaeologists recognise late, rather than early or transitional, forms of domestic donkey. This raises the question of fit between definitions of domestication and morphological criteria for identification of early domestic donkeys.

Definitions of domestic animals focus on domestication as a microevolutionary process, with associated genetic change, as well as on the effect of animal domestication on human societies (Clutton-Brock 1992b: 79; Russell 2002). Clutton-Brock (1999: 32) defines a domestic animal as 'one that has been bred in captivity for purposes of economic profit to a human community that maintains total control over its breeding, organization of territory, and food supply.' Not all zooarchaeologists agree with the emphasis on profit. Meadow (1984) stresses a change in focus from the dead to the living animal and Ingold (1980) and Ducos (1989) emphasise the importance of ownership. Genetic change results from intentional mechanisms such as selective breeding, but also unintentionally. Toleration of animals in human environments, protective tending or taming may cause unintentional changes in diet, relations with predators and increases in population densities that result in selection (Clutton-Brock 1992b; Hemmer 1990). Intentional changes such as selective breeding are more likely to occur in the later rather than earlier stages of domesticatory relations. Morphological characters for identification of domestication have more easily been brought to bear on evolutionary changes than changes in property relations, and on the later rather than earlier stages of domesticatory relations.

Following Harris's (1989) concept of a continuum of people-animal interactions, neither protection of wild ass, including taming and protective herding, nor the early stages of domestication, including breeding of donkeys by settled agriculturalists or nomadic pastoralists, are currently recognizable through conventional zooarchaeological measures of size or morphological change. There are, however, several lines of evidence that have potential for recognising earlier domesticatory relations. Geoarchaeological lines of evidence for identification of concentrations of cattle dung on African pastoral sites (Shahack-Gross et al 2003) can be adapted for identification of the dung of African wild ass and used to identify sites where tame *E. africanus* or early domestic donkeys were penned. But if they were kept in small numbers, or for short periods, this method will not be as successful for donkeys as for cattle or horses. The study of pathologies indicative of heavy use may be more successful for donkeys and Early Dynastic Egyptian specimens are currently under investigation. Geometric analyses of long bone shafts also show potential for documenting the work life of wild ass or donkeys. This research is still in the early stages, however, and the current literature reflects significant methodological problems in distinguishing early domestic donkeys from wild ass and the later rather than earlier stages of domestication.

E. AFRICANUS AND *E. ASINUS* IN THE AFRICAN FOSSIL RECORD

A survey of the existing information on the locations and dates of *E. africanus* specimens in Africa shows that sites are distributed from the North African coast across the northern Sahara and through the Horn (Figure 20.2, Table 20.1). There are at least 15 rockshelter sequences from Morocco and Algeria in the ancient range of the Atlas wild ass. Many of these are from classic sequences such as Mugharet el Aliya or el Mouhaad and were studied by famous palaeontologists Romer (1928, 1935) and Arambourg (1931). A concentration of nine sites also occurs from the fringes of the Sahara in Libya to the Nile in Egypt and the northern Sudan. Churcher (1972, 1974, 1986; Churcher and Richardson 1978) is known for his work on material from Kom Ombo and review of African equid fossils. Very few sites fall within the modern distribution of the African wild ass in the Sudan and Horn, including Eritrea, Ethiopia, Somalia and Djibouti. This almost certainly reflects the distribution of research, because the few sites that have been excavated in this region, such as Midishi in Somaliland and Guli Wabayo and Buur Heybe in Somalia, preserve wild ass (Table 20.1).

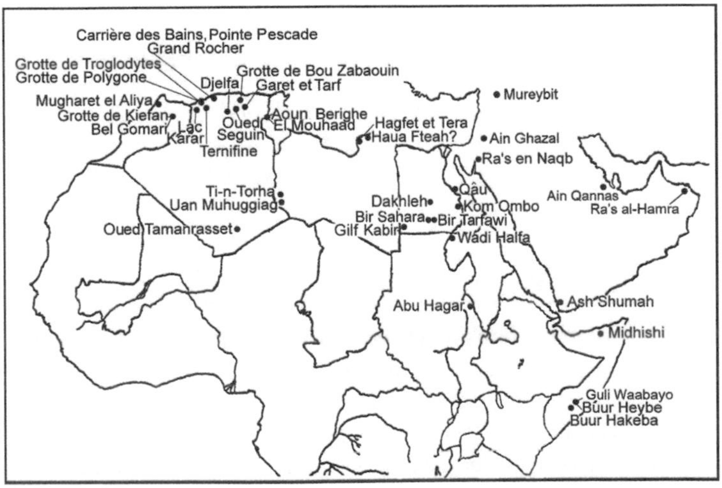

Figure 20.2 Distribution of sites with African wild ass, *E. africanu* (selected from Table 20.1).

Most specimens from African sites reviewed here are not well dated, but range in age from c 40,000 BP to recent. No complete skeletons were recovered from these sites and in general the number of isolated specimens that could be identified as wild ass, and differentiated from zebra, is fairly small (Table 20.1). As a result this review is somewhat tentative and one cannot be certain that specimens from more recent periods might not be attributable to early or large domestic donkey.

In the remainder of this section I review zooarchaeological research on domesticated donkeys (*E. asinus*) from African archaeological sites. During the last fifteen years a number of zooarchaeologists have collected important data on early *E. asinus* in Egypt. Driesch (1997: 31) notes that domestic donkeys occur earlier in Egypt than previously thought and that they are large. The dating of the earliest donkeys lacks precision, but early domestic donkeys are found at six Egyptian Predynastic sites, including El Omari, thought to date to c 4600–4400 BC (c 5700–5500 BP) (Boessneck and Driesch 1998: 99–101), Maadi dating to the first half of the 4th millennium BC (c 6100–5700 BP) (Boessneck et al 1989: 90–92; Bökönyi 1985: 495–98), and the Naqada I settlement at Hierakonpolis (McArdle 1982: 120, 1992: 56) (Table 20.2, Figure 20.3).

Clutton-Brock (1992a: 65; Burleigh et al 1991: 10) has obtained the only direct dates for Predynastic/Early Dynastic equid material from Egypt. This work points out that context cannot be assumed, especially for excavations from the early years of the century. Direct dates from the ass specimens from Badari, considered to date to the end of the Predynastic, showed that the specimen was intrusive. The context-based First Dynasty dates for Tarkan (Petrie 1914) were, however, supported by direct dates, eg c 4390 ± 130 BP, (OXA-566) (Burleigh et al 1991: 10; Clutton-Brock 1992a: 65). Although there are very few AMS dates, it seems clear that domestic donkeys were present in Egypt by c 4000 cal BC (c 5200 BP) (Hollmann 1990: 71) (Table 20.2).

There are relatively few complete skeletons, an exception being the three donkeys buried at Abusir (Boessneck et al 1992: 1–10). Nevertheless, sufficient specimens have been studied to document changes in size through time (Boessneck et al 1992: 2; Boessneck and Driesch 1998: 497; Driesch 1997: 31). Predynastic donkeys from Maadi and Hierakonpolis are large (Boessneck et al 1989: 91; Bökönyi 1985: 497; McArdle 1982: 120, 1992: 56). Donkeys from other Predynastic and Early Dynastic sites such as Elephantine and Buto appear smaller than those from Maadi and larger than Tell e-Dab'a (Boessneck 1976: 22; Boessneck et al 1992: 2; Driesch 1997: 31; Hollmann 1990: 72). The First Dynasty donkeys from Abusir are considerably smaller (Boessneck et al 1992: 2). It is thought that Predynastic donkeys were used for transport (Hollmann 1990: 71). A burned specimen from Maadi suggests that they may also have been eaten, and it is likely that donkeys played a role in Predynastic burial ritual, possibly associated with the god Seth (Hollmann 1990: 71). Continued hunting, or capture, of wild ass is also documented at a number of sites such as Elephantine and Buto (Driesch 1997: 25, 31–32; Hollmann 1990: 71).

Table 20.2 Information on sites with early domestic donkey, *E. asinus* (Figure 20.3)

Site	*Time Period*	*Sample*	*References*
AFRICA			
Egypt			
Maadi	Predynastic, C 4000–3500 cal BC	*E. asinus*; 20 specimens (1985); 113 specimens (1989)	Bökönyi 1985: 495–98; Boessneck et al 1989: 90–92
El Omari	Predynastic, c 4600–4400 cal BC	*E. asinus*; 3 specimens	Boessneck and Driesch 1998: 99–101
Hierakonpolis	Predynastic Naqada I–II, c 3600 cal BC	*E. asinus; fewer than* 20 specimens (1982–1992)	McArdle 1982: 120, 1992: 56; van Neer et al 2004
Buto	Predynastic and Old Kingdom	*E. asinus*; 32 specimens	Driesch 1997: 25, 31–32
Elephantine	Predynastic	12 *E. africanus* and 53 *E. asinus* specimens	Hollmann 1990: 70–73
Tarkan	Early Dynastic, 4390 ± 130 BP (OXA-566, c 2850 cal BC)	*E. asinus*; donkey burials: one cranium, originally 3 skeletons	Petrie 1914; Burleigh 1986: 234; Burleigh et al 1991: 10; Clutton-Brock 1992a: 65; Eisenmann 1995: 10
Abydos	Early Dynastic	*E. asinus*; 10 complete skeletons (2003); two specimens (1911)	Peet 1914: 6; Roessel pers comm 2003
Abusir	First Dynasty	*E. asinus*; burial of 3 individuals	Boessneck et al 1992: 1–10
Tell el-Dab'a	Predynastic–Dynastic	more than 5 individual animals	Boessneck 1976: 21–24; Boessneck and Driesch 1992: 23–24
Sudan			
Kerma	Late fifth–early fourth millennium BP (third millennium BC)	*E. asinus*	Chaix and Grant 1987: 78
Wadi Hariq	3560 ± 150 BP (KN-5318, 1920 ± 200 cal BC)	*E. asinus*; 1 complete skeleton	Berke 2001: 235–256
Shaqadud	Mid-fourth millennium BP	*E. asinus*; 1 os tarsi centrale	Peters 1991: 222
Mahal Teglinos	Early fourth millennium BP	*E. asinus*	Fattovich 1993: 444; Peters in Sadr 1991: 138, 142
Kenya			
Narosura	Third millennium BP	*E. asinus* several teeth	Gifford-Gonzalez and Kimengich 1984: 470; Marshall pers observation 2003
Ethiopia			
Axum	Axumite, c AD 630–770	*E. asinus*; 2 specimens	Cain 2000: 66
ASIA			
Iran			
Tal-e Malyan	c 4200 BP (c 2800 cal BC); Banesh and Kaftari phases (c 3400–1800 cal BC)	Banesh 27 specimens; Kaftari 139 specimens	Zeder 1986

(*Continued*)

Table 20.2 (*Continued*)

Site	*Time Period*	*Sample*	*References*
Iraq			
Abu Salabikh	Sumerian c 3900 BP (BM-1365 A-D, c 2400 cal BC)	metatarsal III	Burleigh 1986: 234; Clutton-Brock 1986; Ducos 1986: 240–42; Eisenmann 1995: 11
Tell Madhhur	Sumerian, Early Dynastic I, c 2200 cal BC	2 skeletons	Burleigh 1986: 234; Clutton-Brock 1986: 210–12;
Syria			
Apamee	c 4300 BP (2300–1800 cal BC)	metatarsal III	Ducos 1986: 240–41; Eisenmann 1995: 11
Tell Brak	c 4000 BP (c 2500 cal BC)	6 donkey skeletons	Clutton-Brock 2003
Israel			
Tell Duweir	Early Bronze Age, early third millennium BC (3000–2850 cal BC)	6 metatarsal III	Ducos 1986: 240–41; Burleigh 1986: 232; Clutton-Brock 1986: 212; Eisenmann 1995: 11
Tell Gat	Early Bronze Age	metatarsal III	Ducos 1986: 240–41
Jericho	Middle Bronze Age, mid-fourth millennium BP (c 1720 cal BC)	metacarpal	Burleigh 1986: 234
Oman			
Maysar-6-25	M-6, Early Bronze Age; M-25, Bronze Age (late 3rd–early 2nd millennium BC)	n = 9, n = 19, metatarsal III, first phalanx posterior (size between wild and domestic)	Uerpmann 1991: 15–16; Eisenmann 1995: 11, 18, 22

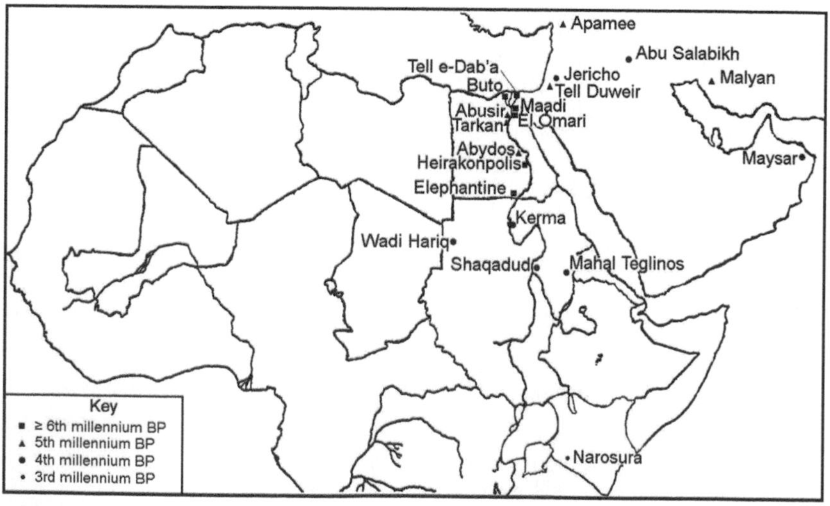

Figure 20.3 Distribution of sites with early domestic donkey, *E. asinus*. (Selected from Table 20.2).

By contrast with the Nile, attempts have not usually been made to distinguish *E. africanus* from *E. asinus* on prehistoric Saharan pastoral sites. Thus at Uan Muhaggiag in Libya, specimen attribution and dating are unclear, but a specimen identified as *E. africanus* (Gautier 1987b: 286, 288) could be early *E. asinus*. The same is true of specimens identified as *E. africanus* by Romer (1935: 173–85) and others from Holocene Capsian sites such as Ain Beida and El Mouhaad in Algeria, Mugharet el Aliya, and other North African sites (Figure 20.2, Table 20.1). Secure identifications of *E. asinus* are limited to the later pastoral sites of Wadi Hariq, dated to c 3460±150 BP (KN-5318) (c 1920± 200 cal BC), where a small complete skeleton was recovered (Berke 2001, pers comm 2003), and a few isolated specimens at Shaqadud at c 3000 BP in the Sudan (Peters 1991: 222) and Narosura at c 3000 BP in Kenya (Gifford-Gonzalez and Kimengich 1984: 470). Sodmein, the only Holocene site excavated in the Red Sea Hills area in Egypt, has not yielded early domestic donkey (Vermeersch et al 1996). Eritrea appears a key area for the distribution and possible domestication of both the Nubian and Somali wild ass, but no long Holocene sequences have so far been excavated in the region. Early donkeys have not been identified, nor has much research been conducted in the areas of primary distribution of the Somali wild ass: lowland Ethiopia, Djibouti and Somalia (see Brandt 1980, 1986; Clark 1954; Guerin and Faure 1996).

E. AFRICANUS AND *E. ASINUS* IN THE ASIAN FOSSIL RECORD

This paper focuses on possible African contexts for the domestication of the donkey, but cannot be divorced from considerations of the possible domestication of donkeys in Asia. Following this review of the age and geographic distribution of ancient *E. africanus* and *E. asinus* in Africa, I examine similar data from Asia.

Over the past 50 years thinking about the wild ancestor of, and geographic location for, domestication of the donkey has changed considerably. The onager, or half ass (*Equus hemionus*) was smaller than the wild ass of Iran and areas to the east and existed in Southwest Asia until the 1930s. It used to be thought that the onager was domesticated by ancient Sumerians and was ancestral to domestic donkeys of western Asia (reviewed in Clutton-Brock 1992a: 87; Zeuner 1963: 367). When crossed with donkeys hemiones do not produce fertile offspring; they are also difficult to tame (Clutton-Brock 1999: 123; Epstein 1971: 396). As a result, since the 1980s zooarchaeologists and epigraphers working in western Asia have argued that *Equus hemionus* was never domesticated (Clutton-Brock 1986: 210–11, 1992a: 37; Postgate 1992: 165–66; Zarins 1986: 189). Researchers have focused instead on *Equus africanus* as the ancestor of the domestic donkey. The question that arises, then, is whether *Equus africanus* might in the past have existed in parts of western Asia, as well as in Africa, and whether it could have been domesticated in Asia.

The possibility of the existence of African wild ass in Asia was raised as early as the 1920s (Antonius 1922: 249; Groves 1974: 115–16, 1986: 45). As Uerpmann (1986: 260) points out, Arabia is geologically part of Africa and often considered part of the African floral and faunal province. It is subject to the same rainfall regime as the Horn of Africa (Cleuziou et al 2002: 11–12), and there are strong biogeographic reasons to think that *E. africanus* was distributed in parts of the Arabian Peninsula.

Evaluating the possible fossil evidence for the presence of African wild ass in Asia has been complicated because of the difficulty of differentiating small samples of possible *E. africanus* from *E. hemionus*. Two other wild equids, *E. hydruntinus*, and *E. caballus*, the horse, were also distributed in parts of western Asia during ancient times. In the 1960s Ducos (1975, 1986) identified a metatarsal from the Jordanian Natufian site of Mureybit as having the size and proportions of *E. africanus* rather than *E. hemionus*. This identification was not widely accepted at the time (Buitenhuis 1991: 41) and there has been debate over whether wild *E. africanus* and *E. hemionus* could have co-occurred in Southwest Asia. Since then, however, a small but growing body of data suggests the possible presence of *E. africanus* at sites in the Levant and the Arabian Peninsula (Cattani and Bökönyi 2002: 44–51; Driesch and Wodtke 1997: 530–33; Uerpmann 1986: 259–61, 1991; Whitcher et al 2000: 43–44) (Table 20.1). Uerpmann (1991: 29–30) argues that well-identified specimens of wild ass are present at four sites, Mureybit and Shams et Din in Syria, Ra's en Naqb in southern Jordan, and Ra's al-Hamra shell midden in Oman. Driesch and Wodke (1997: 530–31) have also tentatively identified *Equus africanus* at the site of Ain Ghazal in Jordan (see Figure 20.2, Table 20.1).

In a posthumous paper, Bökönyi (Cattani and Bökönyi 2002: 34, 44–51) argues for the presence of *Equus africanus* or early *E. asinus* at the shell midden site of Ash Shumah in Yemen. A substantial sample of more than 900 specimens, with ample measurable dental and postcranial specimens, provides strong support for the presence of *E. africanus* on the west coast of the Arabian Peninsula, across the Red Sea from Eritrea, by c 7770±95 BP (c 6684–6475 cal BC). This is the largest known sample of *E. africanus* from any archaeological site in Africa or Asia. More than 90% of the identifiable specimens are attributed to ass; domestic cattle are also present. On the basis of variability in tooth size and enamel patterns, as well as patterns of body-part representation, Bökönyi (Cattani and Bökönyi 2002: 45) suggests that donkeys at this site were in the process of domestication. The body-part data are open to many interpretations, but the dental information is suggestive. Nevertheless, the date of Ash Shumah is earlier than conventionally associated with domestication of the donkey.

Bökönyi (Cattani and Bökönyi 2002: 50; Cleuziou et al 2002: 11–12) notes the similarity of the Ash Shumah fauna, which includes ostrich (*Struthio camelus* Linnaeus, 1758), gazelle (*Gazella* sp.) and cattle, but no sheep (*Ovis aries* Linnaeus, 1758) or goat (*Capra hircus* Linnaeus, 1758), with those of northeast Africa. He points out geological, climatic, and faunal similarities between Arabia and the Horn of Africa as well as mid-Holocene cultural contacts between the regions. Bökönyi argues that in parts of Arabia mobile herders used early cattle to exploit arid lands in a strategy similar to that used in Africa, without dependence on sheep and goat (Cattani and Bökönyi 2002: 50).

Regardless of whether the animals were domestic or not, the identification of African wild ass at Ash Shumah is secure. At most other sites in western Asia small sample sizes present interpretive problems and only one or two specimens are complete enough to make size or morphologically based identifications of *E. africanus*. As a result, there is currently tentative, but not overwhelming evidence for *E. africanus* at sites in the Levant and strong evidence for *E. africanus* or possibly early domestic donkeys in Yemen.

Early Donkeys in Asia

Based on fragments found at archaeological sites it is difficult to differentiate early, large domestic donkey from *E. africanus*, or even from *E. hemionus*. Large samples or complete skeletons are needed for secure identifications. As a result, there are many sites where early donkey may exist in Southwest Asia, but few that are considered reliable. The earliest well-documented domestic donkeys are the easternmost in distribution, dating to c 2800 cal BC in the ancient city of Anshan at Tal-e Malyan in Iran (Zeder 1986) (Figure 20.3, Table 20.2). These animals are slightly smaller than modern African wild ass, but larger than contemporary donkeys of the region (Zeder 1986: 407).

Osteological evidence for early donkeys or donkey/hemione hybrids is also known from Sumerian sites, the best documented being Abu Slabikh (c 2400 cal BC) and Tell Madhhur (Early Dynastic I) in Iraq (Clutton-Brock 1986). Sumerian cuneiform texts dating to c 2600 cal BC provide another line of evidence that suggests the presence of domestic donkeys in western Asia by the late fourth millennium and during the third millennium BC (Postgate 1986: 200; Zarins 1986: 180–89). Postgate (1986: 194, 1992: 166) thinks that these texts may refer to at least three different equids: donkeys indigenous to western Asia, wild onagers 'equid of the desert', and onager/donkey crosses. The wild onagers were reported to have been kept, but used for breeding rather than as draught animals (Postgate 1992: 166; Zarins 1986: 188–89). They are now thought by many epigraphers and zooarchaeologists to have been untameable.

Zooarchaeological evidence from Syria and Israel suggests the possible presence of early domestic donkeys at Apamee (2300–1800 cal BC) (Ducos 1986: 240–41; Eisenmann 1995: 11), Tell Duweir (3000–2850 cal BC) (Burleigh 1986: 232; Ducos 1986: 240–41; Eisenmann 1995: 11), and Tell Gat (Ducos 1986: 240–41). Donkeys are also present somewhat later during Middle Bronze Age Jericho (c 1720 cal BC) (Burleigh 1986: 234) (Table 20.2). Clutton-Brock (1992a: 65) notes that by 2500 cal BC the donkey was widely used and that by 1000 cal BC it was the most common means of transport in Egypt and western Asia. In the less arid regions of Asia and Europe the horse occupied a similar role. The donkey in Asia was, however, primarily used for ploughing or transport of heavy burdens, and not for speed and warfare in the same way as the donkey/onager hybrid and the horse. The donkey played an important role in transport of goods for early cities and in the long-distance trade of the Sumerians (Postgate 1992: 166; Zeder 1986: 372). Early use of donkeys may have influenced urban production and exchange at cities such as Anshan (Zeder 1986: 372).

AFRICAN LINGUISTICS

Historical linguists have contributed considerably to thinking about the origins and spread of food production in Africa, providing archaeologists with hypotheses regarding routes and the sequence of the spread of early Cushitic and Nilotic-speaking African pastoralists, and the origins and spread of African crops such as sorghum (Bechhaus-Gerst 2000; Blench 1997; Ehret 1997, 1998, 2002). Recently Blench (2000) and Ehret (1998, 2002) have analyzed linguistic evidence for the history of the donkey in Africa, focusing on terms for donkeys, African wild ass, mules and zebra, and roots for, variation in, and timing and geographic spread of terms.

Blench (2000: 346) and others (Ehret 2002: 77, 90–91) identify at least three principal base forms for donkey: #kuur., or *Kwer-., #harre-. or *Harr-. and #d-q-r-. or *dakw-. All of these ancient root words for donkey are found in the Afroasiatic language phylum. Lexical terms for donkey appear to have been borrowed from Afroasiatic into other African language phyla such as Niger-Congo or Nilo-Saharan. Niger-Congo speakers are the most numerous in Africa; there are few Nilo-Saharan language speakers and these are distributed in North Africa.

Further information on the likely area of domestication of the donkey in Africa can be drawn from an examination of the distribution of principal base forms within the Afroasiatic phylum. The Afroasiatic phylum is made up of three main families: Omotic, Cushitic-Chadic and North Afroasiatic. Because early splits in the Afroasiatic phylum occurred between Omotic and Cushitic families of the Horn, many scholars (Ehret 1998; Blench 2000 and references therein) think that Afroasiatic originated in the eastern parts of the southern Sahara, or farther south.

Blench (2000: 349, 352) thinks that it is significant that all three branches of the Afroasiatic language phylum have different lexical terms for wild ass or donkey. On the basis of this variation in ancient root words for donkey he argues that donkeys were domesticated more than once around the edges of the Sahara. Ehret (1998: 11, 2002: 77, 90–91) points to the southern Red Sea Hills, the source of the proto-Afroasiatic language, and reconstruction of terms for donkey in proto-Afroasiatic and proto-Cushitic (Blench 2000: 349; Ehret 2002: 90–91). Because Afroasiatic is so ancient and there are proto-Afroasiatic words for donkey, Ehret (2002: 77) suggests that the domestication of the donkey may have occurred in the southern Red Sea Hills area earlier than currently thought, almost as early as the domestication of African cattle. His analysis of the geographic distribution of terms emphasises, therefore, domestication in the southern Sahara or Horn of Africa, but does not rule out additional domestication in the central

Sahara by proto-Chadic-speaking people (Ehret 2002: 77, pers comm 2003). In summary, African linguists concur that at least one and possibly two domestications of the donkey occurred in the Horn and possibly in the south-central Sahara. Their interpretations have not been widely discussed or accepted by Africanist archaeologists.

MODERN DNA

Recent studies of genetic variability in contemporary donkeys and *E. africanus* contribute to considerations of pastoral rather than urban contexts for domestication of the donkey in Africa and to ongoing debates over whether donkeys were domesticated in Africa or Asia (Beja-Pereira et al 2004; Oakenfull 2000, 2002).

Albano Beja-Pereira and colleagues (Beja-Pereira et al 2004) recently analyzed mitochondrial DNA from donkeys from 52 countries in Africa and Eurasia. Mitochondrial DNA is maternally inherited and can be used to provide a particularly interesting record of genetic change through time. They studied a highly variable region, region 1 (HVRI) of the mtDNA control region, sequencing 479 base pairs. Sequences are grouped with one of several equid haplogroups, differentiated by distinctive base pair changes. Subsequently neighbour-joining and other methods are used to construct dendrograms of relationships from genetic distances.

Beja-Pereira and colleagues' phylogenetic analysis suggests at least two clades – one incorporating African wild ass and domestic donkeys, and the other the Asiatic wild asses (Beja-Pereira et al 2004). This study demonstrates unequivocally that the African wild ass, and not the Asian wild ass or hemione, is the ancestor of the domestic donkey. This supports earlier genetic research (Oakenfull 2000) and hypotheses based on Egyptian archaeology and the historic distribution of the African wild ass. It also fits with zooarchaeological and epigraphic data that precluded the onager as an ancestor to the domestic donkey. Finally, this study lays to rest Zeuner's (1963) earlier arguments for domestication of the onager and the onager as an ancestor of Near Eastern donkeys (see also Clutton-Brock 1992a: 87).

Interestingly Beja-Pereira and colleagues' (Beja-Pereira et al 2004) phylogenetic analysis also suggests at least two clades of domestic donkeys, each grouping associated with a different possible wild ancestor. They argue for two separate domestication events from two distinct wild populations or subspecies of *E. africanus*, Nubian and Somali, respectively (Beja-Pereira et al 2004: 1781). Nubian and Somali clades of domestic donkeys occur in both Africa and Asia, with the Nubian clade being somewhat more common in Africa and the Somali clade somewhat more common in Asia. Nucleotide diversity is, however, higher for both lineages in northeast Africa than in Eurasia. As a result, Beja-Pereira et al (2004: 1781) argue that it is likely that both domestication events took place in Africa. They cannot, however, completely rule out domestication of donkeys of the Somali clade in Asia.

SOCIAL CONTEXTS FOR DOMESTICATION AND CREATION OF NEW AGRICULTURAL SYSTEMS: A REVIEW

Historically researchers have focused on urban or peri-urban contexts for domestication and integration of donkeys into mixed agricultural traditions of settled villages and towns of the Mediterranean and western Asia. These agricultural traditions relied upon wheat, barley and pulses and cattle, sheep and goats. Donkeys were used for ploughing, threshing and to pull carts. Donkey caravans were the chief source of land transport for the ancient Egyptians and Sumerians (Brewer et al 1994; Clutton-Brock 1992a; Hassan 1988; Postgate 1992). It is likely that donkeys played a significant role in urban production and exchange in Asia (Zeder 1986: 372). It has also been suggested that donkey transport played a role in the spacing of Egyptian nomes and management of the ancient Egyptian state (Hassan 1988, 1993). Donkey caravans influenced the distribution of goods and the development of trade among ancient cities of ancient Egypt and western Asia. It is known from textual evidence that the Sumerians used donkeys to carry grain over the mountains to Atarra and tin and textiles over the Taurus Mountains (Postgate 1992: 166). The role of donkeys in ancient Egyptian long-distance trade is similarly well documented, with caravans moving between Egypt, the Levant and sub-Saharan Africa (Brewer et al 1994). During the Fifth Dynasty, the caravan master of King Meren-Re, for example, returned from his third trip to Yam with 300 donkeys carrying incense, ebony and grain (Brewer et al 1994: 100).

There has been much less emphasis on the role of the donkey outside of trade. As ethnoarchaeological data show, donkeys played an equally important subsistence role among African herders who lived a nomadic life under arid conditions. Donkeys allowed rapid mobile responses to unpredictable rainfall, flexible management of pasture away from permanent water, residential household mobility and improved health of livestock, women and children. The pastoral hypothesis suggests that donkeys played a significant role in the spread of early pastoral societies in Africa. It is likely that the subsistence role of the donkey was equally important outside of Africa, and that the finding that the donkey is often a woman's animal – especially significant for household tasks, collecting water and firewood – has implications for understanding social roles, the organisation of women's labour and factors affecting population growth in other areas of the world.

DISCUSSION

Prominent in Early Dynastic art and found buried in tombs, donkeys have long been assumed to have been domesticated by ancient Egyptians. It is well known that domestication of the donkey would have had a significant impact on urban commerce and the development of trade in Egypt and western Asia. The ethnoarchaeologically based hypothesis that early Northeast African pastoralists domesticated the donkey in order to facilitate long-distance mobility as the Sahara dried c 7000–6500 BP, suggests a non-urban context for domestication of the donkey in the arid tropics or subtropics. It also has significant implications for the development of pastoralism and the spread of food production in Africa. In the following sections I discuss some of the key themes arising from this hypothesis.

An African Pastoral Context for Domestication of the Donkey?

The geographic location and timing of zooarchaeological evidence for early domestic donkeys is central to testing the African pastoral hypothesis. Support for an African pastoral context for domestication of the donkey would be provided by earlier evidence for domestic donkeys in pastoral contexts of the Sahara, Sudan or Horn than in the Egyptian Nile valley. Early dates for domestic donkey in the Egyptian Nile would not support, nor would they rule out, an African pastoral context for domestication of the donkey. The relative amount of archaeological research undertaken in these regions, sample sizes and preservation issues should be considered.

Zooarchaeologists have recently conducted rigorous metrical and morphological studies of wild ass and donkeys from Egyptian Predynastic and Early Dynastic sites. Scholars working in Egypt now argue that domestic donkeys were present in Egypt c 6000 years ago, considerably earlier than previously thought. Early domestic donkeys are large, difficult to differentiate from African wild ass, and consequently could have been present in Egypt earlier than currently recognised.

In contrast, archaeologists working in the Sahara and the Horn have not focused on zebra or ass in archaeological sites. This is largely because, as predicted by ethnoarchaeological findings, ass are not very common in pastoral sites. The few reported remains are often identified in broad terms as 'equid' or 'cf *E. africanus*'. In cases where *E. africanus* has been identified in mid to late Holocene sites, such as the cave sequences of the Magreb, early pastoral sites of the Sahara such as Uan Muhuggiag, and rockshelters from Somalia such as Guli Wabayo, it is unclear whether specimens should be attributed to *E. africanus* or whether they might be domestic donkeys. Furthermore, except for a few isolated sites, neither the archaeology of the refuge of the Nubian wild ass in the Red Sea Hills area of Sudan and Eritrea, nor the key areas for distribution of the Somali wild ass – lowland Ethiopia, Djibouti or Somalia – have been studied. These gaps highlight areas for future research, but make it difficult to test the pastoral hypothesis with current zooarchaeological data.

The earliest securely documented donkey in a prehistoric pastoral context dates only to c 3600 BP at Wadi Hariq in the northern Sudan. This is c 2500 years after donkeys appear in Egypt. As a result, zooarchaeological data suggest domestication of the donkey by ancient Egyptians. However recent biogeographic, historical linguistic and genetic studies necessitate revision of zooarchaeological interpretations of the domestication of the donkey and point towards the role of ancient African herders in multiple domestication events.

Beja-Pereira et al's (2004) analysis of mitochondrial DNA from modern domestic donkeys suggests domestication of both Somali and Nubian wild ass in Africa, but no evidence for domestication of extinct Atlas or Saharan forms. The authors suggest domestication by African pastoralists. Beja-Pereira et al's (2004) genetic findings fit intriguingly well with Blench's (2000: 349, 352) linguistic arguments for multiple domestication of the African wild ass. Blench thinks that ancient roots in all three branches of the Afroasiatic language phylum suggest likely domestication of the donkey by people speaking languages in the Omotic, Cushitic-Chadic and North Afroasiatic families. Both Blench (2000) and Ehret (1998: 349, 352, 2002: 77) argue on linguistic grounds for at least two possible domestication events around the southern and eastern edges of the Sahara.

It is interesting to consider conditions that might have promoted two domestication events by African pastoralists. The archaeology of early pastoralism in northern Africa suggests that the independent domestication of the Nubian and Somali wild ass and the spread of the idea of domestication among local pastoralist populations are both possible. With increasing aridity from 7000 years ago and increasingly variable rainfall, use of pastures farther from water, and increased mobility of groups of people and cattle, motivation for the domestication of the donkey may have been strong throughout the Sahara and Horn, an area of more than 20 million square miles.

The location and timing of possible domestication of the Nubian wild ass by African pastoralists are limited, however, by the ancient distribution of the Nubian wild ass, the timing of the development and spread of herding societies in northern Africa and the appearance of early domestic donkeys in ancient Egypt. The earliest evidence for early herders in Africa is found 9000–8000 BP in the deserts of southern Egypt and regions to the west (Close and Wendorf 1992; Marshall and Hildebrand 2002; Wendorf et al 2001). Through a combination of movement of herders and contact with local hunter-gatherers, food production spread patchily to the west and then to the south, accelerating as aridity in the Sahara increased from c 7000 BP (Marshall and Hildebrand 2002).

In ancient times, the Nubian wild ass had a broad distribution from northern Eritrea, across the Sudan to parts of the Sahara, and north to Egypt (Groves 2002: 101–03) (Figure 20.1). The linguistic evidence points to the Red Sea Hills for early domestication of the donkey (Ehret 1998, 2002). There is no archaeological evidence for this region, but the first evidence for the appearance of early herders with domestic stock in the adjacent Khartoum area of the northern Sudan dates only to c 6000 BP, at sites such as Kadero (Gautier 1984; Kryzaniak 1991; Marshall 2000). Early domestic stock appears later to the east, between the Nile and the Red Sea Hills, and to the south in Ethiopia, the Horn, southern Sudan and northern Kenya. The earliest domestic stock in the Atbara region of the eastern Sudan dates to around 5th–4th millennium BP (Marshall 2000; Peters in Sadr 1991). Domestic donkeys appear in Egypt by at least 6000 BP. Therefore, if Nubian wild ass were domesticated by herders prior to their appearance in ancient Egypt, domestication must on current evidence have taken place in the central Saharan regions where the earliest pastoralists lived and north of the Sudanese Nile or Red Sea Hills.

The probable area, social context for and timing of domestication of Somali wild ass appear even more tightly circumscribed. It is unlikely that the Somali wild ass were ever found farther north than Eritrea, at about 11.6º N; populations center on northern Somalia (Groves 2002: 102). No long archaeological sequences have been excavated in Eritrea and the culture history of the region is unknown. In Ethiopia and Somalia excavation has been limited, however, early herders and their domestic livestock, ie cattle, sheep and goats, appear significantly later than in regions to the north. Early domestic stock has been identified at c 3500 BP at Lake Besaka, Gobedra, Baati Ataro and Kawlos in Ethiopia (Brandt 1980; Marshall 2000; Negash 2001; Phillipson 1977) and at Buur Heybe in Somalia (Brandt 1986). In these regions, as in the Sudan, the spread of herders appears to result from contact between local hunter-gatherers and early herders, as well as a southward movement of pastoral groups (Marshall and Hildebrand 2002). Consequently, if early herders, rather than hunter-gatherers, of the Horn domesticated the Somali wild ass, it is likely to have occurred after c 4000 BP (uncalibrated). Some archaeologists think the timing and location of domestication is the result of poor archaeological sampling (Brandt pers comm 2002; Phillipson pers comm 2002). If domestication of the Somali wild ass did in fact take place significantly later than domestication of the Nubian wild ass, it could have occurred following contact with donkey-owning herders from farther north or west.

Taken together, archaeological, genetic and linguistic evidence provide tentative support for domestication of the Nubian wild ass by ancient Egyptians and later domestication of the Somali wild ass by African herders. The domestication of the Nubian wild ass by African herders north of the Sahel is not ruled out, however, and nothing is yet known about the possible domestication of the Atlas wild ass. These research developments provide a new framework for thinking about the beginnings of food production in Africa, the spread of African pastoralism, the domestication of the donkey in Africa and multiple patterns of animal domestication and point to many areas for future research.

African or Asian Domestication of the Donkey?

The archaeological and genetic material reviewed in this paper contribute to current debates regarding an African or Asian ancestry of the donkey. Beja-Pereira et al's (2004) modern genetic data rule out the onager as an ancestor of the Asian donkey. The zooarchaeological data suggest, however, that *E. africanus* was present in Asian sites such as Mureybit (Ducos 1975, 1986), Ra's en Naqb (Uerpmann 1991), Ain Ghazal (Driesch and Wodke 1997) and Ash Shumah (Cattani and Bökönyi 2002) and raise the question of an Asian domestication of the donkey. There has been little or no discussion of which subspecies of wild ass might have been present in

Asia (see Eisenmann 1995). On geographic grounds the wild ass of the Levant might be expected to have been more closely related to the Nubian wild ass of the Sahara than to the Somali wild ass of the Horn of Africa. However, donkeys of both the Somali and Nubian clades are found in Asia, and because of the lack of genetic diversity among Asian donkeys, it is much more likely that both Nubian and Somali wild ass were domesticated in Africa (Beja-Pereira et al 2004).

The zooarchaeological evidence tentatively supports the African hypothesis. Metrical and morphological analysis of wild ass and donkey specimens recently excavated in Egypt show that domestic donkeys were present in Predynastic Egypt by c 5500–6000 years ago. Securely identified domestic donkeys in Asia date significantly later, to c 2800 cal BC (c 4200 BP) in the ancient city of Anshan at Tal-e Malyan in Iran (Zeder 1986). Just as in Africa, morphological identification is difficult, and domestic donkeys could have been used earlier than currently thought.

There are hints, such as at Ash Shumah, that donkeys could have been domesticated in Asia, and this is not completely ruled out by the genetic data (Beja-Pereira et al 2004). In order to make the argument that donkeys were domesticated in Asia it is necessary for there to have been a greater reduction in genetic diversity among donkeys in Asia than in Africa. Mechanisms could include less interbreeding among domestic and wild ass, founder effects related to more extensive trade in Asia and eradication of local populations due to disease. If donkeys were domesticated in Asia, Beja-Pereira et al's (2004) data on the distribution of Somali and Nubian haplotypes suggest that it was the Somali, rather than the Nubian wild ass, that was domesticated. From this perspective, it is interesting to consider the Ash Shumah faunal assemblage of the western coast of Yemen, which suggests the possibility of the existence of populations of Somali wild ass in the Arabian Peninsula, across the Red Sea and adjacent to their historic distribution in Eritrea, Djibouti and Somalia. Bökönyi (Cattani and Bökönyi 2002: 45) argues there is evidence for ongoing domestication of African wild ass by c 7000 BP at Ash Shumah. He goes on to propose that this took place among mobile cattle herders, without sheep and goat, and to stress the African nature of this site and pattern of domestication (Cattani and Bökönyi 2002: 46–50). Bökönyi's argument admits a possible pastoral context for domestication of the Somali wild ass in the Arabian Peninsula and provides a way to reconcile aspects of the zooarchaeological and genetic data. At present, however, the evidence for domestication at Ash Shumah is weak.

The proximity of Ash Shumah to Africa suggests an alternative interpretation. If donkeys were domesticated early in the Horn of Africa, as Ehret (2002: 77) suggests on the basis of linguistic data, then animals at Ash Shumah might be early African domestic donkeys. There has been increasing recognition among Africanist and Arabian archaeologists that close ties existed among mid-Holocene hunter-gatherers across the narrow straits of the Red Sea. In particular, there is evidence for importation of obsidian from the Horn of Africa to Yemen during the middle Holocene, and prehistoric pastoral African rock art styles exist in Arabia (Cervicek 1979; Cleuziou et al 2002; Fattovich 1996a). Subsequently, there was well-documented trade between Africa and Asia along this route (Fattovich 1996a, 1996b: 24; Zarins 2002: 420–21). It might make sense that the donkey was imported from Africa to the coast of Yemen as the early engine of this trade.

We are already aware of the problems of recognition of early domestic donkey zooarchaeologically. It is abundantly clear that early donkey are large, and ethnoarchaeological data suggest multiple reasons why selection processes for small donkeys might not have been strong on early pastoral sites. In addition pastoral sites are ephemeral and there are many reasons why, if donkeys were not numerous, they would not be well represented in the faunas from such sites. It also true that the key areas of Eritrea and the Red Sea Hills are archaeologically unexplored. Perhaps we have simply 'missed' early domestication of the donkey by African hunter-gatherers of the Horn. In this case one would have to argue that in addition to Ash Shumah, the specimens identified as *E. africanus* at sites in the Levant, such as Mureybit and Ain Ghazal, were early African donkey. Such has already been argued for later *E. africanus* at the site of Maysar in Oman (Eisenmann 1995). It is interesting that Eisenmann (1995) notes that the Mureybit specimen is small for African wild ass in Africa.

It is time for a major paradigm shift in thinking about the domestication of the donkey. At present it is difficult to reconcile faunal with genetic evidence on this topic and the zooarchaeological data from Ash Shumah, Yemen, with that from other regions. It is also evident that there are significant methodological barriers to zooarchaeological recognition of early donkeys, particularly in the context of mobile herding societies. Additional studies of modern DNA, especially microsatellites from both modern donkey and wild ass, as well as studies of ancient mitochondrial DNA, and further detailed research on the morphology of African wild ass will help to resolve these issues. Genetics may help to resolve ancestor and place; archaeology and zooarchaeology are necessary to clarify social context and timing. Further research is also badly needed on the Holocene archaeology of the eastern Sahara, Horn of Africa, and Arabia.

CONCLUSION

In conclusion, recent genetic studies suggest that a number of large mammals, including sheep, pigs and taurine cattle were domesticated more than once in different geographic regions. Given the very few large mammals, or even plant species, ever domesticated in the world, archaeologists have tended to look upon domestication of plants or animals as rare events resulting from the unusual convergence of particular social and ecological circumstances (Harris 1996: 570). It has long been argued by zooarchaeologists, however, that certain mammals are behaviourally more suited for domestication than others. Many species of bovines, including water buffalo, humped cattle (*Bos indicus*) and unhumped taurine cattle (*Bos taurus*), were domesticated, for example, although antelope and deer were not (Clutton-Brock 1999; Hemmer 1990). *Bos taurus* was domesticated twice, in both Africa and western Asia (Bradley and Loftus 2000; Hannotte et al 2002). For all this there has been little anthropological focus on the occurrence of, or social contexts for, multiple domestication events within one geographic region. In fact, it is only geneticists that have so far identified such events. This is largely because the archaeological record is sparse, providing few data points, and because parsimony has suggested that for a species in a particular region, one domestication event was more likely than two. Genetic findings have prompted archaeologists to reconsider this issue for animals such as pigs, sheep and cattle. Currently, extensive research is being carried out on animals of temperate regions, especially pigs and goats. The donkey appears to be an unusual case, however, with possible multiple domestications within one geographic region. Pastoral perspectives are therefore timely, as they provide a social context in which this event could easily have occurred in the arid tropics and subtropics of Africa and possibly Asia.

Domestication of the donkey has significant implications for the development of African pastoralism, domestication of African plants and the spread of food production in Africa. Integration of the domestic donkey into cattle-based herding systems of the Sahara and Horn of Africa reinforced the mobility of early herders, creating a more productive and sustainable herding system. The increased mobility of early herders generated circumstances that would have reduced continuous directional selection on local plants, contributing to early reliance on domestic herds and late domestication of African plants (Marshall and Hildebrand 2002). Because mobile cattle, sheep, goat and donkey-based herding developed early, food production spread through the northern half of Africa as a result of migration of early herders and through cultural contact among herders and hunter-gatherers, rather than through the influence of settled villagers. Mobile use of the landscape preserved local resources, lessening direct competition and allowing hunter-gatherers to coexist with early herders. It also encouraged continued use of wild plants and patchy spread of food production (Marshall and Hildebrand 2002). Continued use of wild plants and their domestication by later, more settled herders across many regions of the Sahel contributed to the non-centric pattern of plant domestication characteristic of Africa.

Mobile herding systems are still the most efficient form of food production in arid lands of Africa today (McCabe 2004). The domestication of the donkey was a turning point in the organization of movement, land management and labour of ancient herders; it also contributed to the development of distinctively African patterns of food production and plant domestication.

ACKNOWLEDGMENTS

This research was supported by NSF BCS-0447369. I am indebted to Joris Peters, Patricia Moehlmann and Olivier Hannotte for advice in the early stages of this research. I also thank Wim van Neer, Achilles Gautier, Louis

Chaix and members of the Poznan conference 1993 for genial discussions. I am grateful to the Lesorogal family of Maralal for hospitality and information regarding donkeys in pastoral life, to Jean Ensminger for information on donkeys and the Orma, and to Hubert Berke for faunal data from Wadi Hariq. William Stanley of the Field Museum of Natural History, Chicago, and Richard Sabin of the Natural History Museum, London, facilitated study of collections. I acknowledge Rufus Churcher, Juliet Clutton-Brock, Angela von den Driesch, Vera Eisenmann, Achilles Gautier, Colin Groves and Peter Uerpmann for the decades of meticulous scholarship that make such a synthesis of fossil *E. africanus* and *E. asinus* possible. Finally, I am most grateful to Albano Beja-Pereira for collegial donkey-centred discussions over the last few years, to Tim Denham and Luc Vrydaghs for meticulous editing and to members of my department for their enthusiasm for and patience with asses. Any errors are my own.

REFERENCES

Antonius, O (1922) *Grundzüge einer Stammesgeschichte der Haustiere*, Jena: Fischer

Antonius, O (1938) 'On the geographical distribution, in former times and today, of the recent Equidae', *Journal of Zoology* 107: 557–64

Arambourg, C (1931) 'Observations sur une grotte à ossements des environs d'Alger', *Bulletin de la Société d'Histoire Naturelle de l'Afrique du Nord* 22: 169–76

Arambourg, C (1952) *La Paléontologie des Vertébrés en Afrique du Nord Française*, 19e Congrès Géologique International, Monographies Régionales, hors série, pp 1–63, Algiers: Alger

Århem, K (1985) *Pastoral Man in the Garden of Eden*. Uppsala: Department of Cultural Anthropology, University of Uppsala

Barich, B (2002) 'Cultural responses to climatic changes in North Africa: beginning and spread of pastoralism in the Sahara', in Hassan, F A (ed), *Droughts, Food and Culture*, pp 209–25, New York: Kluwer Academic/Plenum Publishers

Bate, D M A (1951) 'The mammals from Singa and Abu Hugar', *Fossil Mammals of Africa* 2: 1–50

Bate, D M A (1954) 'Report on the faunal remains conducted by J. D. Clark from sites in the Somali lands', in Clark, J D, pp 362–65, *The Prehistoric Cultures of the Horn of Africa*, Cambridge: Cambridge University Press

Baxter, P T W (1991) '"Big men" and cattle licks in Oromoland', in Baxter, P T W (ed), *When the Grass is Gone: Development Intervention in African Arid Lands*, pp 192–212. Scandinavian Institute of African Studies Seminar Proceedings no 25, Uppsala: Scandinavian Institute of African Studies

Bechhaus-Gerst, M (2000) 'Linguistic evidence for the prehistory of livestock in Sudan', in McDonald, K C and Blench, R M (eds), *The Origins and Development of African Livestock: Archaeology, Genetics, Linguistics and Ethnography*, pp 449–61, London: University College London Press

Beja-Pereira, A, England, P R, Nuno, Ferrand, Jordan, S, Bakhiet, A O, Abdalla, M A, Mashkour, M, Jordana, J, Taberlet, P and Luikart, G (2004) 'African origins of the domestic donkey', *Science* 304(5678): 1781

Berke, H (2001) 'Gunstraeume und Grenzbereiche', in Gehlna, B, Heinen, M and Tillmann, A (eds), *Zeit-Räume: Gedenkschrift für Wolfgang Taute*, pp 235–56. Archäologische Berichte 14, Bonn: Deutsche Gesellschaft für Ur- und Frühgeschichte

Blench, R M (1997) 'Language studies in Africa', in Vogel, J O (ed), *Encyclopedia of Precolonial Africa*, pp 90–102, Walnut Creek, CA: Altamira Press

Blench, R M (2000) 'A History of donkeys, wild asses and mules in Africa', in McDonald, K C and Blench, R M (eds), *The Origins and Development of African Livestock: Archaeology, Genetics, Linguistics and Ethnography*, pp 339–54, London: University College London Press

Boessneck, J (1976) *Tell el-Dab'a III: die Tierknochenfunde 1966–1969*, Vienna: Verlag der Österreichischen Akademie der Wissenschaften
Boessneck, J and Driesch, A von den (1992) *Tell el-Dab'a VII: Tiere und historische Umwelt im nordost-Delta im 2. Jahrtausend v Chr. anhand der Knochenfunde der Ausgrabungen 1975–1986*, Vienna: Verlag der Österreichischen Akademie der Wissenschaften
Boessneck, J and Driesch, A von den (1998) 'Tierreste aus der vorgeschichtlichen siedlung von El-Omari bei Heluan/UnterÄgypten', in Debono, F and Montensen, B (eds), *El Omari*, pp 99–101, Mainz: Philipp von Zabern
Boessneck, J, Driesch, A von den and Ziegler, R (1989) 'Die Tierreste von Maadi und Wadi Digla', in Rizkana, I and Seeher, J (eds), *Maadi III*, pp 87–128, Mainz: Philipp von Zabern
Boessneck, J, Driesch, A von den and Eissa, A (1992) 'Eine Eselsbestattung der 1. Dynastie in Abusir', in *Mitteilungen des Deutschen Archaeologischen Instituts Abteilung Kairo*, vol 48, pp 1–10, Mainz: Phillipp von Zabern
Bökönyi, S (1985) 'The animal remains of Maadi, Egypt: a preliminary report', in Liverani, M, Palmieri, A and Peroni, R (eds), *Studi di paletnologia in onore di Salvatore M. Puglisi*, pp 495–99, Rome: Università di Roma 'La Sapienza'
Bradley, D and Loftus, R T (2000) 'Two eves for taurus? Bovine mitochondrial DNA and African cattle domestication', in Blench, R M and MacDonald, K C (eds), *The Origins and Development of African Livestock: Archaeology, Genetics, Linguistics and Ethnography*, pp 244–50, London: University College London Press
Brandt, S (1980) 'Investigation of Late Stone Age occurrences at Lake Besaka, Ethiopia', in Leakey, R B and Ogot, B (eds), *Proceedings of the Eighth PanAfrican Congress on Prehistory and Quaternary Studies 1977*, pp 239–43, Nairobi: TILLMIAP
Brandt, S (1986) 'The Upper Pleistocene and early Holocene prehistory of the Horn of Africa', *African Archaeological Review* 4: 41–82
Brehm, A (1915) *Die Säugetiere*, Leipzig: Brehms Tierleben
Brewer, D J, Redford, D B and Redford, S (1994) *Domestic Plants and Animals: The Egyptian Origins*, Warminster: Aris and Phillips
Buitenhuis, H (1991) 'Some equid remains from south Turkey, north Syria, and Jordan', in Meadow, R H and Uerpmann, H-P (eds), *Equids in the Ancient World*, vol II, pp 34–74, Wiesbaden: Ludwig Reichert Verlag
Burleigh, R (1986) 'Chronology of some early domestic equids in Egypt and western Asia', in Meadow, R H and Uerpmann, H-P (eds), *Equids in the Ancient World*, pp 230–36, Wiesbaden: Ludwig Reichert Verlag
Burleigh, R, Clutton-Brock, J and Gowlett, J (1991) 'Early domestic equids in Egypt and western Asia: an additional note', in Meadow, R H and Uerpmann, H-P (eds), *Equids in the Ancient World*, vol II, pp 9–11, Wiesbaden: Ludwig Reichert Verlag
Cain, C (2000) 'Animals at Axum: initial zooarchaeological research in the later prehistory of the northern Ethiopian Highlands' unpublished PhD thesis, Washington University in St. Louis
Cattani, M and Bökönyi, S (2002) 'Ash-Shumah: an early Holocene settlement of desert hunters and mangrove foragers in the Yemeni Tihamah', in Cleuziou, S, Tosi, M and Zarins, J (eds), *Essays on the Late Prehistory of the Arabian Peninsula*, pp 31–53, Rome: Istituto italiano per l'Africa e l'Oriente
Cervicek, P (1979) 'Some African affinities of Arabian rock art', *Rassegna di Studi Etiopici* (Rome) 27: 5–12
Chaix, L and Grant, A (1987) 'A study of a prehistoric population of sheep (*Ovis aries* L.) from Kerma (Sudan)', *Archaeozoologia* 1: 93–107
Churcher, C S (1972) 'Late Pleistocene vertebrates from archaeological sites in the plain of Kom Ombo, Upper Egypt', *Life Sciences Contributions* (Royal Ontario Museum, Toronto) 82: 1–172
Churcher, C S (1974) 'Relationships of the Late Pleistocene vertebrate fauna from Kom Ombo, Upper Egypt', *Annals of the Geological Survey of Egypt* IV: 363–84
Churcher, C S (1986) 'Equid remains from Neolithic horizons at Dakhleh Oasis, Western Desert of Egypt', in Meadow, R H and Uerpmann, H–P (eds), *Equids in the Ancient World*, pp 413–21, Wiesbaden: Ludwig Reichert Verlag

Churcher, C S and Richardson, M L (1978) 'Equidae', in Maglio, V J and Cooke, H B S (eds), *Evolution of African Mammals*, pp 379–422, Cambridge, MA: Harvard University Press

Clark, J D (1954) *The Prehistoric Cultures of the Horn of Africa*, Cambridge: Cambridge University Press

Clark, J D (1976) 'Prehistoric populations and pressures favouring plant domestication in Africa', in Harlan, J, de Wet, J M J and Stemler, A B L (eds), *Origins of African Plant Domestication*, pp 465–78, The Hague: Mouton

Cleuziou, S, Tosi, M and Zarins, J (2002) 'Introduction', in Cleuziou, S, Tosi, M and Zarins, J (eds), *Essays on the Late Prehistory of the Arabian Peninsula*, pp 379–422, Roma: Istituto italiano per l'Africa e l'Oriente

Close, A (1995) 'Few and far between: early ceramics in North Africa', in Barnett, W K and Hoopes, J W (eds), *The Emergence of Pottery: Technology and Innovation in Ancient Societies*, pp 23–37, Washington DC: Smithsonian Institution Press

Close, A and Wendorf, F (1992) 'The beginnings of food production in the eastern Sahara', in Gebauer, A B and Price, T D (eds), *Transitions to Agriculture in Prehistory*, pp 63–72, Madison, WI: Prehistory Press

Closse, K (1998) 'Les ânes dans l'Égypte ancienne', *Anthropozoologica* 27: 27–39

Clutton-Brock, J (1986) 'Osteology of equids from Sumer', in Meadow, R H and Uerpmann, H-P (eds), *Equids in the Ancient World*, pp 207–29, Wiesbaden: Ludwig Reichert Verlag

Clutton-Brock, J (1989) *The Walking Larder*, London: Unwin Hyman

Clutton-Brock, J (1992a) *Horse Power: A History of the Horse and the Donkey in Human Societies*, Cambridge, MA: Harvard University Press and London: Natural History Museum Publications

Clutton-Brock, J (1992b) 'The process of domestication', *Mammal Review* 22: 79–85

Clutton-Brock, J (1999) A *Natural History of Domesticated Mammals*, 2nd ed, Cambridge: Cambridge University Press

Clutton-Brock, J (2003) 'Were the donkeys at Tell Brak (Syria) harnessed with a bit?', in Levine, M, Renfrew, C and Boyle, K (eds), *Prehistoric Steppe Adaptation and the Horse*, pp 126–27, Cambridge: McDonald Institute for Archaeological Research

Coppock, D L (1994) *The Borana Plateau of Southern Ethiopia: Synthesis of Pastoral Research, Development and Change, 1980–1991*. Systems Study 5, Addis Ababa: International Livestock Centre for Africa

Cremaschi, M (2002) 'Late Pleistocene and Holocene climatic changes in the central Sahara: the case study of the southwestern Fezzan, Libya', in Hassan, F A (ed), *Droughts, Food and Culture*, pp 65–82, New York: Kluwer Academic/Plenum Publishers

Cremaschi, M, di Lernia, S and Trombino, L (1996) 'From taming to pastoralism in a drying environment: site formation processes in the shelters of the Tadrart Acacus Massif (Libya, central Sahara)', in Castelletti, L and Cremaschi, M (eds), *Paleoecology*, Colloquium VI: *Micromorphology of Deposits of Anthropogenic Origin*, pp 87–106, Forlì: Edizioni UISPP

Diamond, J (1997) *Guns, Germs and Steel*, New York: Norton

Dijkman, J T (1991) 'A note on the influence of negative gradients on the energy expenditure of donkeys walking, carrying and pulling loads', in Fielding, D and Pearson, R A (eds), *Donkeys, Mules and Horses in Tropical Agricultural Development*, pp 221–22, Edinburgh: University of Edinburgh

di Lernia, S (2002) 'Dry climatic events and cultural trajectories: adjusting Middle Holocene pastoral economy of the Libyan Sahara', in Hassan, F A (ed), *Droughts, Food and Culture*, pp 209–25, New York: Kluwer Academic/Plenum Publishers

Divé, J and Eisenmann, V (1991) 'Identification and discrimination of first phalanges from Pleistocene and modern *Equus*, wild and domestic', in Meadow, R and Uerpmann, H-P (eds), *Equids in the Ancient World*, vol II, pp 278–333, Wiesbaden: Ludwig Reichert Verlag

Driesch, A von den (1997) 'Tierreste aus Buto im Nildelta' *Archaeofauna* 6: 23–39

Driesch, A von den and Wodtke, U (1997) 'The fauna of 'Ain Ghazal: a major PPN and early PN settlement in Central Jordan', in Gebel, H G K, Kafafi, Z and Rollefson, G O (eds), *The Prehistory of Jordan*, vol II, pp 511–56. Studies in Early Near Eastern Production, Subsistence, and Environment 4, Berlin: Ex Oriente

Ducos, P (1975) 'A new find of an equid metatarsal bone from Tell Mureibet in Syria and its relevance to the identification of equids from the early Holocene of the Levant', *Journal of Archaeological Science* 2: 71–73

Ducos, P (1986) 'The equid of Tell Muraibit, Syria', in Meadow, R H and Uerpmann, H-P (eds), *Equids in the Ancient World*, pp 237–45, Wiesbaden: Ludwig Reichert Verlag

Ducos, P (1989) 'Defining domestication: a clarification', in Clutton-Brock, J (ed), *The Walking Larder: Patterns of Domestication, Pastoralism, and Predation*, pp 28–30, London: Unwin Hyman

Dyson-Hudson, N and Dyson-Hudson, R (1999) 'The social organization of resource exploitation', in Little, M A and Leslie, P W (eds), *Turkana Herders of the Dry Savanna: Ecology and Behavioral Response of Nomads to an Uncertain Environment*, pp 69–88, Oxford: Oxford University Press

Edwards, C J, MacHugh, D E, Dobney, K M, Martin, L, Russell, N, Horwitz, L K, McIntosh, S K, MacDonald, K C, Helmer, D, Tresset, A, Vigne, J-D and Bradley, D G (2004) 'Ancient DNA analysis of 101 cattle remains: limits and prospects', *Journal of Archaeological Science* 6: 695–710

Ehret, C (1997) 'African languages: a historical survey', in Vogel, J O (ed), *Encyclopedia of Precolonial Africa*, pp 159–71, Walnut Creek, CA: Altamira Press

Ehret, C (1998) *An African Classical Age*, Charlottesville, VA: University Press of Virginia

Ehret, C (2002) *The Civilizations of Africa: A History to 1800*, Charlottesville, VA: University Press of Virginia

Eisenmann, V (1986) 'Comparative osteology of modern and fossil horses, half-asses, and asses', in Meadow, R H and Uerpmann, H-P (eds), *Equids in the Ancient World*, pp 68–116, Wiesbaden: Ludwig Reichert Verlag

Eisenmann, V (1995) 'L'origine des ânes: questions et réponses paléontologiques', *Ethnozootechnie* 56: 5–26

Eisenmann, V and Beckouche, S (1986) 'Identification and discrimination of metapodials from Pleistocene and modern *Equus*, wild and domestic', in Meadow, R H and Uerpmann, H-P (eds), *Equids in the Ancient World*, pp 117–63, Wiesbaden: Ludwig Reichert Verlag

Epstein, H (1971) *The Origin of the Domestic Animals of Africa*, 2 vols, New York: Africana Publishing Corporation

Fattovich, R (1993) 'The Gash Group of the eastern Sudan: an outline' in Krzyzaniak, L, Kobusiewicz, M and Alexander, J (eds), *Environmental Change and Human Culture in the Nile Basin and Northern Africa until the Second Millennium BC*, pp 439–48, Poznan: Poznan Archaeological Museum

Fattovich, R (1996a) 'The Afro-Arabian circuit: contacts between the Horn of Africa and southern Arabia in the 3rd–2nd millennia BC', in Krzyzaniak, L and Kobusiewicz, M (eds), *Interregional Contacts in the Later Prehistory of North-Eastern Africa*, pp 395–402, Poznan: Poznan Archaeological Museum

Fattovich, R (1996b) 'Punt: the archaeological perspective', *Beiträge zur Sudanforschung* 6: 15–29

Fielding, D and Pearson, R A (eds) (1991) *Donkeys, Mules and Horses in Tropical Agricultural Development*, Edinburgh: University of Edinburgh

Fratkin, E (1991) *Surviving Drought and Development: Ariaal Pastoralists of Northern Kenya*, Boulder: Westview Press

Garcea, E A A (2004) 'An alternative way towards food production: the perspective from the Libyan Sahara', *Journal of World Prehistory* 18: 107–54

Gautier, A (1968) 'Mammalian remains of the northern Sudan and southern Egypt', in Wendorf, F (ed), *The Prehistory of Nubia*, pp 80–99, Dallas: Fort Burgwin Research Center and Southern Methodist University Press

Gautier, A (1980) 'Contributions to the archaeozoology of Egypt', in Wendorf, F and Schild, R (eds), *The Prehistory of the Eastern Sahara*, pp 317–44, New York: Academic Press

Gautier, A (1984) 'Quaternary mammals and archaeozoology of Egypt and the Sudan: a survey', in Krzyzaniak, L and Kobusiewicz, M (eds), *Origin and Early Development of Food-Producing Cultures in North-Eastern Africa*, pp 43–56, Poznan: Polish Academy of Sciences

Gautier, A (1987a) 'Prehistoric men and cattle in North Africa: a dearth of data and a surfeit of models', in Close, A E (ed), *Prehistory of Arid North Africa*, pp 163–87, Dallas: Southern Methodist University Press

Gautier, A (1987b) 'The archaeological sequence of the Acacus', in Barich, B E (ed), *Archaeology and Environment in the Libyan Sahara, 1978–1983*, pp 283–308. Cambridge Monographs in African Archaeology 23 and British Archaeological Report International Series 368, Oxford: British Archaeological Reports

Gautier, A and van Neer, W (1977–1982) 'Prehistoric fauna from Ti-n-Torha (Tadrart Acacus, Libya)', *Origini: Preistoria e Protostoria delle Civiltà Antiche* (Rome) XI: 87–127

Gentry, A, Clutton-Brock, J and Groves, C P (1996) 'Proposed conservation of usage of 15 mammal specific names based on wild species which are antedated by or contemporary with those based on domestic animals', *Bulletin of Zoological Nomenclature* 53(1): 28–37

Gentry, A, Clutton-Brock, J and Groves, C P (2004) 'The naming of wild animal species and their domestic derivatives', *Journal of Archaeological Science* 31: 645–51

Ghuffra, E, Kijas, J M H, Amarger, V, Carlborg, Ö, Jeon, J-T and Anderson, L (2000) 'The origin of the domestic pig: independent domestication and subsequent introgression', *Genetics* 154: 1785–91

Gibson, M A and Mace, R (2002) 'Labor-saving technology and fertility increase in rural Africa', *Current Anthropology* 43(4): 630–37

Gifford-Gonzalez, D P (2005) 'Pastoralism and its consequences', in Stahl, A B (ed), *African Archaeology*, pp 187–224, Oxford: Blackwell

Gifford-Gonzalez, D P and Kimengich, J (1984) 'Faunal evidence for early stock-keeping in the central Rift of Kenya: preliminary findings', in Krzyzaniak, L and Kobusiewiecz, M (eds), *Origin and Early Development of Food-Producing Cultures in North-Eastern Africa*, pp 357–471, Poznan: Polish Academy of Sciences

Groves, C P (1974) *Horses, Asses, and Zebras in the Wild*, London: David and Charles

Groves, C P (1986) 'The taxonomy, distribution, and adaptations of recent equids', in Meadow, R H and Uerpmann, H-P (eds), *Equids in the Ancient World*, pp 11–65, Wiesbaden: Ludwig Reichert Verlag

Groves, C P (2002) 'Taxonomy of living Equidae', in Moehlman, P D (ed), *Equids: Zebras, Asses and Horses*, pp 94–107, Gland, Switzerland: IUCN

Groves, C P and Willoughby, D P (1981) 'Studies on the taxonomy and phylogeny of the genus *Equus*, 1: subgeneric classification of the recent species', *Mammalia* 45: 321–54

Guerin, C and Faure, M (1996) 'Chasse au chacal et domestication du boeuf dans le site Néolithique d'Asa Koma (République de Djibouti)', *Journal des Africanistes* 66: 299–311

Hanotte, O, Daniel G B, Ochieng, J W, Verjee, Y, Hill, E W and Rege, J E O (2002) 'African pastoralism: genetic imprints of origins and migrations', *Science* 296: 336–39

Harlan, J R (1971) 'Agricultural origins: centers and noncenters', *Science* 174: 468–74

Harris, D R (1989) 'An evolutionary continuum of people–plant interaction', in Harris, D R and Hillman, G C (eds), *Foraging and Farming: The Evolution of Plant Exploitation*, pp 11–26, London: Unwin Hyman

Harris, D R (ed) (1996) *The Origins and Spread of Agriculture and Pastoralism in Eurasia*, Washington, DC: Smithsonian Institution Press

Harris, D R and Hillman, G C (eds) (1989) *Foraging and Farming: The Evolution of Plant Exploitation*, London: Unwin Hyman

Hassan, F A (1988) 'The Predynastic of Egypt', *Journal of World Prehistory* 2(2): 135–85

Hassan, F A (1993) 'Town and village in ancient Egypt: ecology, society and urbanization', in Shaw, T P, Sinclair, B A and Okpoko, A (eds), *The Archaeology of Africa: Food, Metals and Towns*, pp 551–69, London: Routledge

Hassan, F A (2002) 'Paleoclimate, food and culture change in Africa: an overview', in Hassan, F A (ed), *Droughts, Food and Culture*, pp 11–26, New York: Kluwer Academic/Plenum Publishers

Heck, L (1899) *Lebende Bilder aus dem Reiche der Tiere*, Berlin: Werner Verlag

Hemmer, H (1990) *Domestication*, Cambridge: Cambridge University Press

Hienleder, S, Lewalski, H, Wassmuth, R and Janke, A (1998) 'The complete mitochondrial DNA sequence of the domestic sheep (*Ovis aries*) and comparison with the other major ovine haplotype', *Journal of Molecular Evolution* 47: 441–48

Higgs, E S (1967) 'Environment and chronology – the evidence from mammalian fauna', in McBurney, C B M (ed), *The Haua Fteah (Cyrenaica) and the Stone Age of the South-East Mediterranean*, pp 16–64, Cambridge: Cambridge University Press

Holl, A F C and Dueppen, S A (1999) 'Iheren I: research on Tassilian pastoral iconography', *Sahara* 11: 21–34

Hollmann, A (1990) 'Saeugertierknochenfunde aus Elephantine in Oberaegypten', unpublished dissertation, Ludwig-Maximilians-Universitaet, Munich

Homewood, K M and Rodgers, W A (1991) *Maasailand Ecology: Pastoralist Development and Wildlife Conservation in Ngorongoro, Tanzania*, Cambridge: Cambridge University Press

Houlihan, P F (1996) *The Animal World of the Pharaohs*, London: Thames and Hudson

Howe, B and Movius, H L (1947) *A Stone Age Cave Site in Tangier: Preliminary Report on the Excavations at the Mugharet el 'Aliya, or High Cave, in Tangier*. Papers of the Peabody Museum of American Archaeology and Ethnology, vol XXVIII, no 1, Cambridge, MA: Peabody Museum, Harvard University

Ingold, T (1980) *Hunters, Pastoralists, and Ranchers: Reindeer Economies and Their Transformations*, Cambridge: Cambridge University Press

Jacobs, A H (1975) 'Maasai pastoralism in historical perspective', in Monod, T (ed), *Pastoralism in Tropical Africa*, pp 406–25, London: Oxford University Press for the International African Institute

Jansen, T, Forster, P, Levine, M A, Oelke, H, Hurles, M, Renfrew, C, Weber, J and Olek, K (2002) 'Mitochondrial DNA and the origins of the domestic horse', *Proceedings of the National Academy of Science* (Washington, DC) 99: 10905–10

Johnson, B R (1999) 'Social networks and exchange', in Little, M A and Leslie, P W (eds), *Turkana Herders of the Dry Savanna: Ecology and Behavioral Response of Nomads to an Uncertain Environment*, pp 89–108, Oxford: Oxford University Press

Jones, P (1977) *Donkeys for Development*, Harare: Animal Traction Network for Eastern and Southern Africa (ATNESA)

Klima, G J (1970) *The Barabaig: East African Cattle-Herders*, New York: Holt Rinehart and Winston

Krzyzaniak, L (1991) 'Early farming in the Middle Nile basin: recent discoveries at Kadero (central Sudan)', *Antiquity* 65: 515–32

Larson, G, Dobney, K, Albarella, U, Fang, M, Matisoo-Smith, E, Robins, J, Lowden, S, Finlayson, H, Brand, T, Willersley, E, Rowley-Conwy, P, Andersson, L and Cooper, A (2005) 'Worldwide phylogeography of wild board reveals multiple centers of pig domestication', *Science* 307: 1618–621

Leslie, P W and Dyson-Hudson, R (1999) 'Peoples and herds', in Little, M A and Leslie, P W (eds), *Turkana Herders of the Dry Savanna: Ecology and Behavioral Response of Nomads to an Uncertain Environment*, pp 233–48, Oxford: Oxford University Press

Little, M A and Leslie, P W (eds) (1999) *Turkana Herders of the Dry Savanna: Ecology and Behavioral Response of Nomads to an Uncertain Environment*, Oxford: Oxford University Press

Little, M A, Dyson-Hudson, R and McCabe, J T (1999) 'Ecology of South Turkana', in Little, M A and Leslie, P W (eds), *Turkana Herders of the Dry Savanna: Ecology and Behavioral Response of Nomads to an Uncertain Environment*, pp 43–66, Oxford: Oxford University Press

MacHugh, D E and Bradley, D G (2001) 'Livestock genetic origins: goats buck the trend', *Proceedings of the National Academy of Science* (Washington, DC) 98: 5382–84

Maloiy, G M O (1970) 'Water economy of the Somali donkey', *American Journal of Physiology* 219: 1522–27

Maloiy, G M O (1971) 'Temperature regulation in the Somali donkey (*Equus asinus*)', *Comparative Biochemistry and Physiology* 39: 403–12

Maloiy, G M O and Boarer, D H (1971) 'Response of the Somali donkey to dehydration hematological changes', *American Journal of Physiology* 221: 37–41

Marshall, F (2000) 'The origins and spread of domestic animals in East Africa', in McDonald, K C and Blench, R M (eds), *The Origins and Development of African Livestock: Archaeology, Genetics, Linguistics and Ethnography*, pp 191–221, London: University College London Press

Marshall, F (2003) '*Equus africanus* and the development of African pastoral societies, New contributions of Africanist archaeologists to understanding the continent's past, II, Abstracts of the African Studies Association annual meetings, Boston, October 31

Marshall, F and Hildebrand, E (2002) 'Cattle before crops: the origins and spread of food production in Africa', *Journal of World Prehistory* 16: 99–143

Masry, A H (1973) 'Prehistory in northeastern Arabia: the problem of interregional interaction', unpublished PhD dissertation, University of Chicago

McArdle, J (1982) 'Preliminary report on the Predynastic fauna of the Hierakonpolis Project', in Hoffman, M A (ed), *The Predynastic of Hierakonpolis: An Interim Report*, pp 116–21. Egyptian Studies Association Publication no 1, Giza: Cairo University Herbarium

McArdle, J (1992) 'Preliminary observations on the mammalian fauna from Predynastic localities at Hierakonpolis', in Friedman, R and Adams, B (eds), *The Followers of Horus*, pp 53–56. Oxford Monograph 20, Oxford: Egyptian Studies Association Publication no 2

McCabe, J T (2004) *Cattle Bring Us to Our Enemies: Turkana Ecology, Politics, and Raiding in a Disequilibrium System*, Ann Arbor, MI: University of Michigan Press

McCabe, J T, Dyson-Hudson, R and Wienpahl, J (1999) 'Nomadic movements', in Little, M A and Leslie, P W (eds), *Turkana Herders of the Dry Savanna: Ecology and Behavioral Response of Nomads to an Uncertain Environment*, pp 108–22, Oxford: Oxford University Press

Meadow, R H (1984) 'Animal domestication in the Middle East: a view from the eastern margin', in Clutton-Brock, J and Grigson, C (eds), *Animals and Archaeology 3: Early Herders and Their Flocks*, pp 309–37. British Archaeological Reports International Series no 502, Oxford: British Archaeological Reports

Meadow, R H and Uerpmann, H-P (eds) (1991) *Equids in the Ancient World*, vol II, Wiesbaden: Ludwig Reichert Verlag

Moehlman, P D (2002) 'Status and action plan for the African wild ass (*Equus africanus*)', in Moehlman, P D (ed), *Equids: Zebras, Asses and Horses*, pp 2–9, Gland, Switzerland: IUCN

Moehlman, P D, Kebede, F and Yohannes, H (1998) 'The African wild ass (*Equus africanus*): conservation status in the Horn of Africa', *Applied Animal Behaviour Science* 60: 115–24

Mohammed, A (1991) 'Management and breeding aspects of donkeys around Awassa, Ethiopia', in Fielding, D and Pearson, R A (eds), *Donkeys, Mules and Horses in Tropical Agricultural Development*, pp 185–88, Edinburgh: University of Edinburgh

Murdock, G P (1959) *Africa: Its Peoples and Their Culture History*, New York: McGraw-Hill

Negash, A (2001) 'The Holocene prehistoric archaeology of the Temben region, northern Ethiopia', unpublished PhD dissertation, University of Florida

Neumann, K (2003) 'The late emergence of agriculture in sub-Saharan Africa: archaeobotanical evidence and ecological considerations', in Neumann, K, Butler, A, and Kahlheber, S (eds), *Food, Fuel and Fields: Progress in African Archaeobotany*, pp 71–92, Cologne: Heinrich Barth Institute

Neumann, K (2005) 'The romance of farming: plant cultivation and domestication in Africa', in Stahl, A B (ed), *African Archaeology*, pp 249–75, Oxford: Blackwell

Nicolaisen, J (1963) *Ecology and Culture of the Pastoral Tuareg*, Copenhagen: National Museum of Copenhagen

Nicolaisen, J and Nicolaisen, I (1997) *The Pastoral Tuareg*, 2 vols, New York: Thames and Hudson

Nordström, H-Å (1972) *Neolithic and A-Group Sites*. Scandinavian Joint Expedition to Sudanese Nubia 3(1–2), Stockholm: Läromedesfärlagen

Oakenfull, E A (2000) 'A survey of equid mitochondrial DNA: implications for the evolution, genetic diversity and conservation of *Equus*', *Conservation Genetics* 1: 341–55

Oakenfull, E A (2002) 'Genetics of equid species and subspecies', in Moehlman, P D (ed), *Equids: Zebras, Asses and Horses*, pp 108–12, Gland, Switzerland: IUCN

Olsen, S L (1996) *Horses Through Time*, Pittsburgh, PA: Carnegie Museum of Natural History

Pao, B and Palin, M (2002) *Inside Sahara*, London: Widenfeld and Nicolson

Peet, T E (1914) *The Cemeteries of Abydos*, part II: 1911–1912, London: Egypt Exploration Fund

Perkins, Jr., D (1965) 'Three Faunal Assemblages from Sudanese Nubia', *Kush* 13: 56–61

Peters, J (1991) 'The faunal remains from Shaqadud', in Marks, A E and Mohammed-Ali, A (eds), *The Late Prehistory of the Eastern Sahel*, pp 197–235, Dallas: Southern Methodist University Press

Petrie, W M F (1914) *Tarkhan II*, London: School of Archaeology in Egypt

Phillipson, D W (1977) 'The excavation of Gobedra Rock Shelter, Axum: an early occurrence of cultivated finger millet in northern Ethiopia', *Azania* (Nairobi) 12: 53–82

Piperno, D R and Pearsall, D M (1998) *The Origins of Agriculture in the Lowland Neotropics*, San Diego: Academic Press

Postgate, J N (1986) 'The equids of Sumer, again', in Meadow, R H and Uerpmann, H-P (eds), *Equids in the Ancient World*, pp 194–206, Wiesbaden: Ludwig Reichert Verlag

Postgate, J N (1992) *Early Mesopotamia: Society and Economy at the Dawn of History*, London: Routledge

Price, T D and Gebauer, A B (eds) (1995) *Last Hunters–First Farmers: New Perspectives on the Prehistoric Transition to Agriculture*, Santa Fe, NM: School of American Research Press

Reed, C A and Turnbull, P F (1969) 'Late Pleistocene mammals from Nubia', *Palaeoecology of Africa* 4: 55–57

Romer, A S (1928) 'Pleistocene mammals of Africa: fauna of the Paleolithic station of Mechta-el-Arbi', *Logan Museum Bulletin* (Beloit, WI) 1: 80–163

Romer, A S (1935) 'Mammalian remains from some Paleolithic stations in Algeria', in Pond, A W, Chapuis, L, Romer, A S and Baker, F C, *Prehistoric Habitation Sites in the Sahara and North Africa*, pp 165–184. Logan Museum Bulletin V, Beloit, WI: Logan Museum, Beloit College

Rosenberg, M and Redding, R W (1998) 'Early pig husbandry in southwestern Asia and its implications for modelling the origins of food production', in Nelson, S M (ed), *Ancestors for the Pigs*, pp 55–64, Philadelphia, PA: University of Pennsylvania Museum of Archaeology and Anthropology

Rowley-Conwy, P (1988) 'The camel in the Nile Valley: new radiocarbon accelerator (AMS) dates from Qaṣr Ibrîm', *Journal of Egyptian Archaeology* 74: 245–48

Russell, N (2002) 'The wild side of animal domestication', *Society and Animals* 10: 285–300

Sadr, K (1991) *The Development of Nomadism in Ancient Northeast Africa*, Philadelphia, PA: University of Pennsylvania Press

Shahack-Gross, R, Marshall, F and Weiner, S (2003) 'Geo-ethnoarchaeology of pastoral sites: the identification of livestock enclosures in abandoned Maasai settlements', *Journal of Archaeological Science* 30: 439–59

Shaw, T (1977) 'Hunters, gatherers and first farmers in West Africa', in Megaw, J V S (ed), *Hunters, Gatherers and First Farmers Beyond Europe*, pp 69–126, Leicester: Leicester University Press

Smith, B D (1998) *The Emergence of Agriculture*, New York: Scientific American Library

Spencer, P (1973) *Nomads in Alliance*, London: Oxford University Press

Starkey, P (2000) 'The history of working animals in Africa', in McDonald, K C and Blench, R M (eds), *The Origins and Development of African Livestock: Archaeology, Genetics, Linguistics and Ethnography*, pp 478–502, London: University College London Press

Uerpmann, H-P (1986) 'Halafian equid remains from Shams ed-Din Tannira in northern Syria', in Meadow, R H and Uerpmann, H-P (eds), *Equids in the Ancient World*, pp 246–65, Wiesbaden: Ludwig Reichert Verlag

Uerpmann, H-P (1987) *The Ancient Distribution of Ungulate Mammals in the Middle East*, Wiesbaden: Ludwig Reichert Verlag

Uerpmann, H-P (1991) '*Equus africanus* in Arabia', in Meadow, R H and Uerpmann, H-P (eds), *Equids in the Ancient World*, vol II, pp 12–33, Wiesbaden: Ludwig Reichert Verlag

van Neer, W, Linseele, V and Friedman, R F (2004) 'Animal burials and food offerings at the elite cemetery HK6 of Hierakonpolis', in Hendrickx, S, Friedman, R F, Cial/owicz, K M and Chl/odnicki, M (eds), *Egypt at Its Origins: Studies in Memory of Barbara Adams*, pp 67–130, Leuven: Peeters

Vermeersch, P M, van Peer, P, Moeyersons, J and van Neer, W (1996) 'Neolithic occupation of the Sodmein area, Red Sea Mountains, Egypt', in Pwiti, G and Soper, R (eds), *Aspects of African Archaeology: Papers from the 10th Congress of the PanAfrican Association for Prehistory and Related Studies*, pp 411–19, Harare: University of Zimbabwe Publications

Vigne, J E, Dollfus, G and Peters, J (2000) 'Les débuts de l'élevage au Proche-Orient: données nouvelles et réflexions', *Paleorient* 25: 5–8

Wendorf, F, Schild, R and associates (2001) *Holocene Settlement of the Egyptian Sahara: The Archaeology of Nabta Playa*, vol 1, New York: Kluwer Academic/Plenum Publishers

Whitcher, S E, Janetski, J C and Meadow, R H (2000) 'Animal bones from Wadi Mataha (Petra Basin, Jordan): the initial analysis', in Mashkour, M, Choyke, A M, Buitenhuis, H and Poplin, F (eds), *Archaeozoology of the Near East, IV*, pp 39–48. ARC-Publicaties 32, Groningen: Centre for Archeological Research and Consultancy, Rijksuniversiteit Groningen

Yalden, D W, Largen, M J and Kock, D (1986) 'Catalogue of the mammals of Ethiopia', *Italian Journal of Zoology/Monitore zoologico italiano* supp 4: 31–103

Yousef, M K (1991) 'Energy cost of locomotion in the donkey', in Fielding, D and Pearson, R A (eds), *Donkeys, Mules and Horses in Tropical Agricultural Development*, p 220, Edinburgh: University of Edinburgh

Zarins, J (1986) 'Equids associated with human burials in third millennium BC Mesopotamia: two complementary facets', in Meadow, R H and Uerpmannn, H-P (eds), *Equids in the Ancient World*, pp 164–93, Wiesbaden: Ludwig Reichert Verlag

Zarins, J (2002) 'Dhofar, frankincense, and Dilmun precursors to the Iobaritae and Omani', in Cleuziou, S, Tosi, M and Zarins, J (eds), *Essays on the Late Prehistory of the Arabian Peninsula*, pp 403–38, Rome: Instituto italiano per l'Africa e l'Oriente

Zeder, M A (1986) 'The equid remains from Tal-e Malyan, southern Iran', in Meadow, R H and Uerpmann, H-P (eds), *Equids in the Ancient World*, pp 366–412, Wiesbaden: Ludwig Reichert Verlag

Zeder, M and Hesse, B (2000) 'The initial domestication of goats (*Capra hircus*) in the Zagros Mountains 10,000 years ago', *Science* 287: 2254–57

Zeuner, F E (1963) *A History of Domesticated Animals*, London: Hutchinson

CHAPTER 21

Using Linguistics to Reconstruct African Subsistence Systems: Comparing Crop Names to Trees and Livestock

Roger Blench

INTRODUCTION

Reconstructing the history of agriculture in Africa, or indeed any area of the world where written documentation is sparse or non-existent, is inevitably a multidisciplinary exercise. Although the volume of archaeobotanical data available for Africa is gradually increasing, coverage remains extremely patchy and concentrates on a few species, notably sorghum, millet and finger millet (see review in Neumann 2003). It is safe to say that most plants cultivated in Africa today are nowhere represented in the repertoire of plant remains recovered from the archaeological record. In the case of domestic animals, particularly cattle, sheep, goats and chickens, the situation is marginally better, with some materials for most species in much of the continent (see individual reviews in Blench and Macdonald 2000). African arboriculture or the intentional planting of trees, an ancient characteristic of many agricultural systems in the Old World, is a poorly understood and little-documented area; linking the sparse archaeobotanical material from trees with present-day management systems has hardly begun. Indirect indicators of agriculture in the past and the present, such as agricultural tools, settlement patterns, field systems and animal pens, remain understudied.

Two tools other than archaeology are available for the reconstruction of agrarian history: historical and comparative linguistics and DNA (deoxyribonucleic acid) studies. The use of DNA to determine taxonomic relationships between or within wild and cultivated crops and trees has yet to be undertaken even for major species. In contrast, studies of nuclear and mitochondrial DNA (mtDNA) in livestock have begun to produce intriguing results (eg Loftus et al 1994; Bradley et al 1994, 1996 on cattle; Giuffra et al 2000 on pigs; Hiendleder et al 1998 on sheep; Luikart et al 2001 on goats). The origins of the domestic dog have recently been the subject of renewed interest (Gallant 2002; Savolainen et al 2002) but many African breeds of ruminant and other minor domestic species are still unsampled. DNA studies tend to show multiple origins for well-known species, with the consequence that classical phenotypic or osteometric work (eg Epstein 1971; Grigson 2000) in archaeozoology must be rethought.

Historical linguistics can be defined as the analysis of the relationship between languages, in particular those assumed to be genetically related and to have 'sprung from some common source'. Historical linguists establish rules that explain how individual languages evolve from this common source through the reconstruction of hypothetical proto-forms. Usually they base their reconstructions on the comparison of two or more languages, but the 'internal reconstruction' of a single language is also possible, using indications such as dialect variation or fossil morphology to create a picture of an earlier stage of a given language. In principle, historical linguistics can provide essentially two sorts of insights relevant to the prehistory of agriculture:

(a) describing patterns of loanwords that track the introduction and diffusion of new cultivated plants and animals, innovative management techniques and related socioeconomic institutions; and,
(b) reconstructing individual lexical items to a hypothetical proto-language that were likely to be known to speakers of that language.

The first author to point to the potential of this method was probably Julius von Klaproth in 1830. He observed that the names for 'birch tree' linked European languages with those of India and therefore had implications for prehistory:

> Il est digne de remarque que le *bouleau* s'appelle en sanscrit *bhourtchtcha*, et que ce mot dérive de la même racine que l'allemand *birke*, l'anglais *birch* et le russe, бере*за* [bereza], tandis que les noms des autres arbres de l'Inde ne se retrouvent pas dans les langues indo-germaniques de l'Europe. La raison en est, vraisemblablement, que les nations indo-germaniques venaient du nord, quand elles entrèrent dans l'Inde, où elles apportèrent la langue qui a servi de base au sanscrit, et qui a repoussé de la presqu'île, les idiomes de la même origine que le malabar et le télinga, que ces nations, dis-je, ne trouvèrent pas dans leur nouvelle patrie les arbres qu'elles avaient connu dans l'ancienne, à l'exception du bouleau, qui croît sur le versant méridional de l'Himâlaya[1] (Klaproth 1830: 112–13).

Max Müller (1864: 222–24) may well have been the first to link etymological data with archaeological finds and, by a fortunate chance, his examples refer to flora and environment. He argued that linguistic interchanges between the names of 'fir', 'oak' and 'beech' in early Indo-European can be interpreted in the light of the changing vegetation patterns deduced from visible strata in Danish peat bogs. Although this type of correspondence is now a longstanding tradition in Indo-European scholarship and is now very much part of the reconstruction of Austronesian prehistory, elsewhere in the world it has had a less enthusiastic reception. In Africa in particular, reconstructed protoforms for the major language phyla are at best the subject of disagreement (see Blench 2002 for a discussion of the controversy over Nilo-Saharan reconstructions) and are often marked by the unwillingness of linguists to engage with archaeobotanical and archaeozoological databases. Although linguistics can provide information on topics on which archaeology has little to say, including social organisation, music, religion and vegetative crops, it can only ever provide relative dates or estimations. Only archaeology can provide absolute dating.

Another aspect of the failure of the two disciplines to engage is the contrast between archaeological visibility and linguistic salience. This works in two ways; something may have high archaeological visibility and low linguistic salience and, conversely, something may be prominent as a reconstruction or a loanword, but be invisible archaeologically. Table 21.1 presents some examples of fields that illustrate the potential mismatch between archaeological visibility and linguistic salience.

Linguistics may also sometimes produce only banal, circular inferences, such that fish names will be salient in fishing communities or that savanna populations will have names for common useful trees. Nonetheless, salience of terms clearly varies from one era of prehistory to another and leaves its traces in vernacular names.

Nonetheless, a body of linguistic evidence for African crops, economic plants, trees and livestock has now been compiled and some examples of the way linguistics, archaeology and genetics can be linked now exist (eg Banti 1993; Blench 1993, 1995, 1998; Blench et al 1997; Heine 1978; Philippson and Bahuchet 1996; Portères 1958; Skinner 1977; Williamson 1993, 2000). But the linguistic evidence shows curiously patchy results; it seems that some categories reconstruct much better than others and that this variation is not necessarily connected with either their salience or their antiquity. An intriguing

Table 21.1 Contrasting archaeological visibility and linguistic salience

Example	*Archaeological Visibility*	*Linguistic Salience*	*Comment*
Fish bones	high	low	Fish species are too numerous and diverse to generate widespread reconstructions.
Tuber crops	low	high	Tubers are not easily identified in African sites with present techniques, although analyses of starch grains may change this. Phytoliths may be valuable in detecting fruits, such *Musa* spp.
Recently introduced crops	low	high	Neotropical introductions have transformed African agriculture, but too recently to be reflected in archaeological materials.
Livestock	high	high	Bones are well preserved and vernacular terms highly salient. The study of livestock is the only area where modern DNA work exists.
Humid zone artefacts	low	specific to individual artefacgs	Acid soils result in poor preservation in humid forest.
Large predators	low	low	Predators are not eaten, hence their bones are rarely found at settlement sites. They are subject to linguistic taboo, hence reconstruct poorly despite high anthropological salience.

asymmetry with important consequences for African economic history is the difficulty of reconstructing crop names compared with terms for domestic animals. Given that dates for agriculture in Africa are highly controversial (see discussion in Neumann 2003), it would be of great interest to establish secure reconstructions for major cultigens such as yam (*Dioscorea* spp.), sorghum (*Sorghum bicolor*), millet (*Pennisetum glaucum*), finger millet (*Eleusine coracana*), fonio (*Digitaria exilis*) and others in the different language phyla. But attempts to do this have been generally unsuccessful, in contrast to livestock, where terms for cattle, sheep and goat have been reconstructed in both the Niger-Congo and Afroasiatic language phyla.

By contrast, tree names remain an underexplored topic. Although some important economic species can be reconstructed to median levels of Niger-Congo, others are only notable for the extreme diversity of their vernacular names. Although the biodiversity of African trees remains to be fully documented, the level of floristic biodiversity is relatively high[2] (Groombridge 1992: 66, Table 8.1). There are perhaps 60,000 species of vascular plants of which 35,000 are endemic (Davis et al 1994). Almost all species of tree are potentially of use; however, many have only limited importance and have not made any impact on the linguistic repertoire of tree names. However, individual species may develop local or zonal importance for a variety of reasons, including their medicinal value, fruit, charcoal, contribution to soil fertility, etc, which in turn mesh with evolving production systems—for example, the capacity to survive bush fires. As they gain salience, strategies for exploiting plant species diffuse and, consequently, a name for a particular species occurs over a wide area. This salience is reflected in the existence of widespread linguistic cognate terms[3] that can be taken to mark the point in the evolution of African language phyla at which human society began to attribute significant economic and cultural value to a particular species. This significance will in turn be interpretable in terms of the archaeobotanical profile of particular regions of the continent. This profile usually does not reflect biogeography, but rather assumed cultural significance, as the examples of African mahogany (*Khaya senegalensis*), shea (*Vitellaria paradoxa*), locust (*Parkia biglobosa*), baobab (*Adansonia digitata*), and silk-cotton (*Ceiba pentandra*) show (Blench in press a).

The situation in Africa contrasts sharply with that of the Pacific. In Oceania, few economic tree species have a 'natural' distribution and indeed indigenous tree floras of individual islands may be quite depauperate (Walter and Sam 1999). Individual species were moved from island to island as the Papuan and Austronesian phyla expanded; the very fact that they were moved guarantees their linguistic salience. Blench (2005) reviews the broader literature on Pacific arboriculture and shows that the rich lexical base available for Pacific languages makes possible a detailed reconstruction of the human restructuring of the tree flora of the region. Reconstructions of tree names have also proven important in the identification of the Algonquian homeland in North America (Goddard 1994).

The movement and manipulation of trees in African history can be divided into general categories that broadly correspond to historical epochs but also to the production systems of particular groups. Arboriculture, defined as the intentional planting of trees, was an ancient characteristic of many agricultural systems in the Old World, but was unknown until very recently in sub-Saharan Africa, with the exception of Ethiopia. Although economic species spread through the opportunistic transport of seeds and the selective protection of individual species, trees and their products played important roles in African subsistence systems because of their relative abundance. Fire is a key element in determining the pattern of African vegetation; species that survive annual burning, such as the locust tree, become more prevalent in savannas with high-density occupations. Only a highly schematic view of the correspondences between production system and the spread of particular tree species is possible (Table 21.2).

Prior to the development of agriculture, foragers intensively exploited a wide variety of fruit trees, including species that are considered of only limited value today. It is generally assumed that LSA (Late Stone Age) foragers were highly mobile and would therefore have actively spread the endocarps of economic fruits. However, this is hard to prove without clearer distributional data and some hypotheses as to the 'natural' environment of particular species. Nonetheless, finds of endocarps, as distinct from the identification of a tree from anthracological (ie charcoal from accidentally or intentionally burnt woody vegetation) data, do suggest human intervention.

This paper explores the conflicts and synergies between archaeology and historical linguistics in reconstructing African agricultural history. It presents examples of the link between reconstructions and dated materials in some major economic species, but also highlights lacunae, noting important species with ambiguous linguistic records.

Table 21.2 A general scheme for determinants of tree salience in African prehistory

Production System	*Characteristic*	*Example Species*	
		English	*Latin*
Forager	Transporting economic fruits	Bush candle	*Canarium schweinfurthii*
Pastoralist	Transporting economic fruits	Baobab	*Adansonia digitata*
Settled agriculture	Bush burning with protection of economic trees	Shea	*Vitellaria paradoxa*
	Selective economic extraction	False locust	*Prosopis africana*
	Ritual prohibitions on cutting	West African ebony	*Diopsyros mespiliformis*
	Trade in economic fruits	Locust	*Parkia biglobosa*
Urbanism	Sale of tree products	Cola	*Cola acuminata*
Colonial era	Intentional diffusion of fruit trees	Citrus	*Citrus* spp.
	Selective economic extraction	Tropical hardwoods	*Milicia excelsa*
Post-colonial era	Agroforestry, plantation economies	Teak	*Tectona grandis*

BACKGROUND TO AFRICAN LANGUAGE PHYLA

African languages are conventionally divided into five phyla:, Niger-Congo, Nilo-Saharan, Afroasiatic, Khoesan and Austronesian (Malagasy) (see Blench 1999). Two of these phyla have significant numbers of speakers outside Africa: Afroasiatic, because of the expansion of Arabic northwards and eastwards into Eurasia, and Austronesian, which is mainly centred on Southeast Asia and Oceania. Using the estimates from Ethnologue (Grimes 2000), the number of African languages spoken today is about 2000. Language numbers are distributed very unevenly across the phyla (Table 21.3).

In the case of Khoesan, numerous languages have become extinct in historic times and only inadequately transcribed data remains. Although Khoesan speakers are predominantly hunter-gatherers, reconstructions of domestic animal names in Central Khoesan are indicative of the date of their interactions with pastoralists, an encounter also reflected in the archaeological record. There are several poorly documented language isolates such as Hadza, Jalaa, Baŋgi Me and Laal (Blench 1999). These current and former hunter-gatherer populations are unlikely to contribute significantly to the reconstruction of the prehistory of agriculture since, as single languages, they do not provide the comparative material that makes the Khoesan data so valuable.

Documentation of African languages is highly variable and is certainly not adequate in the technical field of names for crops, livestock, trees or other fields associated with farming, such as agricultural tools. Linguists are poor botanists (and vice versa) and rarely collect more than the names of a few very common species. As a consequence, the reconstruction of tree names is not well developed in any of the language phyla of Africa. Reference sources such as Burkill (1985, 1994, 1995, 1997, 2000) do sometimes constitute important compilations of vernacular names, but the transcriptions are highly variable in quality and are often difficult to use. Nonetheless, information

Table 21.3 Numbers of African languages by phylum

Phylum	*Number*	*Location*	*Source*
Niger-Congo	1,489	West, central and southern Africa	Grimes (2000)
Nilo-Saharan	80	Southern edge of the Sahara from Mali to Ethiopia; southern extension into Tanzania	Bender (1996)
Afroasiatic	339 (does not include 34 Arabic dialects)	North Africa, Ethiopia; southern edge of the Sahara central Africa and Ethiopia	Grimes (2000)
Khoesan	70	Southwestern Africa; Possible outliers in Tanzania	Güldemann and Vossen (2000)
Austronesian	5 (in Africa)	Madagascar	Grimes (2000)

about the most important species is rich enough to make possible the mapping of linguistic and archaeological data.

CROPS

The Reconstruction of Crop Names

The earliest writing on centres of agriculture and domestication of plants tended to ignore Africa, although Vavilov (1931) identified Ethiopia as a centre of domestication for wheat (*Triticum* spp.) and peas (*Pisum* spp.). The notion that West Africa was an important world centre for crop domestication dates from Murdock (1959), whose proposals have been largely confirmed by later work. The largest language phylum in Africa, Niger-Congo, is generally believed to have originated in West Africa and its speakers would have initiated agriculture by the time the Bantu expansion began in southern Cameroon some 3000–4000 years ago. As a consequence, names for domestic plants that occur in the Benue-Congo languages of Nigeria and have Bantu reflexes can be assigned to the early period of agriculture. We should therefore seek linguistic evidence for the origins of agriculture in West Africa. Ethiopia represents a different agrarian nucleus, with a blend of indigenous species and those brought from the Near East, which are in turn reflected in the predominant Afroasiatic languages.

Many of Africa's indigenous crops remain poorly known and few enter into world trade. Ethnobotanical research into crop plants in Africa has tended to focus on those considered most commercially significant. Thus, although there exists a substantial body of research on the taxonomy and local use of sorghum, plantains (*Musa* spp.) or Guinea yams (*Dioscorea rotundata*), cereals such as fonio and iburu (*Digitaria* spp.), and tubers such as the aerial yam, *Dioscorea bulbifera*, and the Sudan potato, *Solenostemon rotundifolius*, remain almost unknown. This leads to an unbalanced picture of the cultigen repertoire in traditional agriculture and a tendency to underestimate the significance of 'minor' crops in prehistory.

If we depended solely on well-dated finds, our picture of African agriculture would be severely impoverished. The identification of centres of origin for most species is based, not on archaeobotany, but on plant geography and analysis of modern-day cultivars and their wild relatives. Neumann (2003) has reviewed the archaeobotanical evidence for agriculture in Africa, in support of her contention for its late origin. The evidence is best for cereals; vegetative crops such as yams and potherbs are poorly represented or non-existent. It is possible to use phytoliths and starch grains to detect starchy roots, but until now only phytoliths have been adopted and are yet to be widely used (cf Mbida et al 2000, 2001 on *Musa* phytoliths in Cameroon, a report which has remained highly controversial[4]). The dichotomy between cereals and vegetative plants is very marked. With cereals it is possible to compare and contrast linguistics and archaeobotany; with other crops, linguistics is presently the only tool available for reconstruction of their history. As a consequence, agriculture tends to be seen from the perspective of semi-arid regions; better data on forest-zone crops might well transform existing models.

A major difficulty in the reconstruction of plant names in African languages is the transfer of names between wild and cultivated varieties of plants, as the Niger-Congo terms for 'yam' and 'sorghum' illustrate. Yams, ie the Dioscoreaceae, occur throughout all of sub-Saharan Africa between the semi-arid and humid zones. The wild ancestors of the present-day cultivated yams, such as *Dioscorea rotundata*, would have been exploited from an early period, as indeed are many species today, especially in periods of famine. At an unknown time, the cultivated yam was developed from the wild *Dioscorea* through a gradual process of protecting, transplanting and then selecting. Although a reconstruction of something like #-*ji* is reasonable at the level of proto-Benue-Congo (Williamson 1993), this is no guarantee that speakers of this proto-language were cultivating yams, as opposed to simply exploiting wild forms. Therefore, no amount of work on reconstructing the basic lexeme for 'yam' can clarify its relative antiquity in cultivation. Similarly, with sorghum, there is a widespread cognate term in Niger-Congo languages, something like #kVN- (Table 21.6), but archaeobotanical evidence for sorghum (Table 21.5) is persistently late for such a reconstruction to refer to cultivated forms. Failure to recognise this has led to somewhat exaggerated claims about the reconstructibility of both cultigens and, by extension, agriculture.

There is a possible way around this dilemma; the reconstruction of lexical items associated with cultivation (Connell 1998; Williamson 1993). There could, for example, be a specific word for a tool to uproot yams, for seed yam or yam heap. If these words were shown to reconstruct to the same time depth as the yam itself, this would be a good indication of the antiquity of cultivation. Although semantic shift remains a possibility, for example a general word for mound becoming 'yam heap', it is unlikely that the same shift would take place in all groups simultaneously. In the case of the Guinea yam, lexical items associated with its cultivation are not reconstructible to the same level as terms for the plant itself (eg Connell 1998 for the Cross River languages in southeast Nigeria). From this we can conclude that speakers of Niger-Congo languages knew about wild yams and began to exploit them for food long before they adopted current cultivation techniques. Even this strategy is only useful in some contexts; for example, it might seem that looking for reconstructions of words such as 'field' would provide evidence for the relative antiquity of agriculture. But in most African languages, 'field' is simply the same word as 'bush' or 'uncultivated land' and not a distinct lexeme. This is informative about the fuzzy conceptual boundaries of land classification, but not very helpful in uncovering the antiquity of agriculture.

The remainder of this section looks at the evidence for the reconstruction of a sample of African cereal crops—fonio, sorghum, finger millet and wheat—since these can be compared with archaeobotanical data.

Individual Crop Species

Fonio, Fundi, Hungry Rice (Digitaria exilis)

Fonio is a short, grass-like cereal derived from a wild species, *Digitaria longiflora*, cultivated between Guinea and the Nigeria-Cameroon border (Chevalier 1922; Hilu et al 1997; Portères 1955). It is only slightly differentiated from its wild relative, and fonio fields are often invisible to unpractised observers[5]. Its rather disjunct distribution in West Africa at present suggests that it was anciently spread over a much wider area, but that it has yielded to larger crops with higher yields (Figure 21.1). The Arab geographer Al-'Umari, writing in 1337–8, says '[*funi*]...is a downy pod, from which, when crushed, there issue seeds like those of mustard, or smaller and white in colour' (Levtzion and Hopkins 1981: 263). Ibn Baṭṭuṭa, who travelled in Sahelian West Africa a decade later, in 1354, also mentions the cultivation of fonio in Mali. Fonio has been retrieved from the site of Cubalel in Senegal dated to the Late Iron Age, ie last few centuries BC to early centuries AD (Dorian Fuller pers comm). Iburu, *Digitaria iburua*, is a lesser-known relative of fonio confined to central Nigeria.

There is a widespread cognate term for fonio (? #fundi) in West African languages spoken in the heartland area between Guinea and Mali (Table 21.4) where fonio is likely to have been domesticated (Portères 1976: 419–22).

A single cognate term is spread across most of Mande, Atlantic and less commonly in adjacent families, across the core area of fonio cultivation. It has thus been loaned between adjacent branches of Niger-Congo and is not to be reconstructed to any of its proto-languages. Elsewhere in West Africa where fonio is grown, such as central Nigeria, the names are completely unrelated, which indicates that this region was cut off from the main zone of cultivation at an early period (Burkill 1994: 226; Portères 1955, 1976). This evidence suggests that the cultivation of fonio was part of a complex that evolved in the area of present-day Guinea at least 2000 years ago.

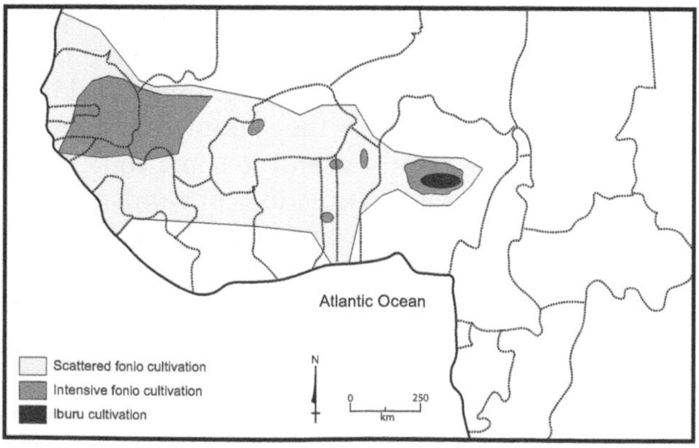

Figure 21.1 Fonio and iburu cultivation in West Africa.

Table 21.4 Cognate terms for fonio in West African languages

Phylum	*Language*	*Attestation*	*Language*	*Attestation*
Niger-Congo	**Mande**		**Atlantic**	
	Mende	**pótέ**	Wolof	**fini**
	Loko	**pénî**	Fulfulde	**fonyo**
	Looma	**pɔdɛ**	Jola-Fonyi	**finya**
	Kpelle	**miniŋ**	Bedik	**fɔndέŋ**
	Jallonke	**fúndéń**	Basari	**funyáŋ**
	Soso	**fundeɲ**	Manjaku	**findi**
	Mandinka	**fíndi**	Kisi	**kpendo**
	Xasonka	**fúndi**	Bulom	**peni**
	Bamana	**fíni**	Balanta	**fénhe**
	Maninka	**fónĩ**		
	Soninke	**fuɲaŋ/fuɲaŋŋe**	**Gur**	
	Bobo	**fẽ̄** *pl.* **fã̄**	Kurumfe	**peŋfe** *pl.* **peŋi**
	Dan	**pɤ̋ŋ̋**	Nawdm	**figm**
	Guro	**fní**		
	Mona	**fĩ̋ĩ̋**	**Kwa**	
	Wan	**fẽ̄ŋ́**	Anufo	**ǹfôni**
	Dogon		**Kru**	
	Dogon	**põ**	Wobe	**pohim**
Nilo-Saharan	**Songhai**			
	Songhay	**fingi**		

Sources: adapted from Vydrine (ined.), Segerer (ined.), Burkill (1994)

Sorghum (Sorghum bicolor)

Cultivated sorghum presents one of the more perplexing problems in African agrarian history (Blench 2003). It is crucial to African subsistence systems in the sub-humid and semi-arid regions of the continent and is embedded in ritual systems, and so would appear to be ancient. The earliest archaeological sorghum which is undoubtedly domesticated (a derived durra/ caudatum type) is from Kawa, Sudan at 400–780 BC (Fuller 2004a). But all attested archaeobotanical materials remain stubbornly recent (Table 21.5) compared with India, where sorghum occurs millennia before confirmed dates in Africa (Fuller 2003a). Archaeobotanical evidence is sometimes hard to read because of the difficulties in distinguishing wild and cultivated races (Neumann 2003: 77).

There is linguistic evidence for a widespread cognate term in west-central Africa, #kVN-, that occurs in a number of distinct language families and phyla (Table 21.6).

Some Niger-Congo families, such as Ijoid and Kru, are not represented in the data because they are confined to the humid zone where sorghum does not grow. The evidence seems to be that the underlying form is widespread, frequently compounded and ancient, but also much borrowed between phyla

Table 21.5 The earliest archaeological records of domesticated sorghum in Africa

Country	*Site*	*Type*	*Date(s)*	*Reference*
Sudan	Kawa	direct AMS	400–780 BC[11]	Fuller (2004a)
Libya	Jarma	direct AMS	110–370 BC	Pelling and Fuller (pers comm); Pelling (2003)
Sudan	Umm Muri	direct AMS	50–230 BC	Fuller (2004b)
Sudan	Jebel el Tomat	direct AMS	245±69 AD	Clark and Stemler (1975)
Sudan	Meroe		20 ±127 BC	Rowley-Conwy (1991)
Nigeria	Elkido		AD 340–430	Magnavita (2002)
Nigeria	Daima		AD 800	Connah (1981)

Table 21.6 Cognate terms for sorghum in West African languages

Phylum	*Branch*	*Language*			
Niger-Congo	Mande	Vai		**ke**	nde
		Mende		**kɛ**	ti
	Atlantic	Fulfulde	ga	**w**	ri
		Konyagi		**ko**	mbo
	Adamawa	Longuda		**kwa**	nla
		Waka		**kɔ**	ŋ
	Kwa	Krobo	ko	**ko**	
	Benue-Congo	Akpa	i	**kwù**	
		Iceve	ì-	**kù**	lé
		Igala	ó	**ko**	lì
		Igbo	o	**kì**	lì
Nilo-Saharan	Songhay	Songhay		**hà**	mà
	Saharan	Kanuri	ngà	**wú**	lì
Afroasiatic	Central	Kamwe		**xà**	
		Bole		**ku**	té
		Dera		**kú**	rè
	West	Mwaghavul		**kà**	s

Source: adapted from Burkill (1994:348 ff.) and personal research

and families, suggesting that sorghum cultivation spread well after the establishment of the main linguistic groups in West Africa.

Philippson and Bahuchet (1996: 103–104) discuss the terms for sorghum in Bantu languages. In much of East Africa, the common term for bulrush millet, ***-bele** (the asterisk is used in historical linguistics to indicate a reconstructed proto-form), seems to have been transferred to sorghum. This implies that sorghum came well after millet was established as a cultigen. Bulrush millet was probably the cereal of the Cushitic speakers who occupied much of East Africa prior to the arrival of the Bantu in the region. Indeed, the Bantu term looks as if it is borrowed from Southern Cushitic (eg Iraqw *balaangw* 'millet' [Mous and Kießling 2004]) or indeed the Eastern Cushitic words associated with cultivation (eg Proto-Sam **bèer* 'garden' [Heine 1978: 46, 54]). To complicate matters still further, many sorghum terms are now applied to maize, which has replaced it as a staple in many areas.

Finger Millet, Ragi (Eleusine coracana)

Finger millet gains its name from the head of the plant which bears some resemblance to a splayed hand. Today it is primarily grown in most regions of eastern and southern Africa to make beer, although it probably played a greater role as a staple in the period before the introduction of maize. The exact area of domestication of finger millet has remained controversial. Because it shows the greatest varietal diversity in India, earlier sources suggested a homeland there. Portères (1951, 1958) inclined to an African origin on the basis of a study of terms in African languages and more recent genetic work has generally supported this view (Hilu et al 1979). Most authors have wanted to assign very old dates to finger millet domestication, despite the sparse archaeobotanical material. Indeed, archaeobotanical records are so far very recent. Boardman (1999 quoted in Barnett 1999) records a first millennium AD find of finger millet near Aksum in Ethiopia. In southeastern Africa, there is a record of cultivated finger millet at Inyanga, in modern-day Zimbabwe, where carbonised seeds are associated with late Iron Age pottery (Summers 1958). Finger millet presumably spread westwards across the centre of the continent in quite recent times, since its western limit is in central Nigeria.

From the point of view of linguistics, finger millet seems to be old in Ethiopia and eastern and southern Africa, but is clearly recent in West Africa. Hausa *támbàà* has been borrowed into many languages of central Nigeria. In Ethiopia, Ehret (1979: 172) notes that Amharic *dagussa* is borrowed from the Agaw languages, suggesting domestication prior to the intrusion of Ethiosemitic. Table 21.7 shows a cognate term recorded in a wide swathe of eastern Africa[6]. The original shape of this word seems to have been something like *#mugimbi*, whence it was also borrowed into Nilotic languages with a loss of the prefix and devoicing of the first consonant.

Another East African cognate term, **-degɪ*, occurs east of the Great Lakes and has been connected to the Indian name *ragi* (Philippson and Bahuchet 1996: Figure 4). But the probable source of *ragi* is proto-South Dravidian **iraki*, 'grain' (Fuller 2003b). The Southern Cushitic languages have a quite different name, **basoróo* (Mous and Kießling 2004) suggesting that they were not the source of the finger millet grown by Bantu speakers.

Wheat (Triticum vulgare); *Durum Wheat, Hard Wheat* (Triticum durum)

Wheat is not indigenous to sub-Saharan Africa, although it has long been grown in North Africa. Soft wheat is the main *Triticum* sp. grown in the

Table 21.7 Cognate terms for finger millet in East African languages

Phylum	*Branch*	*Language*	*Attestation*	*Gloss*
Niger-Congo	Bantu	Swahili	**(m)wimbi**	
		Embu	**ugimbi**	
		Kikuyu	**ugimbi**	
		Chonyi	**wimbi**	
		Sangu	**uwugimbi**	beer
		Sena	**mulimbi**	
		Shona	**mbimbimbi**	bumper crop of finger millet
Nilo-Saharan	Nilotic	Maa	**oloikimbi**	

oases of the Sahara and along its southern margins, from Mauritania to Sudan, as well as in parts of Ethiopia (Chevalier 1932: 75). Wild wheats grow throughout the Near East and are still relatively common today. Wheat grains occur in tombs in Egypt throughout the dynastic period (Darby et al 1977, II: 486). Although the wheats are one of the most common cereals at the oases of the Sahara (Gast 2000), they are only sparsely cultivated farther south. El-Bekri, writing in 1067, mentions wheat at Awdaghost, and Ibn Baṭṭuṭa recorded it at Takedda in the Sahara in the fourteenth century (Lewicki 1974).

Hard wheat was probably replaced by emmer wheat, *Triticum dicoccum*, relatively recently as there is sparse evidence for its presence in the Mediterranean in classical times. Bread wheat (ie soft wheat, or *T. vulgare*) develops from emmer somewhere in the northern Fertile Crescent/South Caspian region. Flotation samples indicate that emmer is in Nubia (3rd cataract) by c 3500 BC and is probably a co-staple with barley at Kerma (Dorian Fuller pers comm). The first Egyptian materials date from the Ptolemaic period (Germer 1985: 212). Durum seems to be a later Roman innovation in Egypt which spread to Libya in the Byzantine period, prior to the Arab expansions (Pelling 2003). Watson (1983: 20) placed its origin in the region between northern Ethiopia and the eastern Mediterranean basin, but it may well be farther north. The linguistic evidence for wheat suggests that everywhere in sub-Saharan Africa except Ethiopia, wheat is a medieval introduction and names in African languages are borrowed from the Arabic, *al qamh*, usually with the article incorporated (eg Hausa *álkámà*). It has been argued that the Ethiopian wheats, *Triticum aethiopicum* Jakubz., are a distinct species (Barnett 1999), but most authors consider it a cultivated variety of *Triticum turgidum*, ie *T. turgidum* conv. *aethiopicum*. If so, then it was introduced from Southwest Asia and diversified in Ethiopia. The widespread cognate term in Ethiopian languages is not adapted from Arabic, with the exception of Oromo (Table 21.8).

The embedding of the #s-n-d cognate term in Cushitic languages strongly supports the introduction of wheat in Ethiopia in the pre-Semitic era.

Table 21.8 #s-n-d, a cognate term for wheat in Ethiopian languages

Phylum	*Branch*	*Language*	*Attestation*	*Gloss/comment*
Afroasiatic	Semitic	Amharic	**sənde, ስንዴ**	
	Cushitic	Oromo	**qamadii**	<Arabic
		Somali	**sarreen**	
		Saho	**sirrey**	
		Beja	**seram/shinray**	
		Sidamo	**sinde**	
		Agaw	**səndayi**	
	Omotic	Wolayta	**sindiya**	<Amharic

Sources: compiled from Lamberti & Sottile (1997), Hudson (1989)

AGRICULTURAL TOOLS

Despite their importance, historical linguists have so far ventured very few reconstructions of African agricultural tools. Indeed, the mapping of existing African agricultural tools and their associated terminology is still in its infancy. There are, however, a variety of ethnological descriptions and overviews which would form useful background material for this enterprise. The German ethnologists took considerable interest in this topic, and Baumann (1944) published a very detailed description of the morphology and distribution of farmers' tools. Some of the descriptions signalled in Raulin (1984) point to the importance of the mapping of tools in the ethnographic record[7]. For example, the sickle used for harvesting cereals is a recent introduction in West Africa, although not of European origin (cf Raynaut 1984:530). In many Nigerian languages, the term for sickle is borrowed from the Hausa *lauje* and it seems likely to have been spread by the Hausa people, based on a North African model. Two edited volumes provide rich material as yet unmined by archaeologists (Seignobos 1984; Seignobos et al 2000).

One database that can be exploited for evidence of the antiquity of agriculture is Bantu Lexical Reconstructions (BLR)[8]. This database lists forms that have been reconstructed in different regions of the Bantu zone, stretching from Cameroon to South Africa and the Kenya coast. Table 21.9 shows all the proto-forms in the database relating to agriculture as well as the zones where they occur. Figure 21.2 shows the location of traditional Bantu zones used by BLR to define the distribution of cognate terms.

The complex of terms associated with farming and cultivation, attested in A and B groups close to the Bantu homeland, argues convincingly that the proto-Bantu had some form of agriculture. There is an intriguing overlap of words for 'hoe', 'axe' and razor', especially partway through the Bantu expansion (C group onwards). This might reflect the period of the introduction of iron tools, when they would still have been rare and expensive and there may have been a tendency to call them by the same name—a type of polysemy that is uncommon today.

Ethiopia seems to have quite a different history from elsewhere. The plough seems to be have been introduced by the Amhara. The Amharic term for

Table 21.9 Bantu reconstructions for tools implying agriculture

Root	Gloss	*form	Zones	Regions
I	hoe, axe	**bàgò**	A J P	NW NE SE
		bògà	A B	NW
II	hoe	**cúkà**	C F G J L M S	NW C NE SE
		kácù	D K L M	NC
		púkà	A J	NW, NC
III	cultivate (especially with hoe)	**dìm**	B C E F G J K L M N P R S	Throughout
	cultivated field	**dìmì**	J L M	NC
	field sp.	**dìmìdò**	J	NC
	cultivated field	**dìma**	J S	NC
	field sp.	**dìmé**	J L M	NC
	farmer	**dìmì**	J L	NC
	work	**dìmò**	C F G H J K L M N S	Throughout
IV	hoe; axe; spear-head; knife	**gèmbè**	C D E F G J M P	NW C NE SE
	shave; cut hair	**gèmb**	J	NC
	razor	**gèmbè**	D F J L	NE
	axe; hoe	**dèmbè**	S	
	axe; hoe	**jèmbè**	E G L M N S	

Source: Bantu Lexical Reconstructions 3 (BLR3)

plough, *marăša*, has been borrowed into all the main languages of Ethiopia. Even where this term is not used, the local terms turn out to be constructs ('hoe of cow', etc) which indicate recent adoption. Barnett (1999: 24) canvasses ideas of introductions from Arabia or Egypt between 3000 and 4000 BP, but the linguistic evidence suggests a more recent date. Neither the design of the Ethiopian plough nor its name points to an external origin; it is likely that it was constructed locally through stimulus diffusion, ie, after seeing a plough elsewhere and designing it for local conditions.

DOMESTIC ANIMALS

This section gives examples of three species of livestock—the camel, the sheep and the chicken—for which the archaeological record is patchy and for which linguistics can make a significant contribution to hypotheses concerning their introduction and spread in sub-Saharan Africa.

The Camel

Camels inhabit much of the desert regions of Africa from Senegambia to the Horn of Africa. They are the typical transport animal of Saharan caravans, but are also increasingly used for agricultural work in sub-Saharan agricultural villages. The one-humped dromedary is originally an Asian domesticate (Epstein 1971; Wilson 1984), although wild camels were known in North Africa in the Pleistocene. Camels were reintroduced from Arabia in the Graeco-Roman period (Bulliet 1990), although bones and dung AMS

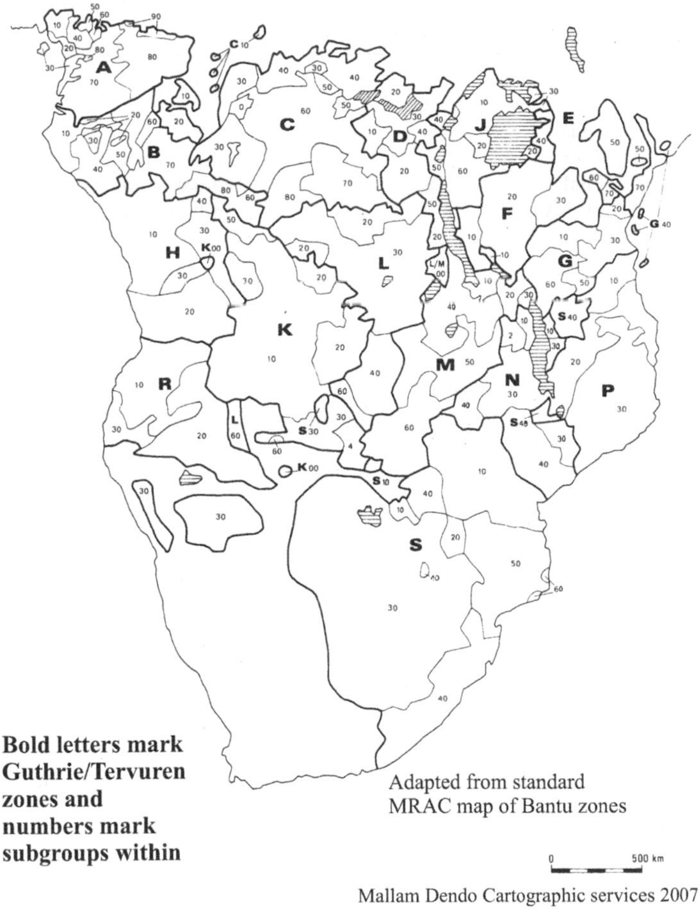

Figure 21.2 Location of Bantu language zones.

dated to 800–1000 BC occur at Qasr Ibrim in Egypt (Rowley-Conwy 1988) and occasional representations suggest that the camel was brought to Egypt as an exotic significantly earlier (Brewer et al 1994: 104). Finds of camel hair and ceramic models of camels confirm that camels were kept sporadically in Egypt, but the introduction of the camel in large numbers may be associated with the Assyrians (c 500 BC).

In the case of sub-Saharan West Africa, the camel is almost certainly more recent. Bones dating to between AD 250 and 400 have been found in the Middle Senegal Valley (MacDonald and MacDonald 2000). Linguistic evidence for the camel in West Africa is reviewed in Blench (1995, 2000). In west-central Africa, there are two sources of words for camel: loans from Berber and from Fulfulde. Versions of Berber **lγm* are common from northern Nigeria to Chad. Skinner (1977: 179–82) discusses the history of the **lγm* consonantal root, which was probably borrowed from Arabic *gml* (also borrowed into English); the Fulfulde term is probably another version of the same root, also adapting Arabic *al-gml*.

Table 21.10 Reconstructed items in proto-Sam showing the antiquity of camel pastoralism

Proto-Sam	*Gloss*
***gaal**	camel
***áùr**	male camel
***hal**	female camel
***ìrbáàn**	milking camel
***qáálìm**	young male camel
***qàálím**	young female camel
***wàdáám**	skin watering bucket
***kor**	camel bell

More problematic is the antiquity of the camel in the Horn of Africa. Archaeological finds of camel materials from this area are summarised in Esser and Esser (1982) and Banti (1993). These authors have argued for a separate domestication in the Horn of Africa, from translocated wild camels of the Arabian Peninsula. There are several studies of the linguistic evidence and terminology in the Horn of Africa (Bechhaus-Gerst 1991/2; Heine 1978). Heine (1981) points to the regular reconstruction of terms connected with camel production, for example the word for 'camel bell' in proto-Sam, ie Somali-Boni-Rendille (Table 21.10).

The camel could therefore have spread across from Arabia in 'pre-Arabic' times and thence up the Red Sea coast to Egypt and North Africa, as well as down the Somali coast and inland to Lake Turkana. The camel is little-known on the Ethiopian Plateau and terms in Cushitic and Omotic languages are loanwords from Oromo.

Sheep

All African sheep ultimately come from outside the continent and all sheep derive from two maternal lines in West and Central Asia (Hiendleder et al 1998). African sheep can be divided into four main types: thin-tailed hair and wool sheep, fat-tailed and fat-rumped sheep (Blench 1993). Wool sheep are only found on the edge of the desert in Mali and Sudan and are probably marginal and late introductions, but hair sheep have a long and complex history in the sub-Saharan region. In Africa, they first occur as domesticates in the eastern Sahara at 7000 BP and at Haua Fteah in North Africa at 6800 BP (Gautier 1981: 336). Muzzolini (1990) reviewed the evidence for sheep in Saharan rock art and his revision of the chronology, placing the first appearance of sheep rather later, at 6000 BP, seems generally accepted. Unfortunately, it is impossible to distinguish sheep and goat bones in most sub-Saharan sites; they are therefore listed together as ovicaprines, despite the two species having rather different histories. Table 21.11 shows selected dates for sub-Saharan African ovicaprines. The complex history of sheep is shown by a widespread and apparently ancient form, *#t-m-k*, which occurs in Afroasiatic, Saharan and Niger-Congo languages (Table 21.12).

Table 21.11 Selected dates for sub-Saharan African ovicaprines

Region	*Location*	*Site*	*Date**
Sahara	Air Massif	Adrar Bous	5000–3350 BC
Sahara	Niger	Arlit	4300–3700 BC
West Africa	Mali	Winde Koroji West	2200–950 BC
West Africa	Mali	Kolima Sud	1400–800 BC
West Africa	Nigeria	Gajiganna	1520–810 BC
Horn of Africa	Ethiopia	Lake Besaka	~1500 BC
East Africa	Kenya	GaJi 4	~2000 BC
East Africa	Kenya	Ngamuriak	1000 BC–AD 0
Southern Africa	Namibia	Falls rockshelter†	190 BC–AD 383
Southern Africa	South Africa	Ma38	AD 200–300

*All dates normalised to a standard format.
†Known to be sheep.

Table 21.12 The *#t-m-(k)* cognate term for 'sheep' across Africa

Phylum	*Family*	*Branch*	*Language*	*Attestation*	*Gloss*
Afroasiatic	Cushitic	East	Oromo	**tumaamaa**	castrate
	Chadic	West	Hausa	**túnkìyáá** *pl.* **túmáákíí**	sheep
		Central	Bade	**taaman, təmakun**	sheep
			Higi of Kiria	**tımbəkə**	sheep
			Tpala	**tə̀mâk**	sheep
		Masa	Masa	**dímíína**	sheep
		East	Mubi	**túmák**	sheep
			Kera	**taaməgá**	sheep
	Berber		Wargla	**adəmmam**	hair sheep
Nilo-Saharan	C. Sudanic	Moru-Madi	Moru	**temélé**	sheep
	Kadu	Eastern	Krongo	**ḍéémà**	female goat
	Saharan		Kanuri	**táma**	female lamb
			Berti	**tami**	lamb
Niger-Congo	Benue-Congo	Nupoid	Ebira Okene	**atɛ́mɛ́**	ewe
	Gur		Kirma	**tumaŋo**	sheep

Source: expanded from Blench (1999)

The linguistic evidence is consistent with the introduction of the sheep some six thousand years ago, probably by Berber populations (Blench 2001). The similarities of names right across the Sahel suggests that the introduction was via a single ethnic group with a common name for sheep. This term would have gradually spread farther south, passing from Afroasiatic and Nilo-Saharan into Niger-Congo.

Table 21.13 Livestock terms in Central Khoesan

Group	*Language*	*Cow*	*Sheep*
Khoekoe	Nama	**koma**	**ku**
Khoe	//Ani	**góɛ̀**	**gû**
Naro	Naro	**góè**	**gǔ**
//Ana	/Ui	**gúè**	**gǔ**
Shua	Cara	**bé**	**gù**
Tshwa	Kua	**dzú bé**	—

A different term, #ku, is reconstructible for Central Khoesan, and this almost certainly is to be correlated with the early dates for sheep in Namibia. Table 21.13 shows the terms for 'cattle' and 'sheep' recorded by Voßen (1996) in Central Khoesan.

All the terms for 'sheep' are cognate with one another, while there are three distinct forms of the names of 'cattle'. Central Khoesan speakers thus had sheep but not cattle when these languages began to diversify, but they acquired (or experienced) cattle after their major division into subgroups. Smith (2000: 226) tabulates the archaeozoological materials from southern Africa, and sheep probably reached this region c 2200 BP. Dates for cattle are consistently later, beginning around the third century AD with Lotshitshi in Botswana (Smith 2000: 225). The sheep kept by Khoe peoples were the fat-tailed race, better known from Arabia and northeast Africa. This evidence is consistent with the idea that these sheep were in the possession of Cushitic speakers practising pastoralism in what would today be Zambia more than 2000 years ago, and that it was there they encountered Khoe speakers and both the animals themselves and the practice of shepherding were transferred. Sadr (2003) has reviewed the evidence for sheep in southern Africa in both rock paintings and excavated sites; he establishes clearly that both sheep and pottery reached the Khoe prior to the incursions of Bantu speakers in the area.

Chickens

Chickens are by far the most important poultry species in Africa, both numerically and in terms of social and economic significance. Despite this, the chicken is an exotic import of relatively recent date. MacDonald and MacDonald (2000), Williamson (2000) and Blench and MacDonald (2001) examine the history of the chicken in Africa in greater detail. In a pioneering study, Johnston (1886) used the names of the chicken in Bantu languages to show that chicken *cannot* be reconstructed to proto-Bantu because of its irregular reflexes. He considered it likely that the chicken was introduced into the Bantu area from the east.

Fumihito et al (1994) argued from mtDNA analysis that the chicken was domesticated just once, from the races of jungle fowl found in northern Thailand. This could fit the archaeological data presented in West and Zhou (1988) for domestic chickens in China as early as 6000 BC. However, more recent analyses (Han Jian-Lin pers comm) have revealed a more complex story. Not only were chickens domesticated twice, once in southwest China/ Thailand and once in northeast India, but there has been regular introgression from wild jungle fowl, *Gallus* sp., in India. The pattern of mtDNA for African chickens suggests at least three distinct introductions: across the Sahara from the Maghreb, to the Horn of Africa and the Kenya coast from India, and a direct introduction on the East African coast of fighting breeds from insular Southeast Asia.

There are several possible routes by which the chicken may have reached Europe and North Africa. There is good evidence for chickens in India, certainly by the time of the Harappan civilization, and they may have spread from the Indus to Mesopotamia in the later third millennium (although perhaps as rarities), and thence to Iron Age Europe and North Africa (Fuller 2003b). Damdama, an aceramic site on the Ganges, has numerous chicken bones as well as rice and barley at around 2500–2000 BC. A diffusion from China across Central Asia, north of India proper, arriving in Europe by 3000 BC, is less well documented, but gains support from the linguistic data (Table 21.14). A much-reproduced painted limestone ostracon from the tomb of Tutankhamun clearly illustrates a cock; several other images suggest the occasional presence of fowl as exotics in Egypt during the New Kingdom (c 1425–1123 BC) (Darby et al 1977, I: 297). However, there is no further evidence in the graphic record until c 650 BC, after which fowl appear in abundance (Coltherd 1966).

Osteological evidence for chicken in sub-Saharan Africa is becoming more common, but is still too sparse to be effectively linked to the mtDNA evidence. Chami (2001) mistakenly identified chicken bones from a Neolithic context on Zanzibar, dated to c 800 BC (Paul Sinclair p.c.). Most finds are from the mid-first millennium AD, with records from Mali (MacDonald 1992), Nubia (MacDonald and Edwards 1993) and South Africa (Plug 1996) all dating to this period. It is hard to know how to interpret this gap. Were the East African finds left by Indian Ocean traders with no implication for the mainland, or is it simply that we have yet to find earlier sites on the continent itself? Many African languages have onomatopoeic words for chicken, usually based on the cry of the cock. Williamson (2000) identifies a number of cognate forms that mirror some of the complexities of the introduction and diffusion of the chicken suggested by the DNA evidence. But one extremely widespread cognate term, *#taxV-*, appears to plot the spread of the chicken from its original zone of domestication to the heart of central Africa. A series of very similar terms forms a chain from Korea across Central Asia to the Near East, North Africa and south to Lake Chad (Table 21.14). This suggests that the chicken not only diffused westward from China as far as central Africa, but it did so *after* the principal language phyla were established.

The astonishing conservatism that permitted a cognate term of the same shape to be retained across as much as 8000 years and virtually the whole of the Old World says something about the economic importance and visual salience of the chicken. Only the term for dog, which has origins in the

Table 21.14 A Eurasian and African cognate term for 'chicken'

Phylum	*Branch*	*Language*	*Attestation*	*Gloss*
Daic	Kadai	Hlai (Li)	**k^{h}ai**	
	Kam-Sui	Dong	**aai**	
		Maonan	**kaai**	
	Tai	Lü (Xishuang Banna)	**kai**	
Miao-Yao	Miao	Laka (Lajia)	**kai**	
	Yao	Mien	**čai**	
Koreanic	Korean	Korean	**ta(r)k**	
Altaic	Mongolic	Buryat	**taxyaa**	
	Tungusic	Manchu	**coko**	
		Hezhen Nanai	**töqo**	
	Turkic	Chuvash	**chax**	hen
		Uyghur	**toxu**	
		Kazakh	**tawɪq**	
Sino-Tibetan	Trung	Nu-jiang	**dɑŋ^{31}gu^{55}**	cock
		Rawang	**tanggu**	cock
Indo-European	Iranian	Sarikoli	**tuxi**	
		Russian	**petux**	
Afroasiatic				
Chadic	Bura-Higi	Bura	**mtəka, təka^{ɣ}**	
		Kyibaku	**ntɨka**	
		Njanyi	**ɗeke**	
	Wandala-Mafa	Dghweɗe	**ɣatukulu**	
		Sukur	**takur**	
	Masa	Masa	**ɬek-ŋa**	cock
	East Chadic	Mubi	**dììk** *pl.* **dàyàkà**	cock
Semitic	Arabic	Classical Arabic	**diik**	cock
	Ethio-Semitic	Harari	**atäwaaq**	
Berber		Awjila	**ṭẹkaʒẹṭ**	
		Tamesgrest	**tekəʒʒit**	
		Tafaghist	**tekəʒit**	
NIGER-CONGO				
Mande		Ligbi	**tùgɔ́**	
Atlantic		Temne	**atɔkɔ**	
East Kainji		Jere	**bètókóró**	

Sources: African language entries from Williamson (2000), Asian data from Reinhold Hahn (pers. comm.)

Eurasian-African #kon-, with a similar distribution and probably even greater antiquity, parallels the importance of the domestic fowl (Sasse 1993).

TREES

Introduction

The reconstruction of tree names is more problematic than either crops or livestock. No terms for tree species in Africa have been reconstructed for the proto-language of any African phylum. This may reflect defective datasets, but this is unlikely in the case of more common species, which are precisely those we would expect to reconstruct. The issue is probably rather different. With such a wealth of species to choose from, only those of marked and widespread economic importance are likely to show up in the linguistic record. Even there, the significance of a particular species can fade in and out. For example, the shea tree is a key species for oil production in much of West Africa proper. However, it occurs as far east as Uganda, but is of little or no economic significance from the centre of Chad eastwards (Hall et al 1996). The merula, *Schlerocarya birrea*, is an important species for beer making in eastern and southern Africa, but of little account in West Africa, despite being present throughout the region. Only where a tree becomes of economic significance over a wide area do vernacular names show widespread distributions. As a consequence, the names of these trees are cognate across patches of Africa where they are salient in the culture, rather than where they are present.

In the case of trees, the archaeobotany of West Africa is in flux. Reviews from the early 1990s, such as Stahl (1993), report species that tend to leave instantly identifiable macro-remains, such as oil-palm (*Elaeis guineensis*), bush candle (*Canarium schweinfurthii*) and nettle tree (*Celtis integrifolia*). More focused archaeobotany and better sieving techniques have begun to produce traces of a much wider range of species, far more consonant with the picture derived from current ethnobotany (Kahlheber chapter 17).

Long-distance trade does not exist in isolation; it acts as a transmission route for the ideologies of the traders. This is particularly true in those parts of Africa where trade was largely in the hands of Islamic merchants. Many economic trees and crops have been spread along these routes. The lexical evidence testifies that dominant trade languages such as Hausa, Kanuri, Songhay, Chadian Arabic and Swahili diffused new plants to remote areas (eg Blench 1998; Blench et al 1997). This worked in several ways. Either a plant could be directly transmitted through the sale of the fruit, or an idea about its use spread through the market. For example, the baobab is indigenous to Africa, as the reconstructibility of a name for the tree itself in some Niger-Congo languages testifies. However, the idea of collecting, drying and crushing the leaves as a soup ingredient is definitely attributable to the Hausa people of northern Nigeria and thus their name, *kúúkà*, is widely spread as a name for the leaves (Burkill 1985: 270–71). In some languages, the Hausa name has actually displaced the original name for the tree itself. This use of the leaves for soup has increased the salience of baobabs in many communities and led village communities to encourage protection of the tree.

Apart from the broad sweep of history, tree salience undergoes considerable local micro-variation, related to the interplay of economic and cultural patterns. Despite the present-day economic importance of shea, *Vitellaria paradoxa*, it may be that techniques for processing their fruits only spread during the last millennium. The shea demands considerable investment in ovens and thus in firewood collection, etc, which is probably only worthwhile when a market opens up and processing can be conducted during the dry season. Neumann et al (1998: 60) report a testa of shea from the medieval village of

Saouga and note that shea butter production was recorded by Ibn Baṭṭuṭa in the fourteenth century, which may in turn be connected with the opening up of long-distance trade routes. However, the shea tree, once predominant as the oil crop of the savanna, has retreated significantly in many regions where the cultivation of groundnut has spread[9]. Once people are no longer willing to process the shea nut, the reasons for protecting the tree itself disappear and its virtue as a wood for carving mortars becomes more apparent.

This section provides samples of reconstructions of two tree species where it is possible to compare the data with an expanded archaeobotanical database compiled by Stefanie Kahlheber. However, it should be emphasised that this barely touches on the material available; the compilation and analysis of vernacular names for trees, with over ten thousand species in sub-Saharan Africa, remains a daunting task.

Oil-palm (Elaeis guineensis)

The oil-palm, *Elaeis guineensis*, remains today the most significant oil crop indigenous to Africa, even if Malaysia has taken over in world production in recent years[10]. Archaeobotanical finds of palm nut husks occur all the way from Liberia to Kenya and also in the Sudan (see review in Stahl 1993). Although the oil-palm is known on the Kenya coast, it is not considered of any economic importance in this region (Maundu 1999). Oil-palms were not cultivated until recently, but protected and allowed to spread by preferential extraction of nearby trees. In many places, the West African humid forest now consists of degraded oil-palm forest with only a few other scattered species (Beier et al 2002).

Palynological data on *Elaeis* pollen exists for Lake Bosumtwi in Ghana (Talbot et al 1984: 185) suggest an expansion of oil-palm at 3500–3000 BP and in the Niger Delta at c 2800 BP (Sowunmi 1985). Whether this can be described as the 'beginnings of agriculture' is highly dubious, but these findings may point to a more intensive local use of the oil-palm. Even this has been questioned; Maley (2001) considers the results from palynology simply as evidence for oil-palm as a pioneer species in natural forest succession stages. Whatever the interpretation, the linguistic evidence *does* point to increased use along the West African coast. While basic terms for 'oil-palm' reconstruct to a deep level in West African languages, terms associated with its processing have a much shallower time depth. The palm nut is partially edible straight from the tree and this must have been known for millennia; pounding, boiling and skimming is almost certainly much more recent. Connell (1998) analysed terms for oil-palm and the nomenclature of processing in the Cross River languages in southeast Nigeria and showed that speakers of Delta-Cross, a hypothetical proto-language spoken in southeast Nigeria some 3000–4000 years ago, were making use of the oil-palm. At least one cognate term is widespread in what is now Nigeria and Cameroon (Table 21.15).

This term is common to the Benue-Congo languages and to Ịjọ, suggesting that the perceived salience of the oil-palm began in the southern humid

Table 21.15 A cognate term for 'oil-palm' in West African languages

Branch	*Group*	*Language*	*Vernacular name*
Yoruboid		Yoruba	**ẹrịn ọ̀pẹ̀**
Edoid		Aoma	**údi**
		Degema	**ìdí**
		Ẹdo	**udin**
Nupoid		Gbari	**èzín**
Idomoid		Idoma	**alǐ**
		Yala	**ali**
Plateau		Koro	**ɛrɛ**
		Ninzo	**iri**
Cross River	Central Delta	Abua	**àlhè**
	Upper Cross	Akpet	**uri**
		Kukele	**ùddì**
		Legbo	**èlì**
		Iyongiyong	**dòré**
Tivoid		Iceve	**ò-vílè**
Bantu		Bafok	**elen**
		Nkosi	**melen**
Ijoid		Kolokuma	**lǐị**

Source: Burkill (1997:354 ff.) and personal field data

forests, perhaps the Niger Delta, and spread outwards from there, probably at a time when the upper limit of the forest was north of its present location.

African Mahogany (Khaya senegalensis)

A tree that is nowadays important in West Africa as a timber tree, the African mahogany must have gained regional importance several thousand years ago, presumably for its medicinal properties. The oil made from its seeds is highly valued and it is often planted around villages as a shade tree. There is a common base term, #-*ko*-, which has an intriguing disjunct distribution, occurring in the Gur languages in Ghana and Burkina Faso as well as in north-central Nigeria. This points strongly to contact between these groups, rather than a reconstructible linguistic root with a great time depth and supports the hypothesis that northern Nigeria was formerly occupied by Gur speakers who were displaced by the Hausa expansion. Table 21.16 shows this form, as well as cognates in both Chadic and Nilo-Saharan languages, with an –m suffix that must have been added at the time of borrowing.

CONCLUSION

One of the striking conclusions from exploring the potential to reconstruct names and terms relevant to African agriculture is their variability. One explanation may be that all the major African cultigens still exist in related wild forms that are still exploited for food. Indeed, the outcrossing of yams,

Table 21.16 A cognate term for mahogany in West African languages

Phylum	*Branch*	*Group*	*Language*	*Vernacular name*
Niger-Congo	Benue-Congo	Plateau	Berom	**cǒ**
			Iten	**ɛho**
			Izere	**kakó**
			Tarok	**ìkò**
		Dakoid	Samba Daka	**nəkum**
	Gur		Gurma	**koka**
			Moore	**koka**
			Dagari	**ko**
			Tayari	**kogbu**
Afroasiatic	Chadic	West	Nimbia	**ágo**
			Miya	**kwə̀m**
Nilo-Saharan	Saharan		Kanuri	**káàm**

the major cereals, pulses and some leafy vegetables with wild and escaped forms is a major problem for plant breeders. As a consequence, the transition between gathering or transplanting uses and cultivation proper is seamless from the terminological point of view. There was little need to adopt or invent a new term to describe an already familiar plant. Linguistic innovations only occurred when technologies began to develop that were related to cultivation and were distinct from wild gathering strategies.

The contrast with domestic animals is evident; none of Africa's domestic animals are indigenous to the continent except the donkey and the guinea fowl. New terms to refer to introduced species such as cattle, sheep and goats are recorded in Niger-Congo and Afroasiatic to high levels of reconstructibility (Blench 1993). Unlike cereals and other domestic plants, livestock are older and are apparently more linguistically stable. Species such as the chicken, introduced more than 3000 years ago, have created a complex trail of loanwords that clearly indicate the routes whereby the bird entered and diffused across the continent.

This chapter has focused on items where common lexical roots are clearly indicative of the salience of a plant or animal species in a particular region. The evidence for crops, livestock and trees suggests a sort of gradient of salience; the more salient a particular species is, the greater the likelihood that widespread cognate terms can be identified. This is in turn connected with the biological diversity of a particular category. Thus, livestock are the most restricted, with less than a dozen domestic animals in use in sub-Saharan Africa. Names for these species are widespread and very conservative. Crops, for which there are perhaps a hundred cultigens across the continent, exhibit a sort of median level of reconstructibility, with a few relatively salient species. Trees, with as many as ten thousand species, provide a wealth of choice and speakers are likely to identify only a few as of sufficient importance to be borrowed and inherited between languages as the terms diversify. This is represented graphically in Figure 21.3.

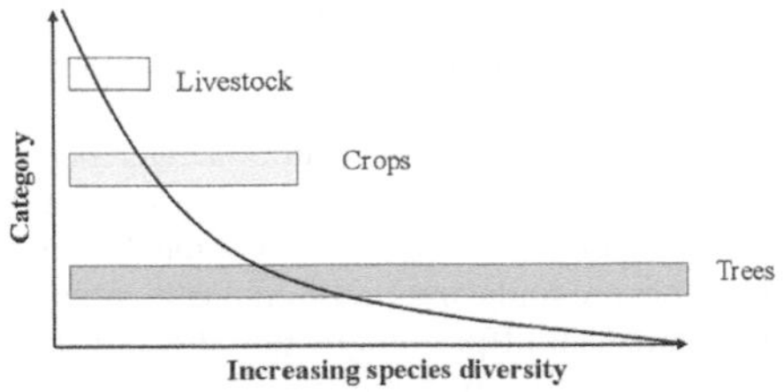

Figure 21.3 Species diversity and reconstruction potential in African languages.

Comparative and historical linguistics remains a mine of little-exploited data. Linguists are not always very accurate in defining technical terms and are prone to ignore history, often through ignorance of archaeoscientific data. Archaeologists are often unwilling to engage with linguistic data, perhaps due to its apparent complexity. But with a topic as important as the origins of agriculture, the opportunities for an interdisciplinary enterprise should be seized.

ACKNOWLEDGMENTS

This paper expands on many themes dealt with in my recent book (Blench 2006) and represents a compilation of data from many sources. I would particularly like to thank Valentin Vydrine, Guillaume Segerer and Kay Williamson (†) for unpublished language materials. Kay Williamson (†) also read and commented on the whole text. Stephanie Kahlheber generously made available her database of African archaeobotanical records (see chapter 17). Han Jian-Lin and Olivier Hanotte at ILRI, Nairobi, kindly gave me access to recent findings of the genetics of domestic animals in late 2004, subsequent to the first version of the paper. Dorian Fuller read the final version before press and kindly corrected and updated the argument in several areas based on recent and unpublished materials.

NOTES

1. Translation by author: 'It is worth saying that the *bouleau* is called *bhourtchtcha* in Sanskrit, and that this word derives from the German *birke*, the English *birch* and the Russian *береза* [*bereza*], although the names of other Indian tree species do not occur in the Indo-Germanic languages of Europe. The likely reason is that the Indo-Germanic nations were coming from the north, and when they came into India they brought the language which became the basis for Sanskrit, thereby pushing down the peninsula the speech-forms of the same origin as Malabar [Malayalam] and Telinga [Telugu]. These peoples did not find the same tree species in their new homeland as those in their former location, with the exception of the birch which grows on the southern slopes of the Himalayas'.
2. Although the diversity is concentrated in a number of 'hot spots', notably Madagascar, the Eastern Arc mountains of Tanzania and the cape in South Africa.
3. The editors suggest I avoid the normal linguistic term 'root' in order to avoid confusion with its botanical sense, hence this chapter adopts this slightly unnatural periphrasis.

4. De Maret (pers comm), one of the co-authors, remarked that it was rejected by referees for several journals because of the difficulties of distinguishing wild enset (*Musa gilletti*, indigenous to Africa) from introduced *Musa* spp.
5. See the website http://fonio.cirad.fr for additional references and more detailed information
6. This is more widespread than indicated in Philippson and Bahuchet (1996: Figure 4).
7. Like so much in the field of material culture, documentation is urgently required, as factory-made tools and tractors are replacing traditional cultivation techniques.
8. BLR3, the third edition, is at http://linguistics.africamuseum.be/BLR3.html
9. Peter Lovett (pers comm) notes that shea production has recently increased again in West Africa due to growing demand from cosmetics companies.
10. Even, regrettably, exporting back to Nigeria palm oil derived from parent material originally brought to Malaysia from Nigeria.
11. Given that this is probably the earliest African sorghum to date, it unfortunately falls within a calibration 'plateau'.

REFERENCES

Banti, G (1993) 'Ancora sull'origine del cammello nel Corno d'Africa: osservazioni di un linguista', in Belardi, A (ed), *Ethno, Lingua e Cultura*, pp 183–223, Rome: Calamo

Barnett, T (1999) *The Emergence of Food Production in Ethiopia*. British Archaeological Reports International Series 763, Oxford: Archaeopress

Baumann, H (1944) 'Zur Morphologie des afrikanischen Ackergerätes', in Baumann, H (ed), *Koloniale Völkerkunde*, pp 192–322, Horn, Austria: Verlag Ferdinand Berger

Bechhaus-Gerst, M (1991/2) 'The Beja and the camel: camel-related lexicon in *tu-beɗauwiɛ*', *SUGIA* 12/13: 41–62

Beier, P, Drielen, M van and Kankam, B O (2002) 'Avifaunal collapse in West African forest fragments', *Conservation Biology* 16(4): 1097–111

Bender, M L (1996) *The Nilo-Saharan Languages: A Comparative Essay*, Munich: Lincom Europa

Blench, R M (1993) 'Ethnographic and linguistic evidence for the prehistory of African ruminant livestock, horses and ponies', in Shaw, T, Sinclair, P, Andah, B and Okpoko, A (eds), *The Archaeology of Africa: Food, Metals and Towns*, pp 71–103, London: Routledge

Blench, R M (1995) 'A history of domestic animals in north-eastern Nigeria', *Cahiers de Science Humaine* (Paris) 31(1): 181–238

Blench, R M (1998) 'The diffusion of New World cultigens in Nigeria', in Chastenet, M (ed), *Plantes et Paysages d'Afrique*, pp 165–210, Paris: Karthala

Blench, R M (1999) 'The languages of Africa: macrophyla proposals and implications for archaeological interpretation', in Blench, R M and Spriggs, M (eds), *Archaeology and Language*, vol IV, pp 29–47, London: Routledge

Blench, R M (2000) 'Minor livestock species in Africa', in Blench, R M and MacDonald, K C (eds), *The Origin and Development of African Livestock*, pp 314–38, London: University College London Press

Blench, R M (2001) 'Types of language spread and their archaeological correlates: the example of Berber', *Origini: Preistoria e Protostoria delle Civiltà Antiche* (Rome) XXIII: 169–90

Blench, R M (2002) 'Besprechungsartikel: the classification of Nilo-Saharan', *Afrika und Übersee* 83: 293–307

Blench, R M (2003) 'The movement of cultivated plants between Africa and India in prehistory', in Neumann, K, Butler, A and Kahlheber, S (eds) *Food, Fuel and Fields: Progress in African Archaeobotany*, pp 273–92, Cologne: Heinrich Barth Institute

Blench, R M (2005) 'Fruits and arboriculture in the Indo-Pacific region', *Bulletin of the Indo-Pacific Prehistory Association* (Canberra) 24: 31–50

Blench, R M (2006) *Language, Archaeology, and the African Past*, Lanham: Rowan & Littlefield

Blench, R M (in press a) 'The intertwined history of the silk-cotton and baobab in West Africa', in Cappers, R T J (ed), *Proceedings of the 4th International Workshop for African Archaeobotany, Groningen 2003* (homepage.ntlworld.com/roger_blench/Unpublished %20Field %20 Materials %20Ethnoscience.htm)

Blench, R M and MacDonald, K C (eds) (2000) *The Origin and Development of African Livestock*, London: University College London Press

Blench, R M and MacDonald, K C (2001) 'Domestic fowl', in Kiple, K F and Ornelas, K C (eds), *The Cambridge History of Food*, vol I, pp 496–99, Cambridge: Cambridge University Press

Blench, R M, Williamson, K and Connell, B (1997) 'The diffusion of maize in Nigeria: a historical and linguistic investigation', *Sprache und Geschichte in Afrika* XIV: 19–46

Bradley, D G, MacHugh, D E, Loftus, R T, Dow, R S, Hoste, C H and Cunningham, E P (1994) 'Zebu-taurine variation in Y chromosomal DNA: a sensitive assay for genetic introgression in West African trypanotolerant cattle populations', *Animal Genetics* 25: 7–12

Bradley, D G, MacHugh, D E, Cunningham, P and Loftus, R T (1996) 'Mitochondrial diversity and the origins of African and European cattle', *Proceedings of the National Academy of Sciences* (Washington, DC) 93: 5131–35

Brewer, D J, Redford, D B and Redford, S (1994) *Domestic Plants and Animals: The Egyptian Origins*, Warminster: Aris and Phillips

Bulliet, R W (1990) *The Camel and the Wheel*, 2nd ed, New York: Columbia University Press

Burkill, H M (1985) *The Useful Plants of West Tropical Africa, Families A–D*, Kew: Royal Botanic Gardens

Burkill, H M (1994) *The Useful Plants of West Tropical Africa, Families E–I*, Kew: Royal Botanic Gardens

Burkill, H M (1995) *The Useful Plants of West Tropical Africa, Families J–L*, Kew: Royal Botanic Gardens

Burkill, H M (1997) *The Useful Plants of West Tropical Africa, Families M–R*, Kew: Royal Botanic Gardens

Burkill, H M (2000) *The Useful Plants of West Tropical Africa, Families S–Z*, Kew: Royal Botanic Gardens

Chami, F A (2001) 'Chicken bones from a Neolithic limestone cave site, Zanzibar: contact between East Africa and Asia', in Chami, F, Pwiti, G and Radimilahy, C (eds), *People, Contacts and the Environment in the African Past*, pp 84–97, Dar es Salaam: DUP

Chevalier, A (1922) 'Les petites céréales', *Revue de Botanique Appliquée et d'Agriculture Tropicale* 2: 544–50

Chevalier, A (1932) 'Les productions végétales du Sahara', *Revue de Botanique Appliquée et d'Agriculture Tropicale* 12: 669–924

Clark, J D and Stemler, A B L (1975) 'Early domesticated sorghum from central Sudan', *Nature* 25(5501): 588–91

Coltherd, J B (1966) 'The domestic fowl in ancient Egypt', *Ibis* 108: 217–23

Connah, G (1981) *Three Thousand Years in Africa*, Cambridge: Cambridge University Press

Connell, B A (1998) 'Linguistic evidence for the development of yam and palm culture among the Delta Cross River peoples of South-eastern Nigeria', in Blench, R M and Spriggs, M (eds), *Archaeology and Language*, vol II, pp 324–65, London: Routledge

Darby, W J, Ghalioungui, P and Grivetti, L (1977) *The Gift of Osiris*, 2 vols, New York: Academic Press

Davis, S D, Heywood, V H and Hamilton, A C (1994) *Centres of Plant Diversity: A Guide and Strategy for Their Conservation*, Cambridge: IUCN–World Conservation Union and World Wildlife Fund

Ehret, C (1979) 'On the antiquity of agriculture in Ethiopia', *Journal of African History* 20: 161–77

Epstein, H (1971) *The Origin of the Domestic Animals of Africa*, 2 vols, New York: Africana Publishing Corp

Esser, M and Esser, O (1982) 'Bemerkungen zum vorkommen des Kamels im östlichen Afrika im 14. Jahrhundert', *Sprache und Geschichte in Afrika* 4: 225–38

FAO (1988) *Traditional Food Plants*, Rome: FAO

Fuller, D Q (2003a) 'African crops in prehistoric South Asia: a critical review', in Neumann, K, Butler, A and Kahlheber, S (eds) *Food, Fuel and Fields: Progress in African Archaeobotany*, pp 239–72, Cologne: Heinrich Barth Institute

Fuller, D Q (2003b) 'An agricultural perspective on Dravidian historical linguistics: archaeological crop packages, livestock and Dravidian crop vocabulary' in Bellwood, P and

Renfrew, C (eds), *Assessing the Language/Farming Dispersal Hypothesis*, pp 191–214, Cambridge: McDonald Institute for Archaeological Research

Fuller, D Q (2004a) 'Early Kushite agriculture: archaeobotanical evidence from Kawa', *Sudan and Nubia Bulletin* 8: 70–74

Fuller, D Q (2004b) 'The Central Amri to Kirbekan survey: a preliminary report on excavations and survey 2003–04', *Sudan and Nubia Bulletin* 8: 4–16

Fumihito, A, Miyake, T, Sumi, S-I, Takada, M, Ohno, S and Kondo, N (1994) 'One subspecies of the red junglefowl (*Gallus gallus gallus*) suffices as the matriarchic ancestor of all domestic breeds', *Proceedings of the National Academy of Sciences* (Washington, DC) 91: 12505–09

Gallant, J (2002) *The Story of the African Dog*, Pietermaritzburg: University of Natal Press

Gast, M (2000) *Moissons du Désert*, Paris: Ibis Press

Gautier, A (1981) 'Contributions to the archaeozoology of Egypt', in Wendorf, F and Schild, R (eds), *Prehistory of the Eastern Sahara*, pp 317–44, New York: Academic Press

Germer, R (1985) *Flora des Pharaonischen Ägypten*, Mainz am Rhein: von Zabern

Giuffra, E J, Kijas, M H, Amarger, V, Carlborg, Ö, Jeon, J-T and Andersson, L (2000) 'The origin of the domestic pig: independent domestication and subsequent introgression', *Genetics* 154: 1785–91

Goddard, I (1994) 'The east-west cline in Algonquian dialectology', in Cowan, W (ed), *Actes du Vingt-Cinquième Congrès des Algonquinistes*, pp 187–211, Ottawa: Carleton University

Grigson, C (2000) '*Bos africanus* (Brehm)? Notes on the archaeozoology of the native cattle of Africa', in Blench, R M and MacDonald, K C (eds) *The Origin and Development of African Livestock*, pp 38–60, London: University College London Press

Grimes, B F (ed) (2000) *Ethnologue: Languages of the World*, 14th ed, Dallas: SIL International (www.ethnologue.com)

Groombridge, B (ed) (1992) *Global Biodiversity: Status of the Earth's Living Resources*, London: Chapman and Hall

Güldemann, T and Vossen, R (2000) 'Khoesan', in Heine, B and Nurse, D (eds), *African Languages: An Introduction*, pp 99–122, Cambridge: Cambridge University Press

Hall, J-B, Aebischer, D P, Tomlinson, H F, Osei-Amaning, E and Hindle, J R (1996) *Vitellaria Paradoxa: A Monograph*, Bangor, Wales: School of Agricultural and Forest Sciences, University of Wales, Bangor

Heine, B (1978) *The Sam Languages: A History of Rendille, Boni and Somali*. Afroasiatic Linguistics 6(2), Malibu, CA: Undena

Heine, B (1981) 'Some cultural evidence on the early Sam-speaking people in East Africa', *Sprache und Geschichte in Afrika* 3: 169–200

Hiendleder, S, Mainz, K, Plante, Y and Lewalski, H (1998) 'Analysis of mitochondrial DNA indicates that domestic sheep are derived from two different ancestral maternal sources: no evidence for contributions from urial and argali sheep', *Journal of Heredity* 89: 113–20

Hilu, K W, de Wet, J M J and Harlan, J R (1979) 'Archaeobotanical studies of *Eleusine coracana* ssp. *coracana* (finger millet)', *American Journal of Botany* 66: 330–33

Hilu, K W, Ribu, K M, Liang, H and Mandelbaum, C (1997) 'Fonio millets: ethnobotany, genetic diversity and evolution', *South African Journal of Botany* 63(4): 185–90

Hudson, G (1989) *Highland East Cushitic Dictionary*. Cushitic Language Studies 7, Hamburg: Buske

Johnston, H H (1886) *The Kili-manjaro Expedition: A Record of Scientific Exploration in Eastern Equatorial Africa*, London: Kegan, Paul and Trench

Klaproth, J von (1830) Réponse à quelques passages de la préface du roman chinois intitulé: *Hao khieou tchhouan*, traduit par M J F Davis', *Journal Asiatique* V: 97–122

Lamberti, M and Sottile, R (1997) *The Wolaytta Language*, Cologne: Rüdiger Köppe

Levtzion, N and Hopkins, J F P (1981) *Corpus of Early Arabic Sources for West African History*, Cambridge: Cambridge University Press

Lewicki, T (1974) *West African Food in the Middle Ages*, Cambridge: Cambridge University Press

Loftus, R T, MacHugh, D E, Bradley, D G, Sharp, P M and Cunningham, E P (1994) 'Evidence for two independent domestications of cattle', *Proceedings of the National Academy of Sciences* (Washington, DC) 91: 2757–61

Luikart, G, Gielly, L, Excoffier, L, Vigne, J-D, Bouvet, J and Taberlet, P (2001) 'Multiple maternal origins and weak phylogeographic structure in domestic goats', *Proceedings of the National Academy of Sciences* (Washington, DC) 98(10): 5927–32

MacDonald, K C (1992) 'The domestic chicken (*Gallus gallus*) in sub-Saharan Africa: a background to its introduction and its osteological differentiation from indigenous fowls (Numidinae and *Francolinus* sp.)', *Journal of Archaeological Science* 19: 303–18

MacDonald, K C and Edwards, D N (1993) 'Chickens in Africa: the importance of Qasr Ibrim', *Antiquity* 67: 584–90

MacDonald, K C and MacDonald, R H (2000) 'The origins and development of domesticated animals in arid West Africa', in Blench, R M and MacDonald, K C (eds), *The Origin and Development of African Livestock*, pp 127–62, London: University College London Press

Magnavita, C (2002) 'Recent archaeological finds of domesticated *Sorghum bicolor* in the Lake Chad region', *Nyame Akuma* (Calgary) 57: 14–20

Maley, J (2001) '*Elaeis guineensis* Jacq. (oil-palm) fluctuations in Central Africa during the late Holocene: climate or human driving forces for this pioneering species?', *Vegetation History and Archaeobotany* 10: 117–20

Marshall, F (2000) 'The origins and spread of domestic animals in East Africa', in Blench, R M and MacDonald, K C (eds), *The Origin and Development of African Livestock*, pp 191–221, London: University College London Press

Maundu, P M (1999) *Traditional Food Plants of Kenya*, Nairobi: National Museums of Kenya

Mbida, C M, Van Neer, W, Doutrelepont, H and Vrydaghs, L (2000) 'Evidence for banana cultivation and animal husbandry during the first millennium BC in the forest of southern Cameroon', *Journal of Archaeological Science* 27: 151–62

Mbida, C, Doutrelepont, H, Vrydaghs, L, Swennen, R, Swennen, R, Beeckman, H, De Langhe, E and de Maret, P (2001) 'First archaeological evidence of banana cultivation in central Africa during the third millennium before present', *Vegetation History and Archaeobotany* 10: 1–6

Mous, M and Kießling, R (2004) *Reconstruction of Proto-West Rift*, Cologne: Rüdiger Köppe

Müller, F M (1864) *Lectures on the Science of Language*, 2nd ser, London: Longman, Roberts and Green

Murdock, G P (1959) *Africa: Its Peoples and Their Culture History*, New York: McGraw-Hill

Muzzolini, A (1990) 'The sheep in Saharan rock art', *Rock Art Research* 7(2): 93–109

Neumann, K (2003) 'The late emergence of agriculture in sub-Saharan Africa: archaeobotanical evidence and ecological considerations', in Neumann, K, Butler, A and Kahlheber, S (eds), *Food, Fuel and Fields: Progress in African Archaeobotany*, pp 71–92, Cologne: Heinrich Barth Institute

Neumann, K, Kahlheber, S and Uebel, D (1998) 'Remains of woody plants from Saouga, a medieval West African village', *Vegetation History and Archaeobotany* 7: 57–77

Pelling, R (2003) 'Medieval and early modern agriculture and crop dispersal in the Wadi el-Algial, Fezzan, Libya', in Neumann, K, Butler, A and Kahlheber, S (eds), *Food, Fuel and Fields: Progress in African Archaeobotany*, pp 129–38, Cologne: Heinrich Barth Institute

Philippson, G and Bahuchet, S (1996) 'Cultivated crops and Bantu migrations in central and eastern Africa: a linguistic approach', *Azania* (Nairobi) 29/30: 103–20

Plug, I (1996) 'Domestic animals during the Early Iron Age in southern Africa', in Pwiti, G and Soper, R (eds), *Aspects of African Archaeology: Papers from the 10th Congress of the PanAfrican Association for Prehistory and Related Studies*, pp 515–22, Harare: University of Zimbabwe Publications

Portères, R (1951) '*Eleusine coracana* Gaertn. céréale des humanités pauvres des pays tropicaux', *Bulletin de l'Institut Français de l'Afrique Noire* 13: 1–78

Portères, R (1955) 'Les céréales mineures du genre *Digitaria* en Afrique et Europe', *Journal d'Agriculture Tropicale et de Botanique Appliquée* 2: 349–86, 477–510, 620–75

Portères, R (1958) 'Les appellations des céréales en Afrique: V. Le millet, teff de l'Abyssinie', *Journal d'Agriculture Tropicale et de Botanique Appliquée* 5: 454–63

Portères, R (1976) 'African cereals: *Eleusine*, fonio, black fonio, teff, *Brachiaria*, *Paspalum*, *Pennisetum* and African rice', in Harlan, J R, de Wet, J M J and Stemler, A B L (eds), *Origins of African Plant Domestication*, pp 409–52, The Hague: Mouton

Raulin, H (1984) 'Techniques agraires et instruments aratoires au sud du Sahara', *Cahiers ORSTOM, Séries Sciences Humaines* XX(3–4): 339–58
Raynaut, C (1984) 'Outils agricoles de la région de Maradi (Niger)', *Cahiers ORSTOM, Séries Sciences Humaines* XX(3–4): 505–36
Rowley-Conwy, P (1988) 'Early radiocarbon accelerator dates for camels from Qasr Ibrim, Egyptian Nubia', *Sahara* 1: 93
Rowley-Conwy, P (1991) 'Sorghum from Qasr Ibrim, Egyptian Nubia, c 800 BC–AD 1811: a preliminary study', in Renfrew, J (ed), *New Light on Early Farming*, pp 191–212, Edinburgh: Edinburgh University Press
Sadr, K (2003) 'The Neolithic of southern Africa', *Journal of African History* 44: 195–209
Sasse, H-J (1993) 'Ein weltweites Hundewort', in Heidermanns, F, Rix, H and Seebold, E (eds), *Sprachen und Schriften des antiken Mittelmeerraums: Festschrift für Jürgen Untermann zum 65. Geburtstag*, pp 349–66. Innsbrucker Beiträge zur Sprachwissenschaft 78, Innsbruck: Institut für Sprachwissenschaft
Savolainen, P, Ya-ping Zhang, Jing Luo, Lundeberg, J and Leitner, T (2002) 'Genetic evidence for an East Asian origin of domestic dogs', *Science* 298: 1610–13
Segerer, G (nd) 'Comparative Atlantic wordlist', electronic manuscript, available from author
Seignobos, C (ed) (1984) *Les Instruments Aratoires en Afrique Tropicale*, Séries Sciences Humaines, XX(3–4). Paris: Cahiers ORSTOM
Seignobos, C, Marzouk, Y and Sigaut, F (eds) (2000) *Outils Aratoires en Afrique: Innovations, Normes et Traces*, Paris: Karthala/IRD
Skinner, N A (1977) 'Domestic animals in Chadic', in Newman, P and Newman, R M (eds), *Papers in Chadic Linguistics*, pp 175–198, Leiden: Afrika-Studiecentrum
Smith, A B (2000) 'The origins of the domesticated animals in southern Africa', in Blench, R M and MacDonald, K C (eds), *The Origin and Development of African Livestock*, pp 222–38, London: University College London Press
Stahl, A B (1993) 'Intensification in the West African Late Stone Age: a view from central Ghana', in Shaw, T, Sinclair, P, Andah, B and Okpoko, A (eds), *The Archaeology of Africa: Food, Metals and Towns*, pp 261–73, London: Routledge
Ṣowunmi, M A (1985) 'The beginnings of agriculture in West Africa: botanical evidence', *Current Anthropology* 26: 127–29
Summers, R (1958) *Inyanga*, Cambridge: Cambridge University Press
Talbot, M R, Livingston, D A, Palmer, P A, Maley, J, Melack, J M, Delibrias, G and Gulliksen, S (1984) 'Preliminary results from sediment cores from Lake Bosumtwi, Ghana', *Palaeoecology of Africa* 16: 173–92
Vavilov, N (1931) *The Wheats of Abyssinia and Erythreae: Their Place in the General System of Wheats* [in Russian]. Bulletin of Applied Botany and Plant Breeding, (Leningrad), supp 53
Voßen, R (1996) *Die Khoe-Sprachen*, Cologne: Rudiger Köppe
Vydrine, V (nd), 'Comparative Mande dictionary', unpublished manuscript, available from author
Walter, A E and Sam, C (1999) *Fruits d'Océanie*, Paris: IRD
Watson, A M (1983) *Agricultural Innovation in the Early Islamic World*, Cambridge: Cambridge University Press
West, B and Zhou, B-X (1988) 'Did chickens go north? New evidence for domestication', *Journal of Archaeological Science* 15: 515–33
Williamson, K (1993) 'Linguistic evidence for the use of some tree and tuber food plants in southern Nigeria', in Shaw, T, Sinclair, P, Andah, B and Okpoko, A (eds), *The Archaeology of Africa: Food, Metals and Towns*, pp 104–16, London: Routledge
Williamson, K (2000) 'Did chickens go west?', in Blench, R M and MacDonald, K C (eds), *The Origin and Development of African Livestock*, pp 368–448, London: University College London Press
Wilson, R T (1984) *The Camel*, London: Longmans

Subject Index

Botanical Index

About the Contributors

HUW BARTON is Wellcome Trust University Fellow at the School of Archaeology and Ancient History, University of Leicester. He is the author of 'The case for rainforest foragers: the starch record at Niah Cave, Sarawak' (*Asian Perspectives* 44: 56–72) and co-editor with Robin Torrence of *Ancient Starch Research*, 2006, Left Coast Press. He specialises in the archaeology of hunter-gatherers in Southeast Asia and the application of starch granule analysis in determining prehistoric diet.

TIM BAYLISS-SMITH is Reader in Pacific Geography at the University of Cambridge, England. His work on the history of Pacific agriculture included research into agriculture in eastern Fiji in the 19th and 20th centuries (co-authored book with Bedford, Brookfield and Latham, *Islands, Islanders and the World*, 1988) and forest uses in western Solomon Islands (with Edvard Hviding, *Islands of Rainforest*, 2000). His collaboration in the Kuk Project with Jack Golson started with fieldwork in Papua New Guinea in 1980–81. He is currently working on the population history of the Solomon Islands.

ROGER BLENCH is the managing director of a consultancy company, Mallam Dendo Ltd, specialising in socioeconomic aspects of conflict, pastoralism, the environment and rural development in Africa and Asia. He has worked with a variety of international agencies including the World Bank, DFID, EC and FAO. He has edited (with Matthew Spriggs) four volumes of *Archaeology and Language* in the One World Archaeology series, books on the history of African livestock (with Kevin MacDonald) and on the linguistic history of Southeast Asia (with Laurent Sagart and Alicia Sanchez-Mazas). His book, *Language, Archaeology, and the African Past* was published by AltaMira in 2006.

TERRY BROWN is Professor of Biomolecular Archaeology at the University of Manchester. His research lies within biomolecular archaeology, the use of biomolecular information, in this case DNA sequences, to address issues relevant to the human past. Primary interests are the origins and spread of agriculture in Southwest Asia and Central and South America. This work combines phylogenetic analysis of DNA sequences from modern crop cultivars with the study of ancient DNA preserved in archaeological remains. He is also interested in the impact of disease on past societies and the application of forensic DNA techniques to archaeological site interpretation.

EDMOND DE LANGHE is Professor Emeritus at the Laboratory of Tropical Crop Improvement, Catholic University of Leuven. He has devoted his life to the study and improvement of banana. While primarily an agronomist, he has also extensively investigated the systematics, physiology, genetics, linguistics and archaeological record of bananas. He was the founding director of the INIBAP under the heading of the FAO and attached to the Laboratory of Tropical Crop Improvement of KU Leuven, which he founded in the 1960s. Currently, his major research interests are *Musa* systematics and archaeology.

TIM DENHAM is a research fellow in the School of Geography and Environmental Science at Monash University. His research builds upon the pioneering investigations of Jack Golson and colleagues and focuses upon early to mid-Holocene plant exploitation and early agriculture in the Highlands of Papua New Guinea. Since 2003, he has published extensively on this research, including articles in *Science, Antiquity, Proceedings of the Prehistoric Society* and *World Archaeology*, and he coedited (with Chris Ballard) a volume of *Archaeology in Oceania* entitled *Perspectives on Prehistoric Agriculture in the New Guinea Highlands.*

GAYLE FRITZ is Professor of Anthropology at Washington University in St. Louis, where she teaches and directs research in the Paleoethnobotany Laboratory. Agricultural origins and dispersals are long-term research interests, with geographical emphases on the eastern United States and the North American Southwest. Current studies focus on precontact landscape management; the adoption of domesticated amaranth north of Mesoamerica; the spread of gourds, tobacco, and other crops; and subsistence change resulting from early contacts between Europeans and Native Americans in the interior Southeast.

JACK GOLSON was Professor of Prehistory in a forerunner of the present Department of Archaeology and Natural History, Research School of Pacific and Asian Studies, Australian National University. His research interests in the prehistory of Australia, New Guinea and the South Pacific in relation to that of Southeast Asia are reflected in volumes where he is a coeditor and contributor: *Aboriginal Man and Environment* (1971), *Sunda and Sahul: Prehistoric Studies in Southeast Asia, Melanesia and Australia* (1977), *Papuan Pasts: Cultural, Linguistic and Biological Histories of Papuan-speaking Peoples* (2005) and a volume in preparation on agricultural origins in New Guinea.

DAVID R. HARRIS is Professor Emeritus of Human Environment at the Institute of Archaeology, University College London. His research interests are comparative ecological, archaeological and archival study of past human subsistence and land use, particularly transitions from foraging to farming and the emergence and spread of agricultural systems. He has written extensively on early agriculture, both theoretically and for different parts of the world. Key publications include *The Origins and Spread of Agriculture* *and Pastoralism in Eurasia* (editor and contributor, 1996) and *Foraging and Farming: The Evolution of Plant Exploitation* (editor, with G C Hillman, and contributor, 1989).

ELISABETH HILDEBRAND is currently Assistant Professor of Anthropology at Stony Brook University. Her ongoing field research includes survey and excavation of rockshelters in southwest Ethiopia and palaeoethnobotanical investigations of early farming in northern Sudan. Recent works include 'Motives and opportunities for domestication: an ethnoarchaeological study in southwest Ethiopia,' *Journal of Anthropological Archaeology* 22: 358–375 (2003), and *Enset, yams, and honey: ethnoarchaeological approaches to the origins of horticulture in southwest Ethiopia*, PhD thesis, Washington University in St. Louis (2003). With Fiona Marshall, she is co-author of 'Cattle before crops: the beginnings of food production in Africa,' *Journal of World Prehistory* 16: 99–143 (2002).

JOSÉ IRIARTE is Lecturer in Archaeology at the Department of Archaeology, University of Exeter. He is a palaeoethnobotanist whose research interests focus on the origins and dispersal of agriculture, human-environment interactions and the emergence of early Formative cultures in lowland South America.

MARTIN JONES is professor and leads a bioarchaeology group at the Cambridge University Department of Archaeology investigating the origins and development of the human quest for food, drawing on a range of techniques from biology to molecular science. Currently his main areas of research involve the tracking of early agricultural expansion through crop genetic markers and the Palaeolithic origins of the modern human meal. His books include *England Before Domesday, The Molecule Hunt: Archaeology and the Search for Ancient DNA*, and *Traces of Ancestry*.

STEFANIE KAHLHEBER is a research associate in the Department of African Archaeology and Archaeobotany at the J W Goethe-University in Frankfurt. Her PhD, completed in 2003, deals with archaeobotanical remains from archaeological sites in Burkina Faso. Currently, she is participating in a multidisciplinary project on ecological and cultural changes in West and Central Africa during the second and first millennia BC. Her main research interests are the domestication history of African crops as well as prehistoric and recent plant use. She has conducted fieldwork in Burkina Faso, Cameroon, Namibia and Nigeria.

ROGER LANGOHR is professor at the Laboratory of Soil Science at Ghent University. His research interests include the earth sciences, archaeo-pedology, soil survey, soil genesis and soil classification, and he has written extensively in these fields. A recent coauthored (with Kai Fechner and Yannick Devos) publication is 'Archaeopedological checklists: proposal for a simplified version for the routine archaeological record in Holocene rural and urban sites in North-Western Europe', in Geoff Carver (ed), *Digging the Dirt: Excavation in a New Millennium* (2004).

GEERTRUI LOUWAGIE is an environmental researcher and received her PhD in geo- and archaeopedology at Ghent University. Her research interests include the earth sciences, archaeology, agriculture and land evaluation (including palaeoland evaluation). A recent coauthored publication (with Christopher M Stevenson and Roger Langohr) is 'The impact of moderate to marginal land suitability on prehistoric agricultural production and models of adaptive strategies for Easter Island (Rapa Nui, Chile)', *Journal of Anthropological Archaeology* 25: 290–317 (2006).

FIONA MARSHALL is a professor in the Department of Anthropology at Washington University in St. Louis. She is an African prehistorian whose research incorporates zooarchaeological and ethnoarchaeogical approaches to the beginnings of food production and development of pastoralism in northeast Africa. Her ethnoarchaeological research among the Okiek also examines taphonomic issues and archaeological signatures of socioeconomic variation among hunter-gatherers. Fiona Marshall has also worked on the archaeology of human origins at Koobi Fora and conservation at the Laetoli hominid footprint site. She is currently leading a multidisciplinary three-year project on African perspectives on the domestication of the donkey.

KATHARINA NEUMANN (Dr. phil nat., J W Goethe–University, Frankfurt, 1988) is a Lecturer of Archaeobotany at the University of Frankfurt. She directs research projects on prehistoric plant use and the Holocene vegetation history of West and Central Africa, with special emphasis on woody plants and the domestication history of African crops. Recent publications include *Woods of the Sahara and the Sahel* (2001); *Food, Fuel and Fields,* coedited with Ann Butler and Stefanie Kahlheber (2003), and 'The romance of farming' in *African Archaeology* edited by Ann Brower Stahl (2005).

RICHARD OSLISLY received a PhD in archaeology from the University of Paris 1 Pantheon-Sorbonne in 1992. Specialising in Central African rain forests, he has been a consultant for ECOFAC, a program founded by the UE, and from 2000–2003 he was in charge of archaeological research for the IRD-Cameroon. Currently he is attached to the Natural History Museum in Paris and is a research associate of WCS-Gabon. He has published numerous papers on the prehistory of Gabon. He is continuing his research on people and the environment in Cameroon and Gabon, and has developed a new research program on Gabon's karst rockshelters.

VICTOR PAZ is the current director of the Archaeological Studies Program at the University of the Philippines. He specialises on macroplant remains determination and is also very interested in larger archaeological questions concerning Island Southeast Asia. He recently edited a festschrift volume in honour of Wilhelm G Solheim II entitled *Southeast Asian Archaeology* (2005). Publications in 2004 include 'Of nuts, seeds and tubers: the archaeobotanical evidence from Leang Burung 1' (*Modern Quaternary Research in Southeast Asia* 18: 191–220) and 'Addressing the redefinition of the Palaeolithic and the Neolithic in the Philippines' (*Proceedings of the Society of Philippine Archaeologists* 2: 1–14).

DEBORAH M PEARSALL received her BA in anthropology from the University of Michigan, where she studied palaeoethnobotany with Richard Ford. During graduate school at Illinois, she worked with South American archaeologist Donald Lathrap, participating in excavations at Real Alto, an early agricultural site in coastal Ecuador. Analysis of macroremains and phytoliths from Real Alto became her dissertation. In addition to continuing to work in Ecuador, Pearsall has conducted palaeoethnobotanical research in Peru, Guatemala, Puerto Rico, US Virgin Islands, Bahamas, Hawaii and Guam. She teaches at the University of Missouri and holds the Frederick A Middlebush Chair in Social Sciences.

LINDA PERRY is a postdoctoral fellow in the Archaeobiology Laboratory in the Department of Anthropology at the Smithsonian National Museum of Natural History. Her current research focuses upon the relationship between food production and social complexity in northwest Venezuela.

EMILE ROCHE is a palynologist at the Université de Liège, Belgium. He obtained a PhD in palynology at University of Brussels (ULB). His research interests include European Tertiary and African Quaternary palynology. He undertook field research from 1976 to 1994 in Burundi, Congo and Rwanda, and in the late 1990s in Tunisia. His research interests include climate change and human impacts on the environment. Publications include 'L'influence anthropique sur l'environnement à l'Âge du fer dans le Rwanda ancien' (*Geo-Eco-Trop* 20 [1–4]: 73–89, 1996) and coauthored (with Marie-Claude Van Grunderbeek and Hugues Doutrelepont) 'Influence humaine sur le milieu au Rwanda et au Burundi à l'Âge du Fer Ancien' in a special issue of *Revue de Paléobiologie* (1984).

MATTHEW P SAYRE is a National Science Foundation Graduate Research Fellow and a PhD candidate at the University of California at Berkeley. He is a paleoethnobotanist specialising in macrobotanical and phytolith analysis. His research focuses on the agricultural, ritual, and domestic practices at the Andean site of Chavín de Huántar.

MARIE-CLAUDE VAN GRUNDERBEEK is an archaeologist and now collaborator at City Museums, Brussels. She studied in archaeology and history of art, Catholic University of Leuven, Belgium. From 1978 to 1987 she undertook field surveys and excavations of Early Iron Age remains in Rwanda, Burundi and Congo. Her research focuses upon Urewe culture, specifically occupation chronology, ceramics and iron-smelting furnaces, as well as upon human-environment interactions. She has published extensively on her research, including 'Essai d'étude typologique de céramique urewe de la région des collines au Burundi et Rwanda' (*Azania* XXIII: 11–55, 1988), 'Essai de délimitation chronologique de l'Âge du Fer Ancien au Burundi, au Rwanda et dans la région des Grands Lacs' (*Azania* XXVII: 53–80, 1992), and (with Emile Roche and Hugues Doutrelepont 1983) *Le Premier Age du Fer au Rwanda et au Burundi: Archéologie et Environnement.*

LUC VRYDAGHS completed his doctor in sciences at the Ghent University (UG), having already completed a degree in philosophy of sciences and one in African civilisation at the Free University of Brussels (ULB). His research is concerned with the phytolith analysis of archaeological deposits produced by agricultural practices in tropical, arid and temperate areas. Currently, he is a scientific collaborator with the Royal Belgian Institute of Natural Sciences, and he is the founding president of ROOTS, a unit specialising in archaeological and palaeoenviromental sciences.

LEE WHITE received a PhD in zoology from the University of Edinburgh in 1992. He is currently a senior conservationist with the Wildlife Conservation Society, New York, and has directed their Gabon Program since completing his PhD. He has published widely on large mammal and vegetation ecology in the Central African rain forests, including writing several chapters in *African Rain Forest Ecology and Conservation* edited by William Weber, Lee White, Amy Vedder and Lisa Naughton-Trevis (2001). He has been working with Richard Oslisly since 1992 on the archaeology and vegetation history of central Gabon.